Springer Aerospace Technology

The series explores the technology and the science related to the aircraft and spacecraft including concept, design, assembly, control and maintenance. The topics cover aircraft, missiles, space vehicles, aircraft engines and propulsion units. The volumes of the series present the fundamentals, the applications and the advances in all the fields related to aerospace engineering, including:

- structural analysis,
- aerodynamics,
- aeroelasticity,
- aeroacoustics,
- flight mechanics and dynamics
- orbital maneuvers,
- avionics,
- systems design,
- materials technology,
- launch technology,
- payload and satellite technology,
- space industry, medicine and biology.

The series' scope includes monographs, professional books, advanced textbooks, as well as selected contributions from specialized conferences and workshops.

The volumes of the series are single-blind peer-reviewed.

To submit a proposal or request further information, please contact:
Mr. Pierpaolo Riva at pierpaolo.riva@springer.com (Europe and Americas)
Mr. Mengchu Huang at mengchu.huang@springer.com (China)

The series is indexed in Scopus and Compendex

More information about this series at https://link.springer.com/bookseries/8613

Florian Sellmaier · Thomas Uhlig ·
Michael Schmidhuber
Editors

Spacecraft Operations

Second Edition

Springer

Editors
Florian Sellmaier
German Space Operations Center (GSOC)
German Aerospace Center (DLR)
Oberpfaffenhofen, Germany

Thomas Uhlig
German Space Operations Center (GSOC)
German Aerospace Center (DLR)
Oberpfaffenhofen, Germany

Michael Schmidhuber
German Space Operations Center (GSOC)
German Aerospace Center (DLR)
Oberpfaffenhofen, Germany

ISSN 1869-1730 ISSN 1869-1749 (electronic)
Springer Aerospace Technology
ISBN 978-3-030-88592-2 ISBN 978-3-030-88593-9 (eBook)
https://doi.org/10.1007/978-3-030-88593-9

This Springer imprint is published by the registered company Springer Nature Switzerland AG
The registered company address is: Gewerbestrasse 11, 6330 Cham, Switzerland

Foreword to the First Edition

Timing is everything—this is especially true for spaceflight operations. 2014 is a special year for the European space community, the year that started with the wake-up of ROSETTA, ESA's comet chaser, with Philae, the German comet lander, which is on its extraterrestrial voyage since 2004. It has been awoken from his hibernation and is providing us with data during its carefully planned first approach on a comet—67P/Churyumov-Gerasimenko.

In November 2014, the mission culminates in the descent of the lander PHILAE to the surface of the comet's frozen nucleus—the resulting measurements may help us answer some of the fundamental questions about the evolution of life on earth. Comets are considered as veterans of our solar system—their analysis provides insights in the early days of our galactic home.

Another fact which makes 2014 quite special for the European spaceflight community is the mission of two Europeans to the International Space Station—never before have we seen two long duration ESA missions within one year. Alexander Gerst has the chance to beat Thomas Reiter's record of logging the longest time in one space mission for a German. Samantha Cristoforetti is only the second ESA female astronaut—and the first Italian woman in space. During their stay in orbit, both will collect data for many months of scientific research and definitely awake the public interest in spaceflight in their home countries.

In this fascinating year falls the publication of the book *Spaceflight Operations*. It discusses important principles and aspects of the operation of space vehicles. Designated experts of the DLR's German Space Operations Center (GSOC), ESA's European Space Operations Center (ESOC) and the University of Southampton have put together a handbook for operations, which provides not only a good overview but also the expert background information, to make the book not only a theoretical description, but a vivid testimonial of many years of experience. Both the GSOC and ESOC spaceflight operations centers were founded in 1967 and GSOC's manned spaceflight history dates back to 1985 with the German spacelab mission D-1.

The authors of this book are involved in many of the most exciting space missions and projects currently ongoing: Columbus and the International Space Station, ROSETTA and the lander PHILAE, TerraSAR-X and Tandem-X, the European Data

Relay System EDRS and the space robotic mission DEOS. They have worked during the preparation and execution phases, acting in their roles as managers, engineers, planners, subsystem specialists and flight controllers. It is inspiring to read their articles and to listen to their "lessons learnt."

It is my desire that the book will provide both an interest and stimulus for future missions—and may help to improve subsequent operations concepts.

Köln, Germany

Jan Wörner
Chairman of the Executive Board of
the German Aerospace Center (DLR)

Preface to the Second Edition

This book originally grew out of the “Spacecraft Operations Course,” a 1-week lecture and exercise series that has been held annually at the German Space Operations Center (GSOC) in Oberpfaffenhofen for the past 21 years. The handout, which was a collection of slides in the very beginning, changed into a book over time. Still, we realized that there is currently no book that deals exclusively with spacecraft operations, so we expanded our project by adding to and detailing the chapters so that we could complete it in book form. As before, most of the chapters are based on lectures from our current “Spacecraft Operations Course.” However, the target audience of this book is not only the participants of the course but also students of technical or scientific courses, as well as technically interested people who want to gain a deeper understanding of spacecraft operations.

Five years after the publication date of the first edition, we felt that a thorough revision and expansion of our book was necessary. This effort resulted in four new chapters: Flight procedures, human factors, ground station operation, as well as software and systems. In addition, some chapters have been extensively supplemented. The entire book has been brought up to date, the language has been revised and we decided to improve its structure: The chapters are grouped into seven parts:

The first part of the book (*Part I*) gives a brief summary of the space segment, introducing the space environment, space systems engineering and space communications. The next four parts deal with the classical fields of space flight operations: The phases of mission operations (*Part II*) are described in chronological order, from preparation to execution and the final evaluation (flight experience). These chapters are now complemented by chapters on flight procedures and human factors. *Part III* addresses ground and communications infrastructure, i.e., cross-mission support services. This part is structured according to the different services. The flight dynamics system (*Part IV*) focuses on attitude and orbit control of the satellite platform, while mission planning (*Part V*) ensures effective payload management and utilization. The last two parts deal with the details of specific mission types: *Part VI* describes the operational tasks of the various subsystems of a classical unmanned satellite in Earth orbit. *Part VII* discusses the special requirements of specific mission types caused by the presence of astronauts, by the approach of a satellite to another target satellite, or by interplanetary cruises and landing operations on other celestial bodies.

We would like to take this opportunity to thank all contributors to this opus: First of all, all of the authors for their contributions, which they provided beside their ongoing operational work. A very big thank you to our design team and editorial office Juliane von Geisau, Yasmin Dorostan, Adriane Woito and Angelica Lenzen, who have been instrumental in supporting us throughout the second edition. Many thanks to Nick Jost, who supported us linguistically as a "native speaker." Thanks also to Martin Peters for his assistance. Last but not least, our thanks go to Pierpaolo Riva at Springer-Verlag for his supervision during the revision of this book. Now, it is time to put the second edition into print—we hope you enjoy reading it.

Oberpfaffenhofen, Germany
February 2022

Florian Sellmaier
Thomas Uhlig
Michael Schmidhuber

Contents

Part I Overview Space Segment

1 Space Environment 3
Adrian R. L. Tatnall and Hauke Fiedler
1.1 Introduction 3
1.2 Launch Vehicle 4
1.2.1 Acoustic/Vibration Levels 4
1.2.2 Static Acceleration 5
1.2.3 Mechanical Shock 6
1.3 Spacecraft Operational Environment 6
1.3.1 Vacuum 6
1.3.2 Solar Radiation Flux 7
1.3.3 Particle Radiation 8
1.3.4 Radiation Belts 8
1.3.5 Atmosphere 11
1.3.6 Space Debris 13
1.3.7 Gravity and Magnetic Fields 15
References 16

2 Spacecraft Design 17
Adrian R. L. Tatnall
2.1 Definition of Systems Engineering 17
2.2 Objectives and Requirements 19
2.3 Design Drivers and Trade-Offs 23
2.4 Concurrent Engineering 25
References 25

3 Fundamentals of Space Communications 27
Felix Huber
3.1 Introduction 27
3.2 Baseband 28
3.2.1 Source Coding 28

3.2.2 Channel Coding 30
3.2.3 Baseband Shaping 32
3.3 Modulation 34
3.4 Carrier 36
3.4.1 Elements of a Space Link 36
3.4.2 Link Budget Equation 38

Part II Mission Operations

4 Mission Operations Preparation 45
Andreas Ohndorf and Franck Chatel
4.1 Introduction 45
4.2 Driving Factors 48
4.2.1 Requirements 48
4.2.2 Cost/Financing 50
4.2.3 Technology/Complexity 50
4.2.4 Schedule 51
4.2.5 Experience 52
4.2.6 Risk 52
4.3 Personnel, Roles and Responsibilities 53
4.3.1 Project Manager 53
4.3.2 Mission Director 54
4.3.3 Flight Director 54
4.3.4 System Engineer 54
4.3.5 Simulation Officer 54
4.3.6 Quality Assurance Engineer 55
4.3.7 Subsystem Engineer 55
4.3.8 Project Office/Project Officer 55
4.3.9 Controlling 56
4.3.10 Configuration Manager 56
4.3.11 Security Officer 56
4.4 Required Data, Products and Tools 56
4.4.1 Test Data and Data Generators 56
4.4.2 Spacecraft Simulator 57
4.4.3 Mission Information Base 57
4.4.4 M&C System Software 57
4.4.5 Ground and Flight Procedures for Nominal and Contingency Situations 58
4.4.6 Operation Support Tools 58
4.4.7 Project Documentation 58
4.5 Activities, Tasks, and Schedule 59
4.6 Review Process 61
4.7 Operational Validation 64
4.7.1 Necessity 64
4.7.2 Operational Validation Organization 65

4.7.3 Points to Evaluate During Simulations and Rehearsals ... 67
4.7.4 Operational Readiness ... 68
4.8 Conclusion ... 69
References ... 69

5 Mission Operations Execution ... 71
Sabrina Eberle, Thomas Uhlig, Ralf Faller, and Michael Schmidhuber
5.1 Introduction ... 71
5.2 Various Phases During Execution ... 72
5.2.1 General Description of the Execution Phase ... 72
5.2.2 Launch and Early Orbit Phase (LEOP) ... 74
5.2.3 Commissioning Phase ... 75
5.2.4 Routine Phase ... 76
5.2.5 End of Mission or Disposal Phase ... 78
5.3 Staffing of the Flight Control Team ... 78
5.3.1 Mission Operations Team Lead ... 79
5.3.2 Subsystem Specialists ... 79
5.3.3 Command Operator ... 80
5.3.4 Planner ... 80
5.3.5 Flight Dynamics ... 81
5.3.6 Ground Data Systems ... 81
5.3.7 Engineering Support Team ... 81
5.4 Interactions Within the Flight Control Team and Flight Procedures ... 82
5.4.1 Interactions Within the Flight Control Team ... 82
5.4.2 Flight Operations Procedures (FOP) ... 82
5.4.3 Anomalies and Recommendations ... 85
5.5 The Mission Type Defines the Operational Concept ... 86
5.5.1 Low Earth Orbit: GRACE ... 86
5.5.2 Geostationary Earth Orbit: EDRS ... 88
5.5.3 Deep Space Missions: Galileo ... 91
5.6 Summary ... 92
References ... 93

6 Flight Experience ... 95
Ralph Ballweg, Andreas Ohndorf, and Franck Chatel
6.1 Introduction ... 95
6.2 Mission Experience: Empirical Data ... 95
6.3 Failure Probability Versus Operational Experience ... 98
6.4 Interpretation of Telemetry ... 100
6.5 Contingency Handling ... 102
6.5.1 Mission Example TV-SAT 1 ... 104

6.5.2 Mission Example: Sun-Acquisition Mode Anomaly ... 109
6.6 Conclusion ... 114

7 Flight Procedures ... 117
Ralf Faller and Michael Schmidhuber
7.1 Introduction ... 117
7.2 General Information ... 118
7.2.1 Basics ... 118
7.2.2 Types of Flight Procedures ... 120
7.2.3 Mission Type Affects Flight Procedures Design ... 121
7.2.4 Flight Procedure Lifecycle ... 122
7.3 Flight Procedure Concepts and Approaches ... 126
7.3.1 Text-Based ... 126
7.3.2 Spreadsheet-Based ... 127
7.3.3 Script-Based ... 129
7.3.4 Commercial Products ... 131
7.3.5 On-Board Control Procedures ... 132
7.4 Development Guidelines and Rules ... 133
7.5 Summary ... 133
References ... 134

8 Human Factors in Spaceflight Operations ... 135
Thomas Uhlig and Gerd Söllner
8.1 Introduction ... 135
8.2 Critical Dependability on Humans in Various Areas ... 137
8.3 Classical Fields of Crew Resource Management ... 138
8.3.1 Communications ... 139
8.3.2 Situational Awareness ... 141
8.3.3 Decision Making ... 141
8.3.4 Team Work and Leadership ... 142
8.4 Translation into Spaceflight Operations ... 143
8.4.1 Flight Control Team Structure ... 143
8.4.2 Operations Products ... 144
8.4.3 Real-Time Decision Making ... 147
8.4.4 Standards for Voice Communication ... 149
8.4.5 Monitoring System ... 153
8.4.6 Commanding and Command Error Data Base ... 154
8.4.7 Training and Simulations ... 155
8.5 Standardization ... 156
8.6 Conclusion ... 157
References ... 158

Part III Communication and Infrastructure

9 Design and Operation of Control Centers 163
Marcin Gnat and Michael Schmidhuber
9.1 Introduction 163
9.2 Infrastructure 164
9.2.1 Control Rooms 166
9.2.2 Public Space in the Control Center 167
9.2.3 Server Rooms and Computer Hardware 168
9.3 Control Center Network 169
9.3.1 Network Topologies 169
9.3.2 Network Technologies 171
9.4 Control Center Software 173
9.4.1 General 173
9.4.2 Space Link Extension Gateway System 175
9.4.3 File Distribution Subsystem 176
9.4.4 Spacecraft Monitoring and Control System 176
9.5 Outlook 178

10 Ground Station Network 181
Marcin Gnat
10.1 Introduction 181
10.2 Station Selection 182
10.3 Station Communication 186
10.3.1 Communication Paths 187
10.3.2 Data Transfer Methods 188
10.3.3 GSN Examples 189
10.3.4 Cloud Based Services 189
10.4 LEOP and Routine Operations 192
10.4.1 GDS Engineering Team (NOPE) 193
10.4.2 Systems Team (Network and Systems Control) 193
10.4.3 Scheduling Office 194
10.4.4 General Tasks Throughout the Project 194
References 195

11 Ground Station Operation 197
Amanuel Geda
11.1 Introduction 197
11.2 The RF Subsystem 198
11.3 The Intermediate Frequency (IF) Baseband Subsystem 198
11.4 Supporting Devices 199
11.5 Telemetry, Tracking and Command Operations 199
11.5.1 Tracking the Spacecraft 202
11.5.2 Ranging 203
11.5.3 Monitoring and Control of Operations 204
11.6 Measurement Campaigns 204

11.6.1 RF Compatibility Test ... 204
11.6.2 LEOP ... 206
11.6.3 In-Orbit Test (IOT) System ... 207
11.7 Space Link Extension (SLE) Services ... 209
11.7.1 Online SLE Service ... 209
11.7.2 Offline SLE Service ... 209
11.8 Summary ... 211
References ... 211

12 Software and Systems ... 213
Markus Hobsch and Michael Schmidhuber
12.1 Introduction ... 213
12.2 Fundamentals ... 214
12.2.1 Telemetry Parameters ... 214
12.2.2 Telecommands ... 214
12.2.3 TM/TC Database ... 215
12.2.4 Monitoring and Control System (MCS) ... 215
12.3 Space-to-Ground Data Streams ... 216
12.3.1 Data Transport ... 216
12.3.2 Frame-Based Telemetry ... 216
12.3.3 Packet Data Structures ... 218
12.3.4 Packet Utilization Standard ... 219
12.3.5 TM/TC and Security Management ... 220
12.4 Monitoring and Control System (MCS) Software ... 222
12.4.1 TM/TC Processing ... 223
12.4.2 Display System ... 225
12.4.3 Offline Data ... 227
12.4.4 Automation of Mission Operations ... 228
12.5 Ops Support Tools ... 229
12.5.1 Procedure Tool Suite (ProToS) ... 230
12.5.2 Operations Support (OpsWeb) ... 230
12.5.3 Predictive Maintenance ... 231
12.6 SW Development and Maintenance ... 232
12.6.1 Multi-Mission Approach ... 233
12.6.2 Standards and Standardization ... 234
12.6.3 Software Procurement ... 235
12.6.4 Software Development ... 236
12.6.5 Maintenance ... 239
12.7 Conclusion ... 240
References ... 240

Part IV Flight Dynamic System

13 Orbital Dynamics ... 245
Michael Kirschner
13.1 Introduction ... 245

13.2 Theoretical Aspects ... 246
13.2.1 Satellite Orbit ... 246
13.2.2 Satellite Velocity ... 249
13.2.3 Orbital Period ... 250
13.2.4 Orbit Perturbations ... 251
13.2.5 Maneuvers ... 254
13.3 Flight Dynamics Tasks ... 257
13.3.1 Mission Preparation ... 257
13.3.2 Mission Execution ... 261
13.3.3 Maneuver Planning ... 263
References ... 270

14 Attitude Dynamics ... 271
Jacobus Herman, Ralph Kahle, and Sofya Spiridonova
14.1 Introduction ... 271
14.2 Disturbances ... 275
14.2.1 Satellite Intrinsic ... 275
14.2.2 External Influences ... 275
14.3 Attitude Determination ... 276
14.3.1 Coarse Attitude ... 276
14.3.2 Precise Attitude ... 280
14.3.3 Ground-Based Attitude ... 285
14.4 Attitude Propagation ... 285
14.5 Attitude Control ... 287
14.6 Tasks of AOCS ... 291
14.6.1 Example 1 ... 292
14.6.2 Example 2 ... 293
References ... 297

Part V Mission Planning System

15 The Planning Problem ... 301
Christoph Lenzen and Sven Prüfer
15.1 Introduction ... 301
15.2 General Overview of a Mission Planning System ... 302
15.3 Techniques for Timeline Generation ... 306
15.3.1 General Considerations ... 306
15.3.2 GSOC Modeling Language ... 307
15.3.3 Application Examples of the Modeling Language ... 313
15.3.4 Templates ... 318
15.3.5 Planning Algorithms ... 319
15.4 Summary ... 322
References ... 322

16 Mission Planning for Unmanned Systems 325
Tobias Göttfert, Sven Prüfer, and Falk Mrowka
16.1 Introduction 325
16.2 Mission Planning System Example 326
16.3 Considerations on Designing a Mission Planning System 328
16.4 Mission Planning at Various Time Scales 330
16.5 Conclusions and Outlook 333
References 333

17 Mission Planning for Human Spaceflight Missions 335
Thomas Uhlig, Dennis Herrmann, and Jérôme Campan
17.1 Introduction 335
17.2 Basic Considerations 338
17.3 Planning Teams 339
17.4 Concept of Crew Flexibility 340
17.5 Planning Phases Overview 341
17.6 Planning Products and Processes 342
17.6.1 Strategic Planning 342
17.6.2 Tactical Planning 344
17.6.3 Pre-increment Planning 346
17.6.4 Increment Planning 348
17.6.5 Real-Time Planning 350
17.7 Planning Tools 351
17.8 Conclusion 352
References 353

Part VI Spacecraft Subsystems

18 Telemetry, Commanding and Ranging System 357
Michael Schmidhuber and Tarsicio Lopez-Delgado
18.1 Definition of Subsystem 357
18.2 Signal Characteristics 357
18.2.1 Frequencies 358
18.2.2 Polarization 358
18.2.3 Side Bands and Side Lobes 359
18.3 Design 360
18.3.1 Subsystem Elements 360
18.3.2 Spacecraft Antenna Layout 363
18.3.3 Redundancies 366
18.4 Monitoring and Commanding 366
18.4.1 Automatic Gain Control (AGC) 366
18.4.2 Loop Stress 367
18.4.3 Lock Status 368
18.5 Operational Situations 368
18.5.1 Acquisition and Loss of Signal (AOS/LOS) 368
18.5.2 Ranging 370

18.5.3 Doppler and Coherency ... 370
18.5.4 Antenna/Transponder Selection ... 371
18.5.5 Polarization Prediction ... 371
18.5.6 Interferences with Another Spacecraft ... 372
18.6 Outlook to Future Developments ... 374
References ... 374

19 On-board Data Handling ... 377
Michael Schmidhuber
19.1 Definition of Subsystem ... 377
19.2 Fundamentals ... 378
19.2.1 Subsystem Elements ... 378
19.2.2 Redundancy ... 380
19.2.3 Telemetry Parameters ... 381
19.2.4 Telecommands ... 381
19.2.5 Spacecraft Database ... 382
19.3 Managing Data Handling On-board ... 383
19.3.1 General ... 383
19.3.2 Safeguard Mechanisms ... 383
19.3.3 Maintenance of the On-board Software ... 386
19.3.4 Execution Management ... 388
19.3.5 Mass Memory Management ... 389
19.4 Summary and Outlook ... 392
References ... 392

20 Power and Thermal Operations ... 393
Kay Müller, Sebastian Löw, and Sina Scholz
20.1 PTS (Power and Thermal System) Design Aspects ... 393
20.1.1 Power System ... 393
20.1.2 Thermal System ... 400
20.2 Operations ... 406
20.2.1 Preparation Phase ... 406
20.2.2 Launch and Early Orbit Phase (LEOP) Operations ... 406
20.2.3 Routine Phase ... 408
20.2.4 End of Life ... 412
20.3 Contingency Operations ... 414
References ... 416

21 Propulsion Operations ... 417
Franck Chatel
21.1 Principle of Propulsion ... 417
21.2 Configurations of Propulsion System ... 420
21.2.1 Layout of a Bi-propellant Propulsion System ... 420
21.2.2 Operational Configurations ... 421
21.2.3 Electric Propulsion ... 422
21.3 Real-Time Operations ... 423

21.3.1 Monitoring During Quiescent Periods 423
21.3.2 Orbit Maneuvers 424
21.3.3 Isolation of the Apogee Engine (Geostationary Earth Orbit—GEO) 427
21.3.4 Autonomous Operations 428
21.4 Offline Operations 428
21.4.1 Preparation and Calibration of Orbit Maneuvers 429
21.4.2 Calculation of the Propellant Mass 430
21.4.3 Calibration of the Center of Mass 431
21.4.4 Estimation of Lifetime 432
Reference 432

22 Attitude and Orbit Control Subsystem Operations 433
Ralf Faller
22.1 Introduction and Overview 433
22.2 Subsystem Description 435
22.2.1 Sensors 435
22.2.2 Actuators 438
22.2.3 On-board Control Unit 439
22.2.4 AOCS Device Combinations and Redundancy 442
22.3 AOCS-Related Ground Operations 444
22.3.1 Basic AOCS Ground Activities 444
22.3.2 LEOP and In-Orbit Tests 446
22.3.3 Contingency Operations 447
22.3.4 Support Tools for AOCS 448
22.4 Experience from Previous Missions 448
22.4.1 AOCS Degradation 448
22.4.2 Strange Wheel Behavior 449
22.4.3 Undocumented AOCS Function 451
22.5 Summary 451
References 452

23 Repeater Operations 455
Jürgen Letschnik
23.1 Introduction 455
23.2 Repeater Subsystem 456
23.2.1 Functions of a Repeater Payload 456
23.2.2 Overview and Layout of a Repeater/Transponder 457
23.3 Repeater Operations 460
23.3.1 Launch and Early Orbit Phase (LEOP) Operations 461
23.3.2 IOT Operations 462
23.3.3 Routine Operations 464
23.3.4 Contingency Operations 468
23.4 Summary 470
References 470

Part VII Special Topics

24 Human Spaceflight Operations 473
Jérôme Campan, Thomas Uhlig, Dennis Herrmann, and Dieter Sabath
24.1 Introduction 473
24.2 Manned and Unmanned Missions 474
24.3 From Satellite to Living Space 474
24.3.1 Advanced Satellite Subsystems 474
24.3.2 Environment Control and Life Support System (ECLSS) 476
24.3.3 Cargo 478
24.3.4 Extravehicular Activities (EVA) 480
24.4 Crew—Another Subsystem to Be Operated 481
24.4.1 Safety of the Crew 481
24.4.2 Crew Health 482
24.4.3 Crew Communication 483
24.5 Ground Support Operations 484
24.5.1 Mission Control Center (MCC) 484
24.5.2 Operations 486
24.6 The Future 487
References 488

25 Operations of On-Orbit Servicing Missions 491
Florian Sellmaier and Heike Frei
25.1 Introduction 491
25.1.1 What is On-Orbit Servicing? 491
25.1.2 Motivations for OOS 492
25.1.3 Space Debris Mitigation 493
25.2 Examples of On-Orbit Servicing Missions 494
25.2.1 Manned OOS Missions 494
25.2.2 Inspection Missions 495
25.2.3 Life Extension Missions 495
25.2.4 OOS Technology Demonstrations 497
25.2.5 Space Debris Mitigation 499
25.3 Challenges in Operating OOS Missions 500
25.3.1 Flight Operation System 501
25.3.2 Ground Data System 502
25.3.3 Flight Dynamic System 502
25.4 Satellite Rendezvous 503
25.4.1 Orbit Mechanics in Local Orbital Frame (LOF) 503
25.4.2 Mission Phases from Launch to Docking 509
25.4.3 Rendezvous Sensors 512
25.5 Satellite Capture 518
25.5.1 Communication Concept 518

25.5.2 Interaction of the Manipulator with the Servicer Platform ... 521
25.5.3 Other Aspects ... 521
25.6 Verification and Test Facilities ... 522
25.6.1 Requirements for Test Facilities ... 523
25.6.2 European Proximity Operations Simulator ... 524
25.6.3 On-Orbit Servicing Simulator ... 526
25.6.4 End-to-End Simulation ... 526
25.7 Summary and Outlook ... 527
References ... 528

26 Interplanetary Operations ... 531
Paolo Ferri
26.1 Types of Interplanetary Missions ... 531
26.1.1 Fly-By Missions ... 532
26.1.2 Orbiting Missions ... 532
26.1.3 Landing Missions ... 534
26.2 The Challenges of Interplanetary Flight ... 536
26.2.1 Trajectory Dynamics ... 536
26.2.2 Energy for the On-Board Systems ... 537
26.2.3 Communications with Earth ... 539
26.3 Mission Control Approach ... 540
26.3.1 Specifics of Interplanetary Flight Operations ... 541
26.3.2 Ground Contact Activities ... 545
26.3.3 Trajectory Determination ... 549
26.4 Special Operations ... 553
26.4.1 Planet Swing-By ... 553
26.4.2 Asteroid Fly-By ... 555
26.4.3 Planet Orbit Insertion ... 556
26.4.4 Landing Operations ... 558
26.5 Conclusions ... 558
References ... 558

27 Lander Operations ... 561
Stephan Ulamec and Paolo Ferri
27.1 Overview ... 561
27.2 Lander Release ... 566
27.3 Various Landing Strategies ... 568
27.3.1 Entry, Descent and Landing Through an Atmosphere ... 568
27.3.2 Landing on Large Airless Bodies: The Moon ... 572
27.3.3 Landing on Asteroids and Comets ... 574
27.3.4 Penetrators ... 576

27.4 Surface Operations 576
27.4.1 Operations of Stationary Surface Elements 577
27.4.2 Rover Operations 579
27.5 Conclusions 583
References 583

Index 587

Editors and Contributors

About the Editors

Florian Sellmaier is the head of business development at the German Space Operations Center (GSOC). He received his Ph.D. from the Institute of Astronomy and Astrophysics/Munich in 1996 and also spent some time at CALTECH in Pasadena/USA. After working in the radar and navigations area and as team lead of the Siemens Competence Center for e-Business Applications, he joined DLR in 2006. As a project manager, he conducted the acquisition and predevelopment of future mission like OLEV, DEOS and EDRS. He is author of several scientific publications in the Astrophysics as well as in the space operations area.

Thomas Uhlig received his physics diploma and his Ph.D. from the University of Regensburg/Germany, where he worked on electromagnetic imaging of nanostructures. After working as scientist at the Argonne National Laboratory in Chicago he changed into the spaceflight business and to the Columbus Control Center in Oberpfaffenhofen, where he first worked as Payload expert on the Space Shuttle mission STS-122, then for several years as Columbus Flight Director. He then took over the training for the European ISS flight controllers, before he was assigned as team lead for the geostationary satellite team at GSOC. In between, he was working as expert for functional safety, first at BMW, then at Heidenhain GmbH before returning to DLR in 2022. He was awarded with the Helmholtz price 2005 of the Physikalisch-Technische Bundesanstalt. He is author of various scientific articles.

Michael Schmidhuber works in the Department for Mission Operations at the German Space Operations Center (GSOC) where he is responsible for satellite operations training. He graduated as an aerospace engineer at the Technical University of Munich in 1994. The training activities comprise not only staff training but also courses for an external audience in the annual Spacecraft Operations Course at GSOC. Also, he is involved in the organization of the bi-annual international SpaceOps Conference.

Contributors

Ralph Ballweg Munich, Germany

Jérôme Campan German Space Operations Center (GSOC), German Aerospace Center (DLR), Oberpfaffenhofen, Germany

Franck Chatel German Space Operations Center (GSOC), German Aerospace Center (DLR), Oberpfaffenhofen, Germany

Sabrina Eberle German Space Operations Center (GSOC), German Aerospace Center (DLR), Oberpfaffenhofen, Germany

Ralf Faller German Space Operations Center (GSOC), German Aerospace Center (DLR), Oberpfaffenhofen, Germany

Paolo Ferri Frankfurt am Main, Germany;
Retired, European Space Agency, ESA-ESOC, Darmstadt, Germany

Hauke Fiedler German Space Operations Center (GSOC), German Aerospace Center (DLR), Oberpfaffenhofen, Germany

Heike Frei German Space Operations Center (GSOC), German Aerospace Center (DLR), Oberpfaffenhofen, Germany

Amanuel Geda German Space Operations Center (GSOC), German Aerospace Center (DLR), Weilheim, Germany

Marcin Gnat German Space Operations Center (GSOC), German Aerospace Center (DLR), Oberpfaffenhofen, Germany

Tobias Göttfert German Space Operations Center (GSOC), German Aerospace Center (DLR), Oberpfaffenhofen, Germany

Jacobus Herman German Space Operations Center (GSOC), German Aerospace Center (DLR), Oberpfaffenhofen, Germany

Dennis Herrmann German Space Operations Center (GSOC), German Aerospace Center (DLR), Oberpfaffenhofen, Germany

Markus Hobsch German Space Operations Center (GSOC), German Aerospace Center (DLR), Oberpfaffenhofen, Germany

Felix Huber German Space Operations Center (GSOC), German Aerospace Center (DLR), Oberpfaffenhofen, Germany

Ralph Kahle German Space Operations Center (GSOC), German Aerospace Center (DLR), Oberpfaffenhofen, Germany

Michael Kirschner German Space Operations Center (GSOC), German Aerospace Center (DLR), Oberpfaffenhofen, Germany

Christoph Lenzen German Space Operations Center (GSOC), German Aerospace Center (DLR), Oberpfaffenhofen, Germany

Jürgen Letschnik Technical University Munich, Garching, Germany

Tarsicio Lopez-Delgado Airbus Defence and Space, Ottobrunn, Germany

Sebastian Löw German Space Operations Center (GSOC), German Aerospace Center (DLR), Oberpfaffenhofen, Germany

Falk Mrowka German Space Operations Center (GSOC), German Aerospace Center (DLR), Oberpfaffenhofen, Germany

Kay Müller German Space Operations Center (GSOC), German Aerospace Center (DLR), Oberpfaffenhofen, Germany

Andreas Ohndorf German Space Operations Center (GSOC), German Aerospace Center (DLR), Oberpfaffenhofen, Germany

Sven Prüfer German Space Operations Center (GSOC), German Aerospace Center (DLR), Oberpfaffenhofen, Germany

Dieter Sabath German Space Operations Center (GSOC), German Aerospace Center (DLR), Oberpfaffenhofen, Germany

Michael Schmidhuber German Space Operations Center (GSOC), German Aerospace Center (DLR), Oberpfaffenhofen, Germany

Sina Scholz German Space Administration, German Aerospace Center (DLR), Bonn, Germany

Florian Sellmaier German Space Operations Center (GSOC), German Aerospace Center (DLR), Oberpfaffenhofen, Germany

Sofya Spiridonova German Space Operations Center (GSOC), German Aerospace Center (DLR), Oberpfaffenhofen, Germany

Gerd Söllner German Space Operations Center (GSOC), German Aerospace Center (DLR), Oberpfaffenhofen, Germany

Adrian R. L. Tatnall University of Southampton, Southampton, UK

Thomas Uhlig German Space Operations Center (GSOC), German Aerospace Center (DLR), Oberpfaffenhofen, Germany

Stephan Ulamec German Aerospace Center (DLR), Cologne, Germany

Abbreviations

3PO	Pre-increment products and WLP officer
A.N.	Ascending node
ABM	Apogee boost maneuver
ACS	Atmosphere control and supply
ACS	Attitude control subsystem
ACU	Antenna control unit
AD/BD	Acceptance mode data/bypass mode data
ADC	Analog-to-digital converter
ADR	Active debris removal
AFD	Automated file distribution
AFW	Astro- und Feinwerktechnik
AGC	Automatic gain control
AI	Artificial intelligence
AIT	Assembly, integration and test
AITV	Assembly, integration, test and validation
ALC	Automatic level control
ALSEP	Apollo Lunar Surface Experiment Packages
AM	Amplitude modulation
AMCS	Attitude measurement and control system
AMF	Apogee motor firing
AND	Alpha-numeric display
AOCS	Attitude and orbit control system
AOS	Acquisition of signal
APID	Application ID
APX	Alpha/X-ray fluorescence
AR	Anomaly report
AR	Atmosphere revitalization
Artemis	Advanced Relay and Technology Mission
ASK	Amplitude shift keying
ASSC	Automatic satellite saturation control
ATHMoS	Automated Telemetry Health Monitoring System

ATM	Asynchronous transfer mode
ATV	Automated Transfer Vehicle
AU	Astronomical unit
AUX	Auxiliary
AWS GS	Amazon Web Service Ground Station
BAPTA	Bearing and power transfer assembly
BAT	Battery
BER	Bit error rate
BLR	Bangalore ground station
BME	Biomedical engineer
BoL	Begin of life
BPSK	Binary phase shift keying
C&DH	Command and data handling
CAM	Collision avoidance maneuver
CAMP	Channel amplifier
CCD	Charge-coupled device
CCS	Central checkout system
CCSDS	Consultative Committee for Space Data Systems
CDR	Critical design review
CE	Concurrent engineering
CEB	Cebreros ground station
CESS	Coarse Earth-Sun sensor
CHeCS	Crew healthcare system
CLCW	Command link control word
CLIP	Columbus lead increment planner
CLTU	Command link transmission unit
CM	Configuration manager
CMC	Communication management center
CMD	Command operator system
CMOS	Complementary metal oxide semiconductor
CNB	Canberra ground station
CNES	Center National d'Études Spatiales
CoG	Center of gravity
COL FD	Columbus flight director
COL OC	Columbus operations coordinator
Col-CC	Columbus Control Center
COMMS	Communications subsystem
COP	Command operations procedure
COP	Composite operations plan
COR	Critical operations review
COSMO	Columbus stowage and maintenance officer
COTS	Commercial off-the-shelf
COUP	Consolidated operations and utilization plan
CPDU	Command pulse distribution unit
CPS	Consolidated planning system

CPU	Central processing unit
CR	Change request
CRAF	Comet rendezvous asteroid flyby
CRC	Cyclic redundancy check
CRM	Crew resource management
CRT	Command ranging and telemetry
CSC	Communication service center
CSRD	Current stage requirement document
CSS	Crew support system
CUP	Composite utilization plan
D/L	Downlink
DARPA	Defense Advanced Research Projects Agency
DART	Demonstration for Autonomous Rendezvous Technology
DC	Direct current
DCU	Diversity combiner unit
DDOR	Delta differential one-way ranging
DEM	Digital elevation model
DEOS	Deutsche Orbitale Servicing Mission
DFT	Data flow test
DH	Data handling
DIM	Digital interface module
DLR	Deutsches Zentrum für Luft- und Raumfahrt (German Aerospace Center)
DMR	Detailed mission requirement
DMS	Data management subsystem
DMZ	Demilitarized zone
DNEL	Disconnection of non-essential loads
DoD	Depth of discharge
DOR	Differential one-way ranging
DPC	Daily planning conferences
DR	Discrepancy report
DS	Daily summary
DSHL	Disconnection of supplementary heater lines
DTM	Digital terrain model
DVB	Digital video broadcasting
EAC	European Astronaut Center
EADS	European Aeronautic Defense and Space Company
EASA	European Union Aviation Safety Agency
EC	Earth center
ECEF	Earth-centered, Earth-fixed
ECLSS	Environmental control and life support system
ECR	Engineering change requests
ECSS	European Cooperation for Space Standardization
EDA	Electronically despun antenna
EDAC	Error detection and correction

EDL	Entry, descent and landing
EDRS	European Data Relay System
EEPROM	Electrically erasable programmable read-only memory
EFN	Electronic flight note
EGS-CC	European Ground System Common Core
EGSE	Electrical ground support equipment
EIRP	Equivalent isotropically radiated power
EMC	Electromagnetic compatibility
EMU	Extravehicular mobility unit
ENVISAT	Environmental Satellite
EO	Earth observation
EOC	End of charge
EOL	End of life
EOM	End of mission
EPDS	Electrical power distribution subsystem
EPIC	European Planning and Increment Coordination
EPT	European planning team
ESA	European Space Agency
ESOC	European Space Operations Center
ESS	Experimental Servicing Satellite
ESTEC	European Space Research and Technology Center
EUROCOM	European spacecraft communicator
EVA	Extravehicular activity
EXOSAT	European X-Ray Observatory Satellite
FAA	Federal Aviation Administration
FARM	Frame acceptance and reporting mechanism
FCLTU	Forward command link transmission unit
FCP	Flight control procedure
FCT	Flight control team
FD	Flight director
FD	Flight dynamics
FDIR	Fault detection, isolation and recovery
FDS	Fire detection and suppression
FDS	Flight dynamics system
FEC	Forward error correction
FGAN	Forschungsgesellschaft für angewandte Naturwissenschaften
FIFO	First in, first out
FM	Frequency modulation
FMECA	Failure mode effects and criticality analysis
FOP	Flight operations procedure
FORDEC	Facts, Options, Risks, Decision, Execute, Check
FOS	Flight operations system
FOT	Flight operations team
FOV	Field of view
FSK	Frequency shift keying

FSS	First science sequence
FTP	File transfer protocol
G/T	Gain over noise temperature
GAM	Gravity assist maneuver
GCC	Galileo Control Center
GCM	Gyro calibration mode
GDS	Goldstone ground station
GDS	Ground data system
GECCOS	GSOC Enhanced Command- and Control System for Operating Spacecraft
GEO	Geostationary Earth orbit
GGR&C	Generic groundrules, requirements and constraints
GMRT	Giant Metrewave Radio Telescope
GMT	Greenwich Mean Time
GNC	Guidance, navigation and control
GOCE	Gravitational Ocean Composition Explorer
GOP	Ground operations procedures
GPS	Global Positioning System
GR&C	Ground rules and constraints
GRACE	Gravity Recovery and Climate Experiment
GRO	Gamma ray observatory
GS	Ground segment
GSCDR	Ground segment critical design review
GSN	Ground station network
GSOC	German Space Operations Center
GSQR	Ground segment qualification review
GSYS	Ground system
GTO	Geostationary transfer orbit
GUI	Graphical user interface
HAMR	High area-to-mass ratio
HCP	Health-check parameters
HEO	Highly elliptical orbits
HF	High frequency
HGA	High gain antennae
HK	Housekeeping
HLP	Horizontal linear polarized
HMI	Human-machine interface
HPA	High-power amplifier
HPTC	High-priority (or high-power) telecommand
HST	Hubble Space Telescope
HTV	H-II Transfer Vehicle
HUDEP	Human Dependability
I/F	Interface
IADC	Inter-Agency Space Debris Coordination Committee
ICD	Interface control document

IDRD	Increment definition and requirement document
IEEE	Institute of Electrical and Electronics Engineers
IEPT	International execute planning telecon
IF	Intermediate frequency
IFM	In-flight maintenance
IMF	Interplanetary magnetic field
IMU	Inertial measurement unit
Inmarsat	International Maritime Satellite Organization
IO	Intermediate orbit
IOAG	Interagency Operations Advisory Group
IOT	In-orbit test
IP	International partners
IP	Internet protocol
IPFD	Input power flux density
IPV	International procedure viewer
IR	Infrared
IRES	Infrared Earth sensor
IRIG	Inter-Range Instrumentation Group
ISDN	Integrated Services Digital Network
ISI	Inter-symbol interference
ISL	Inter-satellite links
ISO	International Organization for Standardization
ISRO	Indian Space Research Organization
ISS	International Space Station
IT	Information technology
ITU	International Telecommunication Union
JAA	Joint Aviation Authorities
JAXA	Japan Aerospace Exploration Agency
JPL	Jet Propulsion Laboratory
JSpOC	Joint Space Operations Center
KSAT	Kongsberg Satellite Services
L2	Lagrangian point 2
LAN	Local area network
LC	Large constellations
LCC	Lander Control Center
LDPC	Low-density parity-check
LEO	Low Earth orbit
LEOP	Launch and early orbit phase
LFSR	Linear feedback shift registers
LGA	Low gain antennae
LHCP	Left-hand circular polarized
LIDAR	Light detection and ranging
LM	Lunar module
LNA	Low-noise amplifier
LOF	Local orbital frame

LOS	Loss of signal
LRP	Long-range planner
LS	Logistics support
LTS	Long-term science
M&C	Monitoring and control
M.A.I.T.	Manufacturing, assembly, integration and test
MAC	Media-access control
MAD	Madrid ground station
MAM	Mapping antenna mode
MAP ID	Multiplexing access point identifier
MASCOT	Mobile Asteroid Surface Scout
MCC	Mission control center
MCC-H	Mission Control Center Houston
MCC-M	Mission Control Center Moscow
MCMUR	McMurdo ground station
MCS	Monitoring and control (sub)system
MD	Mission director
MDS	Mission data subsystem
MEA	Mean error amplifier
MedOps	Medical operations
MEO	Medium Earth orbit
MEP	Mission extension pods
MER	Mars Exploration Rover
MET	Mission elapsed time
MGA	Medium gain antennae
MIB	Mission information (data)base
MIPD	Multi-increment planning document
MIPROM	Multi-increment payload resupply and outfitting model
MiTEx	Micro-Satellite Technology Experiment
MLI	Multi-layer insulation
MMI	Man-machine interface
MMU	Memory management unit
MMX	Mars Moon eXploration
MOD	Mission operations director
MOIS	Mission operations information system
MOS	Mission operation system (flight director)
MOTS	Modifiable off-the-shelf
MPCB	Multilateral payloads control board
MPLM	Multipurpose logistic module
MPLS	Multi-protocol label switching
MPPT	Maximum power point tracker
MPS	Mission planning subsystem
MRO	Mars Reconnaissance Orbiter
MRV	Mission Robotic Vehicle
MSL	Mars Science Laboratory

MTL	Mission timeline
MW	Momentum wheel
NAS	National Airspace System
NAS	Network-attached storage
NASA	National Aeronautics and Space Administration
NASCOM	NASA communications network
NASDA	National Space Development Agency of Japan
NCR	Non-conformance reports
NEAR	Near Earth Asteroid Rendezvous
NIS	Network information service
NMC	Network management center
NOPE	Network operations project engineer/network operator
NORAD	North American Aerospace Defense Command
NRZ	Non-return to zero
OASPL	Overall acoustic sound pressure level
OBC	On-board computer
OBCP	On-board control procedure
OBDH	On-board data handling
OBSW	On-board software
OBT	On-board training
OD	Orbit determination
ODF	Operations data file
OLEV	Orbital Life Extension Vehicle
OMT	Orthomode transducer
OOS	On-orbit operations summary
OOS	On-orbit servicing
OOS-Sim	On-orbit servicing simulator
OPS	Operations
OPS-LAN	Operational local area network
OPTIMIS	Operations Planning Timeline Integration System
ORR	Operational readiness review
ORU	Orbital replacement unit
OSAM	On-orbit servicing, assembly and manufacturing
ROSAT	ROentgen SATellite (X-ray satellite)
OSIRIS	Optical, Spectroscopic and Infrared Remote Imaging System
OSTP	On-orbit short-term plan
P/T	Power thermal
PAC	Packet assembly controller
PC	Personal computer
PCDU	Power control and distribution unit
PCM	Pulse code modulation
PCS	Payload control system
PD	Proportional differential
PDR	Preliminary design review
PDU	Power distribution unit

PICB	Program integration control board
PID	Proportional integral differential
PINTA	Program for Interactive Timeline Analysis
PKF	Pokerflat ground station
PLC	Programmable logic controller
PLL	Phase-locked loop
PLUTO	Procedure language for users in test and operations
PM & P	Parts, materials and processes
PM	Phase modulation
PM	Processor module
PM	Project manager
PMC	Private medical conference
PMD	Photonic mixer device
PMD	Post-mission disposal
PO	Project office/project officer
POIC	Payload Operations and Integrations Center
PPC	Private psychological conference
PPCR	Planning product change request
PR	Planning request
PRN	Pseudo random noise
PROM	Programmable read-only memory
PROP-F	ПрОП-Ф (Прибор оценки поверхности—Фобос) Russian acronym for “Mobile Robot for Evaluation of the Surface of Phobos”
ProToS	Procedure Tool Suite
PSK	Phase shift keying
PTP	Payload tactical plan
PTS	Power and thermal system
PUS	Packet Utilization Standard
PVT	Pressure-volume-temperature
QA	Quality assurance
QAM	Quadrature amplitude modulation
QOS	Quality of service
QPSK	Quadrature phase shift keying
QR	Qualification review
RAAN	Right ascension of ascending node
RAF	Return all frames
RAM	Random access memory
RCF	Return channel frames
RCS	Robotic control system
RDU	Remote data unit
RDUA	Remote data unit A
RF	Radio frequency
RFE	Radio frequency electronics
RGPS	Relative GPS

RHCP	Right-hand circular polarized
RID	Review item discrepancy
RM	Reconfiguration module
RPM	Revolutions per minute
RSA	Russian Space Agency
RSGS	Robotic Servicing of Geosynchronous Satellites
RSM	Requirements/specification matrices
RTG	Radioisotope thermoelectric generator
RvD	Rendezvous and docking
RW	Reaction wheel
Rx	Receiver
S&M	Structure and mechanics
S/C	Spacecraft
S/G	Space-to-ground
SAD	Solar array drive
SAM	Sun-acquisition mode
SAN	Storage area network
SAR	Satellite anomalies reports and recommendations
SAR	Synthetic aperture radar
SBC	Satellite board computer
SBM	Standby mode
SCC	Satellite control center
SCID	Spacecraft ID
SCOS	Spacecraft Control and Operation System
SDH	Synchronous digital hierarchy
SDM	Space debris mitigation
SDN	Software defined networking
SE	System engineer
SEU	Single event upsets
SGM	Safeguard memory
SIM	Simulation officer
SLE	Space Link Extension
SNG	Satellite news gathering
SNR	Signal-to-noise ratio
SO	Security officer
SoE	Sequence of events
SOM	Spacecraft operations manager
SONC	Science Operations and Navigation Center
SOP	Satellite operations procedure
SP	Sun presence
SPACON	Spacecraft controller
SPZBG	Spitzbergen ground station
SRDB	Satellite reference database
SS	Space system
SS	Sun sensor

SSB	SLE switch board
SSCB	Space Station Control Board
SSE	Subsystem engineer
SSIPC	Space Station Integration and Promotion Center
SSMM	Solid state mass memory
SSPA	Solid state power amplifier
SST	Satellite/spacecraft support team
ST	Star tracker
STL	Satellite team lead
STOL	Systems test and operation language
STP	Short-term plan
STRATOS	Safeguarding, Thermal, Resources, Avionics, Telecommunications, Operations, Systems
STS	Space Transportation System
SV	Space vehicle
SVT	System validation test
SW	Software
TAFF	TanDEM-X autonomous formation flying experiment
TAR	Technical acceptance review
TC	Telecommand
TCP	Transmission control protocol
TCR	Telemetry, commanding and ranging
TCS	Thermal control subsystem
TDM	TanDEM-X mission
TDRS	Tracking and Data Relay Satellite
TDRSS	Tracking and Data Relay Satellite System
TDX	TanDEM-X satellite
TEGA	Thermal evolved gas analyzer
TET	Technologie Erprobungs Träger (Technology Experiment Carrier)
THC	Temperature and humidity control
TIM	Technical interface meetings
TLE	Two-line elements
TLT	Test-loop translator
TM	Telemetry
TM/TC	Telemetry/telecommand
TRL	Technology readiness level
TSM	TerraSAR-X mission
TSTD-MPS	TerraSAR-X TanDEM-X mission planning system
TSX	TerraSAR-X satellite
TT&C	Telemetry, tracking and command
TTC-RF	Telemetry, tracking and command at radio frequency
TTTC	Time-tagged telecommand
TWTA	Traveling wave tube amplifier
TX	Transmitter
U/L	Uplink

UHF	Ultra high frequency
UPS	Unified propulsion system
UPS	Uninterruptible power supply
USOC	User Support and Operations Center
UT1	Universal time 1
UTC	Universal time coordinated
UV	Ultraviolet
VC	Virtual channel
VDS	Virtual distributed switches
VLBI	Very long base interferometry
VLP	Vertical linear polarized
VoIP	Voice over IP
VPN	Virtual private network
VSAT	Very small aperture terminal
WAN	Wide area network
WD	Watchdog
WFC	Wide field camera
WHM	Weilheim ground station
WLP	Weekly look-ahead plan
WPC	Weekly planning conference
WPR	Weekly plan review
WRM	Water recovery and management
WSP-C	Weilheim Service-Provider-Cortex
XML	Extensible markup language
XMM-Newton	X-ray Multi-Mirror Mission

Part I
Overview Space Segment

Chapter 1
Space Environment

Adrian R. L. Tatnall and Hauke Fiedler

Abstract The environment in which spacecraft have to function is not only life-threatening for humans but also challenging for the spacecraft itself. To successfully cope with this environment many aspects including acceleration, atmosphere, vacuum, solar radiation and its implications have to be taken into consideration. Such factors are examined in more detail in the following chapter "Space Environment".

1.1 Introduction

The environment in which spacecraft have to function is life-threatening for humans, and we cannot survive there without protection for more than a few seconds. Fortunately, in this respect, spacecraft are generally more robust than humans are and spacecraft can continuously operate in space for more than 15 years on a regular basis. To take Voyager 1 as an example, launched over 40 years ago, the spacecraft is still operational and communicates with Earth from a distance of 22 billion km. It is interesting to question how this longevity can be achieved when maintenance is not an option and the environment, at first sight, appears so unattractive.

To understand these issues, it is important to take a moment to consider what constitutes the space environment, which is definitely different due to the fact that many of the sources of erosion and wear found on Earth do not exist in space.

The space environment is alien, it is remote in the sense that it is difficult and costly to get there. The borderline of space is generally considered to start at the Kármán line at an altitude of 100 km in the thermosphere (Fig. 1.1). The short trip of 100 km represents a major challenge for rockets, and the trip itself subjects the spacecraft to an environment totally different from that to which it is subjected on the ground or in space.

A. R. L. Tatnall
University of Southampton, Southampton, UK

H. Fiedler (✉)
German Space Operations Center (GSOC), German Aerospace Center (DLR), Oberpfaffenhofen, Germany
e-mail: hauke.fiedler@DLR.de

F. Sellmaier et al. (eds.), *Spacecraft Operations*, Springer Aerospace Technology,
https://doi.org/10.1007/978-3-030-88593-9_1

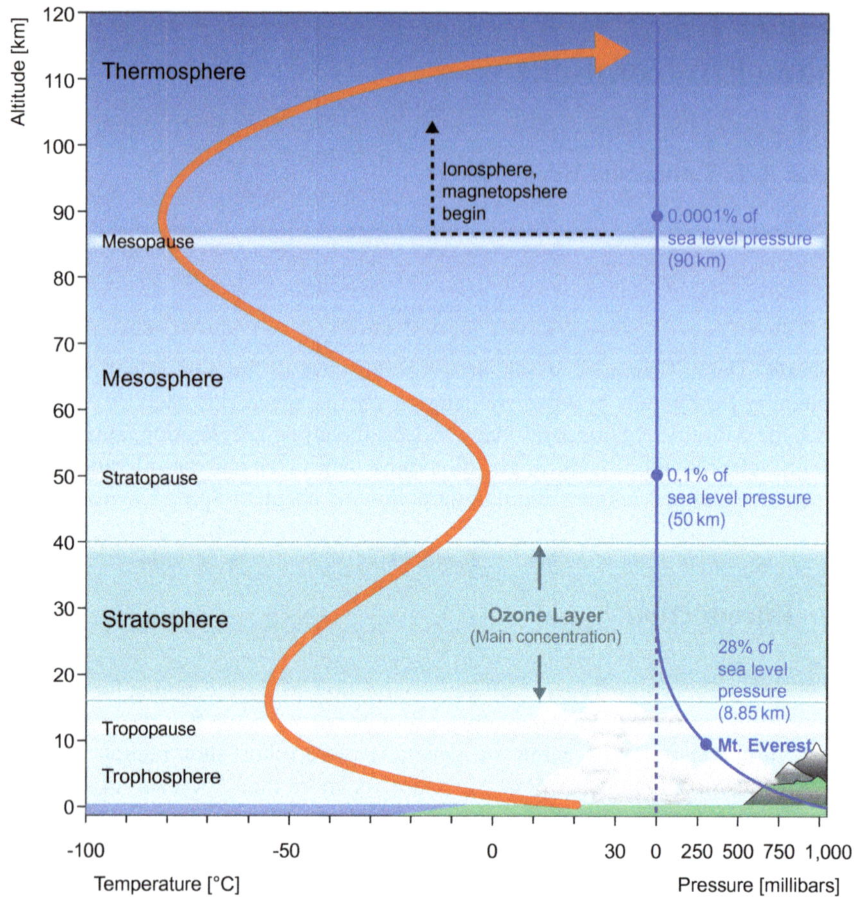

Fig. 1.1 Earth's atmosphere. Adapted from Encyclopedia Britannica online (2012)

1.2 Launch Vehicle

1.2.1 Acoustic/Vibration Levels

Launches are extremely loud, as any spectator will confirm. Acoustics and the vibration levels reach a maximum at the moment of launch when the rocket engines ignite and the exhaust emission is reflected from the ground. As the rocket ascends, the effects governing ground contact decrease, but other mechanical moving parts and unsteady aerodynamic phenomena continue to excite the structure. This structural excitation produces a secondary acoustic field within the structure. As the speed of the rocket increases, the sound field peaks for a second time during the transonic flight phase, which typically occurs just below Mach 1, the speed of sound. The

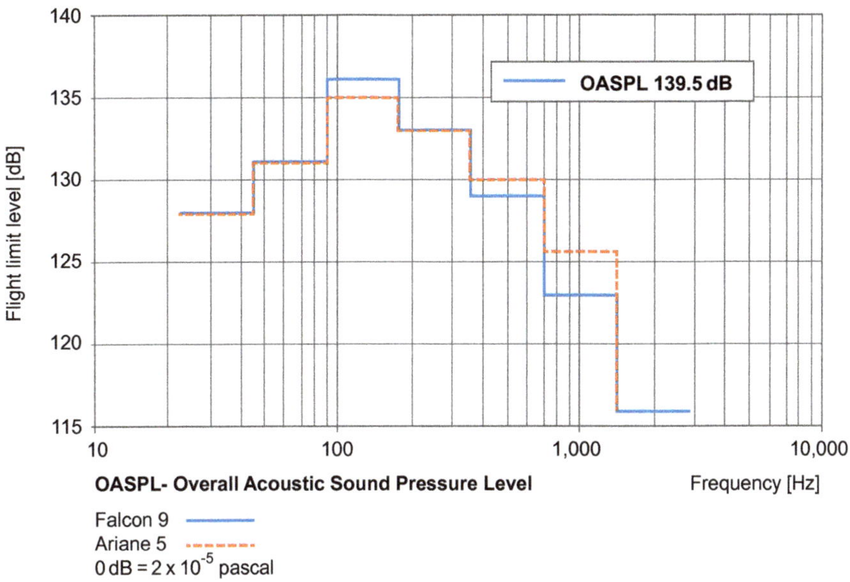

Fig. 1.2 Acoustic sound levels. Data taken from Ariane 5 User's Manual (Arianespace 2011) and Falcon 9 Launch Vehicle Payload User's Guide (SpaceX 2009)

overall levels experienced under the fairing of the Ariane V and Falcon 9 rocket are shown in Fig. 1.2. Acoustic noise has a negative effect on any lightweight structures, and in that context, antenna parabolic reflectors, solar arrays and spacecraft panels are particularly at risk.

1.2.2 *Static Acceleration*

At the moment of launch the rocket mass is at its maximum. The rate of acceleration is correspondingly low as the thrust produced is virtually constant. Rocket acceleration increases as a result of the reduction in the amount of propellant onboard until the solid rocket booster burns out and separation occurs. This gives rise to the distinctive static acceleration profile at launch shown in Fig. 1.3. Since the accelerations vary with time, the effect on the spacecraft is to generate quasi-static loads. These loads determine the major load bearing parts of the spacecraft structure such as the central thrust tube.

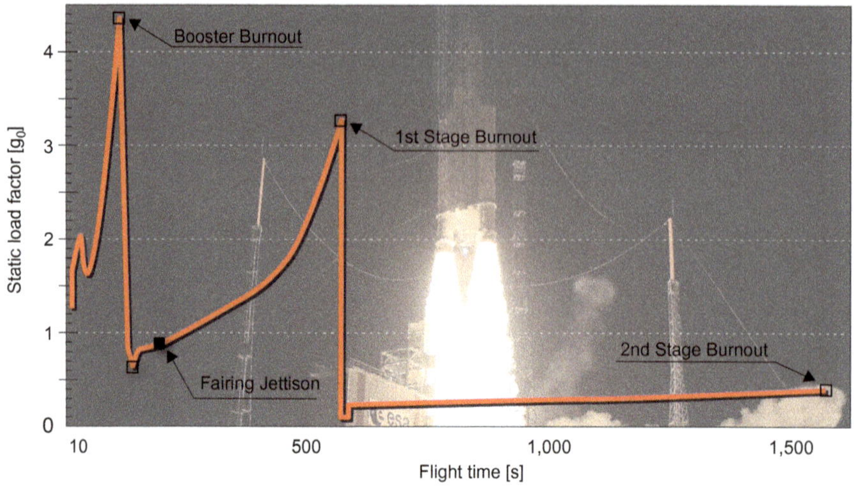

Fig. 1.3 Ariane 5 static acceleration profile (Ariane 5 User's Manual (Arianespace 2011))

1.2.3 Mechanical Shock

A number of events can lead to very high acceleration rates being produced for very short periods of time. These shocks include.

- Ignition and separation of the launch vehicle stages
- Fairing jettison
- Spacecraft separation
- Docking and landing.

Excitation peaks between 1 and 10 kHz for Ariane and Falcon 9. The rate for Ariane 5 is 2000 g_0 and 3000 g_0 in the case of Falcon 9. Despite these very high figures, the transient nature of these loads means that they are of no consequence as regards structural strength, but they are of concern to the functioning of equipment like relays.

1.3 Spacecraft Operational Environment

1.3.1 Vacuum

The ambient pressure by the time a spacecraft reaches low Earth orbit at 300 km is comparable to what could be achieved by using a very good vacuum chamber (about 10^{-7} Pa) on Earth. At an altitude of 800 km the pressure is so low that it cannot be reproduced in a terrestrial environment.

It is therefore important that materials which do not outgas are used in the construction of a spacecraft. Outgassing occurs because the material itself sublimates. Gases are released from cracked materials, or gases that are adsorbed by the surfaces are released in a near vacuum. While this will probably not affect the structural integrity of the spacecraft, it might have an impact on the surface properties and there is always the possibility that the vaporized material will condense if it impinges on colder spacecraft surfaces. It is therefore important that materials such as cadmium, zinc, PVC and many plastics with high vapor pressures are not used. Adsorbed gases can alter the properties of materials. Graphite is a solid lubricant commonly used on Earth, but in space the adsorbed water vapor is lost and graphite is ineffective as a lubricant. Alternatives such as molybdenum disulfide then have to be used. If a vacuum alters the properties of a material which has to be used, residual contaminants can be removed by baking and applying a protective coating or shielding.

1.3.2 Solar Radiation Flux

The spectrum of the radiation from the Sun is approximately that of a black body at a temperature of 5,777 K. This is the temperature of the photosphere, the opaque region of the Sun which is usually considered to constitute its surface. Our eyes have a response which is optimized for the light emitted by the Sun, which peaks at about 550 nm. This radiation is virtually constant and varies by about 0.1% from sunspot maximum to sunspot minimum, although there are seasonal variations in the radiation incident on the Earth which result in deviations of up to 3.3% as the Earth moves in an elliptical orbit round the Sun. The level of radiation we can't see, however, varies to a greater degree. Instead of originating from the photosphere, the source of UV and X-rays is to be found in the outer regions of the Sun, i.e. the chromosphere and the corona.

This can be understood by noting that the temperature increases as the distance from the Sun increases. At a distance of about 2,500 km above the Sun's surface, the temperature in the corona is about a million degrees, and so hot that radiation at X-ray wavelengths is emitted. The variations in the conditions in the corona fluctuate enormously and occur at a rate of seconds to months, which is reflected in the irregularity of the UV and X-rays produced. While terrestrial weather is undoubtedly influenced by variations in solar conditions, but since the overall variations in energy output are very small, it is not easy to distinguish the effects of these variations from the much larger natural variability of our weather. Space weather, however, is dominated by fluctuations in the Sun's output as this has a profound impact on the UV, X-rays and particles impacting the Earth. UV radiation directly affects the materials used in spacecraft, and in particular the solar arrays. The absorption of UV by the cover glass used to protect the solar cell from particle radiation and the slide adhesive can lead to darkening. The effect is twofold: the cell illumination lessens, thus reducing electrical power produced. It also heats up the cell which leads to a reduction in efficiency. By doping the cover glass with cerium oxide UV can be absorbed, thereby preventing darkening.

In addition, solar radiation flux is responsible for the radiation pressure created by absorbed or reflected photons. This is the basis for the force generated from “solar sailing”, a means of controlling or propelling a spacecraft. An experimental spacecraft launched by the Japan Aerospace Exploration Agency (JAXA) called IKAROS successfully used solar sails to fly to Venus (Mori et al. 2009).

1.3.3 Particle Radiation

The Sun emits a continual stream of high-energy particles. These particles are mainly protons and electrons with an energy of 1.5–10 keV. They move at a speed of 400–800 km/s, creating the solar wind that pervades our solar system and extends to the termination shock at 85–95 AU from the Sun. Despite the high speed of the particles in the solar wind, their density when reaching Earth is 5 atoms/cm^3, rising to a few hundred atoms/cm^3 during phases of high solar activity. This exerts a negligible pressure on any spacecraft impacted by the solar wind.

A far greater pressure comes from the light pressure of photons, as described above. While the pressure from the solar wind is negligible, its consequences must be taken into consideration because it has a major impact on the Earth’s environment. The solar wind plasma interacts with the Earth’s dipole magnetic field to form the magnetosphere shown in Fig. 1.4. The magnetosphere’s distinctive asymmetric shape is due to the pressure exerted by the solar wind. On the side facing the Sun, the magnetosphere extends out to a distance of approximately 10 Earth radii under quiet conditions, and in the other direction it extends to several hundred Earth radii. The shape and extent of the magnetosphere depend on the strength and orientation of the magnetic field of the solar wind. This determines the reconnection process of the Earth’s and the Sun’s magnetic field that allows energy and momentum to be transferred from the solar wind into the magnetosphere. It may also result in the acceleration mechanism for the very high-energy particles that can be found in the radiation belts within the magnetosphere.

1.3.4 Radiation Belts

In 1958, the existence of a belt of trapped charged particles around the Earth was confirmed by Explorer 1 and 3 using instrumentation designed by James Van Allen, who had predicted that the belts would exist. The belt detected was the inner radiation belt. In the same year the Soviets—S. N. Vernov and A. E. Chudakov—discovered the second or outer radiation belt. These belts are shaped like a torus and extend from 1,000 to 60,000 km above the Earth. The outer belt is predominantly made up of electrons with a peak at 15,000–20,000 km, whereas the inner belt consists largely

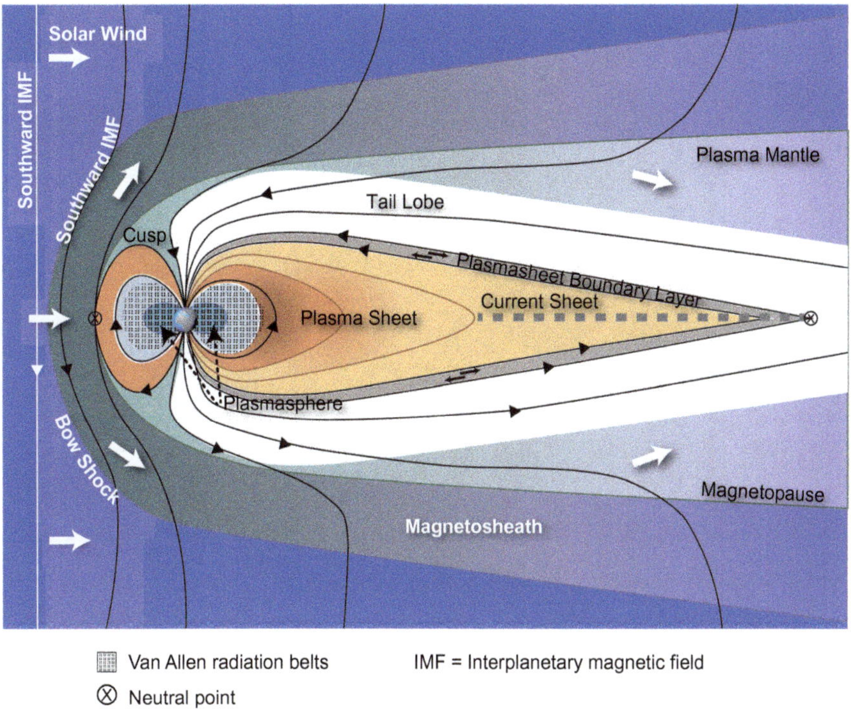

Fig. 1.4 Earth's magnetosphere (Reiff 1999)

of high-energy protons that peak at 3,000 km. Proton energies range from 0.01 to 400 meV and electron energies from 0.4 to 4.5 meV. Both are oscillating between the two Earth poles within one second. They are shown as two distinct belts in Fig. 1.5, but in practice there is no real gap between the belts and they are highly variable depending on solar activity. In September 2012, a third belt was detected further outside, which remained stable for about one month until it was dissipated by a solar flare. It is thought that such temporary belts appear frequently. The location of the radiation belts follows the magnetic field of the Earth and this means that they are not symmetrically placed with respect to the Earth. The axis of this field is offset and tilted with respect to the Earth's rotation axis and so this leads to a location over the South Atlantic where the magnetic field is anomalously low. As a result, the radiation belts are closer to the Earth over this region which is commonly known as the South Atlantic anomaly. A satellite in a low Earth orbit (LEO) orbiting in the South Atlantic anomaly is more likely to encounter energetic particles and hence suffer damage in this part of the world.

A geomagnetic storm is caused by a solar wind shock wave interacting with the Earth's magnetic field. This leads to measurable variations on the Earth's surface in the Earth's magnetic field which are accompanied by increases in charged particles in the radiation belts. These particles are influenced by the magnetic fields and perform

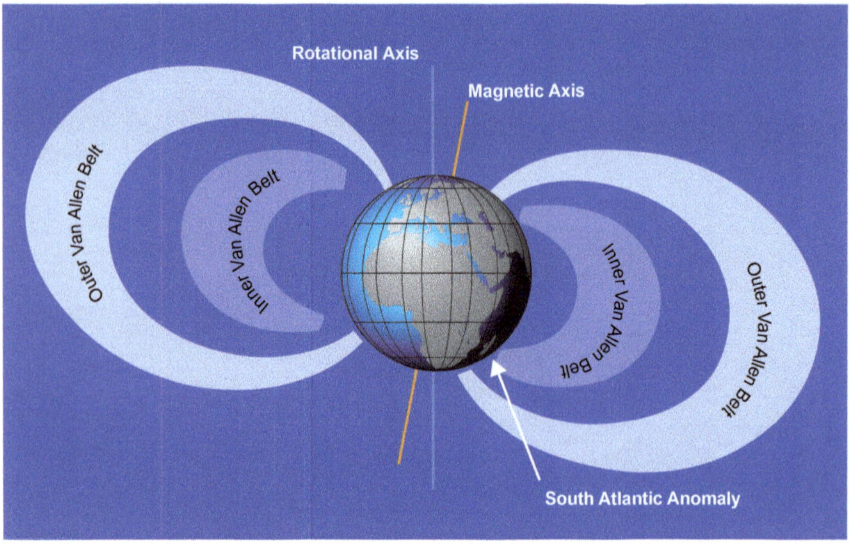

Fig. 1.5 Radiation belts inner and outer Van Allen radiation belts

three types of motions. All particles spiral around the field lines, move down field lines, bounce from one hemisphere to another, and drift around the Earth. This last motion eastward for electrons and westward for protons produces a current known as the ring current, which can be measured by observing the associated magnetic field on the surface of the Earth. It can lead to a decrease of $> 1\%$ in the magnetic field measured at the Earth's surface during a major geomagnetic storm.

The origin of the particles in the radiation belts is solar, terrestrial or cosmic. Particles of solar origin are injected into the outer belts during magnetic storms. It is believed that the protons of the inner belt originate from the decay of neutrons produced when high-energy cosmic rays from outside the solar system collide with atoms and molecules of the Earth's atmosphere.

Radiation effects include total dose effects, e.g. complementary metal oxyd semiconductor (CMOS) problems, lattice displacement damage that can affect solar cells and reduce amplifier gain, single event effects and additional noise in sensors and increased electrostatic charging. The charging of a spacecraft relative to the surrounding plasma does not pose as much of a problem compared to the possibility of increased discharges that can damage equipment and lead to the generation of electromagnetic interference. This has traditionally been thought to be more problematic in GEO (Geostationary Earth Orbit) than in LEO, where the plasma is low in energy and high in density, but nevertheless, in LEO and particularly over the polar regions, high levels of spacecraft surface charging can occur.

1.3.5 Atmosphere

Although the residual atmosphere in LEO is comparable to a very good vacuum on Earth, the resistance of this residual atmosphere has an effect on the satellite motion. Normally atmospheric drag needs to be accounted up to an altitude of 1000 km. The effect of drag on a spacecraft provides the following acceleration a_D:

$$a_D = \frac{1}{2}\rho\frac{A}{m}C_D V_r^2\left(\frac{-\boldsymbol{V}_r}{|\boldsymbol{V}_r|}\right) \tag{1.1}$$

where ρ is the atmospheric density, A the area of the satellite perpendicular to the flight direction, m the satellite mass, C_D the coefficient of drag which is typically ~2.5, and V_r the velocity vector relative to the atmosphere.

As Eq. (1.1) shows, the direction of acceleration a_D is opposite to the velocity vector V_r, so that a_D is rather a deceleration. This deceleration generally decreases the orbit height of a spacecraft, which shortens the lifetime of a mission if the spacecraft is not lifted regularly. The change in height of the International Space Station (ISS) is shown in Fig. 1.6. Reductions in altitude caused by atmospheric drag are compensated by boosts using the station's thrusters. The number of boosts required depends on the atmospheric drag and the permissible variation in height. For a spacecraft like the Gravitational Ocean Composition Explorer (GOCE), which had to be maintained at a constant height and was at a very low altitude in order to measure small changes

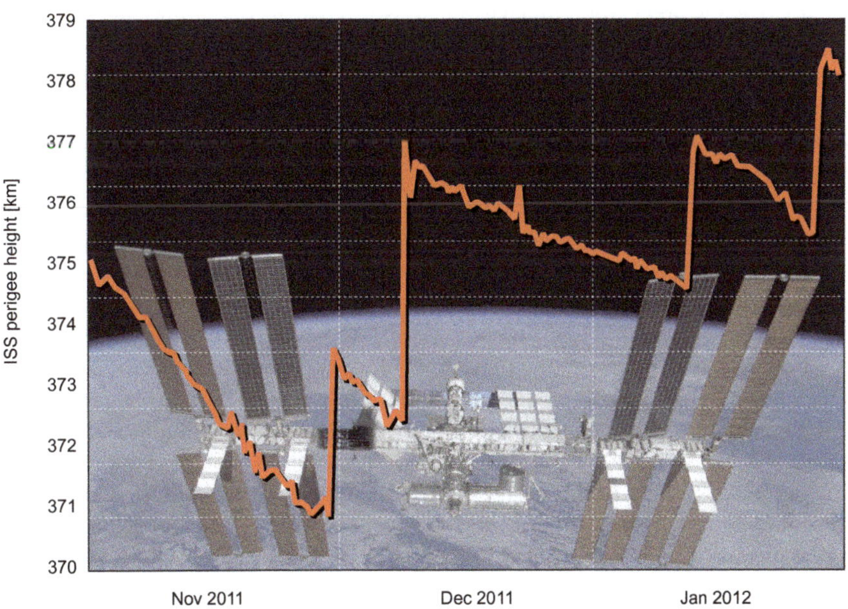

Fig. 1.6 ISS perigee height change

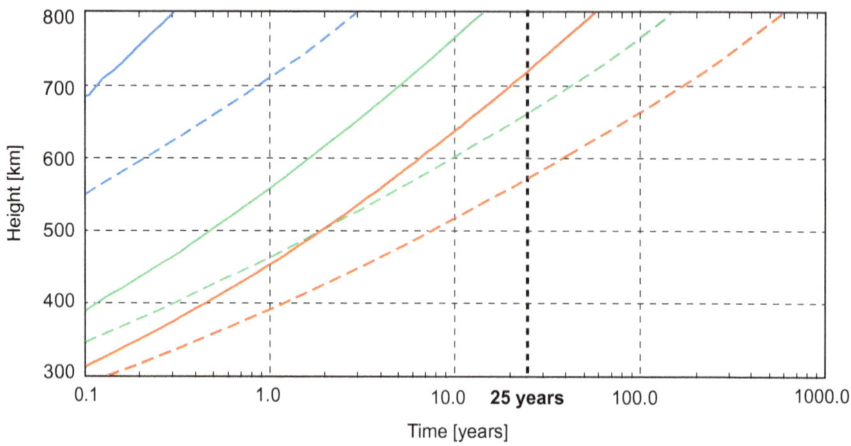

Fig. 1.7 Spacecraft lifetime as a function of altitude for mass-to-area ratio in 200 kg/m^2 (red), 50 kg/m^2 (green) and 1 kg/m^2 (blue) according to Harris-Priester. Solid lines represent high drag, dashed lines low drag. The black thick line marks 25 years

in the gravity field, this meant thrusters had to be used over long time periods. This is why electric propulsion thrusters are applicable. Objects and spacecraft which are not under full control, such as debris, will lose height more quickly when the Sun is active and the atmosphere has expanded.

Additionally, Eq. (1.1) shows, that the acceleration a_D is proportional to the density ρ and indirect proportional to the mass-to-area ratio *m/A*. Since the density of the residual atmosphere decreases with altitude, the lifetime of spacecraft increases with altitude. The lifetime of a spacecraft as a function of altitude and the mass-to-area ratio *m/A* is shown in Fig. 1.7 for a characteristic atmosphere. It shows that spacecraft with a large surface area and low mass are particularly vulnerable to orbit decay.

In order to conform to guidelines suggested by, e.g., the Inter-Agency Space Debris Coordination Committee (IADC) strongly recommended that all spacecraft must either be deorbited or moved into a graveyard orbit within a period of 25 years after the end of the mission. This has resulted in a number of proposals to achieve this by deploying a structure that will greatly increase the area of the spacecraft and hence increase atmospheric drag. Such drag augmentation systems have been adopted, e.g., within ESA's CleanSat program.

However, the residual atmosphere ρ is not only a function of altitude, but also varies with time. The solar activity heats up the upper areas of the atmosphere, which influences the density of the residual atmosphere in the low Earth orbit. Since the solar activity is variable with time, the density also varies. For example, the density of the residual atmosphere at 500 km altitude can change by a factor of 100. The effect of solar activity on the density of the residual atmosphere, and in consequence on the density of space debris, is illustrated by Fig. 1.8. It shows that the reductions

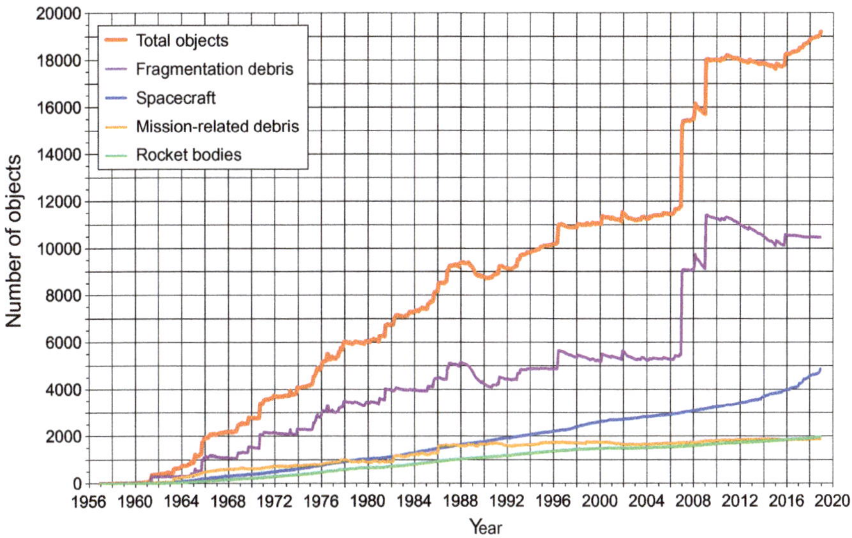

Fig. 1.8 Objects in the Earth's orbit graded by object type (NASA)

in the number of debris occur at times of solar maximum when atmospheric drag is at its greatest, e.g. 1989.

1.3.6 Space Debris

Ever since the Earth has been in existence, it has been impacted by material. The mass flux of this material is currently about 10^7–10^9 kg/year. Much of this material is dust-sized objects called micrometeoroids with a mass of less than 1 g. Their velocities relative to spacecraft average about 10 km/s, and so while they are not likely to cause catastrophic damage to spacecraft they do contribute to the weathering process and can modify material properties. For example, on August 23, 2016, the solar panel of the Sentinel-1A satellite was hit by a millimeter-sized particle. On investigation, it is not clear if the object was a micrometeoroid or man-made space debris. This event had no effect on the satellite's routine operations, which continued normally. Objects larger than 1 g do exist and over the lifetime of the Earth it has been hit by many objects of over 1 km in diameter. It is thought that 65 million years ago a 10 km meteorite hit the Yucatan Peninsula in Mexico and produced a crater 180 km in diameter and probably caused the mass extinction of dinosaurs. Of more concern to spacecraft is the increase in natural debris that occurs when the Earth moves through particles from a comet and a meteor shower can be observed. The Olympus communications spacecraft was damaged by one of the Leonid meteoroids in 1993 and subsequently suffered an electrical failure.

Man-made debris is a growing problem, as illustrated in Fig. 1.8. While the problem has been gradually increasing since man first started launching satellites, it has been exacerbated and highlighted by some recent events that have contributed to the production of large amounts of debris. One example of this was the 2009 satellite collision between Iridium 33 and Kosmos-2251, and further examples are the destructions of satellites in past anti-satellite tests of several nations including US, China and India. The impact of some of these events can be seen in Fig. 1.8. The number of known occurrences is increasing and can be classified as follows:

- Launch and operational debris
- Space vehicle breakup (57 of them deliberate)
- Explosions
- Collision induced (5 to date—latest Iridium 33/Kosmos 2251, 10 Feb 2009)
- Upper stage breakup (largest contribution—Breeze M in 2007, 2010, 2011, 2012)
- Crumbled residue from spacecraft surfaces (paint, MLI, etc.)
- Liquid metal coolant droplets
- Sodium–potassium (NaK) droplets from RORSAT reactor cores
- Solid propellant motor firings
- Anti-satellite test operations (USA: P78-1 Solwind, 13 Sep 1985; China: Fengyun-1C, 11 Jan 2007; USA: USA 193, 21 Feb 2008; India: Microsat-R, 27 Mar 2019).

The only natural current debris sink for LEO is the atmosphere, although a large number of innovative solutions are being considered. These include electromagnetic techniques, momentum exchange methods, remote methods, capture methods, and modification of material properties or restructuring of material. If it is possible to remove debris and to enforce a requirement that re-entry of satellites should be sanctioned up to 25 years after their mission has ended, analyses show that the debris environment could be stabilized. This would involve a 90% post-mission disposal (PMD) which means that, in the case of 90% of the satellite missions, the 25-year rule had been implemented and active debris removal (ADR) of 5–10 large objects per year will be done each over the next few decades. At present, the current PMD rate is below 20% of all missions. The scenario with 90% PMD and the removal of 5–10 objects per year, however, does not consider the possibility of unpredictable events such as the loss of Envisat, an 8 metric ton Earth observation satellite, in April 2012. It is still in one piece but it is out of control and constitutes a definite space debris threat as there is a distinct possibility that it will be struck by other debris, producing thousands of new objects. An analysis of space debris in Envisat's orbit suggests there is a 15–30% chance of collision of its main structure with another piece of junk during the 150 years it is thought Envisat could remain in orbit. Should such a collision take place, a very large debris cloud would be produced in a region of space already full of resident space objects. With the new mega-constellations, like e.g. Starlink or OneWeb, several hundred satellites will be launched into the same orbit. This will change the requirements of PMD dramatically for keeping space sustainable over a long-term timeframe. In that case a PMD of 99% is required, as illustrated in Fig. 1.9.

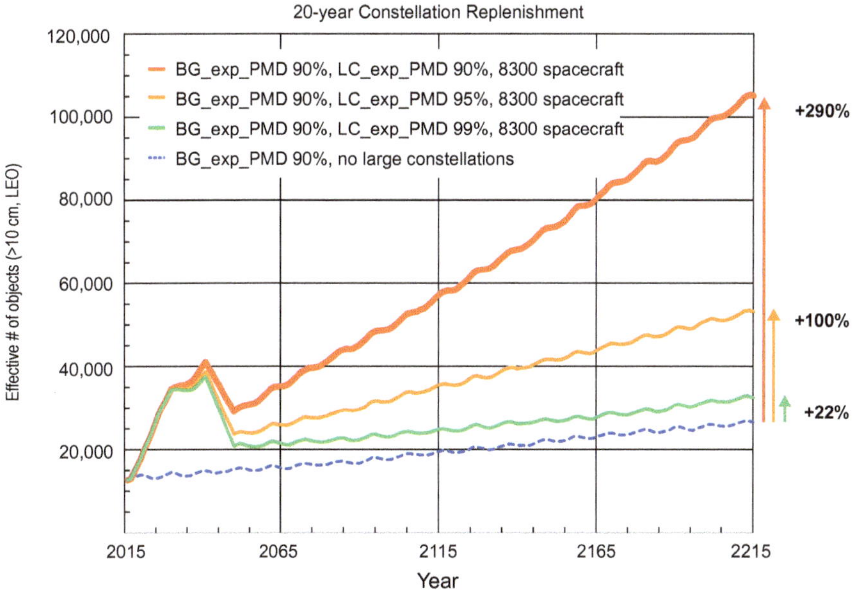

Fig. 1.9 Debris as a function of post-mission disposal of three mega constellations (Liou 2018). The abbreviations stand for background (BG) and large constellations (LC). Both include accidental explosion (exp) and a percentage of successful post-mission disposal (PMD)

In the geostationary region, orbits with low inclinations are relatively stable, thus increasing the negative effect on the space debris situation. After the end of their mission, normally the satellites are raised into graveyard orbits a few hundred kilometers above the GEO. Due to the harsh environment and the fatigue of material involved, the satellites can break up. Some of these newly created items have a high area-to-mass ratio and are called HAMR objects. They establish an eccentricity up to 0.7 as a result of natural forces. These objects dive through the GEO, posing a threat to active satellites. But, there is one possibility to cope with this challenging situation: By increasing the inclination up to 75°, these satellites will be positioned on the decay highway and will, due to natural forces, decay within a few decades in the Earth's atmosphere.

1.3.7 Gravity and Magnetic Fields

In addition to the environmental torques that can be provided by atmospheric drag and solar radiation, there are also gravity gradient torques and magnetic torques. The former are due to the differential gravity forces between the top and the bottom of the spacecraft and can be used to maintain a spacecraft Earth pointing to about ±5°. Magnetic torques are caused by the Earth's magnetic field acting on the residual magnetic dipole moment of the spacecraft. It can be utilized to provide a control torque by generating a controllable magnetic dipole moment on the spacecraft that interacts with the Earth's magnetic field and generates such a torque. In addition to being lightweight, expendable resources are not required. They do need a significant external field, however, and so can only be used for low Earth orbiting missions.

References

Arianespace; Ariane 5 User's Manual; issue 5.1 (2011). https://www.arianespace.com/wp-content/uploads/2015/09/Ariane5_users_manual_Issue5_July2011.pdf

Encyclopedia Britannica Online (2012) Layers of the atmosphere. https://cdn.britannica.com/56/97256-050-4D739762/layers-atmosphere-Earth-yellow-line-height-response.jpg. Cited 14 June 2021

Liou JC (2018) NASA ODPO's large constellation study. Orbital Debris Quart News 22(3)

Mori O, Sawada H, Funase R, Morimoto M, Endo T, Yamamoto T, Tsuda Y, Kawakatsu Y, Kawaguchi J (2009) First solar power sail demonstration by IKAROS. In: International symposium on space technology and science, 9 July 2009

Reiff PH (1999) Live from the Sun; part of the "Passport to Knowledge" series

SpaceX; Falcon 9 Launch Vehicle Payload User's Guide; SCM 2008-010 Rev. 1; Hawthorne, CA (2009)

Adrian R. L. Tatnall is now retired. After obtaining a Ph.D. at Sheffield University in the UK in 1978 he became a senior space systems engineer at British Aerospace at Bristol and specialized in remote sensing instrumentation. He later joined the department of Aeronautics and Astronautics at the University of Southampton and became an associate professor and Head of the Astronautics group. He organized and lectured on spacecraft engineering courses run for industry at the European Space Agency and Southampton for over 20 years. For many years he served on the Space Committee of the UK Royal Aeronautical Society.

Hauke Fiedler heads the Space Situational Awareness team at the German Space Operation Center (GSOC). He studied physics at the Ludwig-Maximilian-University Munich and graduated with his doctorate degree on X-ray binaries in Astronomy. In addition, he was successful in gaining his Master of Space System Engineering from the Technical University of Delft. He played an active role in the TanDEM-X mission in the fields of formation and mission optimization, as well as subsequently combining the TanDEM-X mission with the TerraSAR-X mission. In 2011 he was appointed head of the Space Situational Awareness Team with the main focus on Space Debris research and is currently operating SMARTnet™ in partnership with the Astronomical Institute of the University of Bern.

Chapter 2
Spacecraft Design

Adrian R. L. Tatnall

Abstract In the chapter "Spacecraft Design" the systems engineering process is described, starting from a mission statement and deriving goals and requirements from it. The treatment of design drivers and trade-offs is discussed as well as the use of concurrent engineering.

2.1 Definition of Systems Engineering

Systems engineering requires skills that are traditionally associated with both art and science. Good systems engineering requires the art of technical leadership including creativity, problem solving, knowledge and communication skills, but it also requires the science of system management or the application of a systematic disciplined approach. In this section, the systematic disciplined approach balance is considered in more detail with the emphasis on the methodology of systems engineering, but the main goal of systems engineering is to get the right design. This can only be done using skills that cross traditional boundaries between the arts and the sciences.

The definition of systems engineering is an interdisciplinary approach governing the total technical effort to transform requirements into a system solution. The European Standard for Space System Engineering is described in the European Cooperation for Space Standardization (ECSS) with the document number ECSS-E-ST-10C. The system can be any integrated product or processes that provide a capability to meet a stated objective. This inevitably means that a system can be a subsystem of a larger system and/or a system of systems. A spacecraft is a system, but it is one element of the space mission that will include the launch vehicle and the ground segment and may include other systems such as Global Positioning System (GPS) and a data relay system. The ground segment itself is a combination of systems that is responsible for spacecraft operations and the processing of the data. It is therefore often necessary to consider products at a number of different levels. The boundaries

A. R. L. Tatnall (✉)
University of Southampton, Southampton, UK
e-mail: art4@soton.ac.uk

F. Sellmaier et al. (eds.), *Spacecraft Operations*, Springer Aerospace Technology,
https://doi.org/10.1007/978-3-030-88593-9_2

of the systems engineering discipline and its relationship with production, operations, product assurance, and management disciplines are given in Fig. 2.1, taken from the ECSS-E-ST-10C (2017).

Systems engineering encompasses the following functions:

- **Requirement engineering**, which includes requirement analysis and validation, requirement allocation, and requirement maintenance.
- **Analysis**, which is performed for the purpose of resolving requirements conflicts, decomposing and allocating requirements during functional analysis, assessing system effectiveness (including analyzing risk factors), complementing testing evaluation and providing trade studies for assessing effectiveness, risk, cost and planning.

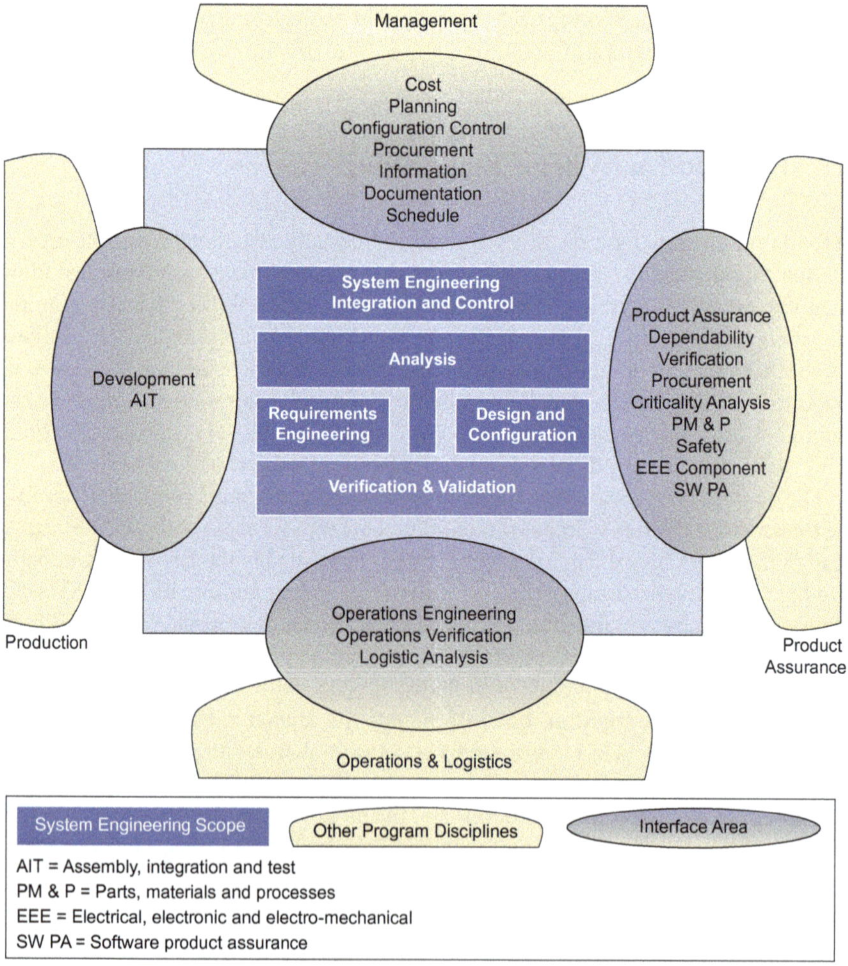

Fig. 2.1 System engineering boundaries (reproduced from ECSS-E-ST-10C; *Credit* ESA)

Table 2.1 Systems engineering techniques (Fortescue et al. 2011)

Requirements identification/analysis	Concept selection
System specification	Budget allocation
Options identification	Performance analysis
Mission assessments	System optimization
Trade-offs	Interface specification
Feasibility assessment	System definition
Cost comparison	Cost estimation

- **Design and configuration** which result in a physical architecture, and its complete system of functional, physical and software characteristics.
- **Verification**, whose objective is to demonstrate that the deliverables conform to the specified requirements, including qualification and acceptance.
- **System engineering integration and control**, which ensures the integration of the various engineering disciplines and participants throughout all the project phases.

These functions require the techniques defined in Table 2.1 to be used.

2.2 Objectives and Requirements

The starting point for the mission is the mission statement: A document established by the customer, which reflects the user needs. It is often a single line that describes the mission, e.g. John F. Kennedy in 1961 said that "this nation should commit itself to achieving the goal, before this decade is out, of landing a man on the Moon and returning him safely to the Earth". The mission objectives are derived from this statement and qualitatively define what this mission should accomplish.

The mission requirements are the top-level requirements in all aspects of the mission. They are usually quantitative in nature, specified by the customer or user, and they are an assessment of the performance required to meet the mission objectives. For the spacecraft system design these requirements are translated into engineering parameters. This translation can be complex, depending on the particular application. The requirements drive the rest of the design and determine all aspects of the mission. They are the single biggest cause of project problems.

For a communication spacecraft, the translation between the user requirements and the engineering requirements is relatively straightforward, since the user coverage and data requirements can readily be used to define the satellite parameters. On an Earth observation and science spacecraft, however, they can be considerably more complex. For example, on the Gravity Recovery and Climate Experiment (GRACE) mission (Wiese et al. 2012), the user requirements on geophysical parameters, such as an ice sheet changes into an instrument specification, have to be translated into measurements of the gravity field and ultimately to the measurements of changes in the speed and distance between two identical spacecraft. In this case, the process involves assumptions about other related parameters such as the level of processing required and GPS data. At the start of the design process, it may not be clear to what extent the requirements are driving the design. So, it is an essential part of the space system engineering process that the requirements are re-evaluated when there is a clearer understanding of the impact they have on the spacecraft design.

This iterative process is essential to ensure that the most relevant and realistic requirements are used for the spacecraft design. There are plenty of examples where the engineering requirements became "tablets of stone" at the start of the design and the overall system suffered because of unwillingness to question them as the design has evolved. It is always necessary to define how much quality is needed or how much "science" is enough in order to hold down mission costs and avoid unnecessarily restrictive requirements. While Augustine's law that "the last 10% of performance generates one-third of the cost and two-thirds of the problems" is an oversimplification, it does encapsulate the problem of overspecification. In other examples, technological constraints, such as the inability to space qualify critical parts or processes, may dictate a revision of requirements. The importance of the requirements should not be underestimated. Relatively little of the total project budget is spent on requirement analysis and initial design, but it does determine the cost commitment for the rest of the program. The later the change in a requirement, the greater the cost impact on the program as a whole.

Figure 2.2 shows how it is necessary to expand these top-level requirements into specifications covering the entire range of system and subsystem engineering parameters. It also shows the importance of establishing, in parallel, budget data.

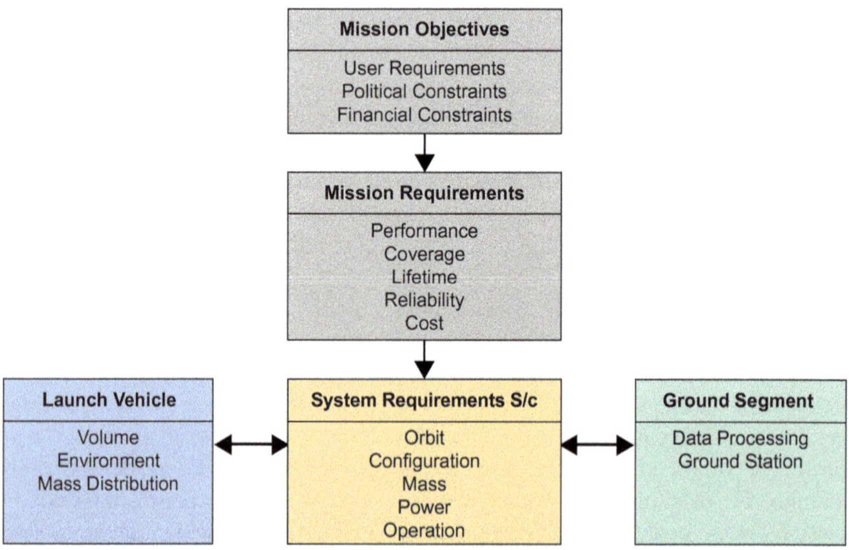

Fig. 2.2 Objectives and requirements of a space mission

Table 2.2 is a checklist of the full range of parameters that are likely to be specified in later, more detailed phases of a program.

There are many systems options that have to be considered in the early design phase of a mission. These include the type of orbit, the launcher, the propulsion system, the type of spacecraft configuration and the attitude control concept.

Example "Astronomy Mission"

The choice of orbit for an astronomy mission is a good example of the kind of choices that have to be made. This highlights some of the key points that must be considered in concept selection and optimization.

Figure 2.3 is a tree diagram showing the possible orbits about Earth and Sun which could be adopted for an astronomy mission. The mission names of spacecraft flown or due to be flown for the different orbits are shown.

It is clear that the choice of orbit for this class of mission is determined by a large number of factors but there is often an overriding consideration. For example, NASA's major observatories—Hubble Space Telescope (HST) and gamma ray observatory (GRO)—had to be in a circular low Earth orbit (LEO) in order that they could be launched/serviced by the Space Transportation System (STS)/Shuttle and the Tracking and Data Relay Satellite System (TDRSS) could be used for data retrieval. As far as the science is concerned these orbits are far from ideal. They suffer from regular eclipse periods and the scope for uninterrupted observation is very limited. Without the constraints of a Shuttle launch, two of ESA's astronomy missions, Integral and the X-ray Multi-Mirror Mission (XMM-Newton), selected highly elliptical orbits (HEO) which can provide long periods of uninterrupted observation away from trapped radiation in the Earth's proton and electron belts. More recent missions, such as GAIA, HERSCHEL and First/Planck, have selected orbits around a spot about 1.5 million km from Earth in the direction away from Sun, known as the L2 Lagrangian point. In this orbit, advantage can be taken of the fact that the benign thermal and radiation environments are ideal for long-distance observations. In addition, by careful choice of the particular orbit around the L2 point, it is possible to have continuous solar power and a continuous communications link. Other spacecraft, such as NASA's Kepler spacecraft, are in orbits around the Sun trailing Earth so that a star field can be observed continuously for several years. The importance of the various factors varies with each mission and the current technology.

It is now a common feature of spacecraft that they reuse existing designs of spacecraft equipment. This can offer very significant savings compared to new developments, e.g. the satellite bus used for Venus Express was almost a copy of that used for Mars Express which in turn was based on the Rosetta bus. The reuse of existing designs and hardware must be treated with caution. Qualification by similarity is a legitimate process but there have been notable failures in the past that have been due to this approach. Examples include the first Ariane V failure because of software inherited form Ariane IV, and the loss of the Mars Observer because of the over reliance on hardware qualified for near-Earth missions.

Table 2.2 Checklist of system requirements, adapted from Fortescue et al. (2011)

Checklist of system requirements	
Mission requirements	**Environmental requirements**
Launch windows Orbit (transfer; operating nominal and back-up) Operations mass Launch and early orbit phase Operational phase End-of-life Lifetime (Retrieval/repair/re-supply) Autonomy Reliability/availability Ground segment	Ground activities Launch and ascent conditions Transfer and operating orbit environment (Reentry, descent)[a] Structural/thermal inputs, loads, ranges Environmental protection Cleanliness/contamination Electromagnetic compatibility (EMC) DC magnetic fields Radiation Spacecraft charge Atomic oxygen[a] Autonomy
Physical requirements	**PA requirements**
Axes definition Configuration constraints dimensions Mass properties Internal torques Disturbances Power/energy	Reliability Availability Maintainability Safety test philosophy Parts, materials, processes Cleanliness Storage, handling, transport Configuration management Software
Performance requirements	**AIV program requirements**
Orbit maintenance Ranging accuracy Timing accuracy Pointing accuracy Measurement accuracy Stability Pointing range Slew rate Data rate Data storage capacity On-board processing Link budget margins Telemetry/telecommands Strength/stiffness Thermal control Reliability	Schedule Model philosophy Safety test philosophy Ground segment equipment requirements Facilities usage
Cost constraints	

[a]For some missions only

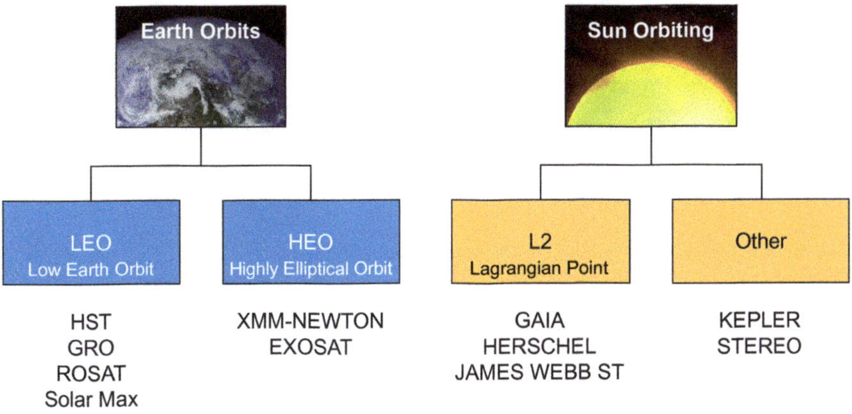

Fig. 2.3 Orbit options for astronomy missions

2.3 Design Drivers and Trade-Offs

The purpose of the satellite bus is to provide the support required for the payload to ensure that it can operate in the required orbit and environment. This makes the payload, in most cases, the single most significant driver of the satellite design.

Power, heating and cooling, structure and communication are all provided to ensure that the payload can operate satisfactorily and relay its data back to ground. The propulsion, attitude and orbit control system (AOCS) and the mission analysis provide the means of getting the payload into the right position to make its measurements. In the case of the Gravitational Ocean Composition Explorer (GOCE), shown in Fig. 2.4, the spacecraft has to fly at a constant, very low altitude of 260 km in order to measure very small changes in the gravity field (Wiese et al. 2012). The effect of the residual atmosphere is very significant and so a main design driver is to minimize air drag forces and torques. Consequently, the satellite body has an octagonal prism shape with two long, fixed solar array wings fitting the launcher fairing dynamic envelope. This requires triple-junction GaAs solar cell technology to generate the maximum power. It also requires an electric propulsion system to ensure the orbit altitude is maintained with the most efficient use of the propellant.

While there may well be a key technological design driver, in a typical space mission there are several factors that need to be considered to determine the optimum mission. A trade-off study is an objective comparison with respect to a number of different criteria and is particularly useful if there are a number of possible design solutions. It is common to make use of trade-off tables to "score" the alternative options in early concept studies. Major evaluation criteria for such trade-offs include:

- Cost, which is generally a dominant factor
- Satisfaction of performance requirements (for example, image quality in an astronomy mission)
- Accommodation of physical characteristics, notably mass, size and power which, in turn, impact on cost and feasibility

Fig. 2.4 GOCE spacecraft (*Credit* ESA)

- Availability of suitable hardware technology and timescales for any predevelopment
- Compatibility with launcher, ground segment and other system elements and the complexity of interfaces
- Flexibility to encompass alternative mission options
- Reliability and availability.

Evaluation criteria should be selected that discriminate between the options. If some of these criteria are considered more important than others, then a weighted trade-off can be performed. The process is shown in Fig. 2.5, adapted from the National Airspace System (NAS) system engineering manual (National Airspace System 2006). Regardless of whether a trade-off is weighed or not it should only be used as a guide. It is impossible to guarantee that a trade-off is entirely objective and that the evaluation criteria are exhaustive and independent. Cost, for example, is influenced by all the criteria above and its use as an independent parameter is highly questionable. Numeric results are useful but may well give a false sense of accuracy and so should be used carefully.

Whereas some factors can be evaluated numerically, many other factors that need to be considered rely on engineering judgement. In addition, quantitative values attributed to factors, can often not be made with sufficient confidence to allow a particular solution to be selected from a number of options. In this case, there are often a number of viable solutions.

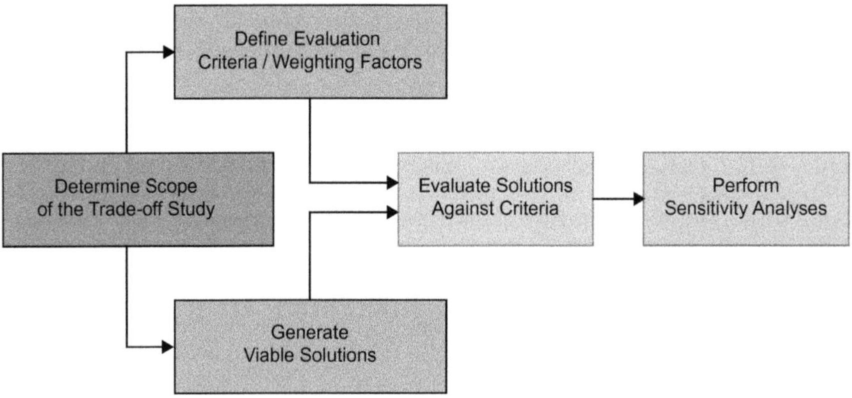

Fig. 2.5 Trade-off process

2.4 Concurrent Engineering

Concurrent engineering (CE) is a relatively new design tool developed to optimize engineering design cycles. It relies on the principles that all elements of a product's lifecycle should be considered in the early design phase and that the design activities required should occur at the same time or concurrently. While system engineering has always recognized the value of this approach, the enabling factor for the CE approach has been the rapid development of information technology (IT). Concurrent engineering has enabled design iterations to be performed much quicker and it has enabled the designer to be more closely involved in the design process.

ESA's concurrent engineering facility at ESTEC (Netherlands) has achieved the following:

- Studies have been performed in 3–6 weeks rather than 6–9 months.
- Cost has been reduced by a factor of 2.
- Overall improvement in the quality of the studies by providing consistent and complete mission designs.

There are many concurrent facilities around the world, and they have become an integral part of the early design phase of a space mission.

References

ECSS-E-ST-10C (2017) Space engineering; system engineering general requirements. https://ecss.nl/standard/ecss-e-st-10c-rev-1-system-engineering-general-requirements-15-february-2017/
Fortescue P, Swinerd G, Stark J (eds) (2011) Spacecraft systems engineering, 4th edn. Wiley
National Airspace System (NAS) (2006) System engineering manual; version 3.1. Federal Aviation Administration ATO Operations Planning

Wiese DN, Nerem RS, Lemoine FG (2012) Design considerations for a dedicated gravity recovery satellite mission consisting of two pairs of satellites. J Geodesy 86:81–98. https://doi.org/10.1007/s00190-011-0493-8

Adrian R. L. Tatnall is now retired. After obtaining a Ph.D. at Sheffield University in the UK in 1978 he became a senior space systems engineer at British Aerospace at Bristol and specialized in remote sensing instrumentation. He later joined the department of Aeronautics and Astronautics at the University of Southampton and became an associate professor and Head of the Astronautics group. He organized and lectured on spacecraft engineering courses run for industry at the European Space Agency and Southampton for over 20 years. For many years he served on the Space Committee of the UK Royal Aeronautical Society.

Chapter 3
Fundamentals of Space Communications

Felix Huber

Abstract This chapter covers the basics of communication with a spacecraft. After an overview, baseband, modulation and carrier aspects are discussed. In the baseband section aspects of source coding, channel coding and shaping are presented. Afterwards, different modulation methods are shown. Finally, carrier aspects like the link budget equation and an example for a link budget calculation are presented.

3.1 Introduction

Radio communication with a spacecraft has to deal with the fact that there are large distances between transmitter and receiver, possible low elevation angles (Fig. 3.10) resulting in a substantial attenuation by the atmosphere, and large Doppler shifts due to the orbital velocity of the satellite. Moreover, the ionosphere reflects or absorbs certain frequencies that are thus unusable for space communications.

A reliable communication is one of the most important components for the operation of a spacecraft. It is necessary to control satellites and bring payload data to the ground. Communication links are either implemented as direct space-to-ground links or as inter-satellite links (ISL).

This already results in a number of requirements:

- The connection for controlling a spacecraft, i.e. the reception of telemetry data and the transmission of telecommands (TM/TC) can be achieved with a small bandwidth, but should be as robust as possible.
- Data downlinks usually require a high bandwidth.
- Direct space-to-ground links must also consider the attenuation by the Earth's atmosphere.

Space communication is realized by electromagnetic signals, usually radio frequencies (RF). In some cases, higher frequencies in the infrared (IR) or optical

F. Huber (✉)
German Space Operations Center (GSOC), German Aerospace Center (DLR), Oberpfaffenhofen, Germany
e-mail: felix.huber@DLR.de

F. Sellmaier et al. (eds.), *Spacecraft Operations*, Springer Aerospace Technology,
https://doi.org/10.1007/978-3-030-88593-9_3

Table 3.1 Space communications radio frequencies

Band	Frequency range (MHz)	Application
VHF	150	Voice
P-band/UHF (lower part)	300–3000	Military satellites
L-band/UHF (upper part)	1215–1850	GNSS, satellite telephony
S-band	2025–2400	TM/TC
C-band	3400–6725	Future LEOP TM/TC
X-band	7025–8500	Payload, deep space
Ku-band	10,700–14,500	TV, routine TM/TC
Ka-band	18,000–35,000	Relay
V-band	37,500–50,200	Inter-satellite links

range are also used. Higher bandwidths can be achieved with higher frequencies. ISL can also be realized at frequencies that would be completely absorbed by the Earth's atmosphere. The relationship between RF bands and the application is shown in Table 3.1.

Two aspects of communications have to be considered:

- Baseband—user aspect
- Carrier—service aspect.

Both aspects can be handled more or less separately. The path of the signals is shown in Fig. 3.1.

3.2 Baseband

The range from signal source to channel coding and from channel decoding to signal presentation is called baseband (see Fig 3.1). Signal sources can be discrete values such as switch on–off or pressure and temperature values on the satellite side, i.e. telemetry (TM) or telecommands (TC) in case of a ground station. Non-digital signals have to be converted into a serial digital signal first in a process called "source coding" that is described in the following subsection.

3.2.1 Source Coding

Sensors convert physical properties such as pressure or temperature into a normalized electrical voltage, such as 5 V. Next, the voltage is sampled at discrete time intervals

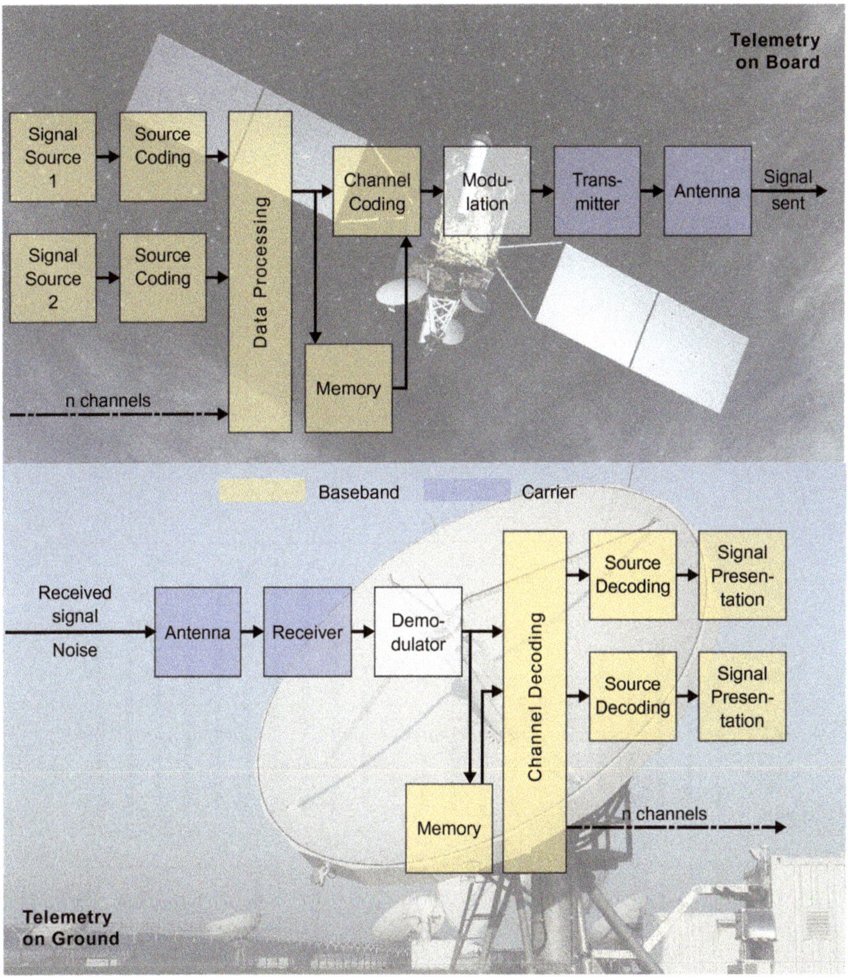

Fig. 3.1 Telemetry signal processing

(sampling) and converted into a binary number by an analog-to-digital converter (ADC) (discretization) (Fig. 3.2). If the signal is to be recoverable without losses, it has to be sampled at a speed twice as fast as its bandwidth (not the highest frequency!): this is called the Nyquist theorem. The number of steps that the ADC can create (quantization) has an influence on the rounding errors that occur when the nearest value has to be chosen. This quantization noise can be made smaller with smaller steps at the cost of a higher data rate that is needed for transmission, thus a trade-off has to be found. The resulting binary numbers are transmitted as a stream of binary pulses, referred to as pulse code modulation (PCM).

If the signal has a bandwidth higher than what the Nyquist theorem allows for, it has to be filtered before being fed to the ADC or frequencies outside the allowed

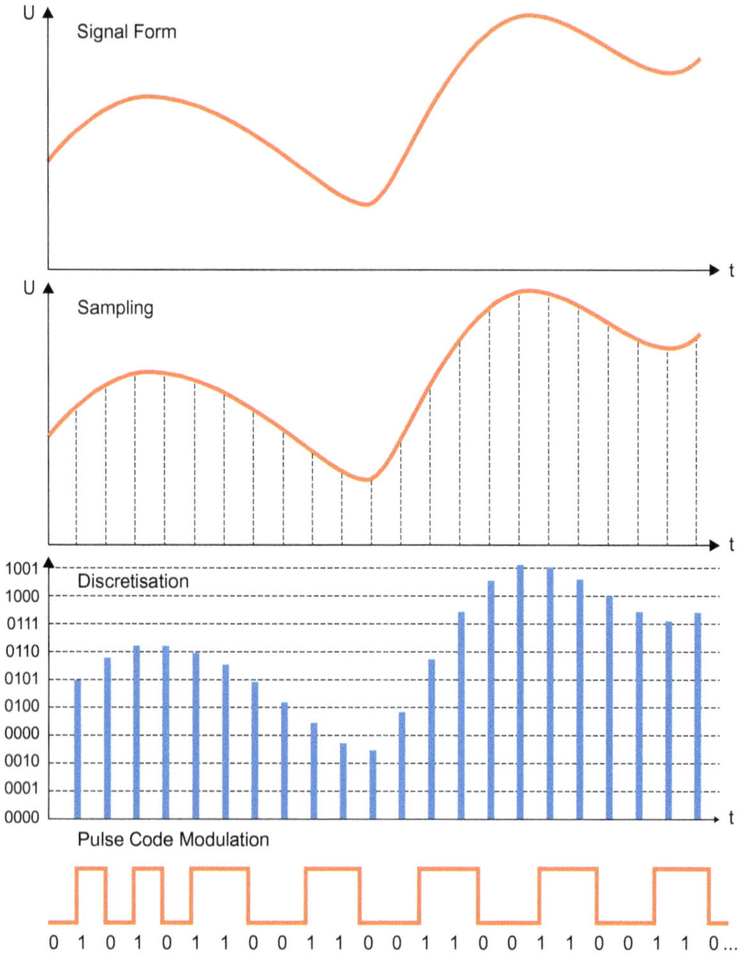

Fig. 3.2 Source coding of a signal

bandwidth will be mapped onto the desired range. This phenomenon is called aliasing and causes a heavy distortion of the signal.

In a next step, various sources have to be combined (multiplexing), formatted (range indication, sequential numbering) and perhaps stored for later transmission.

3.2.2 Channel Coding

Before the digital data can be sent over the air, precautions have to be taken for errors that can occur during the transmission. In this process of channel coding, check sums are added to the data packets like the cyclic redundancy check (CRC), i.e. the "inner checksum", and the resulting serial data stream is run through a convolutional coder ("outer checksum") that adds more check-bits in order to recover the distorted bits later after reception. While the CRC only allows for the detection of bit errors, the convolutional coding has enough information to correct bit errors without requesting a retransmission of the data, hence the name "forward error correction" (FEC). This error correction capability is achieved at the cost of a lower net bit rate, referred to as rate ½, rate ¾, etc. encoding. It should be noted that the bit error rate could also be lowered by using a smaller bit rate, but the FEC achieves a better lowering of bit errors when compared to the net bit rate; this is called the coding gain.

The final PCM is sent out as a sequence of equally long pulses. However, this sequence of pulses, called non-return to zero (NRZ), can create a direct current (DC) offset of the average voltage fed to the transmitter that cannot be handled by the system (Fig. 3.3). Therefore, the PCM code has to be converted into a DC-free signal, for example by a bi-phase (Bi ϕ) coding: the signal is multiplied with a square

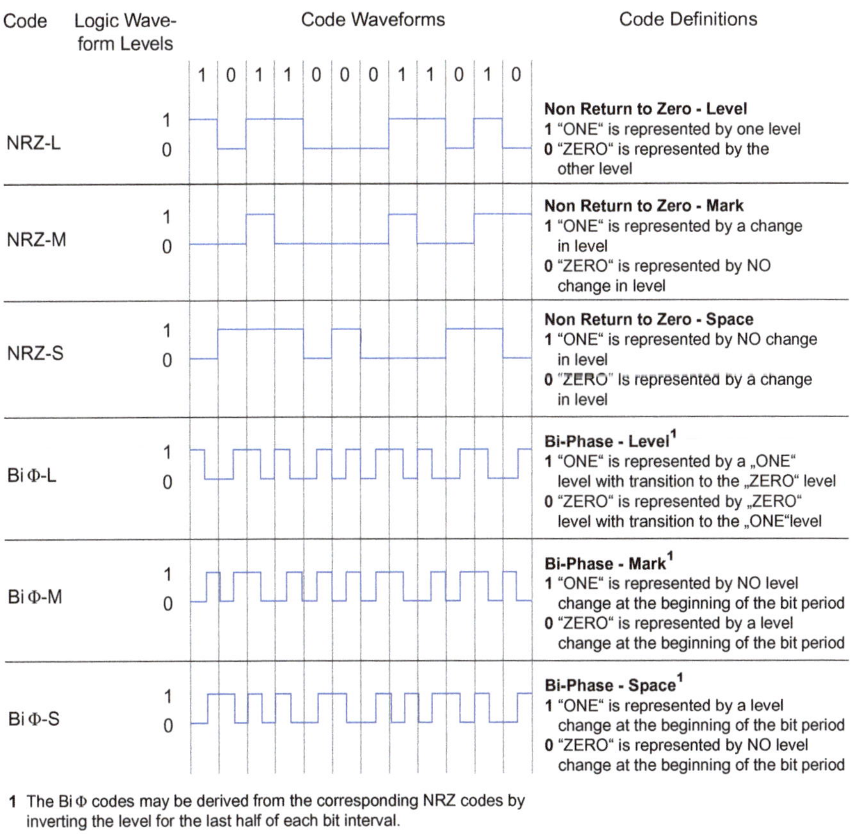

Fig. 3.3 Channel coding according to the telemetry standard IRIG 106

wave carrier of the bit rate which removes the DC offset; however, this is at the cost of a higher bandwidth.

It should also be noted that there are two definitions of bi-phase depending on whether a real multiplication is used or an exclusive-or logic, with the latter being the inverted signal of the first method.

Another critical effect is that if there is an imbalance of zeroes and ones over a certain period of time, a temporary DC offset is generated and the center frequency will shift causing signal losses due to the bandpass filtering at the receiver. A self-synchronizing scrambler is therefore used to smear out periodic patterns that can occur in the data stream. This process, which is also referred to as energy dispersal, creates a uniform spectrum by toggling the PCM bits with a pseudo random number pattern using linear feedback shift registers (LFSR) that have mathematical properties of almost pure randomness. Since the mathematical law of the random numbers is known, the receiver can undo the process of randomizing and recover the original bits.

Hence, the transmitted serial signal has no block structure anymore and therefore synchronization markers have to be added so that the receiver can determine the start of a frame. These synchronization words are known as Barker codes and have a pattern that has a low cross-correlation since the Barker data pattern could also occur anywhere in the data stream and should not trigger the frame detection. Since the frame length is fixed and known to the receiver, it can check for the regular appearance of the Barker codes and determine the start of a frame.

3.2.3 Baseband Shaping

The rectangular pulses occupy a large bandwidth due to their steep edges. If this spectrum is bandwidth-limited due to filtering in the signal path, the shape of the pulse gets distorted and spreads over its bit cell time into adjacent cells causing bit errors. This phenomenon is called inter-symbol interference (ISI). In order to prevent ISI, the signal would have to be filtered with a bandpass filter with a brick wall characteristic. However, such an ideal filter would have an impulse response that spreads over ±infinity with a non-causal behavior and cannot be reached in reality. A more practical approach uses the shape of a raised cosine as the filter transfer function which also has an infinite pulse response, but the corresponding sin(x)/x shape decays faster at the cost of twice the bandwidth. In practical implementations, a linear mixture of both extremes is used, described by a roll-off factor α, where $\alpha = 0$ corresponds to the rectangular filter transfer function and $\alpha = 1$ to the raised cosine shape with an occupied bandwidth of $(1 + \alpha)$ symbol rate. Typical implementations use α between 0.2 and 0.5.

This filtering can be performed at the analog baseband signal where the raised cosine shape has to be approximated by real circuits or in the numerical domain where the pre-calculated impulse responses are superimposed over several bit cells. Since we have a non-causal filter, the output of the bits has to be delayed using a history shift register in order to make signal causal again. This superimposing of the bits is spread over ±4 to ±8 bits, meaning the ideal impulse response is cut off after a certain time, leading to a negligible distortion of the signal.

The resulting signal has zero crossings independently of α always after the bit cell time T, meaning that there is no ISI (see Fig. 3.4). Since we have a symmetric filter, the resulting pulses are also symmetric. A superposition of randomly selected bits leads to a pattern that has the shape of an eye, hence the name "eye pattern". It can be used to judge the quality of the received signal: the eye pattern has to be wide open in the center where the detection of the bits takes place. There should be no zero crossings in the middle as this indicates ISI (see Fig. 3.5).

One more optimization can be performed in order to maximize the signal-to-noise ratio: The transmitter filter and the receiver filter should have the same conjugate complex transfer function. Since the raised cosine function is symmetric, these transfer functions are the same. However, the eye pattern requires the raised cosine shape at the bit detection in order to avoid ISI; therefore, the filtering is shared between the transmitter and receiver by using the square root of the raised cosine. Such a root-raised-cosine filter system is optimal for both ISI and noise.

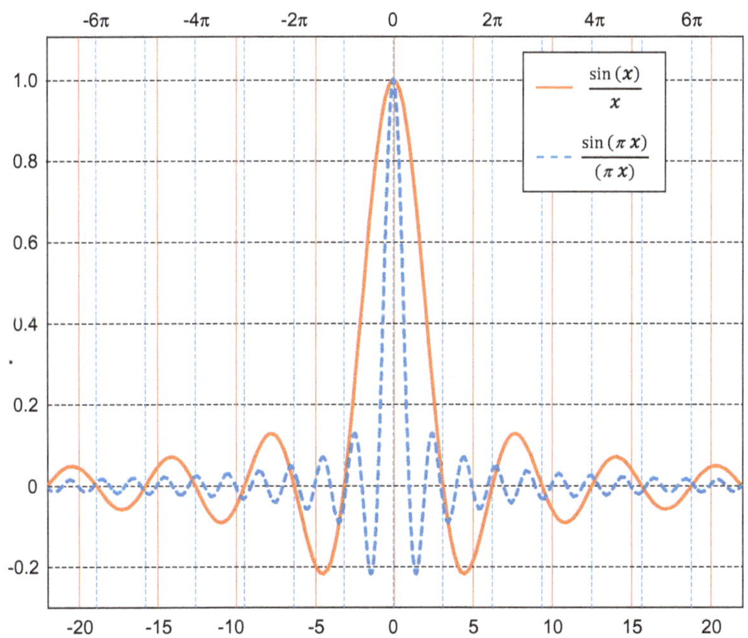

Fig. 3.4 Impulse response of a raised cosine filter

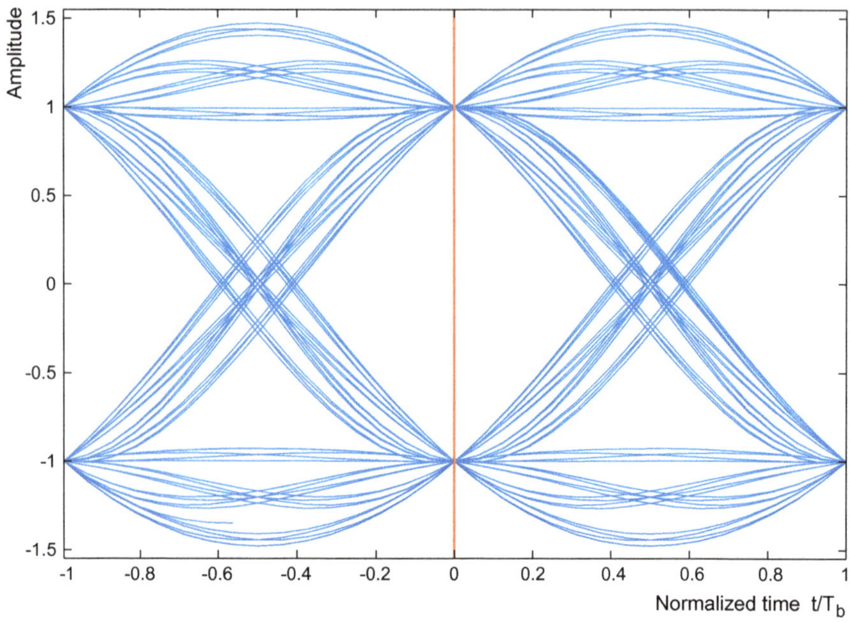

Fig. 3.5 Eye pattern of shaped signal: no zero crossing should occur in the center

3.3 Modulation

Modulation is the process of applying a (coded) signal onto a higher frequency carrier. A radio frequency signal has the form:

$$U(t) = A_c \cos(\omega_c t + \varphi_c) \tag{3.1}$$

with the three parameters amplitude A_C, frequency ω_C and phase φ_C that can be influenced by the modulating baseband signal. If the baseband signal is of analog type, these changes are named amplitude modulation, frequency modulation and phase modulation, respectively.

The modulation of the carrier converts its single frequency into a band of frequencies that is at least twice as wide as the modulating signal—in case of amplitude modulation (AM)—or even more—in case of wide band frequency modulation (FM) (Fig. 3.6).

In the case of a digital PCM signal that has only discrete values, the modulated signal also only takes on discrete values. In this case one speaks of "keying" instead of modulation (Fig. 3.7), since in the beginning of radio communications, the Morse code was generated by pressing the transmission key (beep beep beep, beeeeep, beep beep …). One should note the phase jump after the first bit of frequency shift keying (FSK). These jumps cause side lobes in the spectra and should be avoided by proper design of the frequency switching circuit.

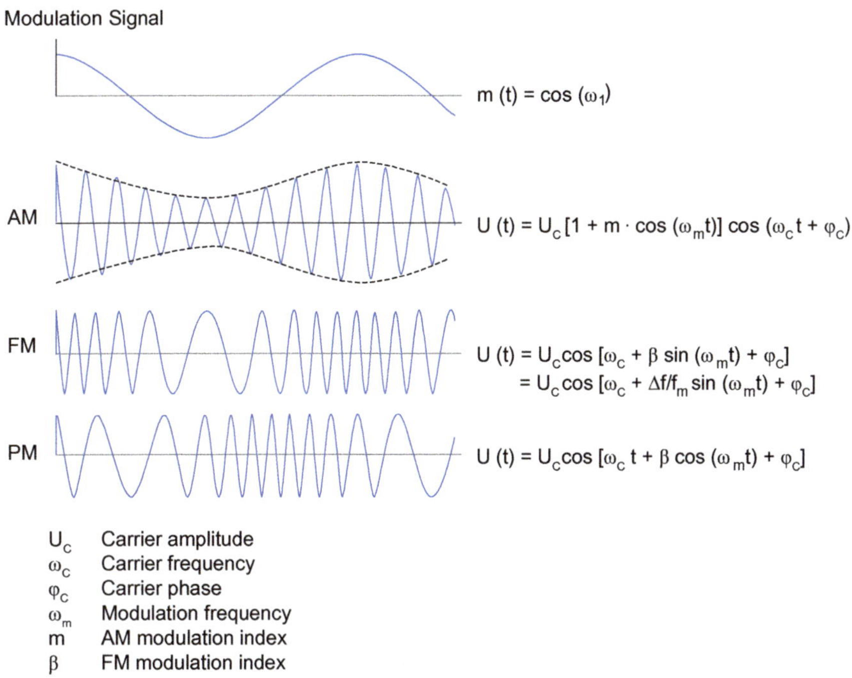

Fig. 3.6 Analog modulation forms

Fig. 3.7 Keying of a digital signal

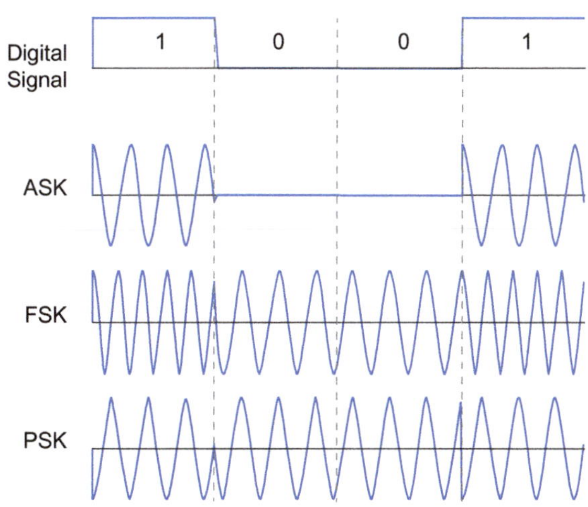

AM Amplitude shift keying ASK
FM Frequency shift keying FSK
PM Phase shift keying PSK.

In the case of PSK, several options are possible, depending on the number of phase values that the signal takes. In the case of only two (0°, 180°) it is called phase reversal keying or binary PSK (BPSK), with four values it is called quadrature PSK (QPSK), since two carriers (sine and cosine) are used. A combination of AM and PSK is also possible. This is called quadrature amplitude modulation (QAM), and allows for more bits to be transmitted within a symbol at the expense of a higher noise sensitivity. All n-ary PSK modulation schemes suffer from an n-ary ambiguity that has to be resolved by the data synchronization mechanism or by using differential phase encoding.

PSK is used in space communications for its noise immunity and better bits/energy ratio despite the high complexity of the electronics.

3.4 Carrier

The carrier, i.e. the electromagnetic wave is realizing the free space transmission. The carrier frequencies used in space communication are either at radio frequencies (see Table 3.1) or in the optical range (e.g., at 1064 and 1550 nm).

3.4.1 Elements of a Space Link

Power Amplifier

Transmits the signal with an average power P_T. Peak power levels can cause a distortion of the signal that has to be accounted for by lowering the input signal (input back-off).

Antenna

Directs the signal into the desired direction by the use of dipoles, horns and reflectors. Since the dimension of the antenna is in the order of the radio signal's wavelength, a Fresnel diffraction occurs that creates a main beam and unwanted side lobes. These side lobes direct energy to unwanted areas in the case of a transmission and collect additional noise in the case of a receiving antenna (Fig. 3.8).

The angle at which the power is at a level of half the maximum value (−3 dB) is called half power beam width and is given approximately by 70° λ/D (in degrees), D: Aperture diameter. The maximum value of the main beam as compared to a theoretical point-like isotropic radiator is called the antenna directivity and its practical value including the efficiency η called "antenna gain" is given by:

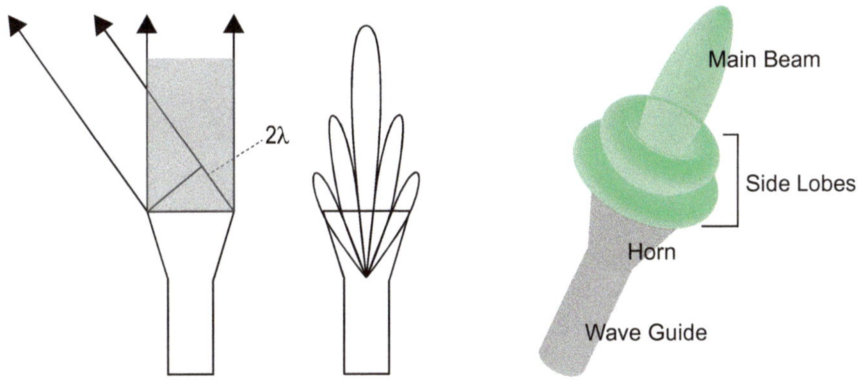

Fig. 3.8 Antenna side lobes

$$G = \frac{4\pi A}{\lambda^2} \cdot \eta = \frac{4\pi A_{eff}}{\lambda^2} \tag{3.2}$$

A: (effective) aperture area. In the case of a dipole, an effective aperture area can be defined as an area perpendicular to the electrical field lines that still has an influence on the field by "capturing" the field lines onto its surface. The gain can be seen as the solid angle into which the antenna concentrates the signal compared to a full solid angle.

The product of transmitter power P_T and antenna gain G is called "equivalent isotropic radiated power" (EIRP) and is the power that an isotropic transmitter would have to transmit in order to create the same power flux density at the receiver.

In the case of a receiving antenna, the effective "capture" area for the incoming signal can be calculated from the above equation if the antenna gain is known.

Noise

All warm bodies transmit thermal electromagnetic noise according to Planck's equation. These sources can be Sun, Moon and Earth's ground, atmosphere and clouds, but also galactic sources. In the case of the radio frequency bands, the spectral density of this noise is constant and given by

$$N_0 = kT_S \tag{3.3}$$

where k is Boltzmann's constant and T_S is the system noise temperature as the sum of all natural noises as seen by the antenna and additional artificial signals such as other transmitters or devices, e.g. human-made noise. Since this noise is purely stochastic, it cannot be removed from the received signal and this limits the sensitivity of the system.

Fig. 3.9 Link geometry

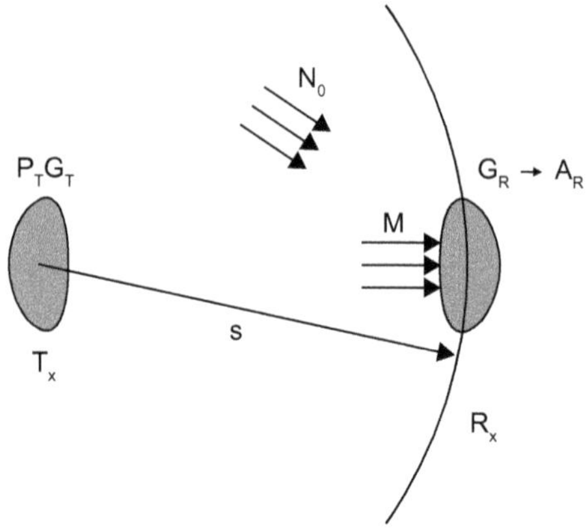

Receiver

The receiver itself creates additional noise in its amplifiers, further decreasing the system sensitivity. In a properly designed system, only the first (low-noise) amplifier contributes substantially to the system noise.

3.4.2 Link Budget Equation

The performance of the radio link is given by the ratio of the wanted signal (Carrier) and unwanted signals (noise) that has to have a certain value in order to recover the data bits without error (Fig. 3.9).

The transmitter of power P_T and antenna gain G_T is assumed to be in the center of a sphere of radius s. P_T is the isotropic radiated power and thus uniformly illuminates the sphere's surface $4\pi s^2$. If the transmitter antenna has a gain G_T, the power flux density at the receiver is therefore:

$$M = \frac{P_T \cdot G_T}{4\pi s^2} = \frac{EIRP}{4\pi s^2} \tag{3.4}$$

If the receiving antenna has a gain G_R, its effective aperture area is

$$A_R = \frac{\lambda^2 G_R}{4\pi} \tag{3.5}$$

and it "captures" at total carrier power C of $M{\cdot}A_R$.

$$C = M \cdot A_R = \frac{P_T \cdot G_T}{4\pi s^2} \cdot \frac{\lambda^2 G_R}{4\pi} \tag{3.6}$$

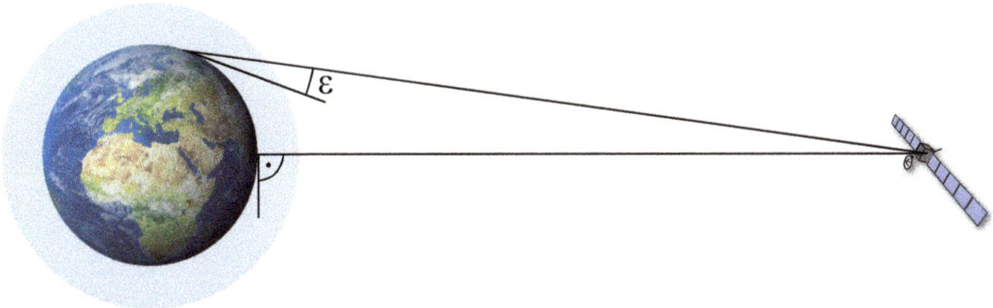

Fig. 3.10 Communication between ground and space at different elevations ε. The damping due to prolonged signal path at low ε is included in the atmospheric attenuation L_A. The free space loss is included by L_S

Reordering and adding an additional atmospheric attenuation L_A due to prolonged signal path at low elevations, rain, snow, etc. leads to:

$$C = P_T G_T \left(\frac{\lambda}{4\pi s} \right)^2 L_A G_R = P_T G_T L_S L_A G_R \tag{3.7}$$

where L_S is called free space loss even though no energy is lost but only diluted over a growing sphere surface area. The reordering of the terms was only done in order to match their position in the transmission path: transmitter, space and receiver.

If the bit rate of the signal is *R*, the time per bit is 1/*R* and the received energy per bit is *C*/*R*. Thus, we finally have the sought ratio of bit energy versus noise power density:

$$\frac{E_b}{N_0} = \frac{P_T G_T L_S L_A G_R}{k T_s R} \tag{3.8}$$

This Link Budget Equation is usually given in a logarithmic scale using the pseudo unit decibel dB:

$$\frac{E_b}{N_0} = EIRP + L_S + L_A + \frac{G_R}{T_s} + 228.6 - 10 lg R \tag{3.9}$$

where 228.6 is the logarithm of Boltzmann's constant and G_R/T_S (the so-called figure of merit) describes the quality of the receiving system. It describes the radio link on an overall power level but does not consider the type of modulation and nature of the noise coming from other possible transmitters.

Depending on the modulation and coding used, the required E_b/N_0 varies from 1–2 dB for turbocoded PSK to 8–10 dB for uncoded FSK. An additional 3 dB are needed in the case of a non-coherent demodulation.

Table 3.2 Link budget summary TerraSAR-X S-band DL with WHM S69 as Rx

General parameters		
Downlink frequency	2280.0 MHZ	
Slant range (@Elevation of 5°)	2078 km	
Free space path loss	166 dB	
GND station WHM (S69, 15 m Antenna)		
Gain	48.4 dBi	
T_{sys}	148.3 K	
G/T	26.7 dB	
Boltzmann constant	−198.599 dBm/kHz	
S/C TeraSAR-X S-band		
Gain	−10 dBi	
Transmitter output power	25.7 dBm	
S/C EIRP	15.7 dBm	
Link budget		
Atmospheric loss	2 dB	
Receiver implementation loss	1 dB	
Total received power	**−104.8 dBm**	Before the LNA
Noise power density RF, N_0	−176.888 dBm/Hz	N_0 in S-band, RF stage
Noise power density IF	−86.5 dBm/Hz	N_0 at the input of the IF receiver
Channel Gain (S-band → IF)	90.388 dB	
Received Power at IF (70 MHz)	−14.5 dBm	
Available C/N_0	**72.0 dB Hz**	
Available E_b/N_0 for a data rate of 1 Mbps, suppressed carrier modulation (BPSK)	12.04 dB	
Required E_b/N_0 for BER = 10^{-6}, without channel coding	10.5 dB	This is the theoretical BER versus E_b/N_0 for AWGN Channel
Link margin without channel coding	1.54 dB	This is the minimum expected as an Elevation of 5° assumed, maximum slant range
Link margin with Reed-Solomon code (K = 223, N = 255)	5.66 dB	Note: RS Code is currently off by TS-X in S-band; this is just an example

dBm (decibel milliwatt) is the unit of power level which describes the ratio of a power compared to the reference power of 1 mW (milliwatt)

In the case of orthogonal signals, other transmitters would not even affect the bit detection even though they bring noise power into the receiver as long as they don't saturate the amplifiers. This is a permanent source of faulty system design (Fig. 3.10).

Table 3.2 shows an example for a link budget calculation using Eq. (3.9).

Felix Huber is the director of the German Space Operations Center (GSOC). He graduated as an aerospace engineer in 1990 at Stuttgart University and received his Ph.D. on antimatter propulsion in 1994 after a guest stay at CERN/Geneva. He was head of a private research institution where he designed and built several space-born experiments including the first commercial experiment on board the International Space Station (ISS) until his appointment of a professorship at the University of the German armed forces in Neubiberg on space operations. He has contributed to several books on aerospace engineering and communications.

Part II
Mission Operations

Chapter 4
Mission Operations Preparation

Andreas Ohndorf and Franck Chatel

Abstract This chapter describes the tasks and activities required to prepare for mission operations. The success of a space mission depends not only on a properly designed and built space segment and the successful launch via a launch segment. It also depends on the ground segment and successful mission operations carried out by a team of experts using the mission ground segment infrastructure and processes. Its organization and design, as well as the assembly, integration, test, and verification (AITV) are therefore as important as the respective activities of the space and launch segment. In this context, a ground segment consists of a ground system, i.e. infrastructure, hardware, software, and processes, and a team that conducts the necessary operations on the space segment.

4.1 Introduction

The success of a space mission depends not only on a properly designed and built space segment and the successful launch via a launch segment. It also depends on the ground segment and successful mission operations carried out by a team of experts using the mission ground segment infrastructure and processes. Its organization and design, as well as the assembly, integration, test, and verification (AITV) are therefore as important as the respective activities of the space and launch segment. In this context, a ground segment consists of a ground system, i.e. infrastructure, hardware, software, and processes, and a team that conducts the necessary operations on the space segment.

This chapter describes the tasks and activities required to prepare for mission operations. It specifies the questions that automatically arise when analyzing the requirements of a mission and how these questions are answered with a design based on available resources and considering project-specific constraints. It is organized as follows:

A. Ohndorf (✉) · F. Chatel
German Space Operations Center (GSOC), German Aerospace Center (DLR), Oberpfaffenhofen, Germany
e-mail: andreas.ohndorf@DLR.de

F. Sellmaier et al. (eds.), *Spacecraft Operations*, Springer Aerospace Technology,
https://doi.org/10.1007/978-3-030-88593-9_4

- First, we introduce mission operations preparation in general, using several examples of past space missions and emphasizing the eminent importance of this phase.
- Then, the input to the overall preparation phase, or the driving factors that influence the design of the ground segment of a particular mission, are explained in Sect. 4.2.
- The team organization, i.e., who needs to do what and when, is described in Sect. 4.3.
- Section 4.4 describes data, products and tools required for effective mission preparation.
- Individual activities, tasks and deliverables are explained in Sect. 4.5.
- Section 4.6 addresses the proven concept of reviews and, in particular, the reviews that are conducted during the preparation phase.
- Section 4.7 describes the operational validation that proves that teams and ground system are able to perform the mission.

When asked what "mission operations preparation" is or means in plain English, the following definition could be given:

> Mission operations preparation includes all activities related to management, development, testing, integration, validation, organization, training, certification, and documentation of the ground segment of a space project. The result of a successful mission operations preparation is a ground segment that is ready for launch.

The duration of this project phase can vary widely. Table 4.1 provides examples of past or current missions and the durations of the operational and preparation phases. These examples cover a range of different mission types, such as Earth-bound satellites in low Earth orbit (LEO), medium Earth orbit (MEO), or geostationary Earth orbit (GEO), interplanetary missions to the Moon, planets, and other solar system celestial bodies, and deep space science missions for solar observation or outer solar system exploration.

The selected examples show that defining a general rule for the duration of operations preparation is nearly impossible. Several interacting factors contribute to space missions in general and ground segment systems in particular. These systems are:

1. at least one mission control center (MCC),
2. a ground station network (GSN), through whose antennas the ground segment communicates with the space segment,
3. a flight operations team (FOT) or flight control team (FCT) that plans and executes the operations of the space segment within its parameters.

The size, dimension, and complexity of each system depends on a number of interacting parameters and constraints, e.g. mission objectives, technical developments, project phase, schedule and budget. The quest for an optimal solution therefore inevitably becomes a search for a compromise acceptable to the customer. This is the task of the mission designers or, more specifically, of the people responsible for the design of the ground segments.

In general, the same (or at least similar) activities and tasks must be carried out in each space project, although very different mission-specific requirements

Table 4.1 Preparation and total project duration of selected space projects

Mission	Type/purpose	Orbit	Spacecraft lifetime (years)	Preparation (years)
TerraSAR-X	Earth observation, science	LEO	5 +	4
TanDEM-X				
Envisat	Earth observation, weather forecast	LEO	10	12
GPS	Navigation	MEO	12 +	22
Galileo	Navigation	MEO	12 +	20
Eutelsat W24	Communication	GEO	15 +	2
GRACE	Science	LEO	12	3
Voyager 1	Outer solar system exploration	Deep space	3 (primary mission) 52 (power limit)	7
Apollo	Human exploration (Moon)	Moon	14d	8
Cassini	Interplanetary	Saturnian system	20	10
Huygens	Exploration			
Ulysses	Deep-space Sun observation	Deep space	17	5 (excluding delays[a])

[a] With a project start in 1979, ULYSSES was scheduled for launch in February 1983 (Wenzel et al. 1992), but was postponed to May 1986 and again, due to the Challenger accident, to October 1990

must be met. The European Cooperation for Space Standardization (ECSS) has issued a set of management and technical standard documents to harmonize the management of European space projects. According to this ECSS phase model (ECSS-E-ST-70C), mission operations preparation is covered in project phases C "Detailed Design" and D "Preparation", as shown in Fig. 4.1. They are located between preliminary design and mission execution (see also Chap. 5).

The result of successful mission preparation is an integrated, validated, and ready-to-launch ground segment.

A generic example of a ground segment, its subsystems, and the data flows between them is shown in Fig. 4.2. It consists of a GSN with three ground stations and an MCC. The main systems of the MCC are the ground data system (GDS), the flight dynamics system (FDS) and the flight operations system (FOS).

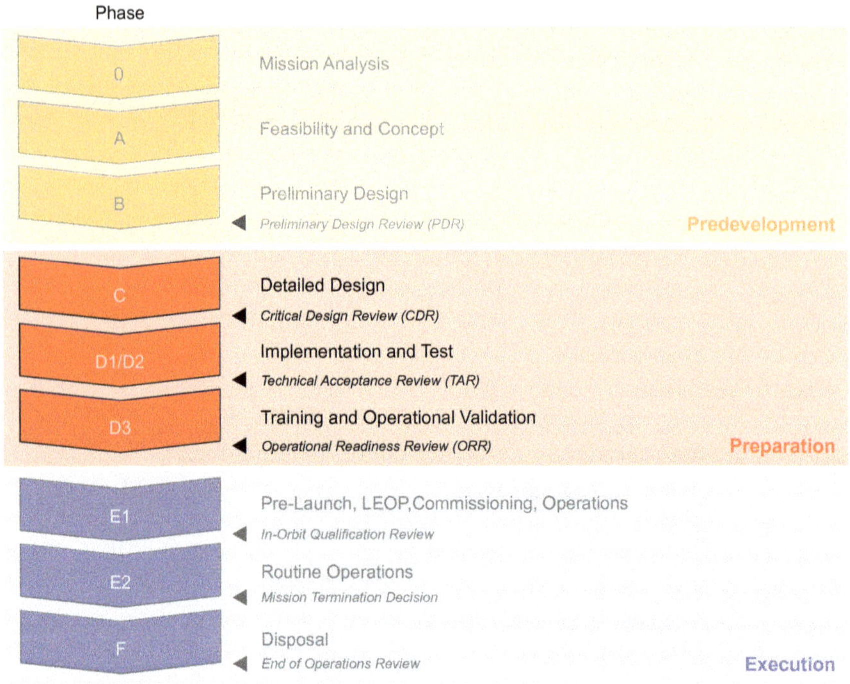

Fig. 4.1 ECSS phase model

4.2 Driving Factors

4.2.1 Requirements

The specific requirements of a space mission determine the technical design of the respective ground segment. Their formulation should follow the rules of requirements engineering and in particular those of ECSS (ECSS-E-ST-10C, ECSS-E-ST-10-06C). These are:

- Performance: Requirements shall be described in quantifiable terms.
- Justification: Each technical requirement should be justified along with the responsible entity.
- Configuration management: Each technical requirement shall be under configuration control.
- Traceability: Each technical requirement shall be traceable backward and forward.
- Unambiguity: Technical requirements shall be unambiguous.
- Identifiability: Each technical requirement shall be identified in terms of the relevant function, product, or system. The identifier shall be unique and reflect the type and life profile situation.

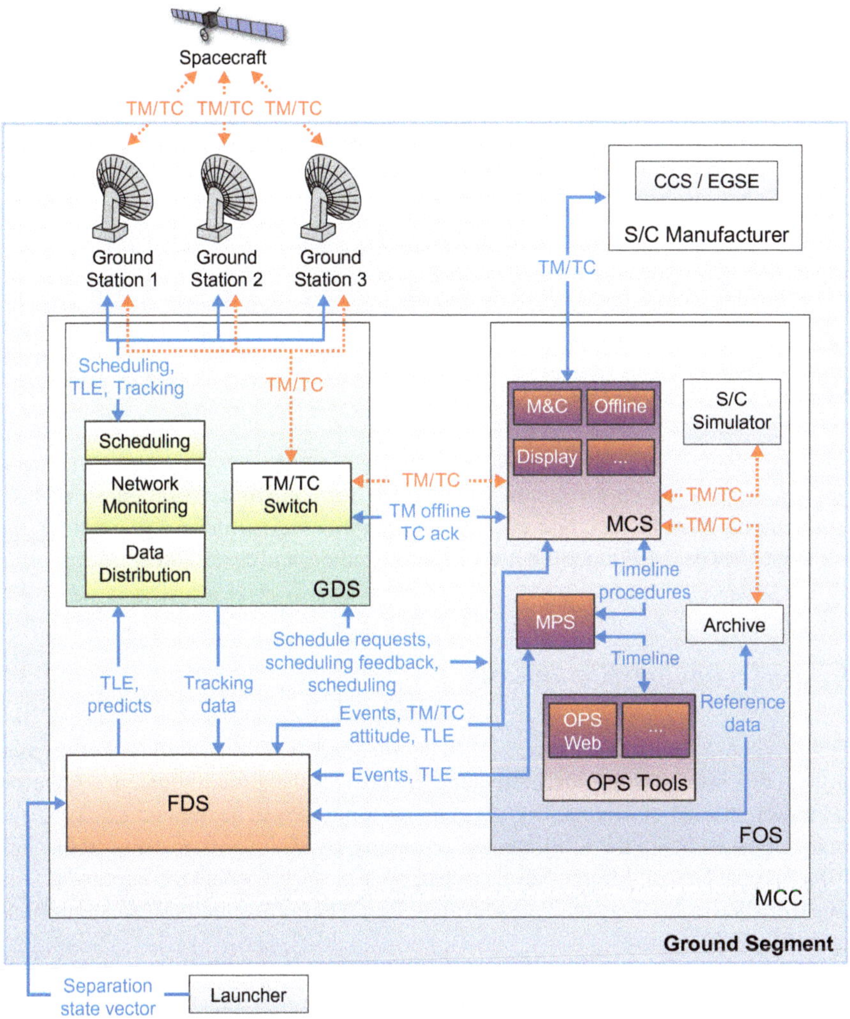

Fig. 4.2 Generic example of a ground segment

- Singularity: Each technical requirement shall be specified individually, i.e., it shall not be a combination of requirements.
- Completeness: Technical requirements shall be self-contained.
- Verification: Technical requirements shall be verifiable by one or more approved verification methods.
- Tolerances: Tolerance shall be specified for each parameter or variable.

These requirements must be analyzed by the responsible engineers and operators (see Sect. 4.3) and answered with a detailed design. Ideally, this concept answers each requirement in the best possible way and the realization takes place within the given project schedule and cost estimate. However, in reality compromises have to be found, otherwise cost and schedule overruns are to be expected.

Every unanswered or partially answered requirement, i.e. a requirement that is not fully met by an appropriate design, or a requirement whose addressing requires an unjustifiably high financial effort, is discussed with the customer. A temporary deviation or non-compliance with a requirement is documented by a so-called waiver if accepted by the customer and thus officially approved by the customer. Yet, a waiver is not an instrument for documenting persistent non-compliances; instead, a requirement change should be considered.

4.2.2 Cost/Financing

Cost-effective design is always required, because most projects do not have the financial resources to develop solutions specifically for a single mission. Nevertheless, requirements must be met, and when multiple options exist for implementation, the one that offers the best tradeoff between risk, schedule, and cost is likely to be chosen.

4.2.3 Technology/Complexity

Technical complexity affects the cost, schedule, and also the overall risk of a space project; this is true for the space segment, but also for the ground segment. However, the required level of complexity depends on a number of factors and drivers, the most important being the fulfillment of requirements. This is often achieved through different concepts of different levels of technical maturity, which are expressed with nine so-called technology readiness levels (TRL). The corresponding TRL definitions according to ECSS are:

- TRL 1: basic principles observed and reported
- TRL 2: technology concept and/or application formulated
- TRL 3: analytical and experimental critical function and/or characteristic proof-of-concept performed
- TRL 4: component and/or breadboard validated in the relevant environment
- TRL 5: component and/or breadboard critical function verification in a relevant environment
- TRL 6: system/subsystem model or prototype demonstrated in the relevant environment (ground or space)
- TRL 7: system prototype demonstrated in a space environment
- TRL 8: actual system completed and flight-qualified through test and demonstrated (ground or flight)
- TRL 9: actual system "flight-proven" through successful mission operations

For each subsystem of the ground segment and for each component, the design must be evaluated for alternatives. These alternatives must then be evaluated for their impact on schedule, cost, and risk before a decision can be made on the technology to be used.

This will be explained using the mission planning subsystem (MPS) of an Earth observation (EO) satellite as an example.

Let's assume that this EO mission requires a defined duration from the receipt of an image order to the delivery of the processed image to the ordering customer of less than a defined number of hours or days. Between these two points in time, several activities are to be carried out. These are scheduling the next available contact for uploading the image acquisition commands, predicting the next opportunity to observe the desired point on the Earth's surface, the next downlink opportunity, and the processing of the transmitted raw image data. If the required maximum duration between receipt of the order and delivery of the resulting image is, say, seven days, the necessary activities can be carried out manually or semi-automatically. However, the shorter the required time frame becomes, the more likely it is that a fully automatic planning system will be chosen. The development of an automatically operating MPS is in itself a complex project and requires a significant funding and time investment. It should therefore be developed as a generic system that can be used for multiple missions. For later missions, it may therefore be practical to use an existing MPS, even if it has more features than needed. The advantage of reusing an existing system is that, due to its higher TRL compared to a new development, testing and validation activities are lower.

4.2.4 Schedule

The project schedule affects the options for the design, implementation, test, and validation of a ground segment in several ways. First, when the schedule is tight, i.e. when there is not enough time to develop and test project-specific solutions, proven technologies of higher TRL must be used. This means, for example, reusing existing software and accepting possible drawbacks of design concepts tailored to a different mission.

Second, the project schedule also influences the validation and training concept. Since there is hardly ever enough time for training for every possible and foreseeable situation or emergency, the FOT must focus on the most severe ones during the training phase. Another influencing factor results from due dates of deliveries by the customer. A major one is the delivery of the spacecraft simulator. It should be done as early as possible, which of course clashes with the fact that the spacecraft design is often not ready. In the case of a series of satellites of the same type, e.g. a constellation of navigation system satellites, this is true only for the first one. However, without the timely delivery of a satellite software simulator, FOT training and validation activities will be hampered. Therefore, the delivery of this important item should be contractually fixed if possible.

4.2.5 Experience

The short version of this factor is: "Whatever it is, it requires less effort to do it a second time." A company or space center with decades of mission operations experience can more easily tackle future space missions of a similar nature than a new competitor. The effort is less because many concepts, processes, and tools already exist in flight-proven configurations. However, this depends heavily on the nature of the mission; it cannot be generalized because space mission requirements can vary widely. For example, human spaceflight missions have very high safety requirements, while the cost-effective maintenance of a constellation of 20–30 satellites of a global satellite navigation system may have completely opposite requirements. Interplanetary deep space missions differ from Earth-bound satellite projects. Therefore, a control center specialized on Earth observation is not the first choice for a science mission to one of Jupiter's moons.

Note that experience may significantly contradict customer requirements. Early feedback of recommendations to the customer is an important task of the control center, as it enables the search for other solutions and the reduction of cost and risk.

In addition to general experience, the time since last mission of the particular type plays a role, as experience and skill decreases with each year that one does not operate the particular type of mission. Ten years after the last mission of a particular type, one can assume that the experience is more or less gone or no longer applicable and must be acquired anew, usually with the same effort as when preparing for a mission type for the first time.

4.2.6 Risk

A specific risk analysis must be conducted for each space project, specific to the respective segments, i.e. including the ground segment. Risks are primarily related to cost, schedule, and mission requirements. Of course, the overall risk for successive missions should be as low as possible; however, it rarely reduces to zero. Minimization here comes from reducing the risk of each system and subsystem to be ready by launch. Preplanned time buffers and milestones should therefore be part of any project schedule to account for delays in preparation.

Minimizing risk is often the reason why aerospace engineering tends to be conservative when it comes to deploying new technologies. Flight-proven, reliable systems and processes are often preferred over new technologies.

4.3 Personnel, Roles and Responsibilities

Preparing to operate a space project requires the organization and assignment of roles and responsibilities. A generic project structure is shown in Fig. 4.3, with the three main branches: management, system engineering, and operations engineering. Specific roles are assigned to each branch.

The decision on which role to assign to which team member must be made by the project manager (PM) and is specific for each project. For example, for smaller projects the PM may not need a dedicated project officer (PO) for project organization and document management. In addition, it makes sense to combine the roles of flight director (FD) and system engineer (SE) for smaller projects; however, the corresponding workload of larger projects should be split between two people.

4.3.1 Project Manager

The project manager (PM) is responsible for the organization and overall management of the project. The PM is the point of contact for the customer and appoints the FD and the SE. This role is usually assigned to an experienced engineer who has preferably been either flight director or system engineer on a previous space project. Later, during the operations phase, he may additionally assume the role of mission director (MD).

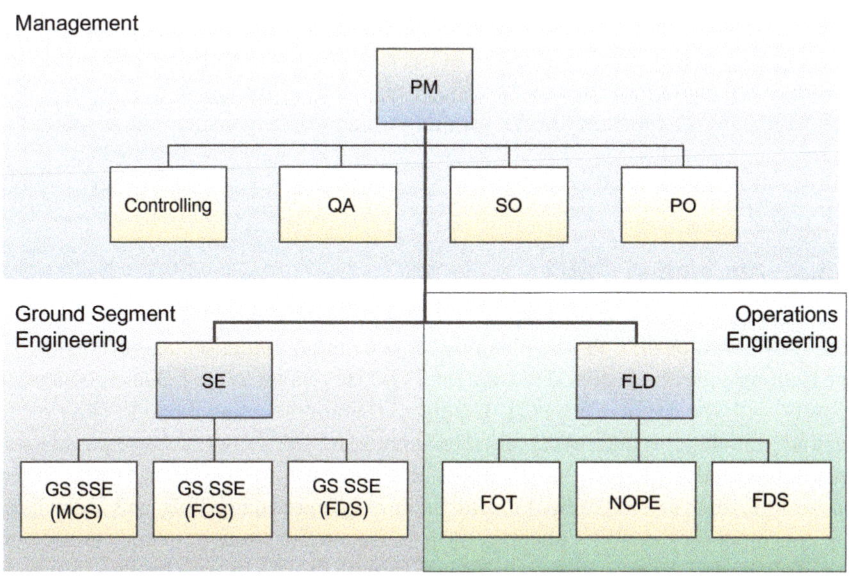

Fig. 4.3 Project organization and roles of a generic ground segment

4.3.2 Mission Director

As the supervisor of the flight director, the system engineer, and the satellite or spacecraft support team (SST) leader, the MD has overall responsibility for the mission execution phase and is accountable to the customer once mission operations have commenced.

4.3.3 Flight Director

The FD is responsible for the preparation and the execution of mission operations. He defines size, composition, qualification, and the training concept of the FOT. He is supported in this task by the simulation officer (SIM). The FD develops the operations concept, formulates low-level technical requirements resulting from that concept, and supervises the FOT during operations. The FD works in close cooperation with the project system engineer and reports to the PM.

4.3.4 System Engineer

The SE is responsible for ground system engineering and defines the technical concept of a mission's ground segment derived from the mission requirements. Together and in close coordination with the FD, the SE defines the specification of the ground segment systems and their subordinate components. He supervises the development of new and the adaption of existing components. He is responsible for the technical implementation of the ground segment, i.e. for integration, testing, and validation. In this function, the SE reports to the project manager and supports the FD.

4.3.5 Simulation Officer

The SIM is responsible for the planning, organization, execution, and evaluation of the training activities required to train the FOT. He reports to the FD and cooperates closely with the SE and the FOT. Common training activities include class room lessons, simulations, and rehearsals. The input to a SIM's work is the required level of training and capability, which needs to be defined by the FD. Organizational constraints, such as availability of infrastructure, required data, tools, and specialists, are also input for the SIM's planning. After a training measure has been carried out, the SIM prepares an evaluation report together with the FD, expressing the success or failure of the respective training measure as well as eventually necessary repetitions or follow-on measures.

4.3.6 *Quality Assurance Engineer*

The quality assurance (QA) engineer is responsible for ensuring that the project complies with internal and external quality standards, i.e., he/she shall monitor the project from a quality and product control perspective and supports the PM, SE, and FD throughout the life of the project. External standards are ISO or ECSS standards, such as ISO 9001 "Quality Management," ISO 27001 "Information Security," or ECSS-Q-ST-10 "Product Assurance Management".

4.3.7 *Subsystem Engineer*

A subsystem engineer (SSE) is responsible for the operations of a specific satellite subsystem, such as the data handling subsystem. Sometimes the operation of multiple subsystems is combined; the attitude and orbit control subsystem (AOCS) or the power and thermal control subsystem (PTS) are common examples. An SSE must learn the functionality of the respective subsystem, know the telemetry to monitor that subsystem, and train to apply subsystem-specific procedures to control its functions, depending on the current situation and intent. They also work with the mission information (data)base (MIB) to validate and optimize the performance of the MCS.

In addition to the FOT SSEs, assembly, integration and test (AIT) activities are carried out by ground system SSEs prior to launch. These are specialists in specific ground system components, e.g. networks, infrastructure, communication, server integration and configuration, security, software, ground stations, etc. The primary subsystems involved are the ground data subsystem (GDS), the mission control subsystem (MCS), and the mission data subsystem (MDS). The respective SSEs of these subsystems support the SE and the FD from project phase B through launch.

4.3.8 *Project Office/Project Officer*

A project office/project officer (PO) may be necessary for larger projects because the organizational workload becomes too large to be handled by the PM alone. The PO covers documentation management and team organization and provides general support to the PM, e.g. for organizing reviews.

4.3.9 Controlling

A controller supports the PM with contractual and financial aspects of the project throughout the project lifecycle. He provides reports and overviews over the project budget at regular intervals and upon request of the PM.

4.3.10 Configuration Manager

The configuration manager (CM) develops a project-specific configuration management plan in phase C and monitors the implementation of this plan in the subsequent phases.

4.3.11 Security Officer

The security officer (SO) supports the PM in all security related issues, e.g. access control concepts, encryption, clearances or classification of documents. This may be relevant for military satellite projects, while it is practically irrelevant for scientific satellite projects. Whether or not a project requires a dedicated SO is the PM's decision.

4.4 Required Data, Products and Tools

Testing the ground segment with its subsystems and components prior to launch requires mission-specific tools and data. For example, validation of the developed flight procedures will be severely hampered without a software simulator that emulates the behavior of the spacecraft and its intended environment, i.e. the conditions in space. Therefore, the following tools or deliverables are required for mission preparation.

4.4.1 Test Data and Data Generators

Specific sets of telemetry data are required to test the monitoring and control (M&C) system, including the mission information (data)base (MIB), processing chains (main and backup), and a potential archiving process. Test data becomes eminently important when no satellite simulator is available. Test data may also be needed for other interfaces like file data deliveries.

4.4.2 Spacecraft Simulator

A spacecraft simulator is an essential component for mission preparation but is not always provided by the spacecraft manufacturer, whether due to cost or time constraints, or both. When this is the case and a simulator is not available to the ground segment, remote access to engineering models or the spacecraft flight model should be provided to the ground system team. However, access to the flight model may provide fewer testing and validation opportunities since the spacecraft is still on the ground. A high-fidelity software simulator provides a representative model of the spacecraft, the spacecraft subsystems, and the physics of the spacecraft environment. The implementation of a satellite simulator can be purely software, but it can also be combined with real spacecraft hardware, such as an engineering model of the on-board computer. Such simulators are called hybrid simulators.

Early deployment of a stable, high-fidelity satellite simulator simplifies mission preparation by allowing early checks of commands and telemetry data. It also enables early familiarization of the FOT with the spacecraft in addition to the spacecraft user manual. The creation and validation of flight procedures is a third mandatory activity during mission preparation that benefits from the early availability of a spacecraft simulator.

4.4.3 Mission Information Base

The MIB contains the definition of the commands, the command parameters, the telemetry, and the location of the telemetry data in the downlink data stream. It is therefore an essential input for validation of the M&C software and also for procedure development and validation. In a preliminary version, it should be delivered in time for the ground segment validation phase. At the end of this validation phase, the MIB should also be final.

4.4.4 M&C System Software

The spacecraft is monitored via telemetry monitoring and display software and commanded via command software. Together, these software packets form an M&C system. It is an essential component of the FOS and is preferably written generically so that only customizations are required for individual missions. This facilitates validation because only the modifications need to be extensively tested. The functionality of the entire M&C system is then verified during validation of the MIB. Simulations and rehearsals are also used to validate the M&C system.

4.4.5 Ground and Flight Procedures for Nominal and Contingency Situations

Control of the spacecraft is basically possible using the M&C software and the validated MIB. However, logical work flows, branching, and timing constraints cannot be described with a simple list of commands. A proven concept to mitigate this shortcoming is the use of validated procedures. They are developed for ground and for flight processes and greatly improve mission operations as they reduce the risk of operational mistakes. A procedure describes a validated workflow step-by-step, along with the required initial conditions, commands to be sent, expected space segment response, timing conditions, and explanatory comments. Such procedures are primarily developed for routine operations, such as activating or deactivating a component or subsystem on board. Nevertheless, it is important to cover possible contingency situations with appropriate procedures as well. The safe mode crash is a prominent example of a contingency situation, and the appropriate procedure should describe the analysis and recovery actions to return the spacecraft to normal operating mode.

4.4.6 Operation Support Tools

A spacecraft must be monitored and operated. However, the amount of attention from the ground depends on the mission. There are space projects that are operated around the clock, such as human spaceflight missions, and there are satellites with a single ground station contact per day. Therefore, the operation of each mission requires a tailored set of operational tools to make the operation as robust as possible. Such tools include anomaly tracking tools, sequence of events, telephone lists, shift plans, minutes, recommendation handling tools, links to documentation, procedure lists, etc. The operation support tools should be organized so that they are easily accessible from any control room position, such as a mission-specific web page.

4.4.7 Project Documentation

For safe and robust mission operations, the FOT must know the functionality of the space segment and be trained for typical or likely situations that will occur. This is facilitated by training and technical documentation. This includes, for example, the spacecraft user manual, the ground segment design description, and operational documentation. These documents shall be available in a timely manner prior to the training and operational validation phase and shall be accessible throughout the operational lifetime.

4.5 Activities, Tasks, and Schedule

There are a number of tasks to be accomplished during the preparation phase for mission operations. At the beginning of the preparations phase, they are primarily technical in nature, such as the integration and testing of ground segment systems and components such as software, computers, and networks. Later, when the technical work is completed, the composition, training, and certification of the FOT becomes the dominant activity. The following description provides an overview of the tasks of each subphase, the requirements to accomplish these tasks, and the expected results of each subphase.

The first of these phases is the project phase C, called "Detailed Design", and it is used to finalize the ground segment design. If technical developments have a long development time (long-lead items), they already start in this phase. The key roles in this phase are those of PM, FD, SE, and QA. They coordinate the design and documentation activities and are supported by SSEs and specialists, such as flight dynamics experts and network specialists. They all use the design documentation from phase B and elaborate the developed concepts to final fidelity. Interfaces are fully defined and appropriate test approaches and plans are written. The result of this phase is a detailed design description, including all internal and external interfaces as well as test plans and schedules.

Project phase C ends with a critical design review (CDR), in which the ground segment provider presents the developed design to the ground segment customer.

The following Phase D, called "Production, AIT and verification", includes three subphases: the development and procurement of the ground segment systems (D1); the assembly, integration, and test of these systems (D2); and the verification and operational validation of the ground segment (D3). In the D1 subphase, the ground segment systems, subsystems, and components are procured or manufactured, including functional and interface testing. The supervision of these activities is the primary responsibility of the SE, supported by QA. In subphase D2, the ground segment is assembled and integrated. The control room is integrated and any required infrastructure changes are implemented. Networks and automated transfer services are configured and computer hardware is integrated into the operational environment. These activities are again supervised by the SE, supported by QA and the FD. The following activities are the content of subphase D2:

- RF compatibility test (six months to 12 months before launch, preferably with RF components of the flight-model)
- Functional and performance testing of internal and external interfaces
- Functional and performance tests of all ground segment subsystems
- Functional testing of the whole ground segment
- FOT assembly and initial training (classroom instruction, manual study, etc.)
- Initial flight and ground operations procedures
- Validation of the MIB
- Preparation of test and validation reports
- Compilation of results into a report summary.

Subphase D2 ends with the ground segment qualification review (GSQR) and a critical operations review (COR). Both reviews can also be combined into one review.

Ground segment operational validation is the primary content of subphase D3. The FD, SE, QA, and SIM roles are primarily involved in the validation. Intensive flight operations procedures (FOP) training, finalization of flight and ground procedures, and validation must be achieved. In this process, certification of the FOT's readiness for the upcoming launch will be achieved through simulations of increasing complexity. The results of these simulations will be documented with appropriate reports for the customer. These reports and any report summaries are reviewed during the operational readiness review (ORR), the final review prior to launch.

Essential documents resulting from the preparation phase are listed in Table 4.2, together with the review for which they must be available (see also Sect. 4.6). According to ECSS (ECSS-E-ST-10-06C), these documents belong to five levels: space system (SS), ground segment (GS), ground system (GSYS), logistics support (LS), and operations (OPS). The letter "A" designates a document issue for approval by the customer and "F" designates a final document issue approved by the respective supplier and for the customer's information. Note that the documents listed above

Table 4.2 Deliverables per review milestone according to ECSS

	Domain	GSCDR	GSQR	COR	ORR
Space-to-ground ICD	SS	A			
Space segment operability requirements document	SS	A			
CFI and services requirements document	SS	A			
GS engineering plan	GS	F			
ICDs for external and internal entities	GS	F			
GS design definition file	GS	F			
GS design justification file	GS	F			
GS configuration management plan	GS	F			
GS AIT plan	GS	F			
GS verification plan	GS	F			
GS configuration status report	GS		F		
GS integration and test reports	GS		F		
GS verification reports	GS		F		
Ground systems user and maintenance manuals	GSYS		F		
Logistics support plan	LS		F		
Mission analysis report	OPS	F	F		
Operations engineering plan	OPS	F			
Operational validation plan	OPS			F	
Operations training plan	OPS			F	
Mission operation plan	OPS			F	A
Operational validation reports	OPS			F	F

are not necessarily separate documents. The relevant information may be embedded in other documents. In addition, the actual contract may define only a subset of this list as deliverables.

4.6 Review Process

A review is a formal project milestone. Successfully passing this milestone indicates that all measures have been taken to complete a specific predefined work package, project phase, or subphase. This means that the project is paused to examine, evaluate, and assess the project status and decide on whether or not to proceed to the next phase. The status is presented with appropriate documentation.

Several reviews are possible during mission preparation, although not all are necessary or required for each mission. The project-specific tailoring required will depend on the particular type of mission and should be specified in the contract covering Phases C and D. This subsection lists the possible reviews during these mission phases, describing their content, common and essential requirements, and when they take place.

ECSS provides for several major reviews, which are described below. These reviews are ground segment critical design review (GSCDR), GSQR, COR, and ORR. Additional internal reviews are possible, such as test readiness reviews or simulation readiness reviews, and need to be included depending on the complexity of the particular project.

Although each of these reviews covers very different topics or elements of the ground segment, they generally follow common principles in timing and organization. However, customization is possible to meet the needs of a project and also for a particular review. Therefore, a complete description of the review in terms of "who, how, and when" should be written. This so-called review procedure is communicated in a timely manner to the "review team," which should consist the project team, i.e. the space segment and the ground segment engineers as well as the review board. The board should preferably consist of external experts in relevant project areas. These experts may come from other companies, research laboratories, test facilities, or agencies. The more diverse the knowledge assembled on the review board, the better.

A review starts with a presentation of the current project status. Venue of this presentation is often at the customer's site. Each project stakeholder gives a brief overview of the current status. This presentation allows the project to describe and explain the specific boundary conditions or constraints under which the current status was achieved. After the presentation, the so-called review data package is given to the review team for examination and evaluation. The duration of this phase should be chosen depending on the size of the data package. In practice, however, this phase takes between two and six weeks.

While studying the review data package documents, all members of the review team should document any concerns that arise via a so-called review item discrepancy

(RID). The structure and organization of RIDs is the responsibility of the review board lead and must be described in the review procedure. At least the following information should be provided for each item:

1. *Item*: Identifies the item of the data package for which the remark is valid. It may follow a predefined nomenclature or scheme, but the document number, page number, section number, and/or line number will also suffice.
2. *Observation*: This comment describes what the reviewer noticed. Examples include unclear statements, incorrect conclusions, inadequate descriptions, inconsistent analysis, and also typographical errors.
3. *Concern*: The reviewer must express the concern that arises from his observation, e.g. an increased risk of failure of a component due to inadequate testing.
4. *Recommendation*: This is a description of the corrective measures or activities required to resolve the observed problem, e.g. extending a test campaign.

It is the responsibility of the project management and the reviewer to determine an appropriate set of RID data items for the particular project. For example, to help organize the review process a "criticality" criterion with possible values "low", "medium", and "high" (or "major" and "minor") helps to group the RIDs and to focus on the important ones first.

The review period, i.e. the time during which the review team can provide RIDs, is limited to approximately 75% of the total review period. The RIDs are then provided to the project team for response. The team then designates a responsible person. This person first decides whether or not the RID observation is warranted, i.e., whether the RID is accepted or rejected. The accepted RIDs are then analyzed and responded to. The response is usually an action, the provision of more information, or a correction to the existing information. This must always be reflected in the updating of documents, e.g. in the form of updates or new issues of the project documentation.

The final review stage includes a so-called RID discussion and closeout, which results in a review report. During the RID discussion, which takes between 2 and 3 days, the project team presents and defends its RID responses to the RID owners. Depending on the responses, the review board then decides whether to pass or fail the review and documents its decision with the review board report. A passing review is often synonymous with a formal "go" to the next phase of the project.

The specific reviews of the mission preparation phase are explained below. These should each be organized and carried out according to the generic description provided. However, project-specific changes are always permitted, but should be coordinated between the project partners involved. If possible, the changes should not be so extensive that the character of a review is significantly changed. In general, and if applicable, the ground segment reviews can be conducted together with the space segment reviews, e.g., by combining the critical design reviews of both segments into one system CDR.

According to ECSS (ECSS-M-ST-10-01C), the reviews during preparation phase are (Table 4.3):

Table 4.3 Reviews during preparation phase

Ground Segment Critical Design Review (GSCDR)	
Date	End of project phase C "Detailed Design"
Objective	customer acceptance of the detailed ground segment design
Precondition	design complete, justified, and documented
Content	documentation providing description and justification of the ground segment design but also test and training specification as well as interface definitions
Chaired by	ground segment customer
Ground Segment Qualification Review (GSQR)	
Date	During phase D, at the end of ground segment AIT and verification (D2)
Objective	to ensure that the ground segment conforms to the technical requirements and that all conditions are met for proceeding with the operational validation phase (D3)
Precondition	ground segment AIT and verification has been finished, i.e., the ground segment is technically ready for usage
Content	test documentation, e.g. reports and report summaries of AITV activities on various levels
Chaired by	ground segment supplier
Critical Operations Review (COR)	
Date	During phase D, after completion of operational validation
Objective	to ensure that all mission operations data has been validated and that all documentation is available to start the training of an operational validation phase
Precondition	passed GSQR and finished validation of operational data
Content	test reports
Chaired by	operations customer
Operational Readiness Review (ORR)	
Date	end of phase D, after completion of operational validation, often 3–6 weeks before launch
Objective	to ensure full readiness of the ground segment for in-orbit operations, and to authorize its utilization for space segment in-orbit operations; to ensure validation of all procedures and readiness of the FOT
Precondition	FOT training finished; operations procedures validated
Content	documentation describing content, course, and results of operations team training, simulations, and rehearsals
Chaired by	operations customer

4.7 Operational Validation

4.7.1 Necessity

Operational validation is necessary to ensure that the ground segment as a whole, including the operations personnel, is capable to carry out the mission. The demonstration of the ground segment's readiness for the mission shall be reviewed during the operations readiness review (ORR). The operational validation needs to be scheduled in phase D3 according to the ECSS mission phase model.

All the elements of the ground segment are specified and verified on unit and system level. The same applies to interfaces between these elements. Flight operations procedures (FOP) are validated against a simulator and operations personnel is trained. But even if all these individual activities were perfectly executed, it would not guarantee a successful mission.

There are also pragmatic reasons to conduct operational validation. The first reason is the end-to-end execution of all mission activities. A typical example is the execution of an orbital maneuver, which is planned by the flight dynamics engineers using their software (see Chap. 13) and producing input for operations and mission planning. Here, multiple aspects can be covered: The processes leading to the inputs to the flight dynamics system (FDS) by the spacecraft engineers, the publication of the computed data (maneuver overview, planning of events covering both ground stations and satellite), the review of the maneuver data by the manufacturer and operations team, the commanding of the parameters, the monitoring of the execution as well as the provision of the post calibration. All of that is executed with a timing which is close to the one encountered during the mission. The format of the products and the correct transfer between the entities should, at this time, already be verified by a dedicated verification. As can be seen in this example, the operational validation is much more than just a test: It brings all stakeholders together. They should then implement the processes documented in the corresponding procedures and handbooks and learned during the training.

Another aspect of operational validation is the control center environment. The mission itself is executed in a setting that is significantly different from the environment found in a clean room or development area in which the assembly, integration and test (AIT) is performed. The software tools are often different and external personnel need time to get used to the mission operations environment. Access to the mission documentation, log-in credentials, phone numbers, security and safety rules, extraction and transfer of telemetry data or the usage of voice protocol are topics that fall under the new environment. The familiarization with the mission environment is of course not limited to the control center itself; the same holds true for the in-orbit test (IOT) ground station. A further, interesting aspect of the validation is the reaction of the ground segment to the load on the system. Simulation and rehearsals are often the first opportunities to have all designated personnel in the control room, which may result in a slow response time for the delivery of telemetry data or products. Timelines of operations and the feasibility of the sequence of events (SoE) are important artefacts to analyze at this stage.

Finally, simulations and rehearsals are the occasion to bring together all the actors of the mission. Personnel from the customer, the AIT team, the control center and the ground stations interact with the system developed for the mission and carry out operations as planned for the flight. In some cases, the mission may involve several control centers around the world. Although training sessions have been held beforehand and despite having experienced personnel on console, the first simulations are very often a shock, as each team needs a forming phase first and requires an adaptation to the actual mission. The usual issues discovered during the operational validation phase range from missing interfaces, insufficient visibility (e.g. of a display page or information exchange) and incompatible formats to unannounced FOP modifications or ground segment adaptations. In the end, the mission shall be carried out by a *combined* team—operators and manufacturers—whose proficiency needs to be demonstrated during the operational validation. In this sense, the operational validation contributes to the training program. The final rehearsals shall demonstrate the readiness of the combined team for the mission as well as its proficiency to cope with unplanned situations.

It shall be stressed here, that operational validation should not be confused with a system validation test (SVT). The objective of the latter is to assess the compatibility between the ground segment and the satellite at commanding and telemetry level. The compatibility at radio frequency (RF) level is the topic of the RF compatibility test, which complements the SVT. Both tests are performed earlier, at the end of the ECSS D2 phase with real flight hardware, whereas the operational validation is performed using a simulator instead of the spacecraft. However, it is possible to take advantage of the SVT for the ORR to validate some FOPs whose performance is not supported by the simulator. Critical operations related to the mission safety shall also be performed on the flight hardware during the SVT to obtain better confidence.

4.7.2 *Operational Validation Organization*

As shown in the introduction (see Fig. 4.1), operational validation takes place after the qualification phase, during which all elements are tested separately against their specifications. The operational validation should also be completed at the ORR, at which point the operational readiness of the ground facilities and personnel should be assessed.

The operational validation usually takes the form of a series of simulations and one or two rehearsals. Rehearsals are different from simulations only in their position at the end of the phase. A mission can be divided into several thematic clusters like the initial acquisition, an orbit maneuver, the deployment of appendices (solar panels, antennas), or an antenna mapping test. After identifying these blocks and depending on their duration and relationships, simulations are organized to perform them under conditions as close as possible to the mission. Apart from being performed within the real control room environment that will be used during the mission, with the qualified software pieces and the trained operations personnel, an important aspect of a simulation is its timelines. A simulation should be performed as far as possible under real-time conditions, which allows the validation of the timeline. Often, a compromise needs to be found between being realistic and the duration of the simulation. For example, extended idle sequences can be compressed or skipped. Also, orbit maneuvers limited to ten minutes usually allow the same level of validation as realistic ones lasting an hour.

Operation validation is usually performed using a spacecraft simulator rather than the real hardware. Although the flight model would guarantee the most realistic behavior, there are several scenarios which can only be exercised with a simulator: Conducting realistic operations requires simulating the environment (position of spacecraft, Sun, Moon, spacecraft attitude, instrument field of view, ground station visibility, etc.) and its interactions with the spacecraft (e.g., control law to point a celestial body or when firing an apogee engine). Furthermore, the flight model is barely accessible due to all the integration and test activities in a phase when the spacecraft is being prepared for shipment to the launch site.

Validating operations at the ground stations is difficult since no signal is actually received. Actions like pointing the antenna, sweeping the ground station signal or setting the polarization have no effect.

The organization of a simulation or rehearsal requires a lot of preparation work. The flight director writes the validation plan. In preparation and execution, he is supported by the simulation officer. Apart from setting up the simulator in the right configuration for the chosen block of activities, it is also necessary to pay attention to align the orbit parameters in the simulator and in the flight dynamics system with the ones computed in the mission analysis. Also, the monitoring and control system (MCS) needs to be configured appropriately to support a simulation with a simulation time in the future (for example six months to one year ahead) and still handle correctly received time stamps of command acknowledgments or time-tagged commands. The alignment of the on-board software (OBSW) version in the simulator with the satellite reference database (SRDB), the flight operations procedures and the display pages shall be ensured as well.

The operational validation shall address simple and severe anomalies together with the processes to detect and resolve them. This aspect depends a lot on the capabilities of the simulator to inject failures and model realistically the spacecraft behavior. It has to be ensured that a recovery scenario is available in order to continue the simulation timeline.

When correctly set up, the operational validation is a very rich phase which brings several issues to light. This ranges from incorrect or incomplete implementations to planning or timing aspects. A careful schedule is therefore necessary to ensure that enough time is provided between each session to implement the corrective actions agreed among the combined operations team.

It is sometimes difficult to make clear to managers of teams having different deadlines (e.g. shipment versus simulations) and whose schedules are tight, that the operational validation is no loss of time. In the end, the operational validation increases the safety of operations and can help to prevent mishaps during the actual mission.

4.7.3 Points to Evaluate During Simulations and Rehearsals

During the actual execution of the operational validation phase, the simulation officer (more on this position can be found in Sect. 4.3.5) evaluates the capabilities achieved by the combined operational team and the readiness of the ground segment to support the mission. The execution of a rehearsal is rarely perfect, and the evaluation must strike a balance between the aspects related to the safety of the spacecraft and those related to the optimization of operations. What still needs to be achieved must be weighed against the time remaining until the ORR and the training or improvement opportunities.

Anomalies should be introduced with care, because much can be assessed during nominal operating situations as well. It is important to ensure that the available telemetry display pages provide sufficient information to track operation and possibly identify the anomalies that will be addressed in the failure mode effects and criticality analysis (FMECA).

The individual training lessons of the operations team formally do not belong to the validation, but may be carried out in parallel and are entangled to some degree. Some training sessions are a prerequisite for the start of the operational validation. One objective of operational validation is to accurately evaluate whether each individual has the skills and knowledge required for the mission. However, it is possible to repeat or refine some training sessions, depending on the deficiencies found in the simulations. From this point on, the training will be more personalized, depending on the skills achieved by each individual.

Another very important aspect is to ensure that the latest available documentation is accessible from the control room and has been used as input for the various operational tools used during operations.

Finally, the simulation officer ensures that the processes necessary for normal operation (e.g. briefing, voice protocol, logging, etc.) are understood and followed by the integrated operations team.

During nominal phases, the simulation officer shall also keep an eye on the ground system load. This allows ensuring that the resource sizing is correct and that no data is lost or delayed, especially in conditions similar to the ones encountered during the mission. The extraction of telemetry data and its transfer to various tools used for operations shall be assessed as well. Incompatibilities or errors, such as in format or unit, are common even after the qualification of the ground system has been completed. Satellite manufacturers have their own tools, in some cases they bring their own PC hardware, and their access from the control center is often a request arising from the first simulations.

The simulation of anomalies shall be carefully planned by the simulation officer. The expected reaction of the operations integrated team needs to be clear before the ingestion of error in order to evaluate the reactions. An evaluation of the time required to detect the anomaly, its reporting and the implemented contingency procedures needs to be made against the expected reaction. It is also important to assess how the processes (e.g. recommendations, anomaly briefings, reporting to management) to handle anomalous situations and their documentation were implemented. A debriefing with the participants is necessary to collect their comments and debate on the positive and negative points noted during the simulation.

Rehearsals are particular simulations organized shortly before ORR to demonstrate the readiness level achieved by the integrated operations team. That team shall later support the decision in the ORR to declare the ground segment and the operation teams fit for flight. Such rehearsals shall cover the nominal mission and well as anomalous situations. The introduction of at least one really severe anomaly is necessary to push the operations personnel out of their comfort zone. Experience from previous missions should play an important role in the selection of the anomalies.

The fact that an anomaly case may seem improbable should not prevent from rehearsing such a situation. Should the same situation happen in flight, which happens indeed more often than one might want to admit, there would be no refusal to deal with it, and so is the logic for rehearsals as well. What is important is to use the processes and the tools available to handle the situation. It is possible that an improvement of processes or tools will be proposed after coping with such a situation. It is the experience of the simulation officer in picking a suitable situation which makes up a successful rehearsal.

4.7.4 Operational Readiness

At the end of the operational validation phase, the simulation officer needs to be in the position to report the readiness for the mission. Unless the last rehearsal was extremely unsuccessful, it is very unlikely that a launch will be postponed because of the operational validation. The issues discovered during the simulations should be corrected for the following ones and so the final rehearsals are usually successful. The remaining issues constitute the risk that is taken when the launch would occur with a ground segment in the state presented at the final rehearsal. A big benefit of the operational validation phase is the awareness of the integrated operations team of its own weaknesses.

The ORR is the last and final review of the mission operations preparation before launch. It gives the clearance for the following launch and early orbit phase (LEOP), during which the ground segment must prove itself in real orbital operations. However, a number of final actions must be performed before launch, including regular technical checks of the ground segment elements. Depending on the mission and the control center that conducts the operations, a so-called system freeze is also worth considering. This means that from a defined point in time, changes to technical systems affecting the ground segment of the mission are only allowed under strict configuration control. This is particularly recommended in a multi-mission environment. The organization of the system freeze must then be coordinated with the control center management and eventual other projects. A system freeze can end after the LEOP.

4.8 Conclusion

If preparation for operations has been successfully completed, real mission operations will likely proceed smoothly with few, if any, unexpected or unprepared contingencies. However, a ground segment and the FOT can hardly be perfectly prepared and trained for every type of spacecraft malfunction or non-nominal behavior. A high degree of flexibility and improvisational skills are therefore essential to maximize the chances of mission success.

References

ECSS-E-ST-10C. Space engineering: system engineering general requirements, 2017
ECSS-E-ST-10-06C. Space engineering: technical requirements specification, 2009
ECSS-E-ST-70C. Space engineering: ground systems and operations, 2008
Wenzel KP, Page DE, Smith EJ (1992) The Ulysses mission. Astronomy and astrophysics supplement series, vol 92, no 2, p 207

Andreas Ohndorf works and has been working in the Mission Operations department of the German Space Operations Center (GSOC) in several roles: as Flight Director, as Ground Segment System Engineer, and as Satellite Project Manager, and team lead. Graduated in Aerospace Engineering at the University of the Armed Forces of Germany in 2001 in Munich, he served as a Technical Officer in the German Air Force for 12 years before joining DLR in 2008. He received a Ph.D. in Aerospace Engineering from Delft Technical University in 2016 and currently leads the Low-Earth-Orbit Satellite Missions Team in the Mission Operations department. Since 2021 he leads the ground segment department of DLR's Responsive Space Cluster Competence Center.

Franck Chatel was Flight Director for the H36w-1 and EDRS missions within the department mission operations at the German Space Operations Center (GSOC). He graduated in aerospace engineering at the Institut Supérieur de l'Aéronautique et de l'Espace (ISAE) in Toulouse (France) and joined DLR in 2001. He was involved as propulsion and AOCS sub-system engineer, later as flight director, in different LEO missions (BIRD, SAR-Lupe, GRACE) and GEO missions (EUTELSAT W24). He supported the early phase of the GALILEO project at spaceopal and DLR GfR before moving to his current position in flight control operations at EUMETSAT.

Chapter 5
Mission Operations Execution

Sabrina Eberle, Thomas Uhlig, Ralf Faller, and Michael Schmidhuber

Abstract This chapter describes in detail the tasks of mission operations execution phases. The basics concerning the different phases during LEOP, commissioning, routine and disposal phase are explained. The differences are described as well as the necessary team members and the support from other teams. Various examples and procedures are discussed, and the transition between the phases is presented in detail. Finally, some examples of different missions for LEO, GEO, deep space missions and human spaceflight are given. This chapter will mainly concentrate on operating unmanned spacecraft from ground, but special aspects of human spaceflight missions will be mentioned where relevant.

5.1 Introduction

Although the preparation phase of a space mission can exceed the duration of the actual mission execution phase significantly, the execution can be considered as the most important phase, because here the spacecraft is fulfilling subsequently all of its mission objectives. In this phase, a dedicated flight control team is overseeing the operations of the satellite. This chapter briefly discusses this phase of the satellite's lifetime and highlights specific processes and setups.

The chapter is mainly focused on satellite operations. However, some aspects of human spaceflight are also highlighted. For more details of the latter, Chap. 24 can also be consulted.

S. Eberle (✉) · T. Uhlig · R. Faller · M. Schmidhuber
German Space Operations Center (GSOC), German Aerospace Center (DLR), Oberpfaffenhofen, Germany
e-mail: sabrina.eberle@DLR.de

F. Sellmaier et al. (eds.), *Spacecraft Operations*, Springer Aerospace Technology,
https://doi.org/10.1007/978-3-030-88593-9_5

5.2 Various Phases During Execution

The execution phase can be broken down into different periods, which can be distinguished by their special operational requirements.

The first part of a mission is called the launch and early orbit phase (LEOP). It is followed by the commissioning phase, in which the spacecraft as well as the payload on board are prepared for nominal operations. The actual mission goals are then accomplished in the routine phase, which is in most cases the longest phase. The end of mission (EOM) of a satellite is followed by the disposal phase, which ensures that the satellite is either parked in a dedicated graveyard orbit or is destroyed by a controlled reentry into the Earth atmosphere. This phase is also called the de-orbit phase.

The execution phase is of course strongly dependent on the mission goal. A scientific low Earth orbit (LEO) mission with experiments on board for only a few months or one year needs to be prepared in the same way as a LEO mission with an expected lifetime of 5–10 years. On the other hand, geostationary satellites (especially the commercial communication satellites) often have lifetimes of around 20 years. Interplanetary missions in comparison have very long execution phases, because it normally takes the spacecraft a long time to get to its destination in the first place (Figs. 5.1 and Fig. 5.2).

5.2.1 *General Description of the Execution Phase*

The above mentioned phases include some common tasks, that will be highlighted in the following section. All spacecraft have implemented a command and telemetry interface. Therefore, they can be controlled by telecommands from ground and allow

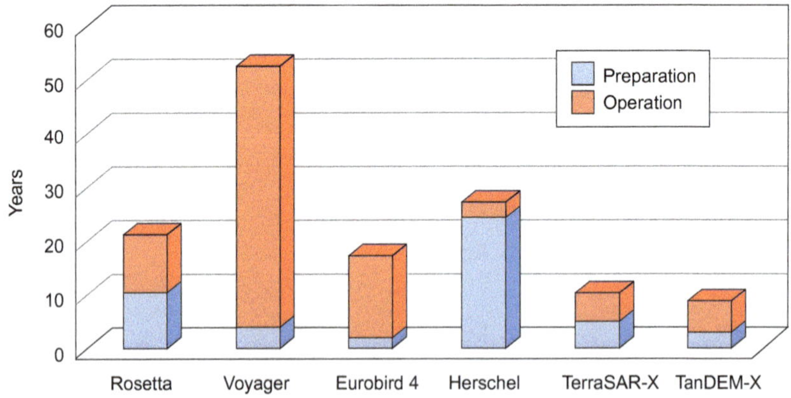

Fig. 5.1 The variable durations of the mission phases using the example of different missions

Fig. 5.2 Overview of the differing lengths of mission execution phases on the basis of different mission types. The preparation phases are assumed here to be roughly of equal length

insight into their internal status via the data they send back to the ground stations. This telemetry not only comprises the data from the corresponding payload, but also information about important parameters of the satellite's subsystems, such as temperatures, currents, status reports from the software or event messages triggered by off-nominal conditions on board. These data—sometimes also referred to as health and status or housekeeping (HK) data—need to be monitored on ground. Commanding of the spacecraft is also a standard task in the execution phase. Commanding encompasses both payload operations and control of the satellite's subsystems.

Not only the internal performance of the satellite requires surveillance, the orbit and the attitude situation also need to be monitored closely. Active adjustments of orbit and attitude are also required. These maneuvers are described in more detail in Chap. 22.

During all execution phases, on-board maintenance activities like software updates or recalibrations of instruments may take place. Their needs are identified during the spacecraft design phase and are usually defined in a maintenance plan that lists all those activities together with the corresponding timeframes when they need to be conducted.

Satellites are usually designed to be highly autonomous. One of the major reasons for involving human control teams in operations is the handling of unexpected situations in the ground or in the space segment. These events are usually called anomalies or contingencies, depending on their severity. Here, humans need to be involved to analyze the sometimes very complex situation, and to put together either troubleshooting plans to identify the root cause of the problem, or to resolve the issue via corrective actions.

Failures can sometimes be prevented, either through preventive measures that can be part of the regular maintenance activities, or through a detailed short- and long-term analysis of the spacecraft parameters. Tendencies and trends can be observed here, that may lead to the decision to take countermeasures to prevent e.g. further degradation of components or subsystems.

5.2.2 Launch and Early Orbit Phase (LEOP)

The LEOP starts after the satellite is launched and released from the carrier. The satellite then has its first time in orbit, in which very specific operational requirements need to be fulfilled, which will be discussed in more detail here.

Launch operations are usually done by a dedicated team in the responsibility of the launch provider. They hand over the satellite to the satellite control center after it is released from the upper-stage and free-floating. The first major milestone for the controllers is then to establish the first contact with the satellite. Since the ascent phase of the carrier rocket and the release process itself are associated with some uncertainties, the position and orbit of the new satellite is not exactly known. Therefore, the first acquisition may involve some search activities of the ground stations. As soon as a radio link is established, the position and orbit can be determined more accurately and a first checkout of the essential components of the spacecraft is started. In many cases, there are also a few very important configuration steps which need to be performed as soon as possible: The radio link with the satellite is dependent of the satellite's attitude, which determines the orientation of the antennas. Therefore, the spacecraft must be given attitude control. It is also crucial to ensure power generation capability on board, since the launch phase is usually only supported by the satellite's batteries, whose capacity is limited. This could encompass deployment of the solar arrays and some reconfigurations of the power distribution system, proper setting of the battery charge regime, and the switch-on of some vital subsystem components.

Such measures can ensure the survival of the satellite in the harsh space environment described in more detail in Chap. 1. One of the next steps is to bring the satellite to its final destination, be it a dedicated orbit or a specific position in the GEO. This may require either several extensive maneuvers with the satellite's propulsion system, or just minor corrections of the orbital parameters. In many cases the maneuvers in this phase of the mission are the main active changes of the satellite's orbit over its entire lifetime.

The LEOP is probably the most critical phase of the entire mission: After a very demanding ascent in terms of vibrations, acceleration, temperature changes, mechanical stresses or sound levels, the spacecraft is exposed to the "real" space environment for the first time. In this way, the satellite becomes "unknown" to the operations team and will unveil new characteristics and behaviors in this phase—which, in most cases, leads to surprises which could lead to a requirement for a redesign of the already prepared operations concept and its corresponding procedures and processes.

Unlike the routine phase, which usually contains a series of repeating and well-understood mission activities, many of the LEOP activities are singular or even irreversible events. The latter may be due to the fact that the developer decided in the design phase that only the transition of a given piece of equipment to its nominal ops configuration should be implemented, but not the transition back into its launch configuration (i.e. only the deployment of solar arrays, payload antennas or instrument booms, but not the retraction).

Events in the LEOP might be time-critical, either in the sense that they need to be performed in a very specific time frame (like orbital maneuvers) or that a strict temporal relationship to other activities exist (i.e. a "thermal clock" for equipment which requires heating after a certain time to prevent degradation or even damage).

Since the LEOP contains many transitional states of the on-board configuration, the level of on-board automatisms is usually considerably lower than for the routine phase, for which a well-defined configuration of the satellite can be assumed. This also limits the capabilities of "self-healing" fault detection, isolation, and recovery (FDIR), making the satellite more vulnerable.

All these reasons mentioned above lead to the requirement to have a good, almost permanent "visibility" of the satellite during the LEOP in order to have the chance to detect and intervene quickly in case of problems. Therefore, multiple ground stations are usually involved to ensure good coverage. It also requires a high level of redundancy in the ground system to cope with problems in this essential part of spacecraft operations. This is different from the routine phase where only one or a small number of ground stations are used due to cost constraints, resulting in a very limited contact time with LEO spacecraft.

This setup of multiple ground stations, potentially owned by different entities, including the coordination of them introduces an additional level of complexity.

5.2.3 Commissioning Phase

The LEOP is followed by the commissioning phase. The transition between them can sometimes be smooth. In this phase the satellite is ready to be used; it flies in its designated orbit and its survival is assured. Now, extensive testing of its platform and payload can be started. This involves checkouts on subsystem level as well as on an integrated level.

In spaceflight it is common to follow a concept of high redundancies. Many subsystems have redundant components—critical ones have even more than one level of redundancy. Elements are called "hot redundant" if the redundant part is already active and thus could take over the function in a very short period of time without interrupting operations. "Cold redundancy" means that the redundant element must first be activated in the event of a fault, which results in a certain latency.

During the commissioning phase, the redundant elements are also checked to ensure that their performance is sufficient for the function to be considered fully redundant or to be able to tune the ops concept accordingly in case of a degraded performance of the redundant component. Testing a hot redundant component is often unnecessary, since for the "hot case" the device is already active and some information about its performance is at hand already.

Redundancy testing is important for reducing operational risks. However, it also poses a certain risk in itself: The spacecraft is brought from a good and reliable configuration into a configuration which involves a not-yet-tested component; also, the transition process between the nominal and the redundant element can be considered as a more vulnerable phase of the satellite, since it is originally only foreseen to be executed in a contingency case where the increased risk of switching to another device is justified. For these reasons, it is attempted to avoid unnecessary switching processes and to bring them to an absolute minimum by means of sophisticated checkout sequences.

As already mentioned, not only the subsystems of the satellite platform but also the payload components are tested during the commissioning phase. This subphase is referred to as in-orbit test (IOT). Depending on the payload and its purpose, it may be necessary to execute special configuration procedures and perform calibration runs. The latter may require additional ground support or equipment not needed during the routine phase.

For geostationary communications satellites, the antenna typically needs to be positioned to direct the antenna beam to the selected area, the solar panels need to be activated to rotate with the Sun, and the payload itself needs to be launched.

In most cases, flight hardware must be proven to meet its specifications or requirements from the design phase. This may have technical or even contractual implications. Therefore, there might be respective test objectives which must be met during the commissioning phase as well.

All of the above tasks, typical for this phase of mission, require the presence and participation of the appropriate experts, be it from the companies involved in the construction of the components, from the expert teams within the flight control team, or from the side of the payload users, the experimenters, or scientists.

Sometimes a mission control center only conducts the LEOP and commissioning phase and transfers operations to a routine operations center for the comparatively easy routine phase. This transition is called the "handover". The LEOP control center is still in standby as a back-up for some time after the handover. After verifying that the routine operations center can operate the spacecraft trouble-free, the LEOP control center ceases operations. This scenario is quite common when the satellite manufacturer offers an in-orbit turnkey delivery to the customer or when the routine control center does not have the experience or the resources available to conduct a LEOP (large control room and access to global ground station network, etc.). Often the LEOP control center will function as a back-up for the routine control center also during routine operations in case a severe contingency or anomaly happens. They can take control of the spacecraft to solve the problems and return the satellite after bringing it into a stable configuration again.

5.2.4 Routine Phase

When the commissioning phase could be successfully completed, the routine phase can be initiated. Satellite operations are now usually linked to routine processes, telemetry is observed and analyzed as already described, and planning, as described in Chap. 16, governs the day-to-day tasks of the satellite. Payloads are operated to achieve mission objectives, whereas subsystems are operated to support payload operations and to ensure the well-being of the entire spacecraft.

Planning also encompasses the management of the limited resources available on the spacecraft, like electrical power or also fuel for orbital maneuvers.

The steady-state character of the routine phase also allows reducing the manpower to a minimum; the experts only need to be activated "on demand" and could be assigned to other projects. Nevertheless, they still monitor their subsystem on a daily or at least weekly basis, dependent on the complexity and flexibility of the subsystem. Especially the thermal system trend analysis is a very important instrument because the system is rather indolent and it takes a while for temperatures to change, so it is important to keep an eye on the long-term behavior. Other subsystems' parameters, like attitude and orbit control (AOCS), are monitored by their specialists more often. All subsystem engineers (SSEs) analyze trends in order to prevent foreseeable contingencies or errors during the whole routine phase.

In addition to telemetry monitoring, weekly team meetings are usually held to discuss special mission topics, events that have occurred, and upcoming actions. Special but expected events include orbit correction maneuvers calculated by the flight dynamics team (FDS) or antenna tests at ground stations. For geostationary satellite missions, orbit correction maneuvers are usually very predictable and follow a certain repetitive pattern. For LEO missions, not many maneuvers are usually required and are performed only every few months, depending on the mission. Further routine tasks beside the weekly team meetings are the monthly reporting to the customer, the maintenance of the change control, and the continuous training of the team members to keep them up-to-date and trained especially during a long routine operations phase, like 15 years or more.

Unexpected events are, for example, collision avoidance maneuvers, a switch to a redundant on-board component or a software upload, normally provided by the satellite manufacturer. Daily routine operations, e.g., dumping telemetry (downloading stored telemetry) and uploading the timetable for the next payload operations, can be taken over by command operators (spacecraft controllers—SPACONS), who are not required to have in-depth knowledge of the satellite's subsystems. In case anything unforeseen happens, the operator immediately contacts the flight director or the relevant SSE.

In routine operations, the ground station network, which is required to maintain the contact to the satellite, is reduced significantly. In many cases a single ground station can serve this purpose, depending on the orbit parameters. In that case, with LEO satellites, the contact to the satellite is reduced to a few passes per day, which then have to suffice to downlink the payload data, to gain insight into the satellite's health and status parameters, and to uplink commands. These are mostly "time-tagged," meaning that they are not immediately executed, but only at a well-defined time during the following orbit(s). Increasingly, highly automated planning engines on board of the satellite can take over control on-board and execute payload tasks autonomously.

5.2.5 *End of Mission or Disposal Phase*

There are some actions to be completed before the satellite mission can be declared to be ended. This is required by international policies or agreements with the customer. All these tasks are summarized in the disposal or decommissioning phase.

According to ISO 24113:2019 (2019), it has to be ensured that the satellite is not posing a risk for future spaceflight missions or for anybody living on Earth and—especially for the case of the very crowded GEO—its position in the orbit needs to be freed for possible successors. This can be achieved by two means: In case of LEO satellites, the orbital maneuvering system is used to change the spacecraft's trajectory in a way that it enters the Earth atmosphere in a controlled manner and is then destroyed by the thermal energy in which the immense kinetic energy of the object is converted to during reentry. This reentry should take place no later than 25 years after the end of the mission. In case the satellite is on a GEO, it is brought into a so-called graveyard orbit, a trajectory in which the satellite does not interfere with operational satellites for many decades or even centuries. This orbital region lies a few hundred kilometers above the GEO. Each mission is committed to leave enough fuel in the tank, so the spacecraft can be maneuvered either into the Earth atmosphere or the graveyard orbit. Therefore, the amount of fuel (see Chap. 21) has to be calculated very thoroughly for each maneuver during the routine phase to ensure that not too much fuel is left, which would cost the mission valuable lifetime, and not too little, so the final maneuver can be executed completely.

For the de-orbiting itself, normally the payload will be switched off and the satellite will be brought in a safe configuration (Skalden 2013). The systems are passivated by shorting the batteries and emptying the tanks to reduce the danger of explosions.

5.3 Staffing of the Flight Control Team

Each mission has different requirements for the flight control team (FCT) or flight operations team (FOT) composition—and different control centers follow slightly different philosophies. However, some elements and some considerations have general validity and are presented below.

The various functions represented in a flight control team are often referred to as "consoles", "subsystems engineer" or "positions". They are interconnected by modern voice communication systems and use dedicated tool suites to spread information within the team, to document decisions which have been made, to record the shift events in a dedicated shift diary, to command the spacecraft, and to monitor its telemetry.

5.3.1 Mission Operations Team Lead

A complex, multi-member team requires a clear hierarchy, a coordination function, and a decision-making process that allows quick reactions to sudden situations. Therefore, in all flight control team setups a team leader function is given. Nomenclature may vary, typically terms like "mission ops team lead," "spacecraft operations manager (SOM)", or "flight director" are used. Flight director (FD) is used in this book.

The person in this position has full responsibility for all operations conducted by his or her team, which specifically involves all commanding of the spacecraft. Therefore, the flight director is also the final authority for all decisions and has to approve all commands which are sent to the vehicle.

In day-to-day operations, his authority is usually only limited by the operations documents which define the operational envelope for the satellite—and under certain emergency circumstances, he may also decide to violate those. It is important for the flexibility and adaptability of operations to equip this person with extensive authority.

Depending on the project setup, the authority of the flight director may be limited to real-time processes only. In these cases, there needs to be another authority which is not part of the flight control team, but provides the team and the flight director in particular with management directives if needed. This position is often called mission director.

The FCT has full responsibility for the satellite during operations. During critical operation phases like the LEOP or special tests during the commissioning phase, an industry team will assist the FCT. If there are any non-nominal situations which were not described in the handbooks and are not covered with procedures, the industry team may help find a solution to return the spacecraft to the nominal configuration. However, the flight director has the overall responsibility of the operations. The SSEs of the industry team advise their corresponding partners of the FCT. The team lead of the industry team is often called the "satellite team lead" or "STL". This person will directly communicate with the flight director.

Representatives of the customer will have a number of console positions in the control room so they can monitor the operations. They communicate within their own team on a dedicated voice communication system loop and, of course, cannot send commands. If and how the customer is involved in operations and decision-making is dependent on the mission.

5.3.2 Subsystem Specialists

All subsystems of a spacecraft are usually reflected as "positions" in the flight control team. This ensures that the team has sufficient expertise and manpower to focus on their subsystems and to decide in critical situations where a deviation from the standard processes is required. The subsystem specialists monitor and analyze the data of their respective subsystem, and ensure that possible anomalies are detected and resolved. The level of responsibility again depends on the overall concept: The flight control team might be empowered and able to bring the spacecraft back into a fully nominal configuration. Or they may just conduct a first contingency response that puts the satellite into a safe mode to have enough time for the further analysis of the problem. This is then discussed and forwarded to engineering support teams, who finally provide the team with advice on how to recover the anomaly and resume nominal operations. The industry team normally will assist the flight control team.

The subsystem specialists support the flight director as the final decision-making instance to make decisions in real time.

For human spaceflight missions, the crew can be considered, in a first approximation, as (an) additional subsystem(s). Therefore, additional positions in the flight control team like "spacecraft communications" or "medical operations" are available here (see Chap. 24).

5.3.3 Command Operator

In most of the teams the actual commanding activity is performed by a dedicated command position. This ensures a good coordination of the overall command activity because it is performed in a centralized manner. The command operator or SPACON (spacecraft controller) takes instructions from the flight director—and from the flight director only! In that way the flight team is relieved from the technical aspects of the MCS and the communication with the network operator (NOPE) and the ground stations (see Chap. 10).

In routine phases, the presence of the flight control team in the control room can be reduced to only the command operator, who receives pregenerated and preapproved command tasks from the subsystem specialists and the flight director. The operator then prepares the command sequence, uploads it, and checks its successful execution. If he detects any anomaly, he can alert the flight director or the subsystem specialists who are usually on call for that purpose. His autonomy is usually constrained to well-understood and strictly defined situations.

5.3.4 Planner

The scheduling of the often very complex activities and a quick and profound reaction in case of malfunctions to ensure that the mission can continue under the new boundary conditions requires the existence of a planning function in the flight control team. The planning concepts for satellites and human spaceflight operations are described in more detail in Part V. The output of the mission planning team, the timeline, is provided as ready-to-send telecommands and is part of the daily command stack. Conventional geostationary communication missions usually don't need no dedicated mission planning. The planning tasks are distributed within the FCT.

5.3.5 *Flight Dynamics*

The orbit maneuvers which have to be performed are calculated and initiated by the flight dynamics team. These may be normal orbit maintenance maneuvers to keep the satellite on its nominal orbit, or an unscheduled maneuver like a collision avoidance maneuver. These maneuvers have the highest priority and all other planned tasks will be canceled and rescheduled in order to prevent the satellite from a possible collision with another spacecraft or impact by an uncontrollable object. The planning and development of an orbit maneuver is described in more detail in Chap. 13.

The flight dynamics team also provides other orbit related information: ground station contact times, S/C sensor usability prediction, maneuver calibration, and the collision risk estimation.

5.3.6 *Ground Data Systems*

To communicate with the spacecraft, the data links from the control center to the ground station and on to the satellite and back have to be established and maintained. This is done by the ground data systems team. Communications between the control room and the ground stations are handled by a special network communications operator. He will also inform the flight director about difficulties or changes regarding the antennas. The communications concept is described in more detail in Chap. 10.

5.3.7 *Engineering Support Team*

Essential support to the flight control team comes from the engineering support team, which in most cases is staffed by experienced engineers of the satellite supplier companies. They have the expert knowledge to analyze problems which are beyond the knowledge and the expertise of the flight control team. During critical phases (e.g. LEOP), representatives of this team need to be present in the control room; in routine phases they can be contacted remotely if needed. They can be considered the second, and in many cases also the last "line of defense". Special attention has to be given to the fact that industry experts naturally move on in their careers and the original knowledge fades over time. For long-term missions the project manager and the flight director need to secure the knowledge necessary and eventually bring this to the attention of the customer.

5.4 Interactions Within the Flight Control Team and Flight Procedures

5.4.1 Interactions Within the Flight Control Team

A typical scenario with a full-fledged flight control team is described here for better illustration of the baseline concept.

During the acquisition of signal (AOS), one of the subsystem components on board shows a high spike in the current of one of the power conditioning devices. The values are sent down to the control center, where the reading appears on the telemetry displays of the flight control team. The out-of-limit condition might be automatically detected by the ground software and the team could have been alerted via a visual and possibly also an audible alarm.

Either the flight director now prompts his team, or the responsible subsystem specialist proactively approaches the flight director and provides some information about the fault signature he has seen, an ad-hoc analysis about the root cause, and, with reference to the ops documentation, a recommendation on how to react in this specific case. The flight director might involve other affected disciplines or might consult the support assistance team, in case it is available, and then base his decision for the problem response on the information he gathered. He would then advice the command controller to prepare the corresponding telecommands and send them to the vehicle in close coordination with the corresponding ground station. The success of the commanding and of the problem resolution approach is then verified by the command operator and/or the subsystem specialist. Based on the results, further steps of recovery, some troubleshooting measures, analysis by further experts, or documentation of the anomaly will follow. All rules for interactions between the various positions are usually described by ops documents, which are also the foundation for the work and responsibility sharing within the team.

5.4.2 Flight Operations Procedures (FOP)

Safe and reliable operations of a spacecraft in orbit require sufficient knowledge about how to fly the spacecraft. Detailed information about the spacecraft itself and the ground system used for the operations is provided by handbooks, telemetry and telecommand databases, and other reference lists, but the basis for the operations is built on the so-called FOPs (more details in Chap. 7).

Definition and Applications

FOPs are a prepared, tested, and validated set of work instructions that list in the very detail all activities and checks to be performed for a specific purpose. This includes the exact sequence and timing of the different steps, complemented with comments about the activities and go/no-go criteria for critical events.

Depending on the mission characteristics, e.g., operating a manned or unmanned spacecraft, long (in GEO) or short (LEO) ground station contact times, the kind of monitoring and control system (MCS), level of spacecraft autonomy, or mission budget, different types of FOPs can be used. For manned missions with astronauts in the loop, it may be sufficient to have FOPs in a free text form on paper or screens (see an example of an astronaut checklist on Fig. 5.3). In unmanned missions, most actions are done through telecommands and telemetry. Here, flight procedure concepts are mostly table-based (see Fig. 5.4). This allows an automated and thus, safer and easier processing of the instructions in preparation and shorter execution times. More sophisticated systems are using script languages, which allow partly- or fully-automated execution of procedures by the ground system with only a minimum of supervisory activities by the operations personnel.

FOPs are typically grouped into procedures for nominal and non-nominal activities and tasks. Nominal procedures might be used for standard and planned situations (e.g. boost maneuver during LEOP), while non-nominal, often called contingency procedures, are prepared for anomalies and trouble-shooting (e.g. no telemetry at signal acquisition).

Flight procedures can be "atomic" or "elementary" and contain only a few instructions around one activity (e.g. switch on of S-band transmitter) or they can be

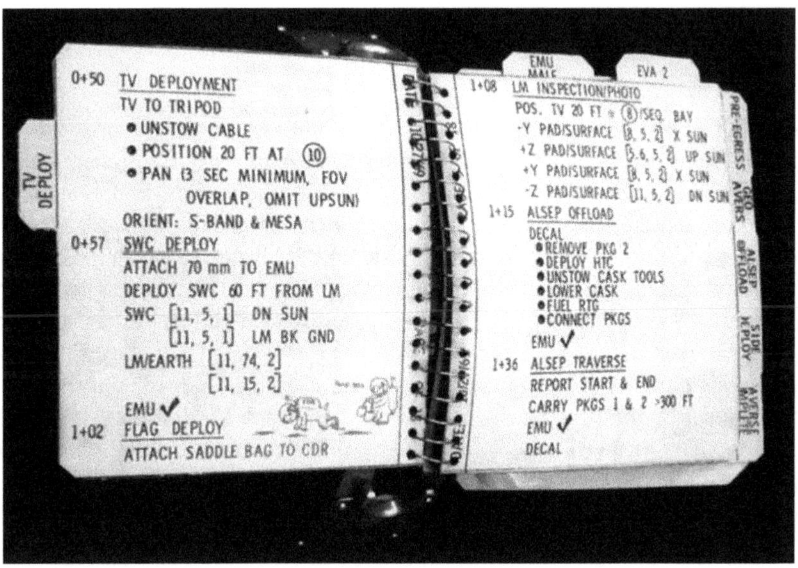

Fig. 5.3 Apollo 12 astronaut cuff checklist, NASA

PX1	NP120		Solar Generator Partial Deployment					TPX-Flight Operations Plan Vol 3		Issue 1.3	
STEP	TIME		ACTIVITY				DISPLAY	TM			
n°	ABSOLUTE	RELATIVE	EVENT DESCRIPTION		CODE	DATA	FMT	CODE	DESCRIPTION	RESULT	
1		T0-8min	**Verify Entry Conditions**								
			Check S/C Status								
			VERIFY:								
			AOCS mode				5200	AS100	AOCS mode	**SAM**	O
			Sun vector S/C x				5200	AD451	Sun S/C x	**0 +/- 0.02**	O
			Sun vector S/C y				5200	AD452	Sun S/C y	**0 +/- 0.02**	O
			Sun vector S/C z				5200	AD453	Sun S/C z	**-1 + 0.02**	O
			Main battery power level				6122	PD300	Bat A PWR	**> 60Ah**	O
			Deployment hinge temperature				2200	TP280	Depl H tmp	**> 10 deg**	O
			IF entry Conditions are not correct								
			THEN postpone deployment								
			ELSE proceed with step 2								
2			**Partial Deployment**								
									UPS: make hardcopy of #4210		O
2,1		T0-5min	**Prepare Pyro System for Deployment**								
			SEND:								
			Activate_Pyro_System		11PAD						[ ]
			Set_Pyro_System		10PAD						[ ]
			Arm_stage			12					
			VERIFY:								
			Pyro system arming				6122	AS230	Pyro Arming	**ACTIVE**	O
			Arming stage selection				6122	AS235	Arm Stage	**12**	O
2,2			Request Go/NoGo for deployment from all subsystems								O
2,3			Request Go/NoGo for deployment from STL								O
2,4			**Deploy Generator**								
		T0-1min	Set TC system in BD mode						**TC operator: verify TC system in BD-mode**		O
			SEND next 2 commands in sequence:								
		T0			20PAD						[ ]
			Fire Pyro		21PAX						[ ]
									TC operator: verify TC system in AD-mode		O

Fig. 5.4 Table-based flight procedure example

encompassing to cover a long complex activity (e.g. in-orbit test of a repeater payload).

FOP Life Cycle

The initial input for the FOP generation is provided by the spacecraft manufacturer, because he has designed the spacecraft and the basic algorithms to operate it. The initial procedures consist of the basic activity flow, the commands to be sent, telemetry checks to be performed, constraints, and basic timing specifications. In the next step, the flight operations specialists complement the procedures by adapting them to the provided monitoring and control system and insertion of additional information, like display page references for telemetry checks or ground-based activities. Finally, the procedures are validated by running them in simulator sessions or comparable test environments, so the correctness of foreseen commands, TM checks, and timing is confirmed. With the release of the validated FOPs, the utilization phase starts. Caused by spacecraft database updates, flight experiences, or changed S/C hardware characteristics, maintenance of the FOP might be required. The procedures need to be updated including additional validation sessions and an official release. A crucial factor is to keep the FOPs under strict configuration control to ensure that only validated and released procedures are used for flight.

Satellites that are built as a series and have a commercial background have the advantage that the flight procedures will be prepared by the manufacturer in a good maturity state. One-of-a-kind missions like scientific satellites or new models will require substantially more work to be done on flight procedure development and the control center may be asked to contribute in that work.

5.4.3 Anomalies and Recommendations

In the previous chapter, flight procedures were introduced as the central element for flight operations. It is obvious, however, that despite optimal mission preparation, a comprehensive set of FOPs, and intensive testing of all space and ground components beforehand, anomalies during the mission cannot be avoided. All kinds of glitches, malfunctions, and mishandling, in the space segment or on ground, might cause a disruption of the running activities and require proper reactions to resume nominal flight operations again.

Anomalies

Anomaly handling is the process of a controlled reaction on problems and anomalies required by common quality management standards. A common approach is to handle spacecraft or ground related issues in separate tracking systems. To process ground related problems, e.g. control room or ground network hardware or software problems, often the established issue tracking system of the ground facility is used by generating observation or discrepancy reports or change requests for logging and troubleshooting. Space segment related problems are covered by the FCT. A work flow has to be established to issue an anomaly, to inform the involved persons for analyzing and problem solving via this anomaly report (AR), and to decide the corrective measure, recorded in a recommendation (see next paragraph). All steps of the process will be logged. The size of the work flow and the number of roles involved in the process depends on the project size and its complexity, but at least the flight director and a responsible subsystem specialist need to be involved.

Recommendations

Recommendations are the controlled way to introduce and process unforeseen and urgent actions or changes to the planned flight operations. A recommendation typically consists of a short description of the context and purpose of the desired action and step-by-step orders to be executed. All kinds of actions can be addressed, such as sending of an additional command and altering a command nominally foreseen in a flight procedure or the execution of a previously unplanned flight procedure. Within the process of anomaly handling, recommendations represent the corrective measure.

Key element for a recommendation is at least a four-eyes principle, i.e. the recommended action needs to be checked and approved by involved engineers and the person in charge of the flight operations (flight director). An example of a recommendation work flow is described as follows: A member of the flight control team prepares the recommendation, which is checked and complemented as necessary by the affected subsystem specialists or by a support team (e.g. during LEOP), and finally approved for execution by the flight director. The recommendation is completed after its execution and the confirmation of the expected results by the SSE and the flight director.

All recommendations have to be noted in a written form, in past-times on paper or nowadays using a dedicated software tool. The steps have to be signed by the corresponding persons/roles. In critical situations, where a quick response is needed, a recommendation may be processed verbally first, but shall be noted and fully logged later.

5.5 The Mission Type Defines the Operational Concept

Basically, there are four different types of missions: LEO satellites, GEO satellites, deep space missions, and human spaceflight. A typical example of LEO satellites are the GRACE twin satellites (Tapley et al. 2004a, b). They were designed to measure the gravitational field of the Earth. They were launched in 2002 and used changes in their relative distance and speed to derive information about local gravity forces.

EDRS is a communication satellite family consisting of the two satellites EDRS-A and EDRS-C. They exemplify geostationary satellites.

Galileo was a deep space probe whose science objective was to explore Jupiter and the Jovian system (Belton et al. 1996). It was launched on the Space Shuttle in 1989 and its lifetime ended with the deliberate entry into Jupiter's atmosphere in 2003. Its mission significantly contributed to our understanding of the Solar system. A GSOC team was located at the JPL to support operations during the mission duration.

For more information on human spaceflight, please see Chap. 24.

5.5.1 Low Earth Orbit: GRACE

The GRACE satellites were orbiting the Earth at an altitude of approximately 430 km in a polar orbit with an inclination of 89°. The orbital period was approximately 93 min (Fig. 5.5). The successor project GRACE Follow-On continues the mission at GSOC.

On LEO missions, one of the typical operations tasks during the routine phase is the "housekeeping" of the spacecraft. The monitoring and control system (MCS) automatically compares most telemetry parameters with predefined limit values and indicates warnings by a yellow highlighting and alarms by a red one (also called soft or hard limit, respectively). Each SSE monitors his subsystem and reports to the flight director in case any of the parameters do not behave as expected. The team does not only react when yellow or red alarm situations are indicated in telemetry, the SSEs also perform long-term monitoring, where data are recorded and plotted over an extended period of time, sometimes years, and evaluated by the experts to make predictions and react in advance to trends and tendencies. During the nights and weekends, command operators will watch the telemetry during the station passes and inform the flight director and the responsible SSE immediately in case any anomalies occur.

Another major task during routine LEO operations is attitude and orbit determination. The flight dynamics team collects the orbit measurement data (typically

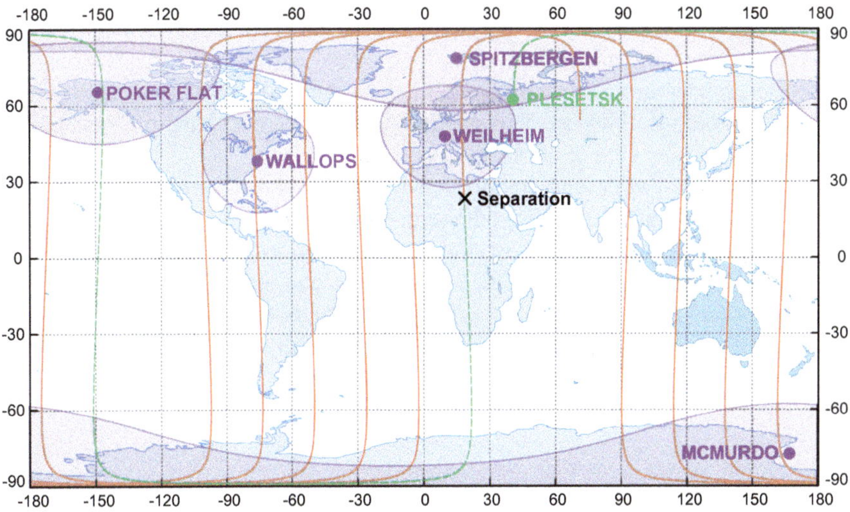

Fig. 5.5 The ground track of GRACE after launch shows that only a few contacts are possible. The green part color indicates the track before first acquisition over Weilheim

GPS measurements), calculates the exact orbit, and generates the information for the next maneuver. The FCT will command the satellite with the orbit maneuver data and execute the maneuver. Afterwards, the flight dynamics specialists will recheck the satellite orbit und evaluate the accuracy of the orbit maneuver. For a detailed description of a maneuver execution, please see Chap. 13.

Payload operations depend on the particular payload of the satellite. In projects like GRACE, there are scientific experiments which have to be commanded and monitored. The recorded data of the satellite payload is dumped over the ground stations and then distributed to the scientists. A dump is the download of a data storage that contains previously recorded telemetry. All these tasks have to be organized and scheduled; this will normally be done by the mission planning tools. For further details see Chap. 16.

Because of the low altitude of the LEO satellites, the contacts with the ground stations are quite short. At an altitude of about 500 km, a ground station normally has contact to the satellite for only around ten minutes. For scientific missions like GRACE, the number of ground stations is limited due to the costs of each ground station contact. As a consequence, the number of ground station contacts with one or two ground stations leads to about five contacts per satellite per day. This is the reason why most of the routine operations are conducted with time-tagged commands. These commands are sent to the spacecraft during the contacts. They contain a specific time stamp, which means they will not be executed right away at reception time, but be stored on board until their intended execution time. This allows the execution of orbit maneuvers, payload-specific actions, or software uploads, which normally take a long time, and are intended to be executed at any time, not only during ground contacts. Without the option of time-tagged commands, operations would be much more complicated and insufficient.

Command execution verification can be done during the next pass, i.e. by indirect verification, via either mechanisms provided by the data handling system (see Chap. 19) or by checking in the telemetry.

During LEOP, the ground station network is, of course, more extensive than during the routine phase for safety, although normally there is still no full coverage. Depending on the mission, four or more ground stations are involved. In case of emergency operations, there will be additional ground stations booked to support the mission and bring it back to normal operations as soon as possible. To make this concept work, ground stations world-wide are usually committed to a spacecraft emergency priority.

5.5.2 *Geostationary Earth Orbit: EDRS*

At the altitude of approximately 36,000 km above the Earth and with an inclination of 0° against the equatorial plane, the geostationary satellite seems to stand still over one location. Thus, with a single antenna located in a suitable region of the Earth (Fig. 5.6), the control center has a 24 h per day visibility of the satellite—it can be operated in real-time without a direct need for time-tagged commanding or telemetry dumping. The two nodes EDRS-A and EDRS-C of the European Data Relay System are located at two positions of the GEO from which they both can be operated through ground stations located in Europe. EDRS-A is a hosted payload on a commercial

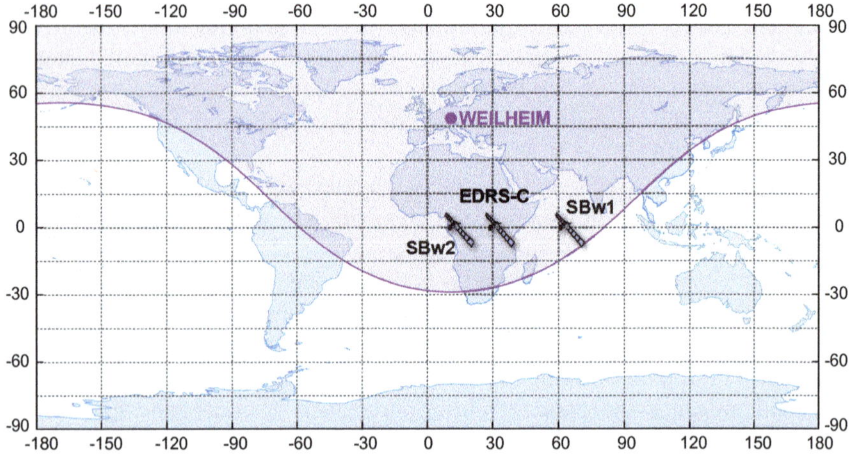

Fig. 5.6 In a geostationary Earth orbit the satellite seems to "stand still" over one location; that means the ground track is only a spot. The shaded area indicates the regions on the Earth from which the satellite can be seen

communications satellite and GSOC is responsible for the payload. EDRS-C is a spacecraft of its own and is under full control of GSOC (payload and platform). The surveillance of both missions is performed with a 24 h a day, 7 days-a-week operator shift concept. The telemetry surveillance and the long-term monitoring of the data are conducted by the SSEs as described in the chapter above. Orbit maneuvers, payload operations, and software uploads can be monitored in real time. The use of time-tagged commands is reduced.

The first days of the GEO-LEOP are not fundamentally different from a LEO-LEOP. The control center takes over the satellite after separation from the launch vehicle and checks out the bus systems. Notable differences from a LEO are the facts that a propulsion system has to be prepared, solar panels have to be unfolded and a longer contact time per orbit is given. Usually the satellite is released from the launcher into a highly elliptical trajectory, the geostationary transfer orbit (GTO). As shown in Fig. 5.7, this orbit has a height above the Earth ranging from 300 km at perigee (point closest to the Earth) up to 36,000 km at apogee (highest point of the orbit).

Variations are possible. Some launchers allow the satellite to be released directly in the GEO, others are using a super-synchronous orbit (apogee higher than 36,000 km height above the Earth) to be able to change the orbit inclination more efficiently.

The orbital period in this phase is about 11 h. Station visibilities are several hours long. Interestingly, the satellite, as seen from the ground station, can change the apparent direction of movement in the sky, as depicted in Fig. 5.7. The general approach is to be able to continuously monitor and control the satellite and therefore to include as many ground stations in the LEOP network as necessary. For critical operations (e.g. for maneuvers) it is advisable to even have redundant stations available. The orbit also dictates that these critical events may be happening at night hours and are possibly not compatible with convenient work shift arrangements.

When the satellite is stable and the main bus components have been checked, orbit maneuvers with the satellite motor are performed around the apogee position. This maneuver sequence (see Chap. 13) will increase the perigee height in several steps to 36,000 km and achieve the desired longitude. Once the spacecraft reaches its final position in the GEO, only one ground station is needed for continuous visibility (Fig. 5.8). In the case of EDRS, additional ground stations are available from which the satellites could be operated in the event of contingency situations. These ground stations have to be tested on a regular basis to make sure that the handover works flawlessly if needed.

In the time when the classic GEO communication satellite was designed, GPS was not available and even later its use was limited, as during large parts of the orbit (above approximately 3000 km height above the Earth) the GPS signals were not continuously receivable. Therefore, range measurements and angle tracking from ground has to be performed. Also, the attitude could usually be determined only with Sun and Earth sensors. This influences operations during eclipse phases and during maneuvers, resulting in costly ground networks, complex activities, and long waiting periods. Modern spacecraft are normally equipped with star sensors which will reduce some of the limitations.

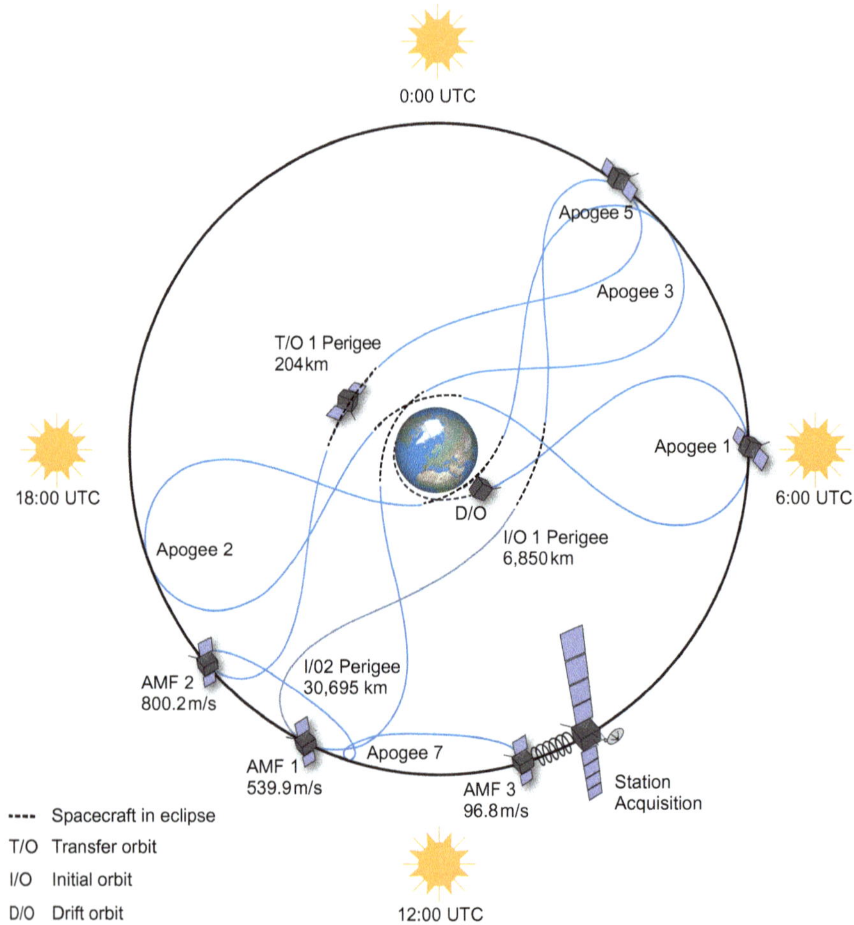

Fig. 5.7 The LEOP trajectory in the Earth-fixed coordinates as seen from the north. This illustration allows seeing the movement of the spacecraft across the sky as seen, e.g., from the ground stations (illustration based on P. Brittinger, H. Wobbe, F. Jochim, DLR)

The routine phase of a geostationary satellite is mostly focused on payload operations. At regular intervals of a few weeks, the satellite's position has to be corrected as perturbation forces influence its orbit. Short boost maneuvers will be executed, which normally do not interfere with payload transmission. Repeater payload activities have to be performed in long intervals. They are described in Chap. 23. Apart from long-term trend analyses of the equipment, the remaining fuel mass has to be calculated (see Chap. 21).

Finally, an aspect that should not be neglected is the fact that GEO missions are different from many other space missions in that they are often highly commercial in nature. Serious business plans and large sums of money rely on a timely service entry. The spacecraft involved are mostly from an established series with proven equipment,

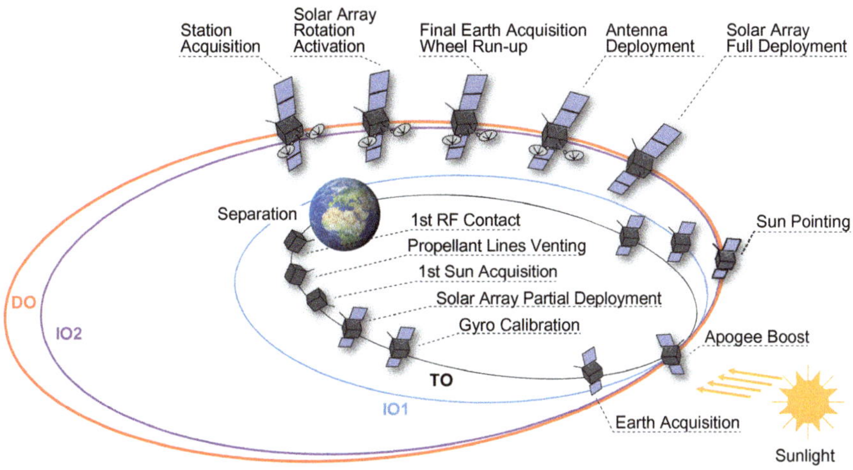

Fig. 5.8 The LEOP orbit of a GEO satellite in inertial coordinates. This view shows the successively larger orbits after the motor firings

but still extremely expensive so expectations from the customer side for system reliability and flawless operations are high. On the positive side, manufacturers often will provide a complete set of documentation including flight procedures along with a convenient full-scale software satellite simulator.

5.5.3 *Deep Space Missions: Galileo*

The typical deep space mission spacecraft is operated outside the Earth's gravitational field at a long distance from Earth. Compared to LEO or GEO missions, it normally takes a very long time for the spacecraft to reach its destination. It took the deep space mission Galileo nearly six years to reach Jupiter's orbit with the goal to study the planet and its moons. Five flyby maneuvers at Venus and Earth were required for the space probe to gain the necessary speed to reach Jupiter. The satellite needs a very high degree of autonomy to detect failures and to autonomously recover from them, because the response from ground can be delayed by several hours due to the radio signal travel time. Because of the geometry of the trajectory, the contact durations are several hours long. Depending on the mission phase, only one ground antenna may be used, which results in daily repeating periods without contact (Fig. 5.9).

One challenge coming from the long transfer times is to keep up the expertise in the operations team, on the manufacturer side and in the scientific community. Projects like Rosetta can span entire careers. It is wise to build up the necessary spacecraft knowledge inside the control center and to preserve access to the engineering model. For more details on deep space missions, please see Chap. 26.

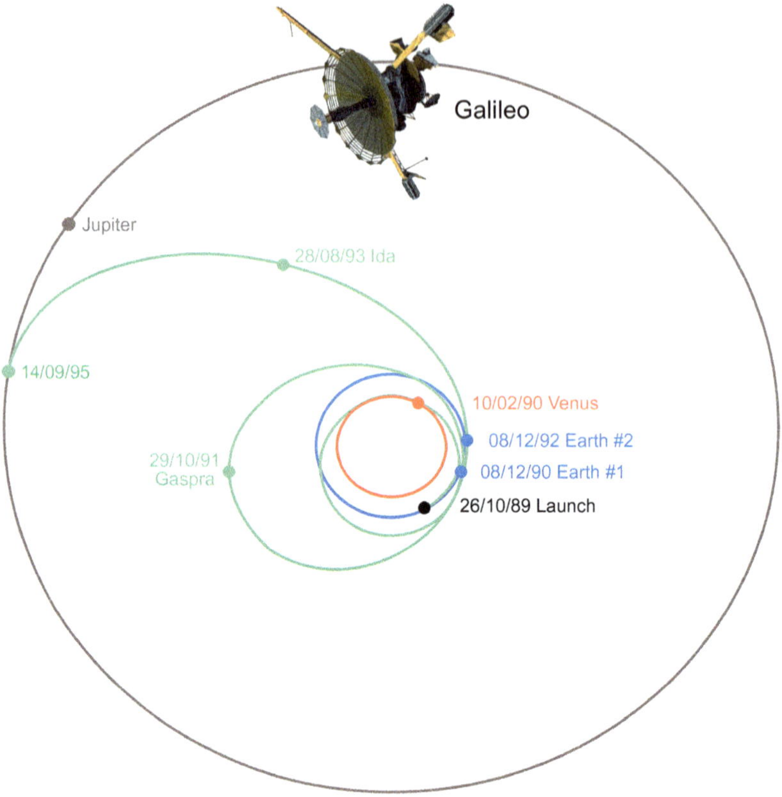

Fig. 5.9 Galileo's long journey to the Jupiter System. In addition to the swing by maneuvers at Venus and Earth, it passed the asteroids Gaspra and Ida on its way

5.6 Summary

The operational concept is very much dependent on the mission type and may vary significantly for the different missions. The composition of the flight control team is in turn driven by the requirements derived from the operational concept. However, some basic commonalities can be deduced. Whereas the operations execution, including shift planning, differs also according to the mission phase, things like a mission control system, a clear responsibility assignation with corresponding processes, e.g. for anomaly resolution, and all necessary operational products, like flight operation procedures, are always in place.

References

Belton MJS, Head III JW, Ingersoll AP et al (1996) Galileo's first images of Jupiter and the Galilean satellites. Science 274

DLR Space Operations and Astronaut Training. Earth observation and science missions—GRACE. https://www.dlr.de/rb/en/desktopdefault.aspx/tabid-12671/4262_read-6309/, cited 14 June 2021

DLR Space Operations and Astronaut Training. Communications missions—SATCOMBw. https://www.dlr.de/rb/en/desktopdefault.aspx/tabid-12673/22115_read-25574/, cited 14 June 2021

DLR Space Operations and Astronaut Training. Science missions—Galileo; https://www.dlr.de/rb/en/desktopdefault.aspx/tabid-12671/4262_read-6338/, cited 14 June 2021

ISO 24113:2019. Space systems—space debris mitigation requirements. https://www.iso.org/standard/72383.html, cited 14 June 2021

Skalden J (2013) Untersuchung von technischen Lösungen zur Passivierung von Satelliten Antriebssystemen. https://elib.dlr.de/82330/, cited 14 June 2021

Tapley BD, Bettadpur S, Ries JC, Thompson PF, Watkins M (2004a) GRACE measurements of mass variability in the earth system. Science 305(5683):503–505

Tapley BD, Bettadpur S, Watkins M, Reigber C (2004b) The gravity recovery and climate experiment: mission overview and early results. Geophys Res Lett 31(9)

Sabrina Eberle is Team Lead for the Geostationary Satellite Missions at the Mission Operations Department at the German Space Operations Center (GSOC). In this group communication satellites are operated and both commercial satellites and military missions are supported. In the past, she was the project manager for SATCOMBw and EnMAP and flight director for SATCOMBw and the SAR-Lupe mission. She was involved as a subsystem engineer in the TerraSAR-X, TanDEM-X and the PRISMA missions. She was also involved in the acquisition of new projects.

Thomas Uhlig received his physics diploma and his Ph.D. from the University of Regensburg/Germany, where he worked on electromagnetic imaging of nanostructures. After working as scientist at the Argonne National Laboratory in Chicago he changed into the spaceflight business and to the Columbus Control Center in Oberpfaffenhofen, where he first worked as Payload expert on the Space Shuttle mission STS-122, then for several years as Columbus Flight Director. He then took over the training for the European ISS flight controllers, before he was assigned as team lead for the geostationary satellite team at GSOC. In between, he was working as expert for functional safety, first at BMW, then at Heidenhain GmbH before returning to DLR in 2022. He was awarded with the Helmholtz price 2005 of the Physikalisch-Technische Bundesanstalt. He is author of various scientific articles.

Ralf Faller is project manager for LEO, GEO, and On-Orbit Servicing projects in the mission operations department at the German Space Operations Center (GSOC). In 1990, he received his diploma in aerospace engineering at the Technical University of Berlin, Germany. He started in 1991 at the DLR Oberpfaffenhofen in the spaceflight dynamics department, being responsible for attitude related software development and mission support and changed to the mission operations department in 1997 where he focused his work on spacecraft flight operations topics. He was involved in several missions as subsystem engineer, flight operations manager and finally project manager.

Michael Schmidhuber works in the Department for Mission Operations at the German Space Operations Center (GSOC) where he is responsible for satellite operations training. He graduated as an aerospace engineer at the Technical University of Munich in 1994. The training activities comprise not only staff training but also courses for an external audience in the annual Spacecraft Operations Course at GSOC. Also, he is involved in the organization of the bi-annual international SpaceOps Conference.

Chapter 6
Flight Experience

Ralph Ballweg, Andreas Ohndorf, and Franck Chatel

Abstract This chapter covers examples of lessons learned at DLR/GSOC, particularly in the course of multiple LEOP phases of communication satellites. It describes the process of dealing with system contingencies, mostly on spacecraft side and wraps up with several spacecraft anomalies and the attempts to deal with them.

6.1 Introduction

The German Space Operations Center (GSOC) was founded in 1967 and operated since then many national and international missions in the human spaceflight and in the satellite area. In this chapter, some of the experiences gathered in these years shall be presented and discussed.

First, some empirical data from a series of almost identical mission is analyzed to show how operational experience is gained by time within a control center. Then, the evolution of failure probability is contrasted with the experience within a team during a given mission.

In another section, we will show how detailed information can be extracted by an experienced flight controller from a single telemetry item.

Finally, contingency operations shall be addressed. After some generic facts about it, we come up with two case studies of real contingency conditions, which were handled by the GSOC teams.

6.2 Mission Experience: Empirical Data

Between 1987 to 2002, GSOC supported in average one launch and early orbit phase (LEOP) of a geostationary communication satellite per year. Among these were two

R. Ballweg · A. Ohndorf · F. Chatel (✉)
German Space Operations Center (GSOC), German Aerospace Center (DLR), Oberpfaffenhofen, Germany
e-mail: franck.chatel@DLR.de

F. Sellmaier et al. (eds.), *Spacecraft Operations*, Springer Aerospace Technology,
https://doi.org/10.1007/978-3-030-88593-9_6

series of six almost identical satellites for the provider Eutelsat from 1990 to 1995 and 1998 to 2002. We will use the first series of those Eutelsat II satellites to demonstrate how learning curves can evolve.

First, the LEOP duration (Fig. 6.1) shall be used as an indicator of experience gain and of improving reliability of both, system and team.

The LEOP operations for a geostationary communication satellite can typically be performed within two to three weeks. These operations include the positioning, configuration and in-orbit testing of the satellite bus. Short LEOPs are desired by the customer in order to bring the satellite into service as early as possible.

Proficiency and reliability of the ground system, its components, and the team have a direct impact on LEOP duration—as experience grows, durations tend to get shorter. Over the course of the six Eutelsat II missions, the LEOP durations were cut from 18 days with Eutelsat II F1 in 1990 to 11 days for Eutelsat II F6 in 1995. Flight F5 was lost due to a launcher failure.

There were several areas where refinements and optimization improved the performance, and therefore led to a shorter duration of the LEOP.

1. Enhanced station acquisition strategies
2. Improvement of procedures
3. Optimization of the sequence of events (SoE)
4. Improvement of hardware and software tools within the control center which allowed, e.g., a faster analysis of data for further processing like ranging data and expedited maneuver calculation.

Another indication of the level of maturity of the operations concept is reflected in the number of engineering change requests (ECR) and non-conformance reports (NCR) that were issued during the mission preparation (Fig. 6.2).

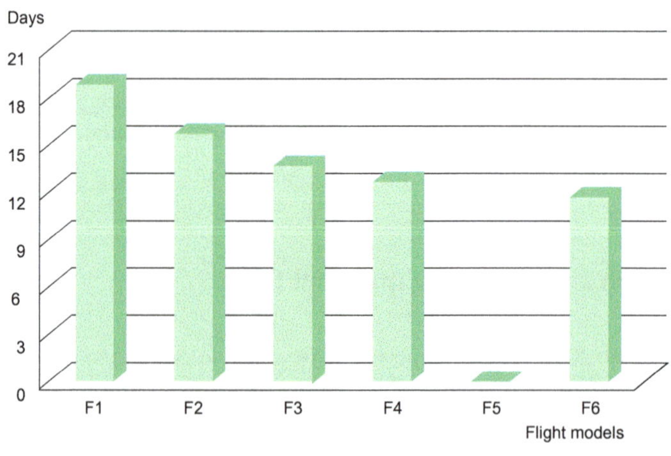

Fig. 6.1 Duration of LEOP for a series of Eutelsat II satellites. Flight F5 was lost due to a launcher failure

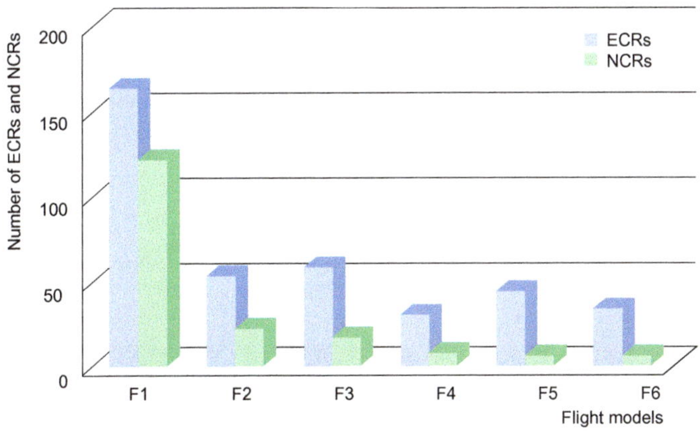

Fig. 6.2 Number of ECRs/NCRs for Eutelsat II series

The change management of the configuration-controlled products during the mission preparation and execution phase is formally managed by ECRs and NCRs:

An ECR is raised whenever it is intended to request a modification on specification level or on the existing configuration.

An NCR is issued whenever a deviation with respect to the specifications is observed or if a subsystem differs from the expected behavior during the mission preparation phase.

The number of ECRs dropped from flight F1 with ~ 170 to ~ 50 for flight F6. The high number of change requests at the beginning is easily explained with the fact that the ground segment had to be configured for a completely new mission. For the next launch, the change requests already dropped to about 60 ECRs because of the gained experience. The slight increase for flight F3 was due to a change of the launcher: While for the first two launches the satellite was mounted on an Ariane, a Lockheed Atlas was selected for the third launch—this launcher placed the spacecraft into a super-synchronous transfer orbit and the launch took place at the Kennedy Space Center in Florida.

Hence, changes in the specifications were driven by the different interfaces to the launcher, the changed launch site, different ground station selection and schedule, and considerable updates to the flight dynamics software to include a perigee maneuver to lower the apogee.

Another small increase can be detected between flight F4 and F5. At that time, GSOC made a change in its control facility. It moved to a different building, and also implemented new hardware with the corresponding operating systems being adapted. The decrease in ECRs from F5 to F6 was not as large as expected since a change in spacecraft hardware resulted in updates to the ground software. Those were in particular in the satellite power subsystem.

During mission execution all deviations from the nominal procedures and actions caused by unexpected and non-nominal satellite behavior are handled by satellite

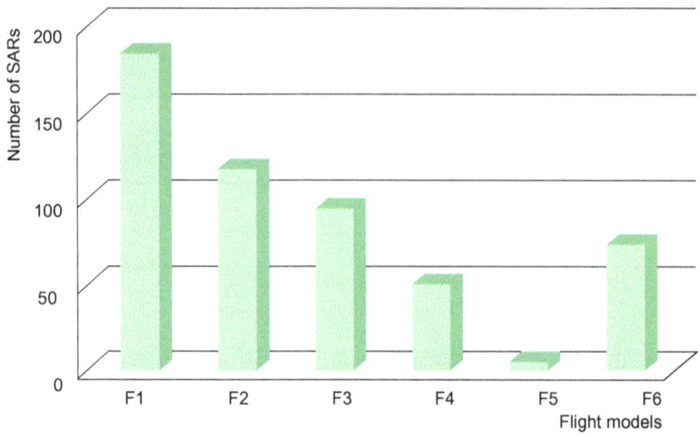

Fig. 6.3 Number of SARs for Eutelsat II series

anomaly reports and recommendations (SAR), see also Sect. 5.4.3. SARs can be issued by any person on the mission control team, control center personnel, as well as representatives of the satellite manufacturer or the end customer. SARs are issued in case of unexpected and off-nominal spacecraft behavior not covered by prepared procedures, online procedure changes, or real-time mission sequence changes.

For the launches from flight F1 to flight F4, one can see a significant decrease in the numbers of SARs from roughly 200 down to 60 (Fig. 6.3). Flight F5, even though it was a launch failure, had some SARs because last minute changes to the database and procedures were introduced, which caused changes to the operational system at GSOC.

Flight F6 had an increase in the SARs because there were several modifications to the spacecraft bus, in particular to the power subsystem.

All those indicators, the duration, the numbers of pre-launch and the number of in-flight discrepancy reports showed a reduction, which can be attributed to the gain of experience, the improved stability, operability and optimization of ground and flight systems, products and procedures.

All "lessons learned" from previous missions pay back in future missions.

6.3 Failure Probability Versus Operational Experience

Not only a series of almost identical missions shows a clear tendency towards more effective operations, also during a given mission there are some observable trends, which shall briefly be discussed in this section.

The typical evolution of the probability of failures shows high values at the beginning of a mission, i.e. the LEOP, with a tendency to decrease and an increase towards the end of mission (EOM) (Fig. 6.4, *dashed line*).

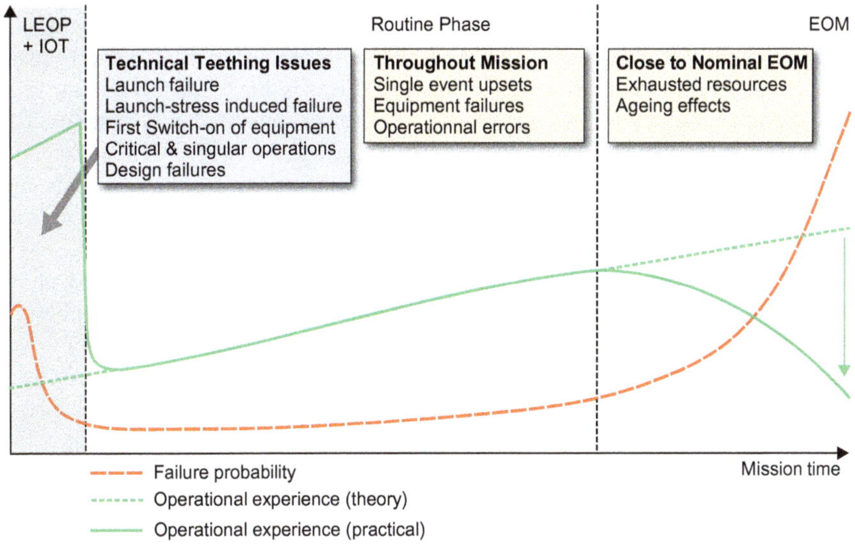

Fig. 6.4 Failure probability versus operational experience

On the other hand, operational experience is growing throughout the following routine phase. However, for long-term missions, this experience might decrease again on the operations as well as at the manufacturer's side towards the end.

Let's first take a look at the failure probability during a mission. It starts out with a rather high likelihood of problems at the beginning; one could call it "technical teething issues". The most likely reasons are:

- Launcher failures.
- Failures induced by stresses during the launch, e.g. vibrations.
- Units or instruments, which experience the space environment for the first time and react differently than expected.
- Operations of time-critical and singular nature are executed. Time-critical events, for example, are the deployment of solar arrays to charge the batteries or the activation of heaters to keep propellant from freezing. One-time-only executions can be the deployment of an antenna or the activation of pyros.
- Design or manufacturing errors are materializing which were not discovered during testing, like faulty Sun sensors.

Once the in-orbit test (IOT) phase is successfully completed, the likelihood of failures drops significantly. The spacecraft is operated in a stable configuration without many changes. From that time on, most problems originate from single event upsets, equipment failures, or in many cases human/operational errors.

Close to the nominal end of the mission, the failure rate increases again due to aging effects on the equipment and exhausted resources, which makes resource management an important issue in the course of operations.

The development of operational experience of the mission control team is shown in Fig. 6.4 with *solid* and *dotted* lines. At the beginning of the mission, there is a team with a high level of basic operational experience. This is due to the fact that the core team members are chosen from staff that has already supported similar tasks. The team has already participated in the preparation phase, has analyzed requirements and databases, was involved in the definition of the mission control system, has written procedures, tested the system, and participated in simulations. In addition, the core team is seconded by experts of the spacecraft manufacturer.

The experience increases during the LEOP, as this is the most demanding phase with all the testing occurring there, and typical problems happening in that phase.

Shortly after LEOP is completed, the expertise level drops significantly (*solid* line). This is caused by several facts: First, the support of the satellite manufacturer leaves the control center once the spacecraft is checked out, then experienced staff might be taken off the team to work on new missions and only a core team is left to support the routine operations phase. In many cases, after check-out a spacecraft will be handed over to a routine control center (in-orbit delivery or turn-key delivery) with the same effect.

Once the steady-state operational setup is established, the experience level gradually increases again.

Once the mission is approaching the end of its lifetime, the experience level of the team starts to decrease again due to natural attrition, team members leaving for other jobs, or because of budget constraints and finances being cut.

It shall be pointed out that this decrease of experience often coincides with the increase of probability of spacecraft failures—for the management of the mission, this is a particular challenge which is not easy to master.

The mentioned effects need to be considered carefully. A manufacturer should be contractually obliged to provide continued support, proficient team members should be made available for a certain time as reference. A spacecraft simulator could help in conducting sessions to train spacecraft contingencies. An efficient training scheme and training documentation should be implemented early in the mission to capture and preserve the knowledge and account for the human factor in operations.

6.4 Interpretation of Telemetry

The status of a spacecraft is represented in its telemetry. Often, only few data items are available, but they still contain valuable information about the situation in space. However, it requires good analytic skills to extract it.

This chapter gives an example how an experienced flight controller can derive information by detailed analysis of telemetry.

As an example, we are using a plot of the spacecraft receiver's automatic gain control (AGC) (Fig. 6.5).

The uplink AGC—a telemetry parameter which is transmitted from the spacecraft to the ground—indicates the on-board measured signal strength of the telecommand

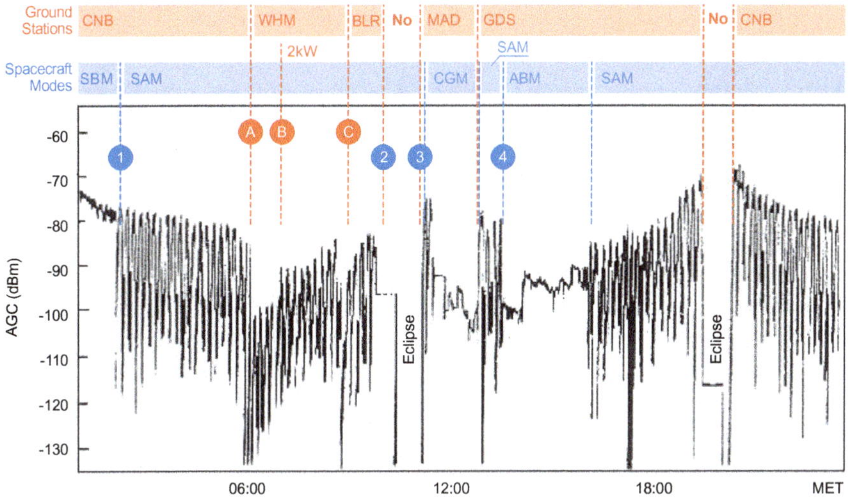

Fig. 6.5 The automatic gain control (AGC) signal during the first day of the LEOP. The corresponding ground stations, the various spacecraft modes SBM (stand-by mode), SAM (Sun-acquisition mode), GCM (gyro calibration mode) and ABM (apogee boost maneuver) are displayed

carrier uplinked from the ground station. The various spacecraft attitude modes, the evolution of the orbit, as well as the ground station coverage are reflected in this AGC plot of the first 24 h in the geostationary transfer orbit (GTO).

At first view, the graph seems to represent a very erratic behavior of the telemetry value plotted. Periods of constant or moderate changes are visible, long time spans of extreme oscillations, and gaps without any telemetry. How can it be interpreted?

Let's first take a look at the spacecraft, its configuration, and the activities performed during the first 24 h:

From the start of the plot during the stand-by mode (SBM), we see a rather stable but slightly decreasing level of the automatic gain control level. Here, the spacecraft is released by the launcher with a predefined attitude at a certain altitude and distance to the ground station. The decrease in signal strength displays the increase in distance between the ground station and the satellite, as it approaches the apogee of the GTO. The decrease of signal strength is therefore related to the increasing distance from the satellite.

At Point **1** the satellite was commanded into Sun-pointing or Sun-acquisition mode (SAM). The satellite is rotating around its z-axis, which is pointing towards the Sun. The oscillation with an amplitude of roughly 25 dBm (decibel milliwatts) is caused by the fact that there is only one receiving antenna, which points—due to the rotation of the spacecraft—sometimes towards the Earth, sometimes away or is shielded by the satellite's structure. This oscillation frequency is therefore a possible means to determine the rotation rate of the spacecraft. The graph also shows a steady decrease in signal strength up until around 06:00, when it starts increasing again.

This is the time when the satellite reaches apogee, the point furthest away from the Earth on its orbit. The overall dip in the AGC level around 06:00 will be explained a little later.

Between point **2** and **3** there is a gap in the telemetry, which indicates a loss of signal. At that time, the satellite is passing through perigee and, due to the low altitude and corresponding high angular speed of the spacecraft, there is no station available to receive a signal.

At point **3**, the signal is acquired again with the spacecraft in SAM, shortly afterwards the satellite is commanded into gyro calibration mode (GCM) which is a 3-axis stabilized mode. Here, the antenna is pointing into a fixed direction. Hence there is no fluctuation in the receive signal strength. The telemetry shows a rather constant value that is only decreasing due to the increasing distance to the ground station. After completion of gyro calibration, the spacecraft is returned to SAM which can be seen by the fluctuating telemetry values.

Point **4**: The spacecraft is configured for the first apogee boost maneuver (ABM). This again is a 3-axis mode with almost constant receive strength. After completion it returns to SAM, interrupted by another eclipse during a perigee pass with no ground station contact.

But there is another way to interpret this plot and receive other information. We will now focus on ground activities. From the beginning to point **A**, we acquired the signal of the spacecraft via the ground station Canberra (CNB), with the AGC level decreasing. At point **A**, the signal drops to a minimum, basically indicating no signal reception by the ground station. At that time there was a ground station handover from Canberra to Weilheim (WHM) with a short interruption in the uplink. The quality of the uplink, i.e. the receive strength, was unsatisfactory, dropping basically down to minimum depending on the spacecraft attitude. So, at point **B**, the uplink power at the ground station was increased from 1 to 2 kW which resulted in a satisfying receive signal strength. Point **C** marks another station handover with a brief interruption of uplink, this time from Weilheim to Bangalore (BLR). Other stations used during this period were Madrid (MAD), Goldstone (GDS), and again Canberra, in that order with the handovers clearly identifiable.

The previous illustration demonstrated clearly what a single telemetry parameter can reveal about the status of the spacecraft and the progressing of operations. It should not be neglected to look at operations from an encompassing view and not only focus on a small detail or device. This can also help in redundant confirmation or failure analysis.

6.5 Contingency Handling

This section covers the aspects of contingency handling during operations. The first level of contingency handling should be covered by the on-board failure detection, isolation and recovery (FDIR) software. The objective of the on-board FDIR, also sometimes known as redundancy management, is the survival of the spacecraft after a single failure for a specified time span without ground intervention. For detailed information on how FDIR is embedded in the data handling subsystem, please refer to the corresponding Chap. 19 about on-board data handling (OBDH). Communication spacecraft are typically designed to survive up to 48 h without ground intervention.

Once the on-board FDIR software has triggered, the detailed failure analysis of the cause of the incident, any recovery actions, like returning to normal mode and restoring the mission, has to be performed by the mission control team on ground.

Due to the increased cosmic radiation, FDIR is sometimes triggered by single event upsets (SEU). Here, the spacecraft can, in most cases, be recovered by simply reestablishing the nominal configuration without any further action.

However, it is a good operational practice not to rely on FDIR mechanisms, but to detect problems before the on-board software intervenes. To support this, the ground system has similar functions implemented to the ones that are established on board. The reaction thresholds are more conservative and allow a detection of an issue before the on-board FDIR kicks in. Often, there are two stages of out-of-limit conditions defined, a warning and an alarm stage. The warning indicates that the situation has to be monitored, but no immediate action is necessary. The alarm stage calls for action—otherwise an instrument or function could be lost.

Another means of detecting satellite issues by ground monitoring is to recognize secondary effects through advanced telemetry analysis. These can be the unexpected changes of telemetry values like temperatures, currents or sensor values even though they still stay within the range of nominal values. Other effects can be attitude perturbations that are not recognized by the on-board software as problems or the loss of up- or downlink, and finally, the long-term analysis of telemetry and consumables. This can give an indication of the remaining lifetime of a unit, instrument, or the mission.

The isolation of problems by ground activities can, of course, be much more subtle than just switching to backup units, which is a common approach of the on-board FDIR. The first step is the in-depth analysis of the problem or failure. Starting with a systematic approach one first identifies the area of the issue: Is it an operator error, is the cause within the ground system, or is it a malfunction within the space system?

Based on this analysis, the corresponding action—in the best case a recovery—can be chosen.

In case the space system is affected, the goal is a return to a normal operational configuration, recover the mission, restore the payload, and activate the nominal equipment again as much as possible.

The way to proceed should be chosen from the prepared contingency procedures. If adaptations are necessary or a fitting procedure is not available, it is recommended to validate this on the spacecraft simulator and together with the spacecraft manufacturer.

Limits within the mission control system might need to be changed, procedures updated, and the FDIR software might need to be rewritten to reflect the operations on a degraded main or the backup system.

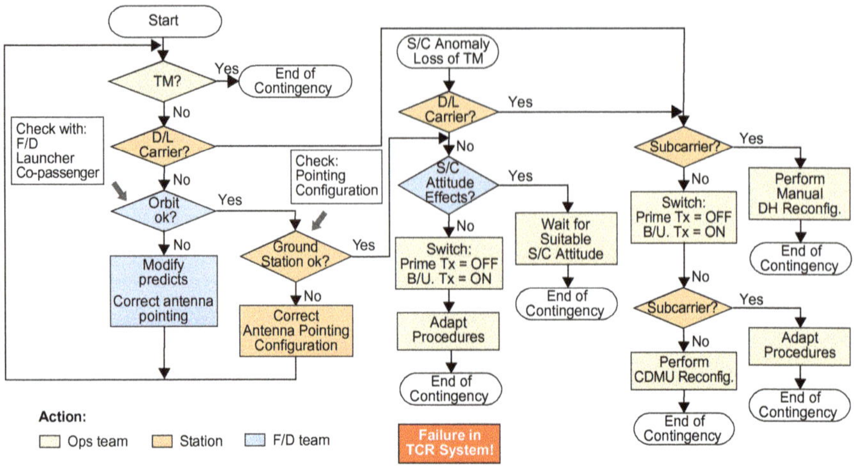

Fig. 6.6 Flowchart for FDIR process in case no telemetry is received

Also, other measures, for example in improved training for operators, new procedures, update of the ground system, databases, or hardware, can be possible consequences.

The baseline for a controlled reaction to system contingencies must be to use verified and approved processes and procedures. Often, the handling of a contingency is too complex to be handled by a single procedure. In this case, flowcharts are useful to guide the operator through a variety of procedures.

In Fig. 6.6, an example of a flowchart is shown. Here, the complex decision-making process in case of a loss of telemetry is captured. The flight controller is guided through the various steps, several decisions paths are displayed. Often, the flowchart is referencing the actual procedures and therefore only provides the logical framework around them.

In the following, two real case studies of contingency handling at GSOC are presented. In the first case, the deployment of the solar array failed, which finally led to the loss of the mission. In the second case, a propulsion subsystem anomaly is discussed.

6.5.1 Mission Example TV-SAT 1

TV-SAT 1 was the first commercial German communication satellite, a joint French-German coproduction. It was a small satellite of two metric tons with five transponders on board, designed for direct broadcasting of TV programs.

TV-SAT 1 was a good example of a mission that went wrong. It provided the mission control teams with unique challenges. The problems GSOC encountered already very early in the mission were:

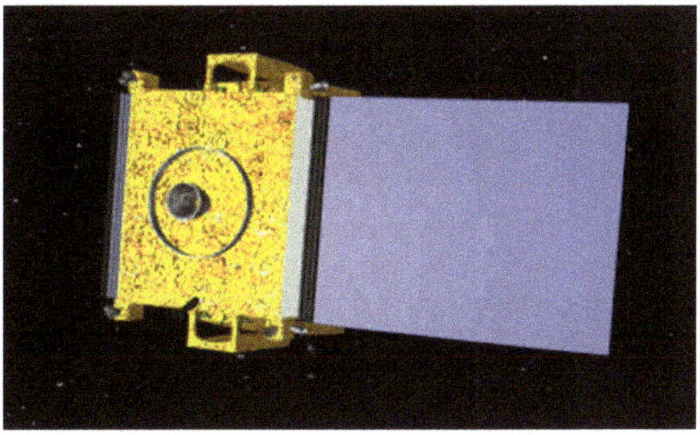

Fig. 6.7 Partial deployment failure of the solar array

- A partial deployment failure
- Gyro failure
- Thruster temperature problems.

In the following, we describe the partial deployment failure in more detail, including the numerous tests to determine the exact cause of the failure. We cover the immediate actions, evaluation of impacts, offline failure analysis, failure investigation, and recovery action attempts.

The problem was caused by one of the solar arrays which failed to deploy at the very beginning of the mission (Fig. 6.7).

Failure Analysis

The partial deployment of the solar arrays was part of the automatic on-board timer function, triggered by the spacecraft separation. An inconsistency was detected by the incorrect status of the deployment microswitch of the "north" panel at the first contact, when the spacecraft was still in eclipse. As first immediate action, the ground database was checked against documentation and with manufacturer experts, whether there was any incorrect bit interpretation. However, no failure in the database was detected.

When the eclipse was terminated, the ground team proceeded with checking the output power of the affected solar array. Unfortunately, the low power levels confirmed the unsuccessful deployment.

When the functioning of the on-board timer was checked, an indication of a failure was discovered. In that situation, the procedure called for sending a manual deployment command sequence. These commands were sent, but no change in the status happened.

Finally, a command sequence was sent to fire the redundant deployment pyros—once again with no effect.

After these first actions, it was concluded that the anomaly was a serious thread for the mission.

To be able to proceed with operations, the impacts of the failure on the mission had to be quickly evaluated. It was found that for the moment, the spacecraft was generally safe from system side; impacts on the subsystems, mainly power, thermal, and attitude, were reviewed and found to be not critical at this mission phase. Apart from some tests it was decided to proceed according to the nominal sequence of events, namely to perform all operations including the first apogee boost maneuver. A number of satellite flight procedures as well as elements of the ground system (e.g. alarm flags) had to be modified and adapted.

An offline analysis by the manufacturer identified a number of more than 50 possible causes for the failure, some of them very unlikely. Test strategies and procedures were defined and developed in order to reduce the number of possible causes. In addition, recovery strategies and procedures were prepared for the different failure scenarios. A review of the failure mode, effects and criticality analysis (FMECA) revealed the blocking of the receive antenna as a fatal mission impact as consequence of the non-deployed solar array.

For the failure investigation new tests had to be defined. Once that was completed, the procedures for those tests in-orbit had to be prepared, validated, executed, and the findings from the tests evaluated.

In the following, some of the tests that were performed are listed:

The first test was to tilt the spacecraft by 45° (Fig. 6.8). The solar array currents were measured to give a rough determination of a possible deployment angle. The expected accuracy was ~ 2°. The result was that the opening angle was less than 2°.

The next test was named "shadowing": The basic idea of this test was to illuminate the panel at low solar incidence angles. Any stirrups holding the panel would thus cast large shadows, which would be measurable from the reduction in the amount of current generated (Fig. 6.9). However, no conclusions could be deducted from this test; it was not sensitive enough to distinguish between the possible cases: No stirrup closed/1 stirrup closed/2 stirrup closed/3 stirrups closed.

This was followed by "current mapping": This test consisted in measuring the power output of the north panel for a variety of solar incidence angles (Fig. 6.10). Power output was expected to vary as the cosine of the angle between the normal to the panel and the solar incidence direction. Any offset in this cosine response could correspond to an opening of the panel. The conclusion from this test was a maximum

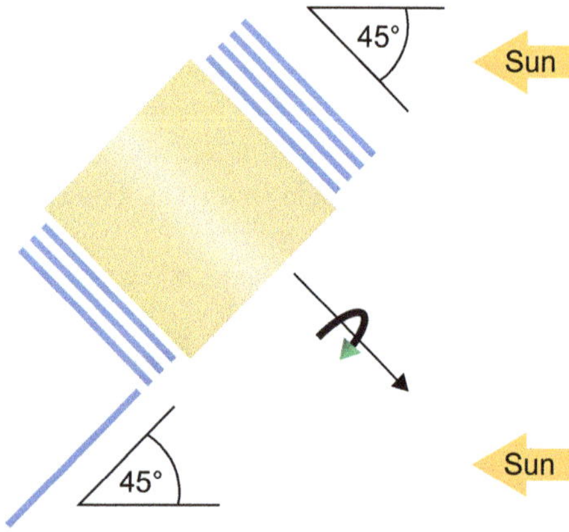

Fig. 6.8 Failure investigation of TV-SAT 1 —tilt S/C by 45°

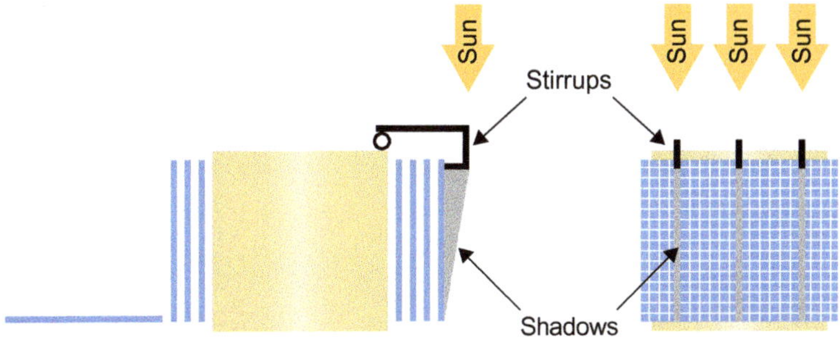

Fig. 6.9 Failure investigation of TV-SAT 1—shadowing test

Fig. 6.10 Failure investigation of TV-SAT 1—current mapping

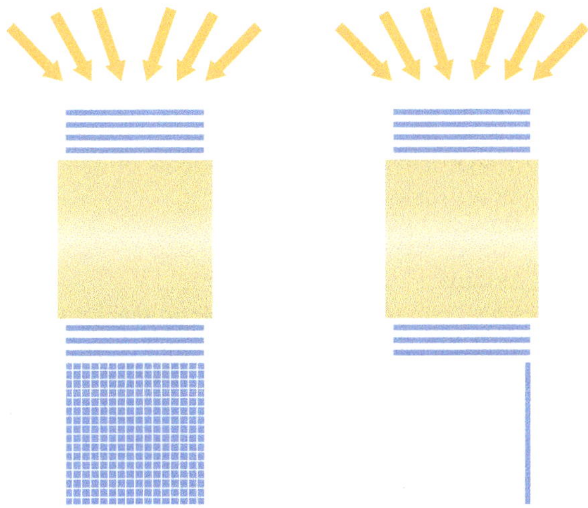

opening of the panel by 0.85°. This test was used in the following after every attempt to fix the problem to find out if they had any effect on the solar array.

Another test was to "shake" the spacecraft using alternating thrust pulses around the spacecraft axes (Fig. 6.11). Purpose of the shaking tests was to determine the resonant frequencies of the north panel, which are different in locked, half-open, and fully deployed position. By shaking the satellite at various frequencies, oscillations in the panel were induced. After stopping the excitation, the continuing oscillations of the panel were measured with the gyros. No resonant frequencies in the expected range could be measured, another indication of a fully blocked array.

Recovery Attempts

As a result of all the tests, the number of possible failure causes could be reduced to 13. A very likely cause for the unsuccessful deployment was a jamming of one or even more stirrups. Different recovery actions were performed, unfortunately all without success:

Fig. 6.11 Failure investigation of TV-SAT 1—shaking test

- Fast spin mode around the satellite's *y*- and *z*-axis in order to exert forces on the stirrups and the panel which could overcome the friction in some failure cases.
- Performing apogee boosts and station keeping boosts in pulsed mode in order to excite resonant frequencies with high amplitudes.
- Exposing the panel and stirrups to alternating hot and cold temperatures.
- The solar array full deployment, the bearing and power transfer assembly (BAPTA) activation, and the shock of the antenna deployment could also overcome some failure cases.

Final Actions

Although it was finally not possible to deploy the solar array, a high amount of operational experience could be gained. For the preparation of the following flight models, preventive actions could be derived for all remaining 13 possible failure modes.

The actual cause was found later: At the launch site it was missed to exchange the transport stirrups with the flight stirrups. As a consequence, payload operations were not possible because the non-deployed solar generator prevented the receive antenna from full deployment. The TV-SAT 1 mission was terminated about six months after launch. Therefore, the satellite was injected into a 325 km over-synchronous orbit by two boost maneuvers. All subsystems were deactivated in order to avoid any risk for other satellites.

The satellite's telemetry transmitter was switched on again after seven years for a short time in order to gain attitude information for the Experimental Servicing Satellite (ESS) study (see Fig. 25.3). The switch on was successful and the satellite signal was acquired at the first attempt.

All following TV-SAT flight models could be operated successfully in orbit.

6.5.2 *Mission Example: Sun-Acquisition Mode Anomaly*

This section provides the description of a severe emergency, which affected a Spacebus 3000 satellite right after its separation from the launcher. This description was reconstructed from operational logs and is presented from the control room point of view.

Anomaly Occurrences and Immediate Actions
After its launch by an Ariane 5 rocket, the satellite of type Spacebus 3000 was acquired with some difficulties over the ground stations of Bangalore and Dongara. The problems with the acquisition of the signal were caused by an incorrect polarization setting at the ground station. The first checkout of the spacecraft revealed, that its systems were working properly. The venting and pressurization of the propulsion system could be performed nominally and allowed the control of the attitude through the reaction thrusters.

At this stage, the next activity was to initiate the Sun Acquisition Mode (SAM) so that the satellite would turn the side towards the Sun, on which the solar panels would be partially deployed (-z_{SC}-axis). After that, the partial deployment of the solar arrays would bring the satellite in a safe state.

However, when the SAM command was sent, it was observed that the thruster valves were closed automatically by the on-board software and no reaction could be detected on the satellite attitude. The mission elapsed time (MET, time counted from launch) was around 02:30 h.

Faced with this obviously anomalous behavior, it was decided by the flight team together with the manufacturer experts to command the thruster valves directly with a dedicated command, instead of using the mode change command, but this action failed again. The next recommendation was to power cycle the electronics controlling the thruster valves and try again the dedicated command to open the valve. This measure failed as well as another trial to command the SAM.

The spacecraft manufacturer then recommended to reconfigure the propulsion system to use the redundant thrusters as a hardware failure of the thruster latch valve was suspected. Once using the redundant thrusters, the dedicated command to open the latch valves for a short duration was sent again and an attitude reaction could be observed this time. After a careful check of the propulsion system, a recommendation was issued to command again the Sun acquisition mode. This resulted in a sharp increase of the rotation rate around the z_{SC} -axis such that an abortion was requested directly over the voice loop due to the urgency of the situation. Aborting the SAM was achieved through two commands for safety reasons and sending them correctly under stress took almost one minute. At that time, the mode was aborted, the satellite rate was so high that the gyroscope measurement was saturated at $-15°/s$ as shown on Fig. 6.12.

Satellite Survival and Saving
After the abortion of the SAM, the anomaly had developed to an emergency where the safety of the satellite was clearly at risk. The battery, which was fully loaded on

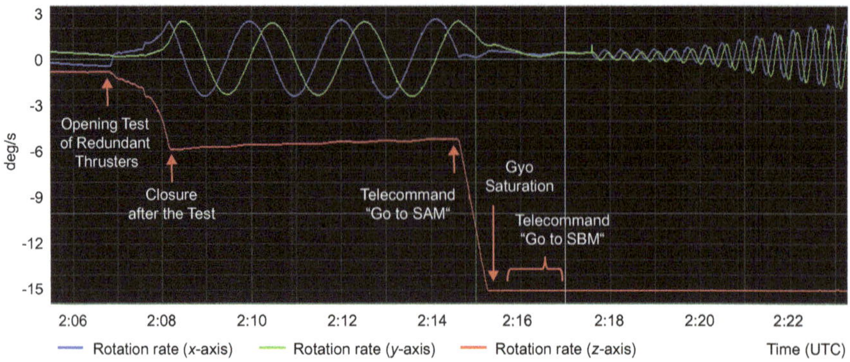

Fig. 6.12 Satellite rates around second SAM activation

the launch pad, was the only source of power for the satellite and the solar panels were still folded on the satellite body. At this point, the satellite manufacturer suspected an incorrect definition of the thrusters in the safeguard memory (SGM) of the attitude and orbit control system (AOCS) to be the root cause. An AOCS expert from industry was summoned at the control center to assess the situation. The MET was around 04:20 at the end of this phase.

The satellite manufacturer team together with the customer representatives agreed on the first course of survival actions. After waiting for the spin to transfer to the main axis of inertia, the spin rate would be reduced by manual opening of the thrusters. As shown in Fig. 6.13, the spin was transferred within three hours from the *z*-axis mainly to the *y*-axis, with a small fraction on the *x*-axis. A comparison between both gyroscope units were performed and showed that their readings were identical, which confirmed their good health status. An estimation of the satellite rotation rate based on the periodicity of the current delivered by the solar panels yielded a value around 30°/s.

The reduction of the satellite spin rate was performed by commanding the thruster opening manually and started around 08:30 MET. At first, only short pulses were

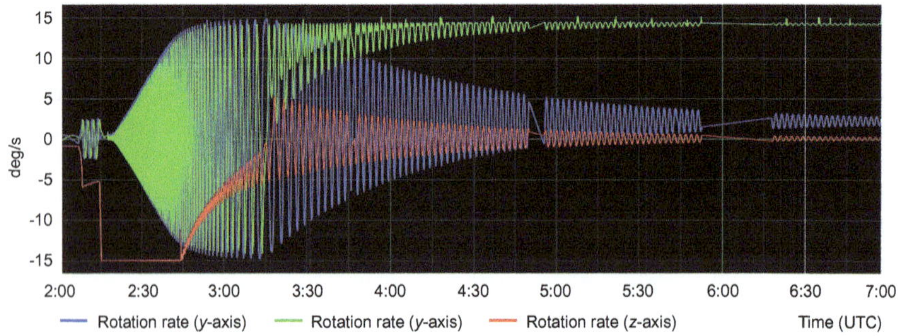

Fig. 6.13 Spin transfer from the *z*- to the *y*-axis

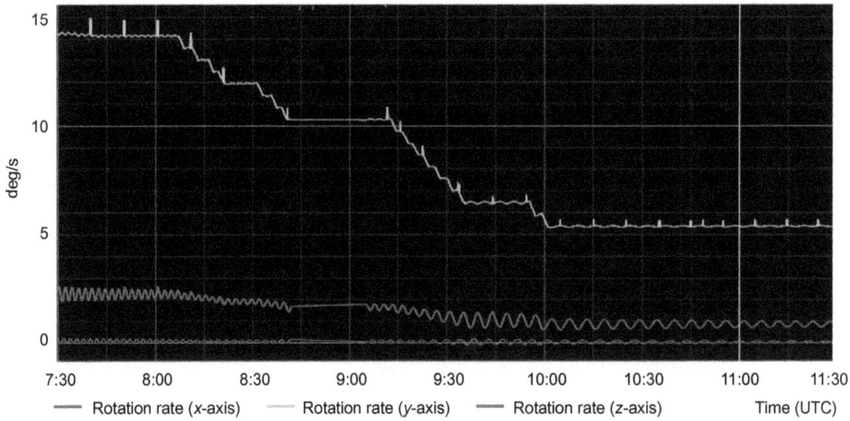

Fig. 6.14 Spin rate reduction by manual thruster pulses

commanded as experiment. After approximately one hour, the confidence in the thruster behavior was high enough to proceed with larger pulses. It took another two hours to bring the spin rate down to approximately 5°/s as displayed in Fig. 6.14.

In parallel, it was necessary to limit the battery discharge as much as possible by reconfiguring the power and thermal systems. This was achieved by increasing the battery load current to take profit as much as possible from the sunlight impacting the folded solar arrays. Any heater not absolutely necessary was switched off and the thermal regulation set points were optimized to reduce the power demand at the cost of cooling down the satellite.

Beside the previous activities to ensure the survival of the satellite, the command history was provided to the experts gathered by the satellite manufacturer, whose task was to investigate the failure, possibly reproduce it on an AOCS ground test bench and provide recovery actions. The content of the SGM loaded on-board was dumped as well to confirm the initial suspicion of the satellite manufacturer team.

This phase was very successful and the all the activities were completed after 12 h of mission elapsed time. The situation was such that all systems of the satellite were affected by the emergency so that the personnel in control room could not be sent home for a rest. It was also necessary to develop and test new procedures to support unforeseen operations. Commanding was rendered difficult by the high rotation rate of the platform. The strength of the signal received from the ground station is measured by the automatic gain control (AGC). The value is nominally around −95 dBm. The plot in Fig. 6.15 shows that the receive signal strength dropped periodically and a command sent in such conditions would have good chances to get lost.

Solar Array Partial Deployment

Around 11:00 MET, the customer's higher management took over from the satellite manufacturer's and customer's combined team. This situation was not foreseen in

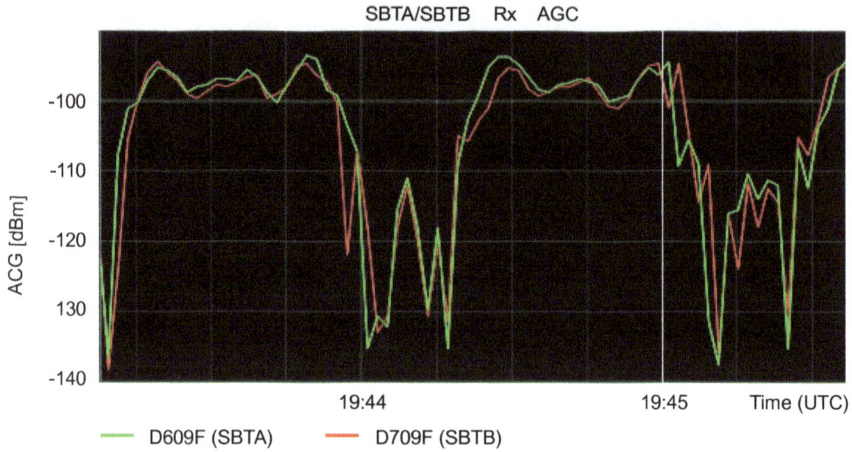

Fig. 6.15 On-board AGC of S-band receiver A (D609F) and B (D709F)

the mission management plan beforehand. Despite the success of the combined team in damping the satellite spin rate, the higher management was more cautious and decided to stop the operations until the situation was fully understood and eventually reproduced on the AOCS ground test bench.

Although the spin rate was almost under control and the reconfiguration of the power and thermal systems slowed down the depletion rate of the battery, its charge level was still dropping. At this point, the combined team decided to go against the higher management decision and gave the order to partially deploy the solar panels on the voice loop, outside of the recommendation process, which is described in Sect. 5.4.3. This deployment involved firing some pyro-elements, as the voltage of the battery cells was already low and dropping.

It is noteworthy that the situation and the way out was fully understood from the perspective of the well-experienced control room team. A further delay of this activity would have critically increased the risk of permanent damaging to the battery. However, the decision to proceed was not taken lightly.

Performing the partial deployment of the solar arrays was a nominal operation procedure, which had been well tested. Its execution was smooth and both solar panels could be partially deployed after 13:45 MET. This action modified the inertia tensor of the satellite and the spin rate was again transferred to the new main axis of inertia as shown in Fig. 6.16.

The new main axis of inertia was such that more Sun light was available on the solar panels and the battery could be charged again.

RAM and SGM Patching

In the meantime, the expert team worked on the assumption that the anomaly was caused by a missing definition of all but two thrusters in the SGM, which resulted in the same missing definition in the random access memory (RAM) when its

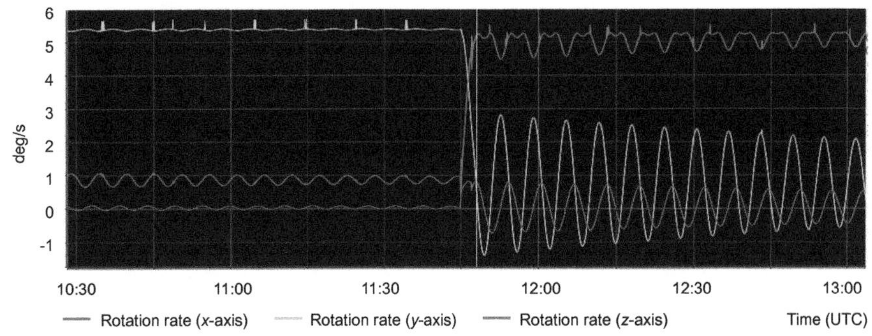

Fig. 6.16 Spin rate around solar array partial deployment

content was initialized from the SGM at boot. The experts were trying to reproduce the observed behavior on the AOCS ground test bench. The proposed correction measures were to load a patch in the RAM, restart the AOCS software and, upon successful recovery, correct the SGM. The mission control team, under the lead of the satellite manufacturer and customer combined team, had to prepare and test the flight procedures to perform these activities. The MET was around 14:00 when green light was given to upload the corrective patch in the RAM. The fluctuating AGC rendered the commanding difficult but the loading could finally be achieved. When time came to activate the patch (MET was around 18:10), higher management decided to stop operations: After such a long shift, it was more than time to give rest to the personnel working in the control room and bring in the backup control team. A complete stop of operations was ordered at MET 21:30.

Operations were resumed at MET 1/08:00 and the prime team was again called to the control room. The decision was taken by higher management to further test the software on-ground, which postponed the operations until MET 1/17:00 when a "go" was given to upload the patch into the SGM this time. However, the loading had to be delayed again by one hour and a half due to a low perigee crossing where the signal of the satellite could not be acquired by the ground station. Once the station was able again to communicate with the satellite, the patch could be loaded and its checksum verified in spite of the fluctuating AGC. Since concerns were raised over the full understanding of the situation, the restart from the SGM was postponed to the day after.

On day 2 of the mission, higher management announced that the root caused was confirmed to be the working assumption by test on the AOCS ground test bench. After the SGM patch performed the previous day, the recovery strategy was to force a reconfiguration with the same equipment pieces by the reconfiguration module (RM). The preparation of these activities and another low perigee crossing shifted the implementation until MET 2/18:00. Although the sequence had been tested and rehearsed on ground, a tense atmosphere could be felt in the control room at the moment of forcing the RM to reboot, which failed. The RM was switched off nominally, but it did not switch on again. The AOCS experts analyzed the situation and concluded that the duration of the OFF command was so long that, due to time overlap, it prevented the execution of the ON command, whose execution was due when the TM was busy. The RM was eventually turned ON by command at MET 2/20:07, which solved the anomaly.

Although the mission was now saved and all involved parties could cheer-up in the control room for half an hour, there was no time to rest. The AOCS had to be reconfigured as well as the heaters. The battery management also needed to be activated and the next low perigee and eclipse had to be prepared. The shift finally ended at MET 3/01:00.

Final Investigations

Despite being involved in the saving of the satellite and having performed shifts way longer than the legal duration without no noticeable loss of performance, the mission control and satellite manufacturer teams were accused of endangering the mission by partially deploying the solar panels without the explicit consent of the higher management. It was finally concluded that this action had probably saved the satellite and that the combined team of satellite manufacturer and customer experts had a better view of the situation when taking the decision.

6.6 Conclusion

The two case studies of contingencies experienced during a mission, the example of a telemetry analysis and the numbers presented in the very beginning of the chapter, prove that experience is a key asset in space operations.

A formal process to gather lessons learned is a valuable endeavor and definitely pays back by improved subsequent missions.

Ralph Ballweg is Project Manager and Flight Director in the geostationary group of the Flight Operations Department of the German Space Operations Center (GSOC). He is responsible for the operations of two communication satellites and two instruments for Laser based communication. He has a degree from the Technical University of Munich in Aerospace Engineering and 25 years of experience in spacecraft operations in Germany and the US.

Andreas Ohndorf works and has been working in the Mission Operations department of the German Space Operations Center (GSOC) in several roles: as Flight Director, as Ground Segment System Engineer, and as Satellite Project Manager. Graduated in Aerospace Engineering at the University of the Armed Forces of Germany in 2001 in Munich, he served as a Technical Officer in the German Air Force for 12 years before joining DLR in 2008. He received a Ph.D. in Aerospace Engineering from Delft Technical University in 2016 and currently leads the Low-Earth-Orbit Satellite Missions Team in the Mission Operations department.

Franck Chatel is Flight Director for the H36w-1 and EDRS missions within the department mission operations at the German Space Operations Center (GSOC). He graduated in aerospace engineering at the Institut Supérieur de l'Aéronautique et de l'Espace (ISAE) in Toulouse (France) and joined DLR in 2001. He was involved as propulsion and AOCS sub-system engineer, later as flight director, in different LEO missions (BIRD, SAR-Lupe, GRACE) and GEO missions (EUTELSAT W24). He supported the early phase of the GALILEO project at spaceopal and DLR GfR before moving to his current position.

Chapter 7
Flight Procedures

Ralf Faller and Michael Schmidhuber

Abstract This chapter describes flight operations procedures which concentrate specific flight operations related knowledge from different resources. After a short introduction, the basics concerning flight operations procedures are explained. Different types of procedures are described as well as the life cycle of a procedure from its creation through the validation process until its utilization. Various technical procedure concepts are discussed, and a simple spreadsheet approach is presented in detail as a practical example. Finally, some general rules and guidelines for setting up and using flight operations procedures are given. This chapter will mainly concentrate on operating an unmanned spacecraft from ground, but special aspects of human spaceflight missions will be mentioned where relevant.

7.1 Introduction

In the normal course of life, our standard knowledge and experience is often not sufficient to bake a cake or to use a new TV set efficiently without additional help. We need clear instructions on what to do. Therefore, cooking recipes are telling us, which ingredients are needed, in what order to mix them together and how long to bake them in an oven to finally get the desired cake. User manuals are providing us with information on how to operate a new TV. Construction manuals are guiding us though the process of assembling the different components to a piece of furniture. All these examples have in common that the user is not having a special education or knowledge. But also, for people with a driving license, cars are equipped with user manuals providing needful information about the car and its functions. Pilots have received a special education to fly airplanes. They have performed intensive training sessions and were certified to fly a dedicated type of airplane, but in spite of this special training and all the experience they might have collected over the years, they are strictly directed by their airlines to follow procedures and checklists for any

R. Faller (✉) · M. Schmidhuber
German Space Operations Center (GSOC), German Aerospace Center (DLR), Oberpfaffenhofen, Germany
e-mail: ralf.faller@DLR.de

F. Sellmaier et al. (eds.), *Spacecraft Operations*, Springer Aerospace Technology,
https://doi.org/10.1007/978-3-030-88593-9_7

takeoff and landing. Hence, manuals and procedures are a central element of their work.

Operating a spacecraft is a similarly complicated process since modern spacecraft are complex machines with more and more sophisticated software. The design of the spacecraft, and consequently the way how to operate them, is defined by the manufacturer. Meanwhile, many different manufacturers are on the market and thus, there are many different ways to operate. In the recent decades of spaceflight, a few standards have been set up, e.g. to harmonize data formats for telemetry and telecommands, but there is almost no standardized way to operate satellites to date. There might be some standards on project or organization level, for instance, there are format standards defined for flight procedures to be run on the ISS, but they are not binding for other human spaceflight missions.

The personnel on ground need a comprehensive background in general flight operations skills and the detailed knowledge of how to operate a dedicated spacecraft. General and detailed knowledge about the spacecraft is usually provided by the spacecraft handbook, but the specific instructions to operate the spacecraft are concentrated in flight operations procedures. They are designed to provide sufficient information for operators to get the planned activities done in a safe and straight forward way.

7.2 General Information

7.2.1 Basics

The flight operations procedures are a prepared, tested, and validated set of work instructions that list, in sufficient detail and chronology, the telemetry checks and the commanding of the respective activities. For unmanned spaceflight missions, where a spacecraft is controlled by an operations team on ground, it is a convenient approach to distinguish between space and ground related activities. In frame of this chapter, procedures in context with the flight operations are called flight operations procedures (FOP). Procedures related to ground system components should be handled separately as so-called ground operation procedures (see Fig. 7.1). It might be the case, that flight and ground procedures have different formats and are developed with different kind of tools.

For manned spaceflight missions with activities, either controlled from ground or performed by the crew on board (or performable both ways), a different nomenclature might be more convenient.

Beside FOP, alternative denotations might be used around the world. ESA is using flight control procedures (FCP) in their standards. Satellite operations procedures (SOP) can be found too.

FOPs mainly consist of commands to be sent to the spacecraft plus the related telemetry checks before and after the command sending. The exact sequence and

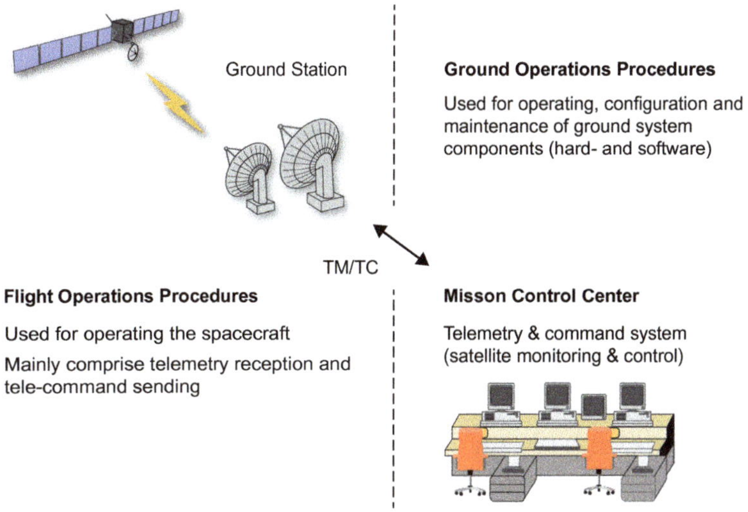

Fig. 7.1 Procedures for spaceflight operations

timing of the different steps are included in the flight procedures. FOPs provide information on which display page a telemetry parameter can be found. They also contain helpful comments/hints about the activities. Decision points, e.g., with some go/no-go criteria for critical events (e.g., boom deployment or pyro firings) can be inserted too, if relevant.

FOPs are specifically designed for use at the corresponding satellite control center (SCC). They are tailored to the dedicated ground and TM/TC (telemetry/telecommand) system software available at that respective control center. Hence, they might have to be adapted when used at a different SCC.

Central input for the development of the flight procedures is the spacecraft related documentation (spacecraft handbook). It is providing specific information about the spacecraft, operational modes and other worth knowing details. All telemetry parameters are listed, as well as all available commands. In fact, the procedures are the specific extract from all available inputs, which are:

- General facts from the handbook about the spacecraft and how to operate it,
- Specific characteristics of the TM/TC system, i.e., the way how the telemetry data are provided and displayed and how telecommands can be sent to the spacecraft,
- All the experience of the operations engineers gathered during previous missions.

The correctness of flight procedures is a critical factor for the flight operations. Thus, only validated procedures shall be used during the mission. They need to be kept under consistent version-control in order to ensure safe and reliable operations and compatibility with the TM/TC databases on ground and in the spacecraft.

7.2.2 Types of Flight Procedures

Depending on the kind of the mission and the complexity of the spacecraft, a significant number of flight operations procedures have to be developed and maintained during the mission lifetime, so it might be a good idea to categorize them with respect to the purpose of the procedure. A common approach is to distinguish between nominal, contingency, and test procedures. This approach is described in more detail hereafter. Nevertheless, such grouping of procedures is an arbitrary decision by the project and is not having any effect with respect to efficiency or reliability of the operations itself.

An alternative example might be the operations in the Columbus module of the ISS, for which a grouping in activation and checkout, nominal, malfunction, corrective and reference procedures have been chosen. There are also maintenance procedures and some for the payloads and experiments.

Nominal Procedures

This group comprises FOPs used in the nominal course of operations, e.g., as foreseen by and listed in the predefined nominal mission sequence of events (SoE), see Chap. 14, or as foreseen in frame of nominal routine operations. Another criterion might be operating of nominal equipment in a nominal configuration.

Contingency Procedures

Flight operations procedures, which are not part of the nominal activities, can be defined as contingency procedures. Typical topics for contingency procedures are operation of redundant equipment, operation in an off-nominal configuration or the transition from nominal to off-nominal states/configurations and back. It might be worth to define dedicated troubleshooting procedures for identification of malfunctions and reconfiguration after anomalies. They speed up recovery activities. In satellite and human spaceflight missions, the standard approach is to only cover single failure deep cases, double failure deep cases are seldom reflected. In general, it is recommended to cover the most probable cases, but finally, the project needs to decide, how many contingency cases are to be covered by respective contingency procedures.

Test Procedures

Test procedures are specially designed flight procedures for tests with the spacecraft, either for on ground tests during the mission preparation before launch, or for in flight tests typically directly after the launch and early orbit phase (LEOP) activities.

For ground tests, the test procedures are specially designed for the use of special ground test equipment on ground and usually cannot be used directly for flight. Thus, for mission safety reasons, it has to be ensured that these test procedures are not accidentally used for flight. A special notation of such ground test procedures is recommended.

For tests in flight during the in-orbit test (IOT) campaign, so-called IOT procedures are used. They are required in the process of the check-out or the calibration of subsystems and its components. Typically, test procedures also try out different configurations in order to demonstrate the readiness and sufficient performance of the spacecraft for the upcoming routine operations phase.

7.2.3 Mission Type Affects Flight Procedures Design

The way to operate a spacecraft is depending on the kind of the mission and its orbit. Satellites on a low Earth orbit (LEO) are designed to operate without connection to the ground most of the time. There is only limited time for interactive operations, when the satellite is in field of view of a ground station. Consequently, flight procedures for LEO missions are typically short with a limited number of steps or sections. For routine activities, they mainly concentrate on sending time-tagged commands to the spacecraft.

Geostationary (GEO) satellite missions are different. Due to the quasi-unlimited contact time, there is no limitation for the activities and consequently for the flight procedure length. It is quite common to design procedures with more than an hour active operations time for the crew on ground. The work on different subsystems is often combined in a single system procedure.

A characteristic aspect of interplanetary missions is the communication delay due to the distance between spacecraft and ground control. Consequently, interactive operations are slowed down or are not possible at all. Commands are sent to the spacecraft and the corresponding reaction can be verified only a significant time later. The real-time telemetry is available with delay only and can hence be compared to recorded and dumped offline data. Commands are sent without immediate confirmation and telemetry is recorded later and available offline.

The following Table 7.1 summarizes the main aspects and characteristics to flight

Table 7.1 Impact of unmanned mission type on flight procedure design

Mission	Characteristics	Impact on procedures
LEO (Low Earth Orbit)	• Short contact times • Most activities happen without ground contact • TM is recorded and dumped	• Short procedures • Mainly loading of time tagged commands, which are executed later
GEO (Geostationary Earth Orbit)	• Long contact times • Most activities are performed with ground contact • No TM recording and dump	• No limitation in procedure length and extend
Interplanetary	• Long response times • No real-time contact • TM is recorded and dumped • All activities performed without ground contact	• Time tagged execution of commands • Only very limited interaction possible

procedure design with respect to the kind of mission.

7.2.4 Flight Procedure Lifecycle

Flight operations procedures are usually designed and prepared during the mission preparation phase, but need to be maintained throughout the whole mission lifetime. Updates due to onboard software changes, changes of configurations, e.g., after major anomalies or loss of equipment might become necessary eventually. Also, experience during the flight operations might be a good reason to update related FOPs. An overview of the FOP lifecycle is shown in Fig. 7.2 and is described in more detail in the next sections.

Basic Procedure Content

In principal, the main input for a flight procedure has to be provided by the spacecraft manufacturer, because he knows how to operate the spacecraft. This input is the procedure backbone, including relevant steps, foreseen TCs to be sent and TM checks relevant at dedicated times. The form in which this input is provided might vary from pure textual description or listings and visualized work flows to already usable procedures.

Development/Adaption

This phase means the transfer of the manufacturer's flight procedure input into a form usable by the satellite control center. The effort for this development process is depending on the received input and the TM/TC system used by the SCC. The SCC also makes suggestions based on earlier projects and own expertise in order to implement useful enhancements. The SCC engineers might propose to break up a complex procedure into smaller parts or, vice versa, they might suggest to combine different FOPs into generic ones (see also Sect. 7.2.2) in order to get procedures which are better to be handled during development, validation, and operations. This is

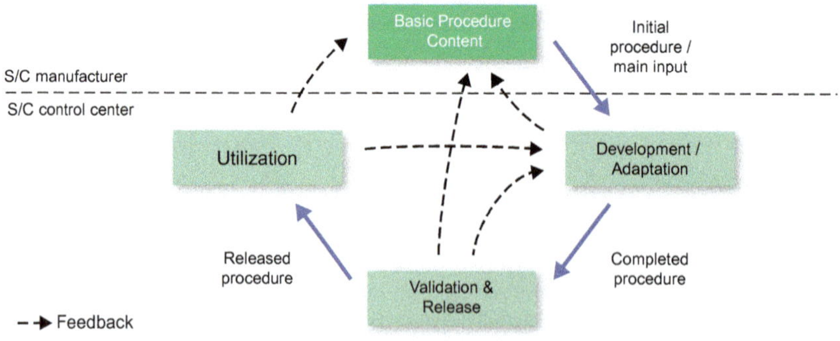

Fig. 7.2 Flight procedure lifecycle

usually done in close contact with the spacecraft manufacturer. In addition, references to display pages, where respective telemetry values can be found, are implemented in the flight procedures.

In an ideal scenario, the manufacturer and SCC are using the same procedure and database formats. These inputs can be easily ingested and only a minimum of adaptation with respect to the transfer is still to be done. The risk of having incompatibilities due to the transfer is minimized, too.

However, more likely is a scenario where the manufacturer uses a different system for development and test of satellite components and system. The SCC needs to spend significant effort on the procedure development. There will be a risk for incompatibilities on SCC side and more effort for test and validation. In addition, procedure redeliveries and updates by the manufacturer might also be provided during the utilization phase, which again will be in a format different to what the SCC is using.

Depending on the project, a manufacturer may only provide the procedure topics and content, e.g., textual information only. In that case, the SCC needs to develop all procedures from scratch. There is a significant risk of incompleteness or incorrectness. Intense testing and validation is even more required. In such a scenario, it is absolutely mandatory to get the confirmation by the spacecraft manufacturer that all inputs have been correctly implemented in the SCC FOPs and that the set of developed FOPs is sufficient to fly the spacecraft.

It was mentioned above that procedures are usually developed during the mission preparation phase. For interplanetary projects, it also might be possible to develop those FOPs, which are only relevant for later operations phases, after launch during the time of the cruise phase.

In general, the spacecraft manufacturer should have a clear view of which procedures are needed for flight and therefore to be ready for launch, but finally, the customer decides what has to be developed and which contingency cases have to be covered by respective FOPs.

Validation and Release

Before a procedure can be used for flight, it has to be ensured that it is correct in terms of doing the right thing and doing the thing right. The syntax with respect to the TM/TC system needs to be checked. The right TCs and TM checks and the compatibility with the current spacecraft database are required. So, with each update of the TM/TC database, the FOPs need to be checked again for correctness and adapted and revalidated if required. The FOP always needs to meet the specifications and has to fulfill its intended purpose.

Therefore, the validation is done within the operational environment using a simulator or real satellite hardware (e.g., an engineering model). Beside the main checks of procedure correctness, provided information about entry conditions and precautions is verified. The procedure flow including all branches has to be validated. The timing of actions, their durations etc., together with checks of referenced display pages and additional information, has to be checked for correctness.

After the validation is passed, the procedure can be released and stated as available for flight operations. A common approach within the final launch preparations is to get an agreement with the manufacturer about procedure versions ready for flight. After launch, releases of updated or new flight procedures are usually done on a case-by-case basis.

Utilization

During the utilization phase, flight procedures are used as planned by the mission SoE or the flight plan. Last minute updates of the SoE or changes due to recommendations (see Sect. 5.4.3) might also ask for execution of additional or alternative procedures at a given time.

Depending on the procedure design approach, a procedure for the upcoming activity might be directly ready for use if it is containing fixed TC and TM parameters only. These FOPs are called static procedures.

It is also possible to set up a procedure as a generic one with variable parts, i.e., the concrete TCs and possibly also the TM checks are finalized shortly before use by ingesting the flexible parts or data (e.g., orbit maneuver parameter, TC execution times, attitude profiles, etc.). The variable inputs can be provided either by the engineer or by automated processes (e.g., mission planning system). Since most spacecraft usually have a high number of parameters and settings to be updated by the respective FOPs, it might be useful to have the latest settings available on ground in form of an image of the configuration, a so-called config mirror. A direct interface between the config mirror and the FOPs might be useful. The correctness of the input needs to be confirmed anyway (e.g., double-check by second engineer). Figure 7.3 summarizes the preparation process. At the end of the preparation, the procedure is finalized, i.e., all required content is implemented.

Before the FOP can be used, some extracts of the procedure, the so-called procedure products, need to be provided in the control room.

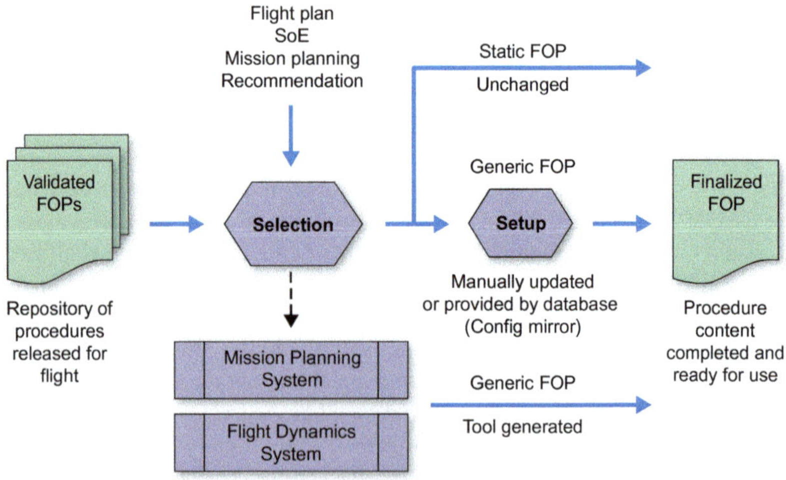

Fig. 7.3 Preparation/selection of flight procedures for upcoming operations

The main product, of course, is the readable part for the operator, which might be purely textual including figures or flow charts. It can be printed out or in an electronic format to be displayed on screens.

Another product is the collection of all telecommands of the procedure (so-called TC stack). These commands are extracted from the FOP and ingested in the TM/TC system. All TCs of the procedure are available in the TC system in the expected order and can be executed respectively. Alternative to TC stacks, script-based files could be used. They are containing TCs and TM checks which allow a semi-autonomous execution of procedures by the TM/TC system (see also Sect. 7.3).

Beside the two main procedure products mentioned above, also some optional items can be set up in order to improve the flight operations process. One optional product is a procedure display page. This display page is directly extracted from the flight procedure and shows all TM parameter checked in the FOP in the same order together with the expected values on a single display page. Such display pages make it easy to follow the activities and consequently improve the flow of operations.

Another optional feature, which improves the flight operations, are so-called configuration checks. Here, the TM/TC system is performing automated TM checks of larger numbers of TM values at a time triggered by the operations team. This feature is supported by some modern TM/TC systems. The corresponding input for the TM/TC system, a so-called configuration check file, is another optional procedure product which can be extracted from the FOP. Figure 7.4 summarizes the procedure products directly provided by or extracted from the FOP.

To summarize, the utilization phase comprises the following: a selection of procedures for the upcoming activity, if relevant, its preparation by ingestion of variable data, the provision of the related procedure products, and then the execution of the

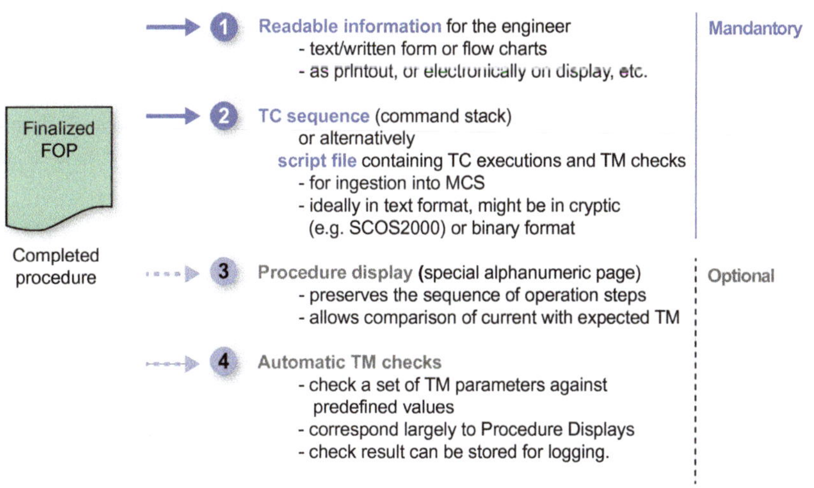

Fig. 7.4 Procedure products extraction

procedure in frame of the mission operations. An archiving of the finalized and used procedure for the purpose of having complete records is recommended.

7.3 Flight Procedure Concepts and Approaches

For the implementation of a flight operations procedure concept in a project, there are different technical approaches possible, from simple self-made solutions to sophisticated commercial tool suites. Influencing factors for the decision are size and complexity of the spacecraft, kind and purpose of the mission, the available TM/TC system, the orbit (LEO, GEO or other), the number of spacecraft to be controlled by the SCC, project budget, among others. The following sections present some examples of technical solutions.

7.3.1 Text-Based

Using purely text-based procedures is one of the easiest technical solutions. All information (orders, etc.) within a procedure is in a readable form in respective text files (e.g., TXT, PDF, RTF, etc.). Pictures, figures or flowcharts can also be inserted to enhance the readability. There is no direct connection between the procedure and the TM/TC system required, so the engineer has to perform all actions, i.e., the selection of required TCs, sending those TCs, and checking of incoming TM. As already mentioned above, the collection of all TCs required by a procedure in TC stacks is a good way to support efficient operations. Thus, the collected TCs can be seen as an annex to the textual procedure.

Text-based procedures have been used successfully for manned flight operations for decades and are still in use on the International Space Station ISS. A special aspect of procedures used for the ISS is that there are two ways to execute them, if technically possible: either remotely from ground, or/and executed by an astronaut in space (an ISS procedure example is shown in Chap. 24).

Alternatively to purely text-oriented procedure concepts, metalanguages like XML (extensible markup language) can be used. Such files support an easy extraction of human-readable text and other procedure products.

Assets and Drawbacks

In general, it is easy to implement this kind of procedure concepts. There are no special costs for software licenses or training of operations personnel on specialized programs. The text-based layout approach provides full flexibility to individual procedure design.

The main disadvantage is the missing connection to the TM/TC system. In case of database updates (changes of TC or TM parameter definitions), each procedure needs to be checked manually for required adaptations here. A general limitation is that more functionality or additional features require external tools to be implemented.

7.3.2 *Spreadsheet-Based*

On a first glance, this approach looks quite comparable to purely text-based procedures. In principle, a table-calculation tool like Microsoft Excel can be used in the same way as a text program, but it has a lot more functionalities. One key advantage is that calculations can be performed within the procedure. This allows, e.g., calculating of times, command parameters or any other data in frame of an upcoming operation. Some programming features are also available to be applied for more complex logic in a procedure. Such logic can also be used to fill the generic parts of a procedure with the specific data for TM and TC, so spreadsheet-based procedures directly support the setup of generic procedures.

An example of a spreadsheet flight procedure for a solar generator partial deployment is shown in the next figures. As you can see here, dedicated sheets are generated for cover and description pages, providing pure text information for the user. The main procedure content is shown in the so-called operations list sheet. All procedure steps with telemetry checks and telecommand transmission steps are listed in chronologic order. On a separate setup sheet, variables (generic parameter) are defined and can be used in the other sheets. In the given example, a parameter for the pyro selection is then used in the operations list sheet. Depending on the selection of this generic parameter, either the prime or the backup equipment telecommands or TM checks are used. This example shows a single procedure to be used for prime or backup configuration. Without the generic feature, two separate procedures would be needed instead. So, the total number of FOPs can be reduced by using the generic approach, but the effort for the generic FOP validation is higher than for the others, because all possible permutations need to be tested (Figs. 7.5, 7.6, 7.7 and 7.8).

In the procedure example, the entry conditions are verified by six TM parameter checks in *step 1*. The TM parameter short code plus a minimum description, the display page number to find it in the display system and the expected value are provided. In *step 2*, two telecommands were sent, and the results are directly checked afterwards in the telemetry.

Assets and Drawbacks

Similar to the text-based procedures, this concept is easy to be implemented. Using such an off-the-shelf office tool does not produce significant cost for software licenses or team training and gives full flexibility for the procedure design. In addition, the advantage of calculation or programming features improves this concept significantly. Such procedure concepts implementing all mentioned features including the implementation of a Config Mirror accessible by the spreadsheet-based FOPs have been used successfully for many kinds of spaceflight missions.

Company logo

Project xyz

Flight Procedure: **NP120**

Solar Generator Partial Deployment

Flight Model	**xyz-1**
Subsystem	**PWR**
Reference Document	**XYZ-Flight Operations Plan Vol 3**
Procedure Issue	**1.3**
Procedure Type	**Nominal**

Generic Procedure - Update Data Sheet!

Fig. 7.5 Spreadsheet FOP example—header page sheets

XYZ-1	NP120	Solar Generator Partial Deployment	Issue 1.3
1	**Purpose**	This procedure is used for the solar generator partial deployment. The procedure allows to perform the deployment with the prime or redundant pyro set.	
2	**Description**	The following step will be performed: 1. Check spacecraft status and configuration 2. Select pyro system and deploy generator 3. Reset pyro system 4. Check spacecraft status after deployment	
3	**Operational Constraints**	It is mandatory to perform the deployment with nominal deployment hinge temperature. If temperature is too low, postpone deployment.	
4	**Related Procedures**	Contingency: CP220 Partial Deployment Malfunction	
5	**Remarks**	none	
6	**Change LOG**	0.1 17.01.2010 RFA - Creation of procedure 0.2 20.01.2010 RFA - TM and TC parameter references completed. 1.0 30.01.2010 RFA - Procedure finalized for validation test. 1.1 17.02.2010 MSC - Step 2.4 revised; one additional check inserted; step 2.6 removed. 1.2 21.02.2010 RFA - Procedure revised according to new TM/TC database 2.3.0 1.3 28.02.2010 MSC - Procedure Released	

Fig. 7.6 Spreadsheet FOP example—description page

Again, a main disadvantage is the missing connection to the TM/TC system and its database. The user either needs to take care of the consistency between procedures and database by entering the data manually or he has to add additional tools.

XYZ-1	NP 120	Solar Generator Partial Deployment	Issue 1.3

Input Data Table

Topic	Defined Name	Description	Value	Comments
General Information	FM	Satellite Short Notation	XYZ-1	
	SUBSYS	Subsystem	PWR	
	PROCNAME	Procedure Short Name	NP120	
	PROCTITLE	Procedure Title	Solar Generator Partial Deployment	
	REFDOC	Reference Document	XYZ-Flight Operations Plan Vol 3	
	ISSUE	Procedure Issue	1.3	
	PROCTYPE	Procedure Type	Nominal	nominal or contingency
Variable Data	Name	Description	Value	Comments
	Psel	Pyro selection	prime	prime or redundant
Calculated Data				

Fig. 7.7 Spreadsheet FOP example—input sheet with generic information

7.3.3 Script-Based

Modern TM/TC systems do not only allow issuing single or bundles of telecommands, they also support TM checks to be done automatically by the ground system against predefined limits. In combination with the option to run batch jobs or other advanced programming features, procedures can be run in an automated way following predefined checks and actions. The ground personnel are only involved where needed. Even the start of a respective procedure can be invoked clock or system driven. Hence, for such a concept, parts of the operations are handled by predefined rules and automated ground processes (semi-automated or supervised operations).

The autonomous flow during the execution of a script-based procedure can be interrupted or paused by inserting break or hold points, where the ground engineer is requested to provide his "go" to proceed, or he is asked to provide some required input. Thus, it can be included into the procedure, how often the automated flow is stopped and how often the ground personnel are actively involved (assisted automation).

The principal mechanism to run flight operations that way was firstly used during pre-launch functional testing, especially for regression testing, but then also practiced for the in-orbit operations. Meanwhile a lot of TM/TC systems allow script-controlled operations. An example of flight procedure concept is based on PLUTO (procedure language for users in test and operations) scripts, which were developed and defined as a standard by the ECSS. Figure 7.9 shows a PLUTO script code example.

XYZ-1	NP 120		Solar Generator Partial Deployment				XYZ-Flight Operations Plan Vol 3		Issue 1.3	
STEP	TIME		ACTIVITY			DISPLAY	TM			
n°	ABSOLUTE	RELATIVE	EVENT DESCRIPTION	CODE	DATA	FMT	CODE	DESCRIPTION	RESULT	
1		T0-8min	**Verify Entry Conditions**							
			Check S/C Status							
			VERIFY:							
			AOCS mode			5200	AS100	AOCS mode	**SAM**	O
			Sun vector S/C x			5200	AD451	Sun S/C x	**0 +/- 0.02**	O
			Sun vector S/C y			5200	AD452	Sun S/C y	**0 +/- 0.02**	O
			Sun vector S/C z			5200	AD453	Sun S/C z	**-1 + 0.02**	O
			Main battery power level			6122	PD300	Bat A PWR	**> 50Ah**	O
			Deployment hinge temperature			2200	TP280	Depl H tmp	**> 10 deg**	O
			IF entry Conditions are not correct							
			THEN postpone deployment							
			ELSE proceed with step 2							
2			**Partial Deployment**					**UPS: make hardcopy of #4210**		O
2,1		T0-5min	**Prepare Pyro System for Deployment**							
			SEND:							
			Activate_Pyro_System	11PAD						[ ]
			Set_Pyro_System	10PAD						[ ]
			Arm_stage		12					
			VERIFY:							
			Pyro system arming			6122	AS230	Pyro Arming	**ACTIVE**	O
			Arming stage selection			6122	AS235	Arm Stage	**12**	O
2,2			Request Go/NoGo for deployment from all subsys							O
2,3			Request Go/NoGo for deployment from STL							O
2,4			**Deploy Generator**							

Fig. 7.8 Spreadsheet FOP example—procedure body showing the first steps of the procedure

Procedure name	Activate Freon Loop

PLUTO script

```
procedure
    main
        initiate and confirm step Heating UP
            declare
                Boolean Status,
                real AvgeTemp units degC
            end declare
            main
                Status := get vadility status of Pump of Freon
Loop1;
                AvgeTemp := get average value of FreonTemp
                    with
                    StartTime := current time ()- 10 min,
                    EndTime := current time ()
                    end with;
                <...>
            end main
        end step
        <...>
end procdure
```

Fig. 7.9 PLUTO script code example from ECSS-E-ST-70-32C

Other script-based procedure concepts use STOL (systems test and operation language) developed by NASA in the late 1970th (Desjardins et al. 1978) or Python.

Assets and Drawbacks

The clear advantage of script-based procedures is the option to perform operations in an autonomous way. This might safe valuable operations time, because such automated checks are performed quicker by the system than by a human operator. In addition, tests on ground in frame of a mission preparation campaign can be performed in an automated way.

The poor readability of script codes and the limitation to integrate additional information for the user within the scripts requires providing such information elsewhere. This can be done by additional text documents, at least for major procedures.

7.3.4 Commercial Products

There are different solutions on the market providing commercial off-the-shelf flight procedure concepts. One example is the mission operations information system (MOIS) of the RHEA Group. MOIS is an integrated suite for writing, managing and testing flight procedures running on a standard office tool set. It is fully database driven, so in case of database updates, the impacted flight procedures are directly flagged as not valid. This supports the consistency between database and procedures. Such commercial tools are customized for dedicated TM/TC systems. Providers of commercial TM/TC systems usually provide own concepts for their flight procedures, e.g., based on scripts.

Assets and Drawbacks

The main advantage of such commercial tools is to have a complete solution for the flight procedure concept and the connection to the TM/TC database. This supports projects with larger numbers of procedures.

On the other hand, such a solution brings along significant costs for software licenses and maintenance, which have to be considered in the project budget.

7.3.5 On-Board Control Procedures

The procedure concepts as described so far have in common, that they are controlled during the execution via ground. Either an engineer or an automated (script-based) process perform TM checks or release commands. During its execution, a communication link to the ground control center is mandatory. As a logical consequence, it would be an improvement to have the procedure and the process to execute it on board. This is the main characteristics of on-board control procedures (OBCP).

An example for such a procedure concept has been developed by ESA and was published in 2010 as a dedicated ECSS standard (ECSS-E-ST-70-01C). This standard also fits to other ECSS standards for data types, packet definitions, etc. (Prochazka 2010).

OBCPs are developed and tested/validated on ground, then uplinked and stored on-board and activated when needed. They allow to send commands and to verify the execution in the same way, as it would be usually done from ground. All spacecraft components, platform and payload, might be operated that way.

OBCP procedures have successfully been flown by various satellite and deep space missions, e.g., Venus Express, GOCE, Sentinel, Herschel, Planck, and Rosetta.

Assets and Drawbacks

On-board control procedures can run without direct connection to the ground. Hence, phases without ground visibility or communication delays due to long distance to a spacecraft do not impose a limiting factor any longer. Less human availability is needed for such an operational concept.

The development of OBCPs is more complex and the effort for its maintenance is higher than for ground-based procedures. The OBCP function can only be used by spacecraft which are specially designed for its use. A later implementation for already flying missions is not possible.

7.4 Development Guidelines and Rules

As described in the sections above, there are different ways and solutions available to implement flight operations procedures in the project, but some general rules should be followed during the development. A good approach to structure a procedure could be as follows:

1. Check current situation against pre-requisites at procedure start.
2. Bring the spacecraft and the ground system into a corresponding state (e.g., configure TM).
3. Perform the action.
4. Check the results.
5. Repeat, confirm, branch as appropriate.
6. Return spacecraft and ground system to the default state or entry state of following procedure, if part of a fixed sequence.

It should be ensured to a have a clear check of the entry conditions at the beginning followed by the activities, and completed at the end by returning either to a default state, or in a configuration adequate to proceed with the next procedure. The main activity block might contain a single task or even more which should be grouped in dedicated steps.

Branching within the procedure is possible but should be handled with care. Too excessive use might make a procedure confusing and hard to be tested and validated. Try to encapsulate activities in complete building blocks (e.g., "configure spacecraft for Earth acquisition"). The start of such building blocks is a suitable branching (e.g., "if wheel unload is skipped, proceed at step 8") or waiting point (e.g., "wait 20 s and then proceed" or "wait until clearance/feedback is available"). If a procedure contains more than ten major steps, consider breaking up the procedure.

7.5 Summary

Flight procedures are the central element for reliable flight operations. They build the backbone for the mission operation execution in providing the specific operations know-how.

The development of the flight procedures during the mission preparation phase and its maintenance over the whole mission lifetime is a major task for the flight operations engineers, which needs sufficient resources allocated in the project budget. Some influencing factors (e.g., customer interest, input from and interface to the spacecraft manufacturer, etc.) have an impact on the FOP related processes in terms of complexity and criticality for the whole project.

Different technical concepts are available for the procedure design, but it needs to be fit to the available TM/TC system. More advanced concepts like the on-board control procedures additionally require a specially designed spacecraft. For the procedure development, basic guidelines shall be established and followed by all involved partners in order to get efficient and reliable products for the mission.

References

Desjardins R, Hall G, McGuire J et al (1978) GSFC Systems Test and Operation Language (STOL) functional requirements and language description. NASA Goddard Space Flight Center

ECSS (2021) European cooperation for space standardization. www.ecss.nl, ESA-ESTEC, ECSS Secretariat, P.O. Box 299, 2200AG Noordwijk ZH, The Netherlands, 14 cited June 2021

Prochazka M (2010) On-board control procedures: autonomous and updateable spacecraft operator onboard and beyond. In: 2010 Workshop on Spacecraft Flight Software (FSW-10), Beckman Institute at the California Institute of Technology, Pasadena, CA (Dec 2010)

RHEA Group (2021) MOIS—mission operations information system. https://www.rheagroup.com/the-new-generation-toolset-for-mission-operations-preparation-and-validation/, cited 14 June 2021

Ralf Faller is project manager for LEO, GEO, and On-Orbit Servicing projects in the mission operations department at the German Space Operations Center (GSOC). In 1990, he received his diploma in aerospace engineering at the Technical University of Berlin, Germany. He started in 1991 at the DLR Oberpfaffenhofen in the spaceflight dynamics department, being responsible for attitude related software development and mission support and changed to the mission operations department in 1997 where he focused his work on spacecraft flight operations topics. He was involved in several missions as subsystem engineer, flight operations manager and finally project manager.

Michael Schmidhuber works in the Department for Mission Operations at the German Space Operations Center (GSOC) where he is responsible for satellite operations training. He graduated as an aerospace engineer at the Technical University of Munich in 1994. The training activities comprise not only staff training but also courses for an external audience in the annual Spacecraft Operations Course at GSOC. Also, he is involved in the organization of the bi-annual international SpaceOps Conference.

Chapter 8
Human Factors in Spaceflight Operations

Thomas Uhlig and Gerd Söllner

Abstract One aspect which is predestined to be overlooked in a highly technical environment like space operations is the "man in the loop", who acts not fully predictable like a computer, but whose actions are partially influenced and affected by completely different, non-technical domains: the psychological or physical constitution of human beings and their interactions within a group. This chapter focuses on factors which minimize errors with potentially severe consequences and which improve the performance of high responsibility teams acting in a complex and dynamic environment. It establishes a link between those factors and operational concepts of spaceflight operations.

8.1 Introduction

Spaceflight is a challenging enterprise at the rim of technical feasibility, and conquering of space has led to a few disastrous events—losses of rockets and spacecrafts, therefore enormous values and, in the worst cases, even to losses of life. Some of these events could at least partially be traced back to human errors. On the other hand, spaceflight has seen as well a number of missions, which only could be concluded successfully and against all predictions without loss of lives or vehicles, because groups or individuals showed extraordinary performance and excellence. Hence, humans can be both: A risk or the only means to handle an unexpected situation. The avoidance and mitigation of risks caused by non-technical factors will be subject of this chapter.

Our starting point is a pictorial principle, which explains, how catastrophic events usually materialize: The Swiss cheese model in Fig. 8.1 shows that they are not caused by one unlucky mishap, but in many cases by a chain of many of them. Under normal circumstances, multiple barriers are established to prevent a critical or catastrophic event. But under certain circumstances, it can nevertheless happen,

T. Uhlig (✉) · G. Söllner
German Space Operations Center (GSOC), German Aerospace Center (DLR), Oberpfaffenhofen, Germany
e-mail: thomas.uhlig@DLR.de

F. Sellmaier et al. (eds.), *Spacecraft Operations*, Springer Aerospace Technology,
https://doi.org/10.1007/978-3-030-88593-9_8

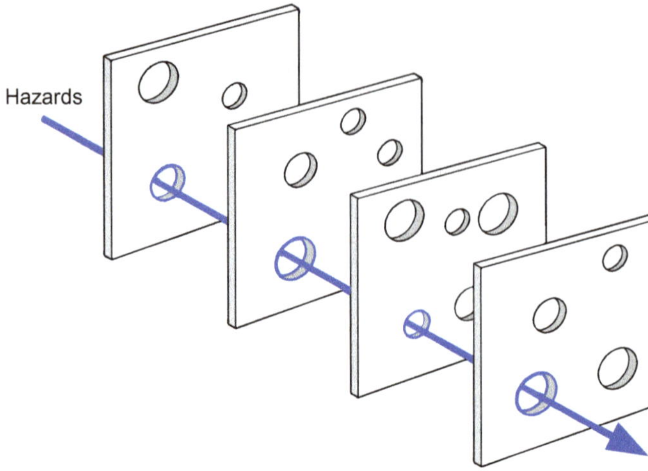

Fig. 8.1 The Swiss cheese model shows how multiple layers, which normally separate us from a hazard to happen, are rendered useless if "aligned holes" open a direct path to the hazard

that all these safety holes get aligned—which then opens the path to a disastrous situation.

Every single barrier usually has a certain permeability—comparable to holes in a cheese: A certain risk must be accepted, otherwise operations would be impossible. However, the combination of multiple layers reduces the risk of a critical/catastrophic event to a tolerable level.

The barriers of the Swiss cheese model in Fig. 8.1 can be classified as:

- Technical safety mechanisms: in space operations critical systems are laid out single- or even dual-fault-tolerant
- Dedicated tools which contribute by their design or their man–machine interface (MMI) to the prevention of failures
- Operational measures or preventive or corrective actions of a team or individuals

In this chapter, we focus on the human aspects and operational measures to prevent hazards. Since we are running operations in spaceflight always as a team, we need to look at the interactions of the team, their processes, and the factors which improve the performance of the team.

The discipline which deals with teams acting in dynamic high responsibility environments is called crew resource management (CRM). The goal of CRM is to reduce failures and hazards introduced by human factors. It optimizes the proper acquisition of information but also team processes and leadership.

This chapter shall provide a first introduction into CRM. It can only touch the surface of this psychological discipline; entire books (e.g. Kanki et al. 2010) and numerous articles have been written about the subject or single aspects of it. It shall create awareness about the importance of psychological and psychosocial factors in a quite technical field like spaceflight operations. Some related aspects can also be found in Chap. 6 within Sect. 6.3.

8.2 Critical Dependability on Humans in Various Areas

In the early years of aviation, planes were flown by one single person, who had to deal with all risks. The increasing complexity of aircrafts, their growth in size and the longer travel distances turned the control of an airplane into a team effort. This also changed the characteristics of an ideal pilot to a fully-fledged team player, especially when the co-pilot turned more and more from a supportive function into a technical partner of the captain. This paradigm shift also opened up a new source of failure: Besides human errors and technical malfunctions, now also team processes could lead to critical situations.

A few catastrophic events enforced counteractions of the aviation industry:

- In 1977, the crash of two airplanes at Tenerife airport resulted in 583 casualties. The root cause of the disaster was—beside other factors, as indicated already in the discussion of the Swiss cheese model before—traced down to communication issues between the pilot and air traffic control, but also the decision-making process in the cabin (Ministerio de Transportes y Communicaciones 1978).
- In 1978, ten passengers died in a crash of an airplane in Portland, USA. Here, the cabin crew was distracted by some minor technical issues while the machine was running out of fuel (National Transportation Safety Board 1979).

In the aftermath of these and some further catastrophic events, and considering the fact that human failure is about four times more often the root cause of an accident than technical reasons (Cooper et al. 1980; Helmreich and Foushee 1993), aviation introduced the concept of CRM, and NASA was already involved in the very first conceptual considerations for it (Cooper et al. 1980). CRM is an approach which acknowledges the fact, that improved group interactions can reduce the risk of human-induced errors in environments, in which those errors can have catastrophic effects. The introduction of CRM in aviation beginning of the 1980's led to a significant reduction in the number of accidents within 30 years (Flin et al. 2002).

The need to know about CRM aspects is not only limited to pilots. When 1988 a Boeing 737–200 lost approximately one third of its casing and one of the identified root causes was traced down to maintenance issues (National Transportation Safety Board 1988), the training on CRM was made mandatory also for ground service personnel. After two aircraft crashes in 1989, which could have been prevented, if information from the cabin would have found its way to the cockpit (Department of Transport 1990; Moshansky 1992), CRM training was also rolled out for flight attendants (Ritzmann et al. 2011). It needs to be stressed that the above-mentioned CRM training in aviation is not complementary or optional, but is required by the different regulations and laws of the corresponding aviation authorities (FAA, EASA, JAA, etc.).

In the meantime, the concept of crew resource management was rolled out also in other areas: Common to all these areas is that teams are acting in high reliability environments (Hagemann et al. 2011). A failure of these teams has the potential of fatal consequences: loss of lives, serious injuries or significant damage to the environment, also with high costs involved. In addition, the complexity of the environment they are acting in is high, which leads to the fact, that the results of taken actions are difficult to predict.

Of course, military operational forces are considered as such teams (O'Connor et al. 2009). Then all kind of emergency response teams have adapted their curriculums to reflect elements of CRM in them: Fire fighters are being taught in it (Okray and Lubnau II 2004) as well as medical response units. Rescue coordination centers, which dispatch and lead those units during larger emergency situations, are highly affected by CRM subjects. CRM also entered medical operation rooms, where a team of experts needs to effectively interact to prevent negative effects on the health of the patients (Rall et al. 2014). Various prominent accidents in the nuclear power or offshore industry, which could be traced back to human factors, also predestine those fields for the introduction of CRM (Gaddy et al. 1992; O' Connor and Flin 2003).

8.3 Classical Fields of Crew Resource Management

Crew (or group) performance and behavioral studies are behind the concepts of CRM. As baseline, it is of course imperative to have a common understanding on the group goals (e.g. safe ops, mission success). What sounds self-evident in the first instance, might be not that clear in all cases. For example, for space operations the interests of agencies, companies and industries, scientists and universities might be diverging.

Group processes are described via a control loop in a model of Foushee and Helmreich (1988).

In that model, the **input factors** provide the framework of the group interactions, the **group process factors** directly control the performance of the group, which generates as an output safe and efficient operations. Before we focus on the group process factors which are directly linked to the operational environment discussed in this book, it is worth having a short look at the input factors, which constitute the baseline for successful and efficient group work.

Any group interaction is influenced by a set of input factors (Foushee and Helmreich 1988):

- *Individual factors* are directly related to the individuals of a group: Their motivation, their emotional state, their knowledge or physical condition have direct impact on the outcome of a group process.
- *Group factors* include the composition of a group, its inner climate, in international efforts like spaceflight often also the various cultural backgrounds or the proficiency in the language.

- *Organizational factors* are giving boundaries, but are crucial for a successful team playing: How is the company/agency dealing with errors? What are the norms within the organization? Is the work done under cost and time pressure or is safety a value on its own?
- *Regulatory factors* describe the boundary conditions which are set up by rules, laws or regulations.
- *Environmental factors* can also influence the group performance. Usually, spaceflight operations are performed in the protected environment of a control room, but even here, temperature or the light conditions can lead to a higher or lower group performance.

All those factors directly influence the group process factors and, thus, the outcome of a group work.

Let's focus on the **group process factors**. For successful and effective group interactions a number of behavioral markers have been identified. These can be clustered in several fields, whereas this grouping is not unambiguous. In this book, we present a selection, which we deem of relevance for the subsequent discussion in the light of spaceflight operations:

- Communications
- Situational awareness
- Decision making
- Teamwork and leadership.

This selection is presented in more detail in the following sections.

8.3.1 Communications

Communication is a key asset in the interaction of humans and therefore of significant importance for CRM. It is defined as an information exchange between a sender and a receiver. In a well-established model (Shannon and Weaver 1949), the communication media is introduced, which is used for information transfer. This is depicted in Fig. 8.2. During the process of information exchange, various portions of the exchange can be subject of a disturbance, which can lead to various effects: For example, the receiver could have a wrong perception about the identity of the sender, the information could be altered or misinterpreted or could get lost. As a result, an inappropriate or no action might be triggered, which can constitute the "key hole" in the Swiss cheese, opening the path to a catastrophic event.

Another important aspect of communication is described by Schulz von Thun's four-sides-model (Schulz von Thun 1981), which is emphasizing, that communication is in most cases not mathematically precise, but leaves room for interpretation, which can, again, be a source of disturbance (see Fig. 8.3). This is in particular true for international collaborations with multicultural participation and non-native

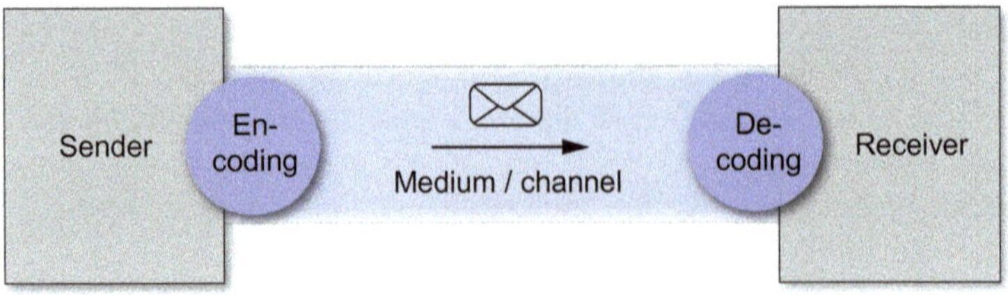

Fig. 8.2 A simple model of communication between a sender and a receiver. The information is transmitted via a dedicated medium/channel. The sender needs to encode the information, the receiver decodes it. All parts of this communication chain can be subject of disturbance

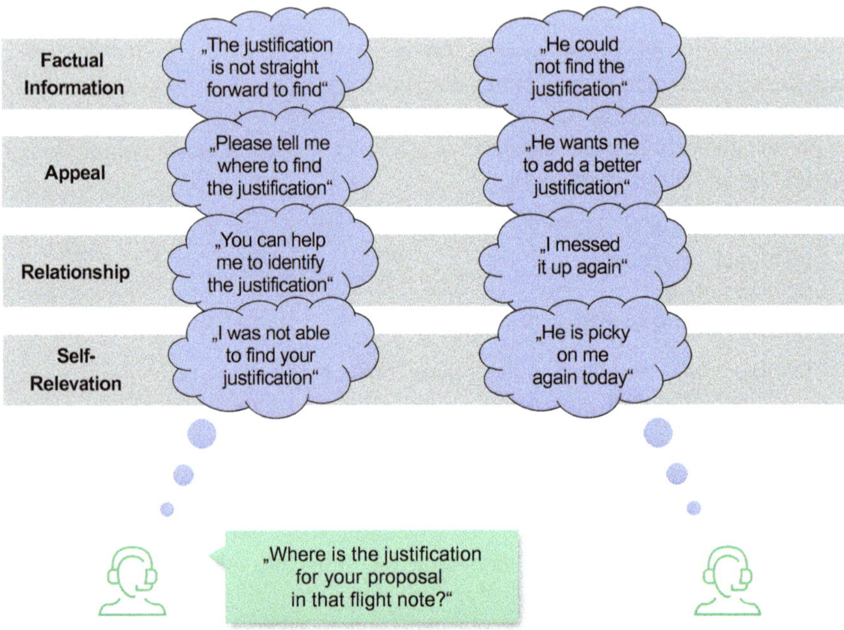

Fig. 8.3 A message has four sides according to Schulz von Thun. One sentence of a sender (green) can have various meanings (blue, left), depending on which "channel" he is broadcasting. The same is true for the receiver (blue, right): He might not be able to identify the correct "channel", which the sender is using. This is a source of misunderstanding and even conflict

speakers, which is often the case in spaceflight projects. However, within the control room it is preferable to always remain on the factual information side.

8.3.2 *Situational Awareness*

Another key variable for successful team performance is situational awareness. It is a quite complex concept and can be seen as the end product of an information collection process. In a commonly accepted approach, situational awareness is divided into three layers: data collection, interpretation and propagation (Endsley 1995).

In the first layer, the relevant elements of the environment are perceived. It can be seen as the data collection process. Here, a normally useful function of the human brain can lead to negative effects. Usually, it is crucial that we can filter out stimuli of our environment, which are not of particular importance in a given situation. We can focus on the essentials and are not disturbed by all the data, to which we are continuously exposed. However, this effect can lead to the fact that we are completely ignoring information which is important in this particular situation. Everybody who has been exposed to a stressful situation knows the effect: The awareness narrows down to a tunnel view. An investigation showed, that ¾ of the failures in aviation, which could be traced back to loss of situational awareness, can be linked with this first layer (Jones and Endsley 1996).

In the second layer, the collected data is interpreted and comprehended. They are incorporated into a mental model of the situation. On that level, the interpretation can suffer from a biased view and result in a wrong mental model.

Different causes for bias are described. Humans tend to focus on the data, which supports their theory of what is going on—and neglect facts, which are contradicting to it (Nisbett and Ross 1980). Often, we are also fixated on a stereotype, which, again, prevents us to have an unprejudiced view on the data (Rasmussen 1987).

In the third layer, a projection of the current situation into the near future is executed. This anticipation is a key feature of situational awareness. It is obvious that this extrapolation is very sensitive. The results of the two precursory layers and any disturbances are also propagated into the anticipation of the situation.

Each of these layers has to be considered to avoid errors due to loss of situational awareness.

8.3.3 *Decision Making*

Communication is elementary to gain situational awareness. Situational awareness is an important part for the decision-making process. A decision is required whenever multiple options are available for action. For a team, a decision-making process needs to be defined. A decision can be made either by a democratic and cooperative process or by an authoritarian approach:

- A democratic and cooperative process, which involves all team members and reflects their individual opinions appropriately, has the advantage that the members feel appreciated and take responsibility for their acting within the team. The satisfaction is usually quite high, since the team's heading direction is the result

of a common effort. However, the disadvantage is that decisions take comparably long.
- In critical situations and for teams acting in high reliability environments (Hagemann et al. 2011), another style is preferred: Here, an authoritarian approach is the most promising concept. The team is set up in a hierarchical order and one or more decision makers are clearly identified, together with a clear scope of their authority which defines who is responsible for which decision-making level. This concept allows quick turn-around times for decisions. Of course, the quality of the decision is very much dependent on the skills of one or a small set of individuals.

The team leader has to have special skills to use the hierarchical setup in a constructive way, involving team members and take decisions where needed. It is advantageous if he/she is able to adapt his/her decision-making style to the corresponding situation.

8.3.4 Team Work and Leadership

The "art of leadership" was already mentioned due to its close relationship to the decision-making process. All above mentioned areas plus disciplines like conflict management, workload and stress management, or group dynamics influence the team performance.

Under ideal circumstances, the team performance leads to an increased outcome compared to the sum of all individual performances.

Successful teams have (implicitly or explicitly) allocated roles, which are taken by their members, in spaceflight operations clearly defined by the control room positions. A role conflict can arise, if the expectations attributed to a certain role are not matching within the team.

Team members not only have roles, but they also have a status, meaning a relative ranking within the group. One of the authors vividly remembers his first simulation run as newly assigned flight director: Leading a team which had an experienced astronaut at the communicator position was a challenge—role and status was perceived as not matching. This could unbalance the team decision making, but can also be seen as a chance by using the availability of high expertise.

On the other hand, it is a task of the team leader to balance out different experience levels. This considers that differently experienced team members need different levels of attention, especially in high pressure scenarios. The balance has to be kept between handling the situation in a proper way and not overloading the individual with too many or too complex tasks.

8.4 Translation into Spaceflight Operations

In the following section, we will discuss how the human factors described above can be addressed in spaceflight operations.

The ISS project as the main current human spaceflight endeavor of the world is used as example in many cases. There are multiple reasons, why this project is a good demonstrator for CRM related principles. The two most important ones are: First, since astronauts and large space borne structures of immense value and high public visibility are involved, the consequences of a failure can be life-threatening for the crew and endangering for a key project of the human spaceflight history. Hence, the flight control teams can definitely be considered as acting in a high reliability environment.

Second, the teams which are interacting and need to deal with a potential problem are much larger, hence group processes and therefore the concepts of CRM play a more significant role compared to a one- or two-person shift for an unmanned mission.

The authors both worked in various functions at the Columbus Control Center (Kuch and Sabath 2008) at GSOC in Oberpfaffenhofen, which is one of the five main ISS control centers (Houston, Huntsville, Moscow, Tsukuba, Munich).

Below, we will now investigate how the different fields of CRM, which were laid down in Sect. 8.3, are incorporated and realized in space operations.

We will discuss the flight control team as a system which supports a quick decision-making process by a hierarchical structure. Then we will focus on operations products and show how they support decision making and increase the situational awareness. This also includes how to establish decisions outside of the nominal envelope of operations and a means to document the decision-making process. The communications approach is introduced.

We will talk about tool support and one dedicated tool which is used for further failure analysis. And finally training and simulations are introduced as an appropriate means to introduce best practices and exercise the corresponding processes.

8.4.1 Flight Control Team Structure

The structure of a flight control team is usually set up in a hierarchical order, which supports an authoritarian leadership style and hence a quick decision-making process with clear responsibilities (Sects. 4.3 and 5.3). The flight director has the full real-time authority, but needs strong rationales to deviate from flight procedures, flight rules, etc., which we will discuss below.

There are cases, where flight directors of different levels of authority are interacting with each other (e.g. the flight directors of the various control centers for ISS operations). There can be flight control team members, who direct another layer of supportive positions outside the main control room.

Usually, the flight directors are responsible for the implementation of the timeline on console which was prepared by the mission planning beforehand. A high-level management layer is taking care for strategic mission objective decisions.

Hence, it needs to be clearly defined where the management responsibility and authority ends and the real-time function of the flight director takes over.

Flight directors are coordinating the technical experts of the mission or spacecraft and rely on their team's expertise. This results in a group decision making process with one final decisive instance.

8.4.2 Operations Products

Spaceflight operations is a strongly regulated business. In many aspects, it can act as a role model for clearly laid down processes, roles, responsibilities and products. The subset of documents which are of relevance on console is subsumed under the term ops products. They describe interfaces, constitute sets of laws, list the steps for performing an activity.

In the following, we will discuss some operations products which are of high relevance for the decision making of a flight control team and support the situational awareness of all team members.

I. **Interface Procedures**

The operations interface procedures define the structure and positions of a flight control team as well as the interaction between the positions of the team, the decision authority or even between different teams (e.g. launch control team and flight control team).

The interface procedures clearly define who has to communicate with whom on what, and which tool or media has to be used for different subjects (e.g., maintenance, anomaly, emergency) or mission phases (e.g., LEOP, routine operations).

This framework provides guidance, defines the team interactions and guarantees a structured approach, which optimizes the situational awareness of the entire team over all mission phases. In very dense situations, e.g. during launch, the communication between positions or centers can be defined down to a single word, which has to be provided.

II. **Flight Procedures**

In spaceflight operations, most—if not all—of the activities performed on-board the spacecraft, either remotely via commands or directly by the astronauts, are executed following dedicated procedures. These provide a sequence of to-be-performed tasks/commands and checks, in some case also with built-in logical decisions or repetition loops.

This is already a major advantage compared to other teams acting in high reliability environments. In other areas, the usage of procedures or checklists is not yet common practice in that generality (Hales and Pronovost 2006), but their importance is acknowledged.

In the space operations domain, these procedures are in many projects called flight operations procedures (FOP, see Chap. 7), in the ISS world the acronym ODF (operations data file) is used (see Sect. 24.4.3). Outside of the spaceflight area, similar products are referred to as standard operation procedures (SOPs) or simply checklists.

An example of an ODF is depicted in Fig. 17.3, an early version of spaceflight procedures is shown in Fig. 5.3.

It is obvious, that a strict adherence to a validated procedure eliminates failures to a high degree. Also, any real-time decision making is minimized, since the standard cases are covered by those procedures.

Procedures for the flight segment are written according to a common layout standard. This allows the user a quick adaptation also to unknown or less frequently used procedures.

Also, the used names for telemetry items or telecommands follows usually a standard, which allows to extract as much information as possible about the item from its name on one hand, e.g., *EPS-PDU3-Outlet-Channel-12* for the 12th outlet channel of the Power Distribution Unit 3 of the Electrical Power Subsystem. On the other hand, care should be taken that names used during operations are not only unique, but also easily to distinguish: *DMS_Tlm_Pkt_Pwr1* and *DMS_Gnd_Tlm_Pkt_Pwr1* are bad examples—the differences between those terms is too small.

The entire content of a flight procedure is validated with desktop reviews and/or simulator validation. This ensures that a verbatim adherence to the procedure does already prevent some errors under normal circumstances.

In addition, procedure displays and command stacks can be used to support the execution of a flight procedure and makes it even less error-prone.

A procedure display consists of all telemetry items which need to be checked in the course of the flight procedure, also the structure of the latter is mirrored in it, as well as the expected values (Fig. 8.4).

A command stack is a previously built and validated set of commands, which corresponds to the commands of a flight procedure in the correct order, which are required to execute that procedure.

Both products reduce even more the possibility of errors in real time, they are only pairwise implemented in the ops environment to ensure consistency at all times. Version control is ensured along with the procedure naming convention.

There can be additional measures in place to ensure full attention of the operations team at critical steps of procedures: In some projects, so-called hazardous commands are identified. Those commands are capable to either generate a critical situation or to reduce the number of barriers to a critical status, if sent under wrong circumstances. If a hazardous command needs to be executed, it requires per process special handling during real-time operations, within the commanding tool or needs special approval. This ensures a good situational awareness of the entire team and enforces an active decision before sending the command.

In summary, flight procedures provide a validated way of operating a spacecraft, and provide clear guidance for decisions.

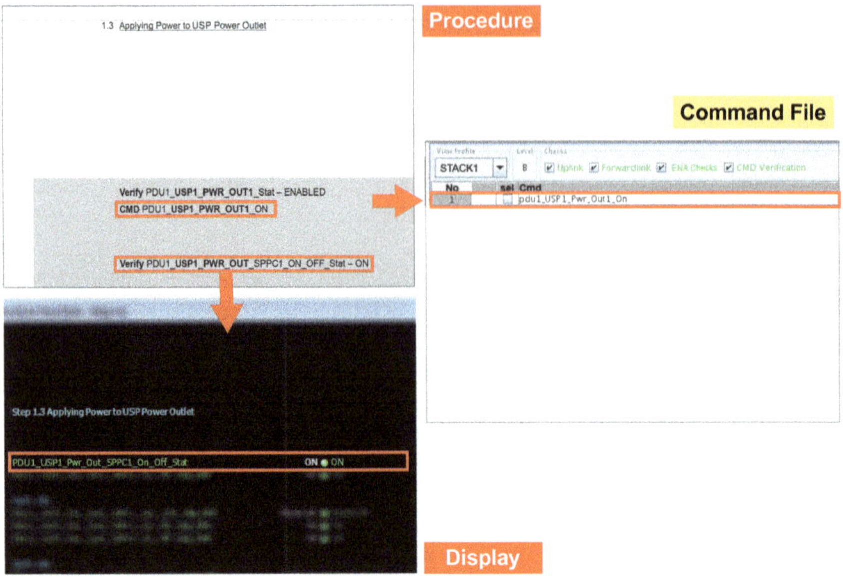

Fig. 8.4 The generation and validation of flight procedures is directly linked to the corresponding command stacks and displays. The products show commands and telemetry items in the correct order and are configuration controlled

III. Flight Rules

Flight rules are ISS operations-specific predefined decisions or rules, comparable to laws, which support the flight control team in real time decision finding.

Flight rules determine general operations principles as well as technical details. They take precedence over individual flight procedures and govern the way how to apply procedures in operations. This becomes important during anomalies where single standard procedures may not be enough to solve the situation. Goal is to keep situational awareness and ease decision making.

As an example, see Fig. 8.5 for a flight rule that specifies the valve operations for an active and passive/redundant water pump in a cooling system.

IV. Flight Notes

The above discussed operations products (flight rules, flight procedures, interface procedures) are normally prepared in the office environment: they are authored, reviewed, iterated until all review findings are worked in and finally approved by the corresponding responsible board. Since this can be a lengthy process, there needs to be a way with a quicker turnaround time in place to allow a fast adaptation in case of failures in the products or a changed onboard configuration/situation.

This process is done via flight notes. In general, those "small documents" are existing with a unique ID and a defined status (e.g. "draft", "in review" or "approved") within a dedicated tool on console. They are used in real time by the operations teams

```
WATER PUMP OPERATIONS

1. IF A WATER PUMP IS ACTIVE, THE ISOLATION VALVE
   OF THE NON ACTIVE WATER PUMP SHALL BE CLOSED
   UNLESS THE NON ACTIVE WPA IS HYDRAULICALLY
   ISOLATED FROM THE COOLING LOOP.

2. THE WATER PUMP ISOLATION VALVES SHALL NOT BE
   OPEN SIMULTANEOUSLY. THIS DOES NOT APPLY.
   A. DURING LOOP RECONFIGURATION
   B. DURING CONTROLLED WATER TRANSFER
   C. IF ONE OF THE TWO WATER PUMPS
      IS HYDRAULICALLY ISOLATED

3. THE WATER PUMP ISOLATION VALVES SHALL NOT BOTH
   BE IN THE CLOSED POSITION AT THE SAME TIME.

Rationale for 1: The Isolation valve of the non-active Water
Pump is closed in order to avoid water backflow and to prohibit
a passive operation/through flow of an inactive pump.
```

Fig. 8.5 A flight rule shows the rule text (capital letters), supported by commentary information (italic letters), which ease the interpretation of the rule

for various purposes, which require coordination, review and approval. They allow to write down forward plans and to detail complex decisions. The flight note tool ensures transparency in data collection and ensures situational awareness; therefore, the team decision making is supported.

Especially during anomalies, a proper coordination is required because not all failures can be considered in predefined procedures. For some operations, the pure equipment saving, e.g. by powering equipment down, is not sufficient. Examples are thermal clocks which require re-powering to avoid damage of equipment or loss of science results.

Note that during time-critical operations, the agreements can also be taken on recorded voice loops and the documentation in a flight note happens "after the fact", which impacts the situational awareness and increases the risk of errors, but support a very quick turnaround. Optimal is a combination of both, i.e., to gather information via the tools and discuss shortly on the loops.

8.4.3 Real-Time Decision Making

Flight rules and flight procedures provide predefined decisions or at least guidance to decide. However, they usually are written "one failure deep" only, which means that the baseline is a nominal situation on board. Otherwise, the content and complexity of flight rules or procedures would increase extremely, if all analysis is also performed in the presence of failures or combination of failures.

In off-nominal situations, it is very likely, that, for example, a flight rule cannot be applied. Hence, the concept of predefined decisions does not work.

In these cases, the flight control team has to perform a real-time analysis of the situation based on the available facts and come to a decision how to proceed.

In the human spaceflight domain, all flight control team members are trained to follow the priority order "crew—vehicle—mission". This means that the uppermost concern needs to be life and health of the astronauts. If this can be guaranteed, then the integrity of the vehicle itself is shifted into the focus of the team. And only if this is ensured as well, then the mission goals are pursued.

What sounds quite logical is in fact not trivial: The teams are mainly focused on the mission goal for almost all of their on-console time, because the crew and vehicle both are usually doing well. Therefore, we saw quite often in training simulations of contingency situations, that flight controllers tend to still be concerned about their mission goals, although the crew and the vehicle have been endangered.

For decision finding, the FORDEC principle, which is well-known also from other disciplines, should be applied (Hörmann 1994): First, collect the **F**acts, which means, try to understand the current situation. Next, evaluate, which **O**ptions are available. That way, a variety of possible forward plans are on the table. Now check the benefits and **R**isks of every option, then **D**ecide. Afterwards, **E**xecute your decision and finally **C**heck the result. In many textbooks, this scheme is depicted as a loop, which is continuously repeated. In the end, the checking of the outcome is, again, an analysis of the situation and a collection of facts—and the start of a new iteration, see Fig. 8.6.

Fig. 8.6 The FORDEC principle should be applied continuously

It is very important to document the facts, the options and risks and finally the decision in a written way, e.g. in a flight note (see Sect. 8.4.2 IV) or in a console log book. Off-nominal situations might lead to investigations afterwards, which are then, in most cases, conducted with all necessary information, sufficient time and the knowledge of the final consequences of the decision. This happens under quite different circumstances than the original decision, which was made under time pressure and with limited data. In such cases, it is important to be able to refer back to a well-documented decision-making summary.

If a contingency lasts longer, those written logs are also a highly valuable source of information for the upcoming shifts. In parallel the situation is described in the shift handover documentation and discussed during handover.

8.4.4 Standards for Voice Communication

It was already highlighted in the previous chapter, that communication is a key aspect for successful group interactions.

In many spaceflight projects, communication is done via a dedicated voice communication system, which links all participants and offers a variety of communication channels which can be activated simultaneously or limited to one or a few, in case of heavy communication traffic.

An example for the voice communication system in the Columbus project of GSOC is shown in Fig. 8.7. On an individual touch screen, each user of the system can select the voice channels he wants to monitor via his headset. One loop can be configured as talk channel on which the user's microphone is patched if the transmit button of the headset is pressed. Dedicated user roles allow a configuration of the available voice loops on the touch screen and to define the rights (no access at all, monitor only, monitor and talk) for every channel. Record/playback functions can be implemented for each user. There can be a replay capability which can be requested on demand by the ground controllers in charge of the voice communication system. In general, a recording/replay capability has been proven to be very helpful, e.g., if a crew call was ambiguous, a replay can help to clarify the situation without having to contact the astronauts again.

In a given spaceflight project, the various voice channels have well-defined functions and scopes as well as different levels of importance, there can be clear rules, which of the voice loops need to be continuously monitored by everyone.

For the ISS project it is commonly agreed, that the space-to-ground loops, which are used for communication between the ISS crew and ground have the uppermost priority and need to be monitored by everyone for situational awareness.

Second priority is with the flight directors' voice loops, where important information is exchanged. Furthermore, every flight control team position has its "prime loop", which is the channel where the corresponding flight controller is reachable. In addition, there are voice loops for specific operational topics or coordination

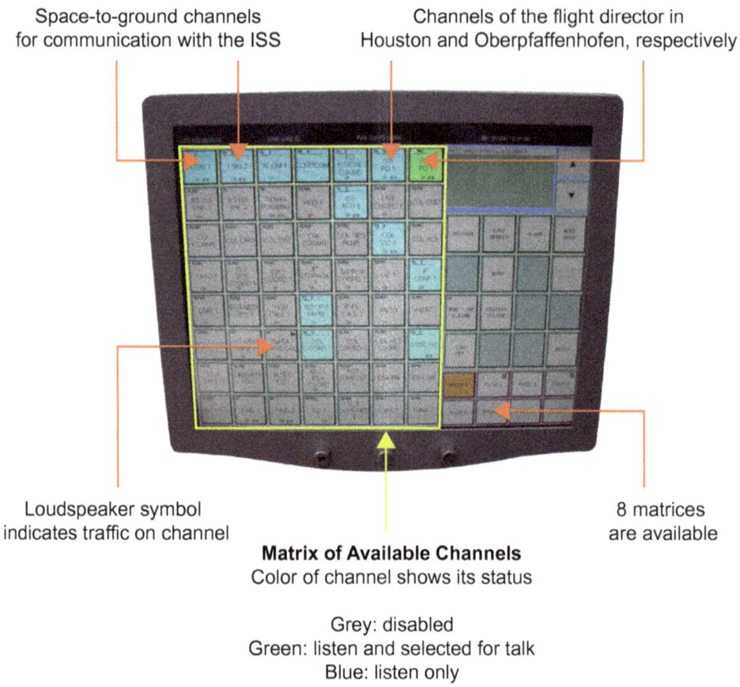

Fig. 8.7 A voice communication system features a set of voice channels, which can be activated (blue) or muted (grey) by the user. Once activated, the voice traffic on the selected channel is audible via the user's headset. One channel can be selected as active talk channel (green): If the headset button is pressed, the microphone of the headset is connected to this channel and the user can talk on it

loops, which can be used for extended discussion without spamming one of the above-mentioned loops.

Depending on the project and the number of participants, the amount of voice loops can be limited to just a few or can involve some hundred loops. In the latter case, all loops do not match on one touch screen, so a preselection must be done for each control room position, which channels are required. In some cases, the voice communication system also allows to switch between multiple pages or matrices.

During communication on the voice loops, a strict aviation-like voice protocol is utilized. This provides a number of advantages: The voice protocol is standardized and uses partially a well-defined terminology and keywords. The keywords are designed in a way that misunderstanding is less likely (e.g., "affirmative" instead of a short "yes"). The international NATO alphabet (Alpha, Bravo, Charlie …) is used, if spelling is required, also for numbers or times a specified format is used to avoid ambiguity.

Table 8.1 provides an overview of the most commonly used "special words" within the ISS project.

Table 8.1 Most commonly used "special words" within the ISS project

"Special word"	Meaning
Affirmative	Yes
Copy	Understand
Go ahead	Proceed with your transmission
Negative	No
Roger	I have received your transmission
Unknown station/station calling	I do not know the identity of the station calling me
Standby	I need to pause for a few seconds either to attend to something else of higher priority or to obtain needed information, but I will respond within one minute
Wilco	I have received your message; I understand it and will comply
All After	I refer to the transmission following "*XXX*". E.g., "*EPIC, say again all after XXX*"
Disregard	Cancel my transmission in progress, or cancel my last transmission
Figure	Numerals will follow, not needed when giving parameter readouts or time
I spell	I will spell the following phonetically. e.g., "*Cislunar, I spell, Charlie India Sierra Lima, Uniform, November, Alpha, Romeo*"
Word after/before	I refer to the word after or before "*XXX*"
Word(s) Twice	Communications are difficult. Transmit, or I will transmit, each word, or group twice
Break (break)	I wish to interrupt a transmission already in progress
On my mark	An event is about to take place. The countdown may start with 10, 5, or 2, but should be **one second** intervals toward zero and should end one second after "*one*" with the word "*mark.*" (e.g., "*On my mark. 3, 2, 1, mark*")
Read back	Repeat all or the specified portion of my last transmission
Say again	Repeat all or the specified portion of your last transmission
Speak slower	You are talking too fast

Voice loop communication is exercised excessively in various training sessions with flight controllers. By generating the awareness for the possible disturbances in communications as described above, the team also started to apply good practices, which were not taught in the classroom lessons: The short words "on" and "off" can easily be misunderstood—with potentially dramatic consequences. Hence, those words are spelled out: "on, O-N", "off, O-F-F". The teams repeat important key information, instead of just replying "I copy" ("I understand"). This introduced a kind of end-to-end protection of the information transfer.

For expected excessive communication with the astronauts, the teams also try to have a video connection to the space station to have an additional visual communication channel with non-verbal cues available if required.

All these measures aim to minimize misunderstandings and make errors on that level more unlikely.

In human spaceflight, the communication with the astronauts on board is deemed the most critical one. There is a generic rule that all other communication has to be silenced, whenever "the station calls".

It is worth to shortly demonstrate the procedure during a space-to-ground communication, since it shows, how different levels of decision making and communication interact with each other. We use an example of a call from ISS to the Columbus Control Center (Col-CC, call sign: "Munich"), see Fig. 8.8. Beside the astronaut

Fig. 8.8 Example of a communication with ISS on voice communication channel "S/G1" and accompanying conversations on other loops

"Frank", there is the EUROCOM involved, which serves as crew communicator. The EUROCOM is usually an astronaut or an astronaut trainer, i.e., a person which is well known to the crew. In this example, the Columbus flight director is a silent listener; since parts of the communication are done on his voice loop ("COL FD 1"), he could however intervene at any time (further details on the setup see Kuch and Sabath 2008).

Two different voice loops are involved in this example: A space-to-ground loop is generically monitored by everyone. For the crew, this is usually the only voice loop accessible to them. The voice loop of the Columbus flight director is available to the entire Columbus project and is monitored by all operational parties of it, but it is neither available at the other ISS control centers nor to the crew.

Astronaut "Frank" initiates the call to ground on the space-to-ground voice channel "S/G 1". Since the entire team is obliged to monitor the space-to-ground loops, they are aware of the crew call. Since the astronaut already indicates, that the subject of his call is related to the payload "Experiment", the responsible payload expert (called USOC OPS) is informing COL FD that he/she is listening. This is done on the "COL FD 1" voice loop, which indicates that the attached payload center is ready for the crew call. Since everyone's awareness is given, EUROCOM can now reply to the crew and asks him to continue. The astronaut's question about the start of the centrifuge removal cannot be answered by Col-CC itself, since the experts are with the payload control center. Hence, the go/no-go-decision is done there and is relayed by USOC OPS to the Columbus flight director on his voice loop. Since the flight director has the final authority for the crew communication, the answer to the crew question needs to be formally approved by him, then EUROCOM can immediately go "on air". As an additional handshake mechanism, the technical expert at the payload center confirms ("good words") the answer provided by EUROCOM to the crew for general awareness on the "COL FD 1" voice loop. In this example, the answer to the crew was trivial, but for non-trivial items to be communicated it is of crucial importance, that the words of EUROCOM directed to the crew are checked and confirmed by the experts: Since EUROCOM is usually not a subject matter specialist, his mental model of what the answer is about could be wrong—and by his rephrasing he could then convey wrong information.

For the communication, also non-verbal aspects like eye contact, gestures, etc. have to be considered. They are elementary to human communication and also help for making crew calls more efficient if all participants are in one control room. The example shows different mechanism which are put into place to ensure that distortions in conversations are avoided or at least unveiled.

8.4.5 *Monitoring System*

Situational awareness in spaceflight operations is usually dependent on the data, which comes from space. When only abstract data like telemetry is available, it is not easy to get an idea of the situation on board. Hence, the tools which display or process these data need to support the process of mental model generation in an optimal way.

To date, the algorithms used here are not too complex, but there are projects which increase the tool-based support for the flight control team. For example, ATHMoS (Automated Telemetry Health Monitoring System) (O'Meara et al. 2016) was developed at GSOC using outlier detection and machine learning. The tool gives early indications of future anomalous behavior of telemetry parameters using past telemetry data and therefore supports the flight controller in grasping the situation on board.

The current telemetry display systems allow a customization of telemetry pages. They can be set up with different focus. It was already mentioned above, that procedure-focused displays support well the execution of flight procedures. The displays can be organized on subsystem level of the spacecraft, they can include graphical elements to allow an easier orientation for the user.

It is quite common to have overview pages, which contain high-level telemetry items and allow to navigate to more detailed pages for in-depth investigations.

In many cases, there are hard and soft limits for each telemetry item defined in the mission database (or even limits for the change rate of the item). The telemetry visualization is then enriched with a check against these limits and corresponding coloring of the telemetry item in case of a limit violation. It is also helpful to logically group limit checks of multiple telemetry items into additional status parameters. If this is done iteratively, the spacecraft's health status can be displayed via a few "summary alarms". If one of those indicates an alarm, the flight controller can then navigate along the "alarming chain" down to the telemetry item which initially went out of limit.

8.4.6 Commanding and Command Error Data Base

One prominent manifestation of errors made by the flight control teams are commands, which were sent erroneously.

In most cases, the negative effect is minor. However, the important step is not to lose the chance to learn from them. A first step towards this goal is the implementation of a command error database.

This database is only focusing on the human factor part of command errors on console, independently of any technical implication of the failure, which is handled via the standard anomaly handling processes.

The error handling culture is important for a successful use of such a tool. In spaceflight operations it should be clear to all parties on console, that no one is blamed for errors. Everybody accepts, that errors are happening as part of human work and is interested to minimize the effect. The quicker an error is reported and the more open potential impacts are discussed, the easier it gets to keep control on the operations. Nevertheless, the collection of errors in a database needs to be done with care. Confidentiality is needed for the database which is read-access restricted to key personnel—ideally, even the names of the flight controllers are not archived, but of course with a shift plan it is easy to reconstruct.

It must be clear that the only goal of the analysis of the errors is to improve the operational handling and be able to identify error patterns, both not linked to individuals.

A few control centers have set up such databases (Harris and Simpson 2016). Here some details of the command error database which was set up at GSOC for the Columbus project.

In the command error entry fields, contributing factors are requested from the operator. These are e.g. rushed, distracted, lost situational awareness, display problem. This information allows analysis when a reasonable set of data is collected.

The goal is to derive potential patterns from the command errors which happened. This could be week days (more errors on Mondays?), shifts (more errors during night shifts?), specific procedures ("easy" procedures or "complicated" procedures), specific periods (handovers, afternoons, biorhythm cycles, etc.). As a consequence, countermeasures could be considered and implemented.

8.4.7 *Training and Simulations*

Training and simulations are imperative elements in improving the team performance of a flight control team and in mitigating the risk of human failures with critical or catastrophic result.

In didactics, the learning objectives are clustered into three major domains (Bloom et al. 1956): The typical technical training is covered by the cognitive domain, practical skills are taught in the psychomotoric domain, and if the goal of training is a change in the mindset or the value system, we talk about the affective domain.

It is important to introduce CRM to the flight controllers not only in the cognitive domain: Here, the required knowledge is provided how to counteract human failures in the different areas which were discussed above. Also, the psychomotoric domain has to be addressed: The flight controllers need to learn the practical techniques (e.g., voice loop communications).

Equally important is the effect on the affective domain: The flight controllers need to acknowledge, that human errors will happen and that they can lead to dangerous situations and actions to protect against such failures. In fact, this attitude is one of the most important non-technical skills of a good flight controller.

An introductory CRM training can be conducted as a dedicated course and is now established in many large control centers.

For example, NASA has a well-established course (O'Keefe 2008; Pruyn and Sterling 2006). Also for the Columbus Control Center, a course on human behavior and performance is provided (Uhlig et al. 2011).

An excellent possibility to practice the various CRM aspects discussed before, are simulations, which are used for exercising the team work (Uhlig et al. 2012). Simulations are executing an ops-like scenario in its typical operations context. A flight control team is working on console with their operational tools, which are connected to a simulator or an engineering model of the flight segment. A dedicated timeline containing a scenario, which was selected according to the learning objectives of the simulation, is executed and corresponding failure scenarios are inserted by a simulation coordinator.

By that, communications by using voice loops, situational awareness and decision making by nominal and off-nominal situations can be trained. A dedicated instructor team monitors the performance of individuals and teams and provides feedback, so that all participants learn from the simulation cases.

8.5 Standardization

In the last years, it was commonly acknowledged that the consideration of the human factor in space operations is an essential component of a successful spaceflight program.

For the ISS project, a dedicated competency model (Bessone et al. 2008) was developed and is now applied at all major control centers for the International Space Station.

More generically, ESA has initiated a Human Dependability Initiative (HUDEP) for the exchange on human dependability topics. Participating parties are space agencies and control centers, universities, space industries as well as other organizations acting in high reliability environments, e.g., nuclear power plant operators.

Therefore, the HUDEP initiative looks at technical systems and projects with the aspect of humans in the loop. This includes project participants from design and test phase, production and maintenance, up to operators and users. The aim is to ensure exchange experiences between all involved parties with the goal to improve spacecraft design, manufacturing and operations.

As a result of the initiative, a Human Dependability Handbook (2015) was developed. It is published as part of the ECSS framework and provides best practices for implementing human dependability concepts in spaceflight. The scope is wider than the operational phase of a mission. It aims to extend the view on all project phases of a spaceflight mission.

The handbook provides a familiarization on human dependability aspects and discusses human performance principles. In detail the handbook covers:

- Principles of human dependability, i.e., human dependability concept and human role in the system
- Human dependability processes, covering human error analysis, human error reporting and investigation
- Implementation of human dependability in the system life cycle

For the last bullet, a systematic approach for the implementation of human dependability in a system life cycle was analyzed. All typical project phases A until F (see Fig. 4.1 ECSS phase model) are considered with this view, including concept and design definition until operations. For each phase, human dependability activities are described in detail with specific items for objective, inputs, tasks and outputs.

8.6 Conclusion

In this chapter we have discussed various aspects of human dependability in high reliability environments. It is an important factor for sustainable success, especially in complex and demanding situations.

Crew resource management (CRM) deals with the interactions within a team. Hence, it is of major importance for the set of flight controllers which runs operations in a control room. Therefore, the principles of CRM are applied in spacecraft operations, both for manned and unmanned spaceflight: Communication is a key success element within a team. It was demonstrated how verbal and written communication can support operations in an optimum way. It is crucial for a flight control team to be aware of the situation at any given point of the mission and to ensure, that a common mental model of the situation on board is shared between the team members. A fast and structured way of decision making is required, the corresponding hierarchical structures need to be established and corresponding processes and supporting products need to be in place. Last but not least, teamwork and good leadership skills of the flight director are of major importance.

The CRM principles need to be respected in project preparation with focus on ops products generation, and exercised during training sessions. The goal is to reduce the risks on console during operations.

By taking CRM topics serious in all project phases, the human factor can be turned from being a risk to becoming a significant gain.

References

Bessone L, Schmidt L, Tomi L, Inoue N, Spychalski A, Sipes W (2008) International space station human behavior and performance competency model. NASA/TM-2008-214775

Bloom BS, Engelhart MD, Furst EJ, Hill WH, Krathwohl DR (1956) Taxonomy of educational objectives: the classification of educational goals; Handbook I: cognitive domain. David McKay Company, New York

Cooper G, White M, Lauber J (eds) (1980) Resource management on the flight deck. In: Proceedings of a NASA/industry workshop

Department of Transport (eds) (1990) Report on the accident to Boeing 737-400 G-OBME near Kegworth, Leicestershire on 8 January 1989. AAR-4/90

Endsley MR (1995) Toward a theory of situation awareness in dynamic system. Hum Factors 37:32–64

Flin R, O'Connor P, Mearns K (2002) Crew resource management: Improving team work in high reliability industries. Team Performance Management. Int J 8:68–78

Foushee HC, Helmreich RL (1988) Group interaction and flight crew performance. In: Wiener EL, Nagel DC (eds) Human factors in aviation. Academic press, San Diego, CA, pp 189–227

Gaddy C, Wachtel J, Swezey R, Salas E (1992) Team skills training in nuclear power plant operations. In: Teams: their training and performance. Ablex, Norwood

Hagemann V, Kluge A, Ritzmann S (2011) High responsibility teams – Eine systematische Analyse von Teamarbeitskontexten für einen effektiven Kompetenzerwerb. Psychologie des Alltagshandelns 4(1):22–42

Hales BM, Pronovost PJ (2006) The checklist—a tool for error management and performance improvement. J Crit Care 21(3):231–235

Harris S, Simpson BC (2016) Human error and the international space station: challenges and triumphs in science operations. In: Proceedings of 14th International Conference on Space Operations

Helmreich RL, Foushee HC (1993) Why crew resource management? Empirical and theoretical bases of human factors training in aviation. In: Wiener EL, Kanki BG, Helmreich RL (eds) Cockpit resource management. San Diego Academic Press

Hörmann HJ (1994) FOR-DEC—a prescriptive model for aeronautical decision making. In: 21. WEAAP-Conference, Dublin

Human Dependability Handbook (2015) ECSS-Q-HB-30-03A

Jones D, Endsley M (1996) Sources of situation awareness errors in aviation. Aviat Space Environ Med 67:507–512

Kanki B, Helmreich R, Anca J (2010) Crew resource management. Elsevier Inc., San Diego

Kuch T, Sabath D (2008) The Columbus-CC—operating the European laboratory at ISS. Acta Astronaut 63(1–4):204–212

Ministerio de Transportes y Communicaciones (ed) (1978) Joint report. K.L.M.-P.A.A., 12 July 1978

Moshansky VP (ed) (1992) Commission of Inquiry into the Air Ontario crash at Dryden, Ontario. Final Report

NASA Space Transportation System/International Space Station (1998) Payload safety review and data submittal requirements. NSTS/ISS 13830 (Rev. C)

National Transportation Safety Board (ed) (1979) Aircraft accident report united airlines—McDonnell-Douglas, DC-8-61. N8082U

National Transportation Safety Board (ed) (1988) Aircraft accident report—Aloha Airlines, Flight 243, Boeing 737-200, N73711, near Maui, Hawaii, 28 Apr 1988. NTSB/AAR-89/03

Nisbett RE, Ross LD (1980) Human interference: strategies and shortcomings of social judgement Englewood Cliffs. Prentice-Hall, NJ

O'Connor P, Flin R (2003) Crew resource management training for offshore oil production teams. Saf Sci 41:591–609

O'Keefe W (2008) Space flight resource management training for international space station flight controllers. In: Proceedings of AIAA SPACE 2008 conference and exposition, San Diego
O'Meara C, Schlag L, Faltenbacher L, Wickler M (2016) Automated telemetry health monitoring system at GSOC using outlier detection and supervised machine learning. SpaceOps Conference, AIAA 2016-2347, Daejeon
O'Connor PE, Hahn G, Salas E (2009) The U.S. Navy's crew resource management program: the past, present and recommendations for the future. In: O'Connor PE, Cohn JV (eds) Human performance enhancement in high-risk environments: insights, developments, and future directions from military research. ABC-CLIO, pp 90–106
Okray R, Lubnau II T (2004) Crew resource management for the fire service. Pennwell, Tulsa, Oklahoma
Pruyn P, Sterling M (2006) Space flight resource management: lessons learned from astronaut team learning reflections. SOL J Knowl Learn Change 7(2):45–57
Rall M, Gaba DM, Dieckmann P, Eich CB (2014) Human performance and patient safety. In: Miller RD et al (eds) Miller's anesthesia. Elsevier, pp 106–166
Rasmussen J (1987) Cognitive control and human error mechanisms. In: Rasmussen J, Duncan K, Leplat J (eds) New technology and human errors. Wiley, Chichester
Reason J (2000) Human error: models and management. BMJ 320(7237):768–770
Ritzmann S, Kluge A, Hagemann V, Tanner M (2011) Integrating safety and crew resource management (CRM) aspects in the recurrent training of cabin crew members; Aviat Psychol Appl Hum Factors
Schulz von Thun F (1981) Miteinander reden: Störungen und Klärungen. Psychologie der zwischenmenschlichen Kommunikation. Rowohlt
Shannon CE, Weaver W (1949) The mathematical theory of communication. University of Illinois Press, Urbana, IL
Uhlig T, Özdemir K, Sabath D (2011) Training concept of the Columbus flight control team. Proceedings of IAC
Uhlig T, Hock D, Özdemir K (2012) Practical exercises as part of flight controller training. Proceedings of IAC

Thomas Uhlig received his physics diploma and his Ph.D. from the University of Regensburg/Germany, where he worked on electromagnetic imaging of nanostructures. After working as scientist at the Argonne National Laboratory in Chicago he changed into the spaceflight business and to the Columbus Control Center in Oberpfaffenhofen, where he first worked as Payload expert on the Space Shuttle mission STS-122, then for several years as Columbus Flight Director. He then took over the training for the European ISS flight controllers, before he was assigned as team lead for the geostationary satellite team at GSOC. In between, he was working as expert for functional safety, first at BMW, then at Heidenhain GmbH before returning to DLR in 2022. He was awarded with the Helmholtz price 2005 of the Physikalisch-Technische Bundesanstalt. He is author of various scientific articles.

Gerd Söllner is Project Manager for the Columbus Project at the German Space Operations Center (GSOC). He has entered DLR in 2001 and has prepared and executed the Columbus 1E mission as Lead Flight Director. Thereafter, he has gained several years of console experience as Flight Director at the Columbus Control Center and has led the Columbus Flight Director Group. In the field of human dependability, he has contributed to the ESA Human Dependability Workshop since 2011.

Part III
Communication and Infrastructure

Chapter 9
Design and Operation of Control Centers

Marcin Gnat and Michael Schmidhuber

Abstract In this chapter we deal with different aspects of the design of a control center. First, the necessary infrastructure is analyzed. Then, the design of the local control center network is examined, followed by the required software. Various aspects of the design for the facility itself (the building), various office and operational subsystems, and IT hardware are discussed.

9.1 Introduction

This chapter describes the design aspects of a typical mission control center (MCC). The German Space Operations Center (GSOC) is used as an example.

The mission control center is, as the name suggests, the fundamental facility of a space mission. It is the central point where all data and management information about the spacecraft is collected (Fig. 9.1). This data is received, reviewed and processed, decisions are made, and in the event of an emergency, the appropriate procedures are performed to restore the mission to nominal conditions. The operation of the MCC is determined by its design, which defines its capabilities, flexibility and robustness. The operation of the MCC is also defined by the staff, primarily the flight operations team, but also by all staff responsible for interfaces and infrastructure. Finally, the design of the MCC must meet customer requirements and provide a safe environment for spacecraft operations. Therefore, this includes not only purely technical solutions, but also the respective environment for the people working there.

A so-called multi-mission operations concept enables greater operational flexibility and easier allocation of new missions. The decision whether the MCC is designed as multi-mission or single-mission should be made early in the design process, as it has far-reaching effects on the overall design, especially on the IT infrastructure and the network. The operational concept (multi-mission or single-mission) has also a major impact on the assignment of personnel, especially the mission operation teams.

M. Gnat (✉) · M. Schmidhuber
German Space Operations Center (GSOC), German Aerospace Center (DLR), Oberpfaffenhofen, Germany
e-mail: marcin.gnat@DLR.de

F. Sellmaier et al. (eds.), *Spacecraft Operations*, Springer Aerospace Technology,
https://doi.org/10.1007/978-3-030-88593-9_9

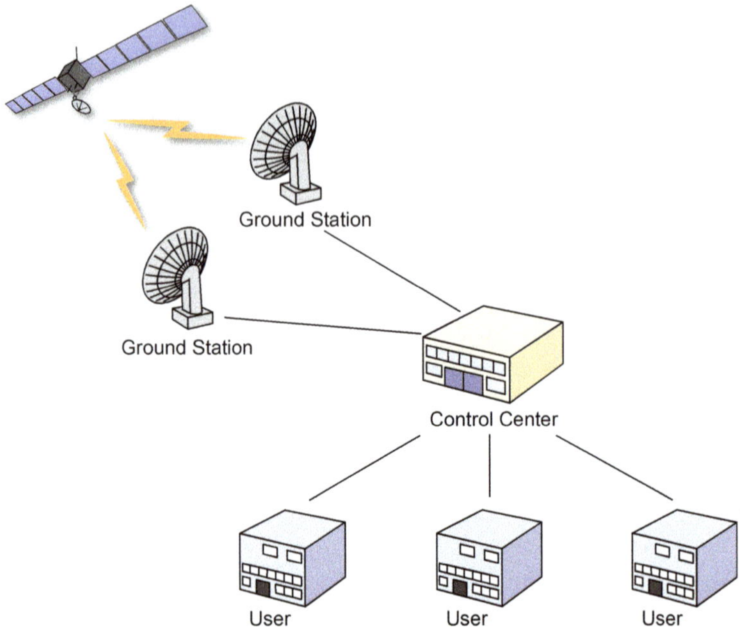

Fig. 9.1 Position of the mission control center in the ground segment

Our focus here is on a multi-mission environment based on the GSOC design. A multi-mission design is usually more complex, which is the price of greater flexibility. However, there are also some situations where a multi-mission design is not appropriate. Simultaneous operations of missions with little or no similarity (e.g. due to different requirements, security aspects) cannot be easily combined in a multi-mission environment. For example, it would be difficult to integrate a scientific mission where data is more or less publicly available, with a military mission with very stringent security requirements.

9.2 Infrastructure

The first task in planning the infrastructure is identifying a suitable location for the control center building. Several important aspects have to be taken into account: Unfavorable geological conditions should be considered, e.g., geologically unsafe zones (earthquakes) and areas with frequent flooding should be avoided if possible. Appropriate measures should be taken to make the control center less vulnerable to natural or technical conditions. This can be achieved, for example, through a concept of redundancy at various levels. A redundant power supply is essential. Uninterruptible power supplies (UPS) can provide constant power to all MCC systems during short power outages or fill a time gap until diesel generators start up. The latter can also provide power for several days. A fully independent backup control center provides maximum redundancy, especially when set up with site diversity in mind. This can range from a separate building providing redundancy in the event of locally concentrated failures, such as a building fire, to a remote location that provides redundancy even in the case of large catastrophic events, such as floods, earthquakes and the like.

In terms of redundancy in the communications infrastructure, it is useful if the control center is located near a major urban center with multiple independent connections to the telecommunications networks. A separate communication antenna on the MCC provides additional independence in case the terrestrial communication lines are interrupted.

The maintenance aspect is also important. Occasionally, it is necessary to replace parts of the equipment like e.g. hard disks, switches, workstations. In most cases, this can be done without affecting operation, but sometimes it is necessary to shut down parts or even the entire MCC. In such a case, planning and coordination is essential. Affected projects need to be informed, maintenance work carefully planned, and backup solutions discussed (e.g. the question, what happens if maintenance takes longer than planned or is unsuccessful). In addition, it should be noted that some equipment can experience significant degradation if its power is repeatedly turned on and off. Many electronic devices are very sensitive to such power cycling. This power interruptions could result in damage to the unit, which has to be repaired by the original maintenance activity. It should also be remembered that cable ducts should be adequately sized when the building is constructed. They should also provide good access for maintenance and expansion work. Finally, things like the sizing of rooms, doors or elevators should also be well planned, because it may well be that the computer racks need to be moved more often than originally expected, especially when the hardware is renewed for the first time.

The MCC facility must meet security standards defined by law, company guidelines and project requirements. An access control system includes the basic technical infrastructure (such as security doors, door key management, corresponding locking policies) as well as more complex elements such as access terminals (with key cards) with corresponding key card management, surveillance cameras in critical rooms and corridors, and alarm systems (intrusion alarm). In addition, security personnel must be available on site at all times. Depending on the characteristics of the projects in the facility, more or less strict visitor control may be implemented. Typically, visitors will not have access to the network and data processing facilities at all, while conference or exhibition areas with satellite replicas on display can be treated as low-security zones.

The facility must be equipped and maintained for the safety of personnel. This includes emergency exits and signage, fire and smoke alarms (possibly connected to the local fire brigade) and various types of fire extinguishing systems. The latter may be essential, especially for larger computer or UPS systems. A central fire extinguishing system may be installed (e.g. using argon inert gas extinguishing or similar systems to prevent damage to the equipment). Finally, fire procedures must be developed and prepared. Especially in case of spacecraft operations, procedures must be precise and should clearly define under what circumstances the control room must be evacuated, when and if running systems should be shut down, and how they are subsequently restored.

9.2.1 Control Rooms

The control rooms are the heart of the MCC. Depending on available resources and needs, there may be multiple control rooms. They may vary in size and serve different purposes. They may be assigned to certain space missions permanently or only for specific phases of a mission such as a launch and early orbit phase (LEOP). Most control rooms will be equipped with air conditioning, not only for human comfort but also for computer hardware. As mentioned earlier, the room must be equipped with emergency exits and provide adequate space for the operating team and additional equipment such as printers, voice and video systems. A copy machine must also be available, but possibly outside the control room itself, as it generates significant noise that may not be desirable inside.

Control rooms should allow for changes in configuration, as space missions may change the layout of control rooms during their operational life. This requires forward planning of elements, such as cabling for network, telephone, voice and power. One of the possible solutions here is a flexible design that takes advantage of the high-capacity network backbone, remote desktop and virtualization. These include hosting virtual machines for multiple projects in the data center, services that enable the deployment of new or cloning of existing virtual machines, load balancing and automated redundancy switching for project systems. In addition, network connectivity is to be established between all control rooms and the data center, including services such as project separation or access control for specific user groups. Finally, the consoles are all identical and allow only remote desktop connection (no local applications). This can be implemented through desktop PCs or zero clients.

The structure of the control rooms described above simplifies many maintenance tasks. Maintenance can also be carried out in parallel with operation without disturbing the actual projects. The advantages can be considered here, but there are also some additional points to consider: The high flexibility and the availability require more detailed planning in order to manage the resources wisely and reach the best cost-benefit ratio. The multi-mission concept also requires a rethinking of the use of control rooms, as they can no longer be occupied exclusively by individual projects.

Consoles should have enough space for work, both digital and traditional, for conventional tasks such as taking notes or viewing documentation. Telephone and voice system should have their place, as well as—indispensable today—several monitors for displaying different systems. Not to be neglected is a comfortable seat, as the operator may be expected to spend the entire shift from eight to ten hours on it. Consoles usually require access not only to the operational systems, but also to the office network (for e-mail, documentation) and the internet [e.g. to allow representatives of the satellite manufacturer or customers to access their company network via virtual private network (VPN)].

Facilities such as restrooms or the coffee kitchen should be located near to the control room to minimize the time spacecraft operators must spend outside the control room.

In addition to the central control rooms for the spacecraft, there are several other rooms that are typically needed during operations. The flight dynamics team typically has its own control room for the LEOP phase to coordinate closely with the flight team and exchange data products quickly. The network and system control room is a communications facility that connects all incoming and outgoing connections to and from the MCC and provides voice communication with the outside world for monitoring and coordinating the ground station network or other key external operational interfaces. Satellite manufacturers or customers may also require special rooms where they can conduct important offline activities in close proximity to mission operations. These rooms may require special access control as well as specific network connections that allow access to either the internet or the operations network.

9.2.2 Public Space in the Control Center

Space missions can attract a lot of public attention, but this should not interfere with operations. Visitor areas with large glass windows allow a direct view into the control room and give a sense of spacecraft operations. For situations or missions that do not want to be publicly visible, it should be possible to cover the windows with blinds or a similar solution (see Fig. 9.2).

Control centers require sufficient office space for both their own staffs and their guests. Depending on the orientation of the control center, different approaches can be taken to the design of areas for public access, exhibition, and catering. Many facilities for military or communications purposes will require only relatively small exhibition and catering areas. Large national control centers that conduct LEOPs and many public relation activities will require large areas for interested members of the public, as well as press and meeting rooms and catering for visitors. Space missions are still on the cutting edge of technology and fascinate many people. A control center provides a unique opportunity to generate public interest in space missions and technologies. It should therefore provide appropriate resources to inform and educate the public.

Fig. 9.2 Control room at the German Space Operations Center (GSOC)

9.2.3 Server Rooms and Computer Hardware

Before we look at the network in more detail in the next chapter, let's briefly review some other subsystems and elements that are no less important but are not as much in focus here.

The server room or data center is equipped with server, routing and switching facilities. When designing the system, special attention must be paid to the servers, since their reliability, flexibility, and capacity are largely determined by the effort required to maintain and expand the system. Until recently, the design was determined by the use of powerful servers. However, this concept was not very flexible; a defective element of the server caused its total failure and the restoration of full operation took a very long time.

With increasing flexibility and redundancy requirements (combined with the increase in the number of servers), the focus shifted to so-called blade servers. They can be tightly packed and support the growth of applications. At the same time, they offer backup options and easy replacement of defective modules. Currently, however, virtualization is the trend. Virtualized application servers are even easier to maintain. They can be seen as a single available space for computing and storage resources that can be used very flexibly. In the days when physical servers were used, running ten applications meant having 20 physical servers (including backup). However, the same situation in a virtualized environment requires only two physical machines (prime and backup), each hosting ten virtual applications. The decision of which technology to use must be made on a case-by-case basis. Each technology has its advantages and disadvantages. For example, virtualization, with all its advantages, is not very suitable for applications with high network traffic since a single physical port of server hardware must be shared. However, the virtualization principle is a prerequisite to further improve the maintainability of control room hardware (control consoles). This is achieved by using thin-client terminals. Since these thin-client terminals do not contain any local hard disks or other moving parts, their reliability and expected service life are significantly higher than those of conventional PCs.

Other benefits include lower power consumption and heat dissipation, which in turn results in significantly lower cost for providing adequate air conditioning for the computer hardware.

Data storage is also an important issue. For most office applications, hard drives from office computers and possibly network attached storages (NAS) may be sufficient. Spacecraft data and documentation requires a different approach, as security, continuity, and collaboration issues must be addressed. Again, several solutions can be considered, ranging from high-capacity NAS-like storage, to storage area network (SAN) for short- and long-term storage, to data safes and long-term archives (in the form of magnetic tapes with automatic update mechanisms). The latest developments in cloud systems and cloud storage can be useful here, and while they may not be suitable for all projects, they can be a real asset for some.

9.3 Control Center Network

In this chapter, we will look at the design, security and maintenance aspects of the control center local area network (LAN). The computer network is the backbone that connects all subsystems within the MCC. It is the connecting element that simultaneously protects certain systems from unauthorized access.

9.3.1 Network Topologies

Figure 9.3 shows the concept of the network connection between the control center and the ground station. Two separate paths can be identified, each tailored to their specific requirements. On the left is the so-called office path, which connects the office LANs of the control center and the ground station (e.g. to allow teams at both sites to exchange documents). This connection is realized with the help of a VPN over the campus network. On the other hand, there is a highly reliable, redundant synchronous digital hierarchy (SDH) connection for the real-time satellite data that links the operational LANs of both sites. The same data router also covers the multiplexed voice over IP (VoIP) traffic on the data link to provide highly reliable voice communication for MCC and ground station operators. The above example shows that there are two independent network branches, the office LAN and the operations LAN. This separation is also implemented in the network structure within the

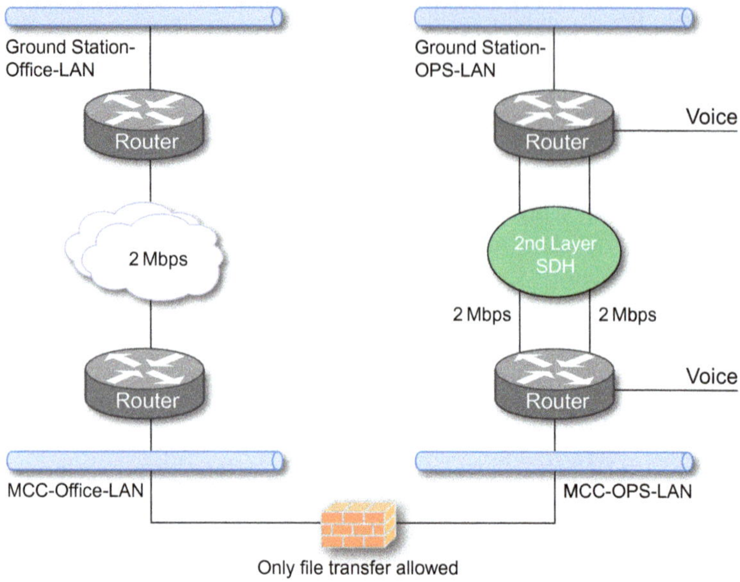

Fig. 9.3 Example of connections between control center and ground station

control center itself. It reflects the solution introduced to meet the security requirements. The operational LAN (also called OPS-LAN) is the highly secure network area. It is physically separated from other networks and only accessible from outside to a very limited extent. File transfers are only allowed through certain file transfer protocol (FTP) servers in the so-called demilitarized zone (DMZ), and real-time connections from ground stations are also only allowed through a firewall to trusted sites with similar high-security operational networks. The aforementioned DMZ is the typical partition between different LANs. It consists of another network section with two access points protected by firewalls. DMZs typically contain only firewalls and FTP servers, but in case of a specific purpose, the outermost DMZ may also contain an application server to provide certain MCC services to the outside.

In modern control centers, where a variety of services are made available to external users, this functionality can become a very serious matter. The Office-LAN is a typical part of the network used in MCC offices and is intended for general office work such as viewing and editing documentation and accessing e-mail. The office LAN has internet access, but this functionality is restricted (for example, the internet can be accessed from the office LAN, but the Office-LAN is not accessible from the internet, so it is only one-way access). The Office-LAN is managed; only registered devices are allowed to access the network and IP addresses are maintained centrally by the MCC network administrator.

Another network shown in the following figures is the so-called ops-support-LAN. This is not necessarily needed for every control center, but contains some supporting systems that require a little more access to the outside world, but are also very important for the operational system (i.e., to provide command files). In the following example, the ops-support-LAN contains flight dynamics and mission planning systems.

Each area has its own service and security segments that host proxies, virus scanners, and authentication, name, time, and file servers. Clients cannot open direct connections to hosts outside their LAN area; connections can only be made through the proxies in the service segment. These proxies are directly connected to the virus scanners in the DMZs, where incoming and outgoing traffic is scanned.

Figure 9.4 shows an example of an MCC network. The real-time TM/TC connections with the ground stations and external partners are shown in the figure below (OPS-LAN). Important operational files are also transported on these network interfaces. To transfer a file from the OPS-LAN to an external customer over the internet, it must pass through several firewalls and DMZs before being made available on an FTP server in the outermost DMZ. This may be quite cumbersome, but it is important for security purposes.

9.3.2 Network Technologies

The MCC network is based on the TCP/IP protocol and the underlying Ethernet. The type of cabling depends on the resources available, but in principle fiber optic cabling offers greater potential for future expansion and is also an important factor in preventing unauthorized eavesdropping. Typically, it also offers higher bandwidths and therefore higher data rates, so future upgrades only require replacing equipment such as routers or switches. Replacing the cabling itself, on the other hand, is very expensive and can require a great deal of effort. Devices with interfaces for fiber optic cabling are also usually much more expensive than conventional cables. Therefore, it can make sense to implement a hybrid solution for office equipment (PCs or laptops) with fiber optic between large hubs and connections to end users via copper cables.

As already mentioned, the control center network forms the backbone of all operating systems and is strictly operation-critical. This requires the appropriate support from specialist personnel. Depending on the size of the control center and thus also of the network and the projects supported, it may be necessary for the corresponding network support personnel to be permanently available, either through shift work or on-call duty.

Another aspect of the network is its maintenance. In many cases, it will be possible to perform maintenance with minimal or no impact on ongoing operations. This can be the case, for example, when equipment needs to be replaced. This can be done in the time between two satellite passes. But even then, and even more if there is a real impact on operations (e.g. a loss of system availability for a few hours), proper planning and preparation is required (e.g. backup arrangements, spacecraft autonomy).

The modern network design pays particular attention to the separation of projects, virtualization, data center functionality, scalability and flexibility. In the future, the

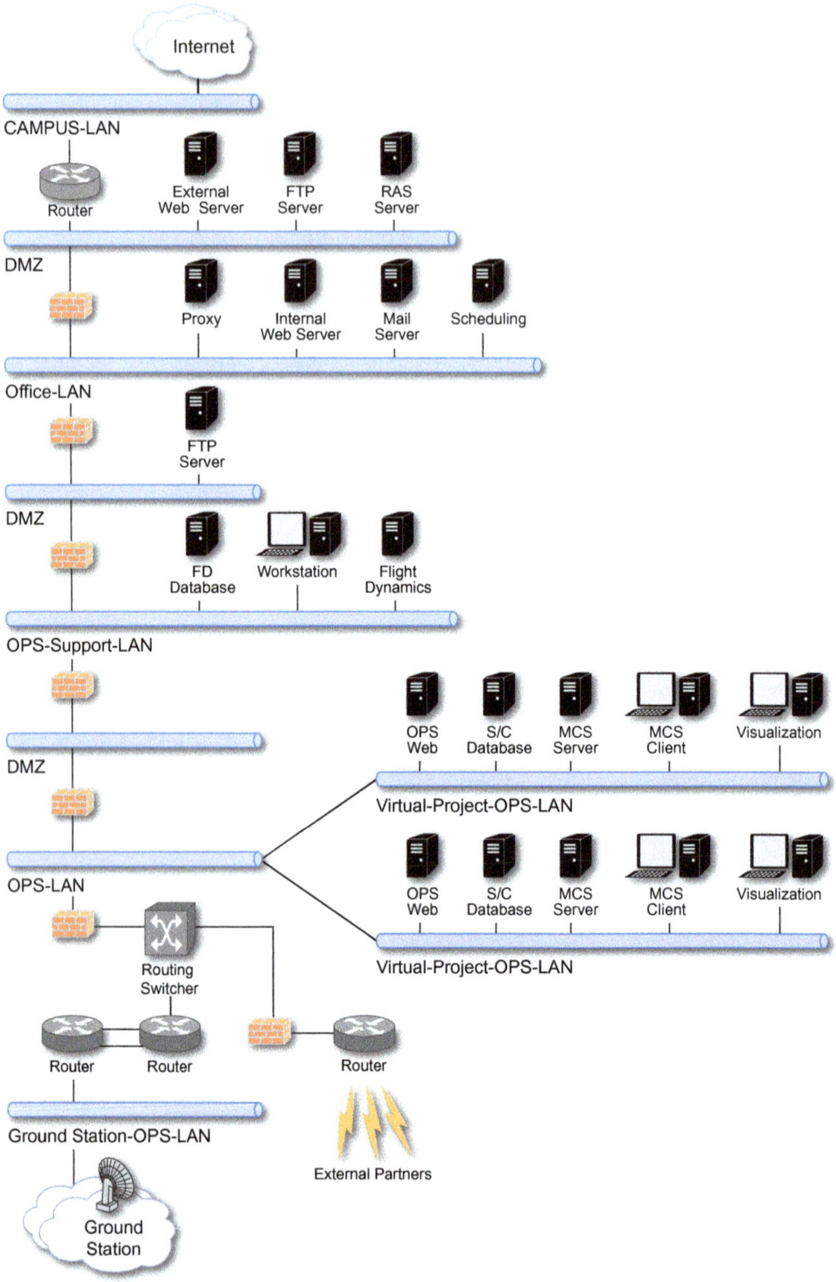

Fig. 9.4 Example of a control center network

system should also be able to form a platform for technologies such as software defined networking (SDN).

The previous Layer 2 design is now implemented by the Layer 3 design. This reduces the number of components to be configured and standardizes and automates various tasks. One important aspect is the establishment of centrally managed virtual switches (so-called virtual distributed switches or VDS), which replace the previous local virtual switches on the individual physical servers. The advantage is that network definitions and properties can be managed and changed centrally instead of making such changes locally on a large number of systems. In addition, the VDS offer significantly more options for error analysis and monitoring in the event of a problem.

9.4 Control Center Software

9.4.1 General

This section presents some prominent examples of specific software used in a control center. Their generic functionality is explained using specific applications as example. Standard programs and software packages such as office software and operating systems are not discussed here.

The software of a control center is specific and often custom-made, although there are a number of commercial software packages on the market that support satellite operations. They are usually not cheap, as there is a vast customer base, and it is always necessary to consider how they fit into the specific environment of the particular control center and satellite project. The software must not only process telemetry or perform orbit calculations, but also provide interfaces to other packages or systems. Some of these interfaces may be proprietary, making it impossible to use off-the-shelf products.

The software of the ground station and the control center, and in particular, their interfaces, can be divided into real-time and offline in analogy to the data flow paths.

Thus, a distinction is made between real-time data flow, which flows from the spacecraft via the ground station to the MCC, and the file transfer, which transports large amounts of data in the form of files. The latter are also referred to as "products". They contain data such as event forecasts, converted telemetry excerpts, inputs from external parties, ground station forecasts, etc. They must be exchanged between internal and external partners. Due to the large number of files, the need for timely delivery and reliability, most transfers should be automated. They are more or less asynchronous and not as time-critical as the real-time data stream.

Real-time data are often referred to as "online" and file transfers as "offline". The diagram in Fig. 9.5 shows these two types of information flow between ground segment subsystems. For both types of data flows, there is special software that generates, processes, transmits, and converts them. Online communication is realized

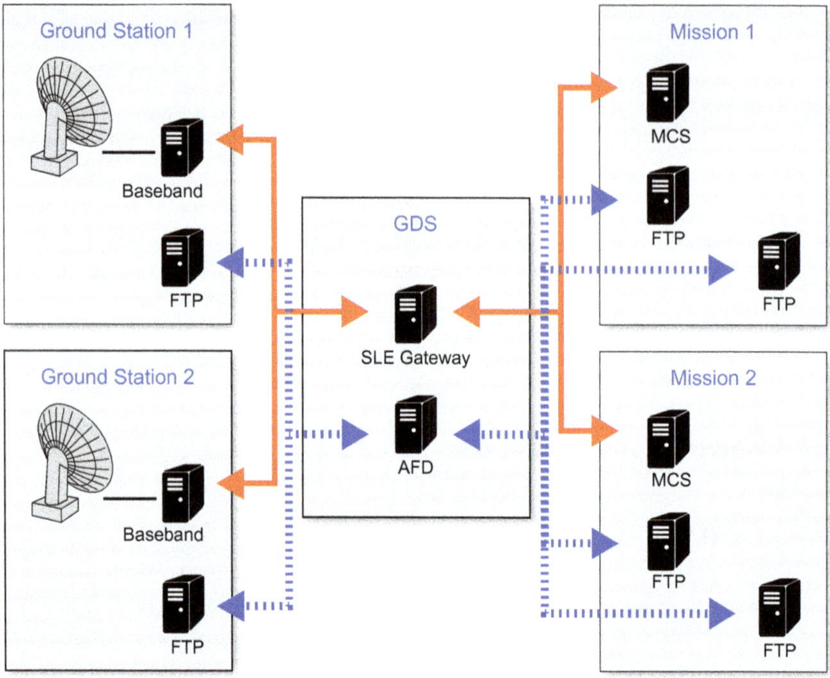

Fig. 9.5 Online (*solid lines*) and offline (*dotted lines*) data transfers

with four main elements in a chain. The baseband software is usually installed at the site of the ground station (see Fig. 3.1). It performs basic tasks at the lowest level, such as frame synchronization, error correction, or time stamping. The service provider delivers the data from the ground station to the appropriate MCC. Currently, the Space Link Extension (SLE) is used in most cases and is described in more detail in Sect. 9.4.2. The service user acts as a counterpart to the ground station on the MCC side. The SLE application software receives SLE data and provides it to the monitoring and control system (MCS) in an appropriate format. Finally, the spacecraft M&C system (also called TM/TC processor) provides the actual data processing and user interface for the flight controllers.

Offline communication also consists of four components. The generation and processing systems produce or use files; these systems may also include the MCS system mentioned above. There are dedicated storage systems that provide both the necessary hardware and the appropriate data management software. In particular, this includes all databases necessary for operation. There may be automated file transfer software. An exemplary implementation, the automated file distribution (AFD) software, is described in more detail in Sect. 9.4.3. Since security plays an important role in expensive and sensitive satellite missions, firewalls and virus scanner software are also used.

9.4.2 Space Link Extension Gateway System

Special communication links must be used to ensure reliable communication with ground stations, some of which are located far away. This includes the communication lines that must be ordered and leased, e.g. dedicated lines with very small aperture terminals (VSAT). In most cases, commercially available lines are being used. Protocols tailored to the specifics of space missions must also be used. One commonly used protocol is Space Link Extension (SLE). SLE is a standard according to the Consultative Committee for Space Data Systems (CCSDS) and is widely adopted by agencies and companies operating ground stations as it provides good interoperability. Unlike previous solutions, it can provide cross-support without the need to customize interfaces for each mission and customer. SLE is based on a client-server architecture and enables the transmission of telecommands and telemetry, which are encapsulated in SLE packets and therefore can be transported over the wide area network (WAN) (Fig. 9.6).

As already mentioned, the SLE is server-client based. The role of the server is taken over by the so-called SLE service provider. The service provider is located at the ground station and provides the services associated with that station upon request. These services are called forward command link transfer units (FCLTU) for telecommand and return channel frames (RCF) or return all frames (RAF) for telemetry. These services are described in detail in the relevant CCSDS standards (see Table 10.2 in the next chapter).

The SLE user is located in the control center on the opposite side of the network. It manages the above services, performs all necessary protocol conversions and acts as an interface to the satellite monitoring and control (M&C) system. For example, GSOC uses the SLE switch board (SSB), which is capable of receiving telemetry

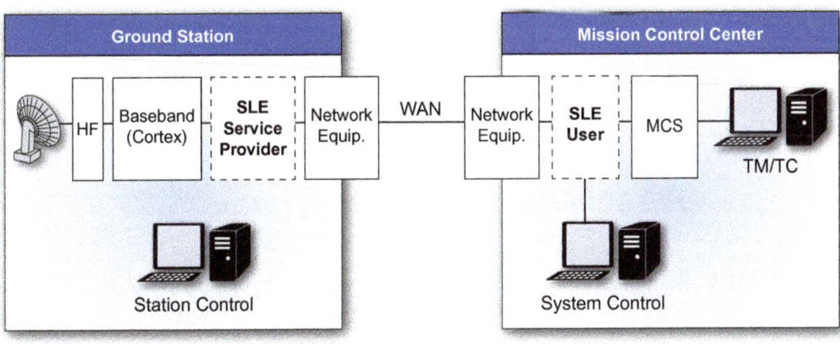

Fig. 9.6 Communication between the ground station and the mission control center over a WAN can be accomplished with the SLE gateway. For this purpose, the ground station equipment and software such as the RF and baseband connect to the SLE service provider connected over the network, and its equipment connects to the SLE user on the MCC side, which in turn connects to the MCC software and hardware such as the monitoring and control software

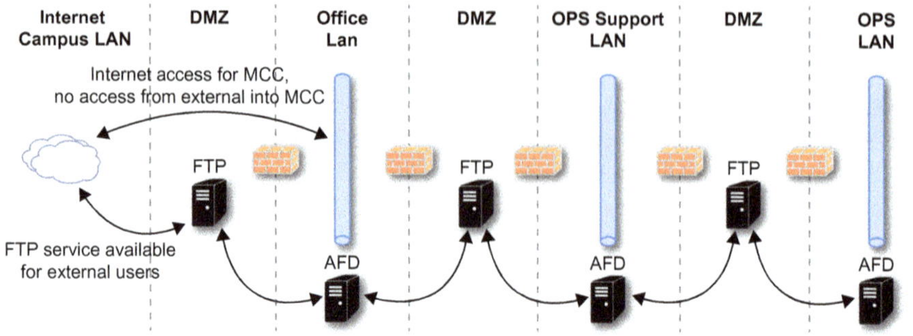

Fig. 9.7 AFD configuration in the GSOC network

flows from different stations in parallel and sending them to different MCS instances. Alternatively, it receives telecommands from the MCS software, converts them to SLE format, and sends them to the appropriate station.

9.4.3 File Distribution Subsystem

File distribution is a key element in the offline communication described above. GSOC uses the automated file distributor (AFD), an open source tool developed by the German Weather Service and being commonly used in various areas. This system is used by all GSOC satellite projects. The system is in a multi-mission environment, so each project defines its own file transfer matrix that serves as input for the AFD configuration. The matrix defines what type of files are to be transferred, from where to where, and how often. Once the configuration is activated, the AFD system starts monitoring the defined directories and performs the transfer completely autonomously. AFD is particularly useful for complex network structures, as shown in Fig. 9.7 for the three LANs used at GSOC.

9.4.4 Spacecraft Monitoring and Control System

As stated above, the spacecraft M&C software package is tasked with the receiving and unpacking the telemetry, processing and displaying the data, processing and encoding the telecommands, and sending them out through its interfaces. It is the central software component used by the flight operations team in the control room of the control center.

In many cases, it is a monolithic application, but designs with separate components for different tasks are also in use. Over the past few decades, a number of different systems have been developed and are available on the market. In the field of M&C

systems, there are now efforts to standardize them. The European Space Agency (ESA), for example, has just commissioned the development of a European Ground Systems Common Core (EGS-CC).

In many control centers and especially in LEOP operations, the mission data must be available at different workplaces at the same time. Therefore, a server/client functionality is usually included.

As shown in Fig. 9.8, there are two basic data flow paths in an M&C system. On both paths, dedicated interfaces are needed to establish communication with the data user interface, which in turn ensures the connection with the ground station. Telecommands, usually grouped in the so-called command stacks, must be processed until they can be sent out over the interface: They have to be encoded and packetized. On the other hand, incoming telemetry must also be processed by the monitoring and control system: The packets must be "opened" and their content must be mapped to the original parameters. This raw telemetry cannot yet be efficiently represented to the flight controllers, since it is only a bit pattern that must be first calibrated to appear, for example, as meaningful physical values such as temperatures or currents.

Finally, there is the possibility to perform an automatic threshold check of selected telemetry parameters: Their values are compared with predefined thresholds, and in some cases more sophisticated mathematical operations are performed. The result of this check can be either just the corresponding telemetry item highlighted on the

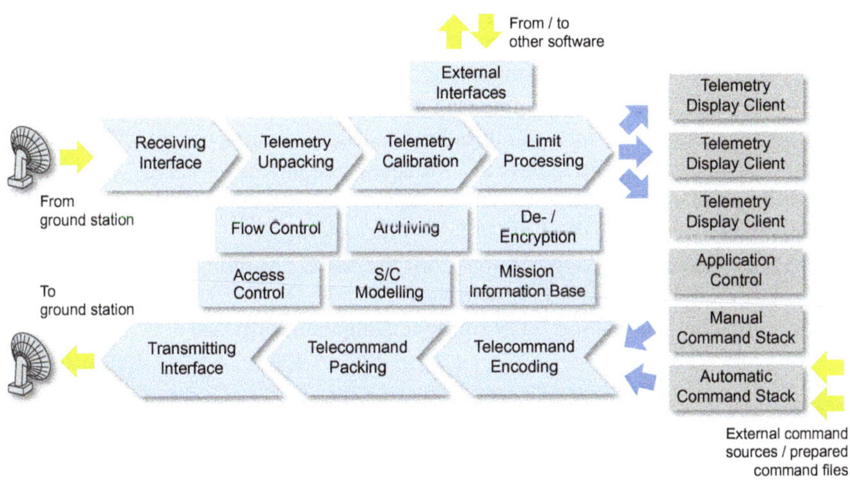

Fig. 9.8 The functional components of a typical monitoring and control system (MCS). Two data streams can be distinguished: Coming from the ground station, some processing steps are necessary before data can be displayed on a telemetry client. After reception at the interface of the MCS, the telemetry packets have to be unpacked, the content calibrated and appropriate limit checks performed. In the opposite way, the telecommands of the so-called command stacks have to be coded, packets generated and finally sent via a defined transmission interface. Other essential components are also listed

display, an alarm for the flight controller, or even an automated, defined response from the system.

The central component of the M&C that enables all the processing described above is the spacecraft's database, also called mission information base (MIB). It contains the definition of the telemetry and the telecommand streams, the calibration information, and the limit definition. Although most M&C systems are developed along standards and intended for use in all types of missions, it usually requires some effort to customize the software for each mission. Ultimately, M&C must reflect the capabilities of the spacecraft's onboard data handling system, as described in Chap. 19.

After the spacecraft is acquired, the satellite begins transmitting telemetry and the ground station antenna receives it. The ground station performs demodulation and decoding; initial error checks and possibly error correction are applied to the data stream. All received data is stored locally in either short-term or long-term archives (depending on the ground station and its capabilities). Furthermore, the data stream is made available to the MCC via the WAN interface. There, the data is received, further decoded, further error correction is performed, and finally every part of the current telemetry is processed, analyzed, stored, and a portion of it is displayed on the spacecraft operators' console for direct viewing.

Commands for the spacecraft go in the opposite direction. Unlike telemetry, which streams almost continuously during contact and often contains redundant or repetitive information, commands are sent only using special operating procedures, as described in detail in Chap. 5. Telecommands are sent from the M&C System in the MCC, packed into appropriate transmission protocols and transmitted over WAN to the ground station. In the meantime, the ground station must have established the uplink, which is defined as a stable radio connection with spacecraft. During this time, the ground antenna transmits and the spacecraft antenna receives. To compensate for frequency fluctuations caused by the Doppler effect, a sweep must often be performed.

The telecommand packets coming from the MCC are modulated to the carrier frequency and radiated toward the satellite.

For more details about the MCS software components, please refer to Chap. 12.

9.5 Outlook

Although the basic function of control centers is the same and remains constant over time, their design depends on evolves with the requirements of the missions for which they are responsible. Some design principles have been maintained since the beginning and are found throughout the world, while new requirements emerge with new technologies. In this way, most control centers are unique.

Cloud systems are designed to be available anywhere, anytime. This has a huge impact on the concept of operations. With processing resources available as a service, you can have virtually unlimited amounts of them and scale as needed. Operators could work from home, sub-system engineers would not need to be physically present during on-call shifts, and the mission management would have 24/7 access to all relevant information. Facilities would be more or less redundant. Essentially, spacecraft operations would take place in a completely virtual environment, much like in a multiplayer online game.

The main technologies (virtualization, containers, etc.) are already known and in use, but scalability can be a problem for government agencies. Large companies such as Google or Amazon have a lot of experience in this area, while agencies would have to make significant efforts to achieve the possibilities of a seamlessly expanding computing capacity while being transparent about the changes to users. And so, one option for agencies would be to use commercial capacities either in whole or in part. This is likely the most cost-effective solution. There are still many questions regarding actual data governance, ownership and security that still need to be resolved for each individual need and situation.

Marcin Gnat received his M.Sc. degree in electrical engineering in 2000 from the Technical University of Koszalin, Poland. From 2001 to 2010 he worked as a Senior Test Engineer in the semiconductor industry at Infineon Technologies and Qimonda, followed by consultant activities in the same area. Since 2010 he works for satellite missions at DLR's German Space Operations Center (GSOC). Since 2017 he is in charge of Ground Data Systems providing communications and ground station solutions to the projects.

Michael Schmidhuber works in the Department for Mission Operations at the German Space Operations Center (GSOC) where he is responsible for satellite operations training. He graduated as an aerospace engineer at the Technical University of Munich in 1994. The training activities comprise not only staff training but also courses for an external audience in the annual Spacecraft Operations Course at GSOC. Also, he is involved in the organization of the bi-annual international SpaceOps Conference.

Chapter 10
Ground Station Network

Marcin Gnat

Abstract The ground station network (GSN) plays a major role in space missions. It establishes links with the spacecraft and with other control centers, supports specific characteristics of the spacecraft and provides the functionality and safety of the mission. By its nature, GSN participates in cross-support activities between different organizations and agencies. The GSN comprises several functional aspects, the communication path between the control center and the ground stations (online data transport, offline data, voice), the management of the stations and their antennas as well as coordination tasks and station scheduling.

10.1 Introduction

The communication with the spacecraft as an essential part of the spacecraft's operation is mainly characterized by receiving telemetry (TM), transmitting telecommands (TC) and tracking. Optimizing the communication link can significantly increase the operability, the outcome and last but not least the safety of the mission. E.g., the contact time can be increased, usually by introducing additional ground stations into the network and selecting optimally placed stations. For some missions (such as LANDSAT, SILEX and Sentinel) the use of one or multiple geostationary relay satellites is a viable option. A well-known service here is the TDRS (Tracking and Data Relay Satellites) program (Stampfl and Jones 1970), which supported already the early Space Shuttle flights and is still the backbone of communications in the ISS project. In the meantime, more and more relay satellites are available, and the market is growing—the European Data Relay System (EDRS) features even terminals for optical communications (Böhmer et al. 2012).

The launch and early orbit phase (LEOP) is a particular case, since several critical tasks depend on a safely established contact—via a connection which is set up for the first time in the life of a spacecraft. Here, GSN has the task of shortening the time

M. Gnat (✉)
German Space Operations Center (GSOC), German Aerospace Center (DLR), Oberpfaffenhofen, Germany
e-mail: marcin.gnat@DLR.de

F. Sellmaier et al. (eds.), *Spacecraft Operations*, Springer Aerospace Technology,
https://doi.org/10.1007/978-3-030-88593-9_10

from the separation of the spacecraft from the launcher to the first acquisition. The initial acquisition station must carry out the first tracking of the satellite to allow an accurate orbit determination, and it needs to receive telemetry in order to assess the condition of the spacecraft after launch. When time permits or if it is required, the first station also performs an uplink to allow time-critical operations such as attitude setting into Sun-pointing mode or unfolding solar panels.

10.2 Station Selection

When designing the GSN, several technical properties, parameters and requirements must be considered. They are usually provided in the form of the Space to Ground Interface Control Document, while some others can be found in the spacecraft design and requirements documents.

The analysis begins with the main mission feature, the orbit. Based on the knowledge about the position and speed of the satellite during the mission phases, we can decide which ground stations potentially can be used depending on their geographical location (see also Sect. 13.3). For Earth-bound missions, the orbit type may change during LEOP, but remains stable during the routine phase. Orbit types are categorized according to altitude (or, in other words, distance from Earth), the inclination of the orbital plane to the Earth equator, and the shape of the orbit path.

The majority of the satellites are in circular orbits. We distinguish between low Earth orbit (LEO) with altitudes of up to 1,000 km, geostationary Earth orbit (GEO) with about 36,000 km and everything in between, called medium Earth orbits (MEO). The orbits can have different inclinations, with GEO usually at zero degrees, many LEOs in polar orbit close to 90° and all other satellites somewhere in between.

And so, the spacecraft flying in LEO with an inclination of about 55° can only be contacted by stations located at a latitude on Earth, which is smaller than that value. Therefore there is clear dependence between spacecraft orbit inclination and selection of the ground stations.

Polar stations are of major importance for Earth observation missions in polar orbits, since they principally allow communication within practically every orbit (see Fig. 10.1) for an example ground station network supporting polar orbits). However, they cannot be used for missions with low inclinations such as GEO. In such a case, the ground antennas must be ideally distributed along the equator (Fig. 10.2).

Other Earth-bound satellites are on highly elliptical (= eccentric) orbits. The significant changes in altitude, ranging from very low (only about 100 km) up to 60,000 km (far beyond GEO) result in a wide variety of spacecraft velocities. The reason to choose such an orbit can be a transfer phase between different orbits (e.g., geostationary transfer orbit, GTO), a footprint optimized for a specific region of the Earth from the ground (e.g., Molniya type orbits) or scientific requirements.

The GSN must be carefully adjusted to the mission. With highly elliptical orbits, a spacecraft is visible in apogee for many ground stations over a long period of time (hours); however, the signal strength is significantly decreased. At perigee, on the

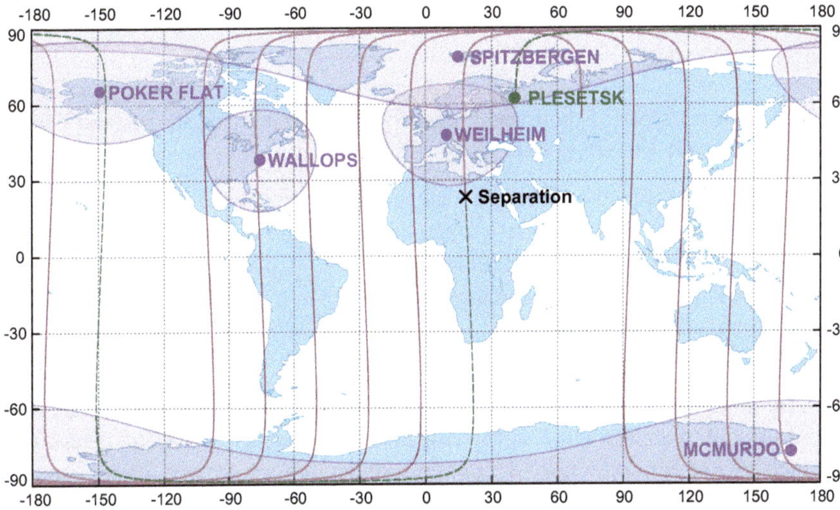

Fig. 10.1 Typical LEOP ground station network for LEO Spacecraft. The ground track of the spacecraft is depicted as well as the footprint of the involved ground stations

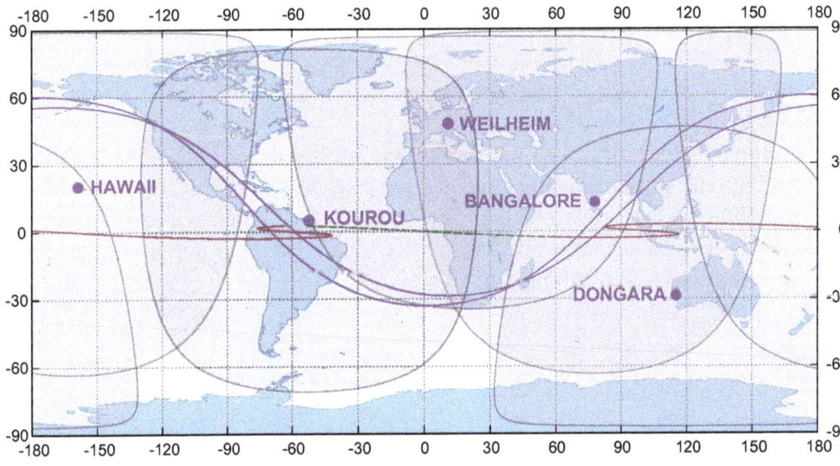

Fig. 10.2 Typical LEOP ground station network for GEO Spacecraft in GTO. The ground track of the spacecraft is depicted as well as the footprint of the involved ground stations

other hand, the spacecraft will be at a very low altitude with extreme speed. The resulting antenna tracking speed is very high and excludes most antennas.

Missions to the Moon, Mars or further in space are called deep space missions, the spaceship is no longer orbiting the Earth. Due to the large distances and the weak signals, the required antennas are larger in diameter to achieve the required signal sensitivity. For such antennas, a high tracking speed is no longer a design requirement, as the target is quasi-stationary in the sky and its movement is dominated by the Earth's rotation speed (Figs. 10.1 and 10.2).

There are also a bunch of technical parameters, which may also influence the station selection. Most of them are known from basic antenna theory and can be found in respective literature, thus we just list them here for completeness:

- Equivalent isotropically radiated power (EIRP)
- Antenna gain and gain-to-noise temperature (G/T)
- Antenna diameter (in case of reflector antennas)
- Available/supported radio frequencies
- Uplink/transmission capability
- Additional services capability (like ranging, Doppler measurements).

These parameters are then used to preselect the antenna. This may be actually refined through the calculation of the so-called link budget. This value determines the space link quality (see Sect. 3.4.2) and is an indication, how much margin remains under different conditions during the mission.

Further parameters that influence the choice of stations are the downlink and uplink frequencies, which are grouped in so-called frequency bands, as shown in Table 10.1. Not all ground stations have antennas that support all possible frequency bands; often a given station serves a dedicated purpose. For example, stations supporting GEO missions typically have antennas with Ku- and Ka-band capabilities, while LEO stations have S- and X-band capabilities. Deep space antennas typically also support S- and X-band frequencies, but with a much larger dish diameter (to enable better EIRP and G/T).

The situation becomes even more complex during the LEOP of geostationary satellites. S-band is traditionally used for payload IOT (in-orbit test) and routine operations before switching to Ku-band or Ka-band. This requires a very demanding planning for the spacecraft itself, the mission operations and the ground segment with various systems and equipment.

Table 10.1 Frequency band assignments as used in space operations

Band	Range (MHz)
L-band	1215–1850
S-band	2025–2400
C-band	3400–6725
X-band	7025–8500
Ku-band	10,700–14,500
Ka-band	18,000–35,000
V-band	37,500–50,200

Since there is no unified frequency band naming convention, the definition of frequency bands may vary across the various publications

Despite wide popularity and low cost of the S-band, there is increasing interest to support LEO satellites with Ka-band since a higher bandwidth is available. Furthermore, in particular GEO missions with long contact times and low signal strengths suffer from an increasingly congested S-band and interferences with other spectrum users such as mobile Internet access. Therefore, the necessary infrastructure for telemetry and telecommand in Ka-band is provided in more and more ground stations.

Another aspect of station selection is bandwidth. Most ground stations support the full available downlink data rate in the specific frequency band. However, the uplink is sometimes restricted. So far, relatively low data rates (between 4 and 20 kbps—kilobits per second) have been used for the uplink. The capabilities and equipment of the existing stations have been designed accordingly. A tendency towards increased uplink data rates can be observed to meet trends such as more frequent software uploads. Currently, not all ground stations can support those.

Other parameters that we will not discuss here in detail, but are worth mentioning, are modulation type, encoding, randomization, space link data format, and finally specific tracking requirements for ranging and Doppler. All those factors are standardized, and the CCSDS (Consultative Committee for Space Data Systems) and ECSS (European Cooperation for Space Standardization) standards listed in Table 10.2 describe them to the full extend.

Full technical compatibility between spacecraft and the station is essential, so it is not enough to rely on standards. Before the spacecraft is launched, a radio frequency (RF) compatibility test (often referred to as RF comptest) is usually performed. This test ensures that the radio interface is working and allows to prepare the station configuration for its use during LEOP. The RF comptest is typically conducted at least six months before the launch, as soon as the so-called "RF suitcase" is available. The RF suitcase contains the flight model of the RF equipment with some parts of the on-board computer (OBC). In some cases, even the entire spacecraft is transported to the ground station for testing. In the latter case, however, a clean room must be available.

It is also important to mention the accuracy of the carrier spacing between signals from multiple satellite antennas or multiple satellites in general, as interferences may occur that render individual or all space links unusable. This can lead to exclusions of up- or downlinks over certain geographical regions or at certain times, which in turn results in specific ground station selection.

A further important aspect for the layout of the GSN is the autonomy of the spacecraft. A satellite with very high autonomy can operate and survive for a long time without commands or surveillance from ground. In such a case the GSN can be very simple for the mission, e.g., with only one antenna and without high redundancy. On the other hand, with low autonomy or critical applications (e.g., precise orbit keeping) the demands on GSN increase. Finally, constraints on board may also require more frequent contact times (e.g., limited data storage on board).

Table 10.2 The most important CCSDS and ECSS standards for space mission communication

Document	Title
CCSDS 401.0-B	Radio frequency and modulation systems—Part 1 Earth stations and spacecraft
CCSDS 132.0-B-1	TM space data link protocol
CCSDS 131.0-B-2	TM synchronization and channel coding
CCSDS 232.0-B-2	TC space data link protocol
CCSDS 231.0-B-2	TC synchronization and channel coding
CCSDS 232.1-B-2	Communications operation procedure-1
CCSDS 301.0-B-4	Time code formats
CCSDS 320.0-B-5	CCSDS global spacecraft identification field: code assignment control procedures
CCSDS 910.4-B-2	Cross support reference model—Part 1: Space link extension services
CCSDS 911.1-B-3	Space link extension—Return all frames service specification
CCSDS 911.2-B-2	Space link extension—Return channel frames service specification
CCSDS 912.1-B-3	Space link extension—Forward CLTU service specification
CCSDS 133.0-B-1	Space packet protocol
ECSS-E-70-41A	Ground systems and operations—Telemetry and telecommand packet utilization
ECSS-E-HB-50A	Communications guidelines
ECSS-E-ST-50-01C	Space data links—Telemetry synchronization and channel coding
ECSS-E-ST-50-02C	Ranging and Doppler tracking
ECSS-E-ST-50-03C	Space data links—Telemetry transfer frame protocol
ECSS-E-ST-50-04C	Space data links—Telecommand protocols synchronization and channel coding
ECSS-E-ST-50-05C	Radio frequency and modulation
ECSS-E-ST-50C	Communications

They are available from the web sites of the organizations

10.3 Station Communication

The selected ground stations are connected to the control center via a communication network. Different levels of network communication characteristics must be considered and used. Here, the decisions made for a dedicated mission are based on some basic requirements such as the bandwidth needed to support the mission, availability at specific locations and the total cost throughout the mission.

10.3.1 Communication Paths

Commonly used are so called leased lines, which basically describes the actual technology used, but their name rather describes the type of use (exclusive use for a particular customer or connection). The backbone technology of these lines is in the hands of the telecommunications provider and the spacecraft control center may have a choice to select between different ones. But even if the selection of the actual WAN technology is not possible, we shall know it in order to assess the quality of the connection, which in turn may have influence on operations. Keywords here are SDH (synchronous digital hierarchy), ATM (asynchronous transfer mode) or MPLS (multi-protocol label switching), the last of which is the current state of the art (Fig. 10.3). Previously often used technology of ISDN (Integrated Services Digital Network) has already been decommissioned.

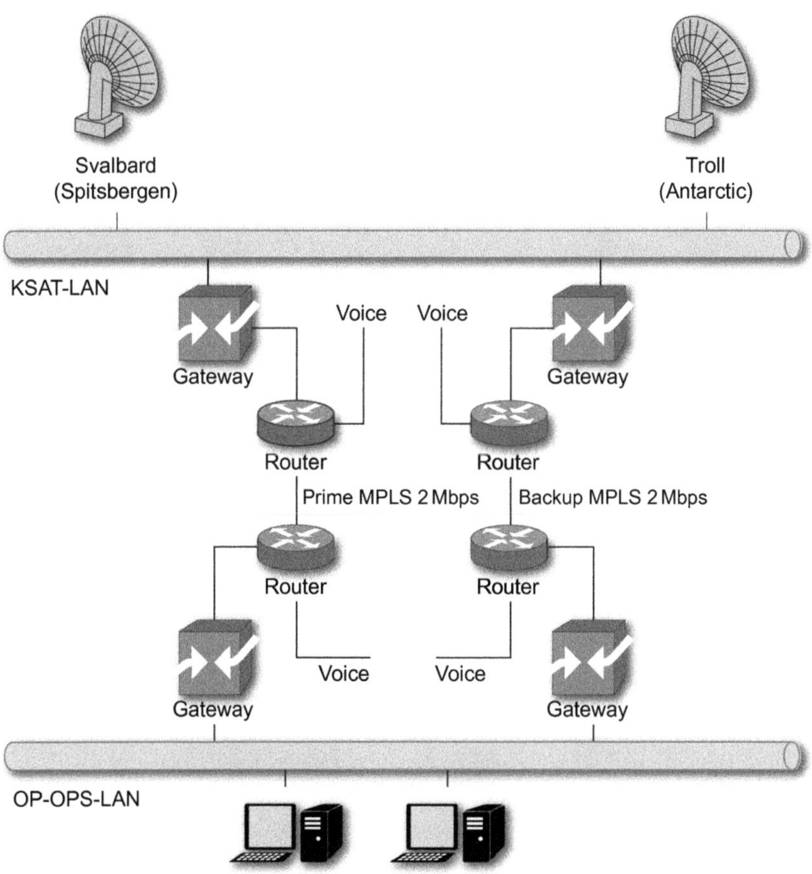

Fig. 10.3 Example of redundant ground station connection. A connection between networks of KSAT (Kongsberg Satellite Services) and GSOC in Oberpfaffenhofen is shown

A specific variant of communication is the so-called roof-to-roof communication, which is typically implemented by VSAT (very small aperture terminal). Here, satellite dishes are installed on the premises of the mission control center (MCC) and its partner center or station and the information exchange is established via a geostationary communication satellite. The advantage is an extremely high flexibility (the MCC can be connected to virtually any point on Earth) at reasonable bandwidths. However, the solution is usually quite expensive (rental of the GEO transponder) and considerable delays in data streaming are introduced.

Although it is not a communication path, but an encryption protocol, a few words about virtual private network (VPN) shall be included here. In most cases, VPN is associated with a connection over the internet. This option may look very attractive because the internet is cheap, offers virtually unlimited bandwidth and is available everywhere. However, also the disadvantages should be considered: the corresponding MCC shares its communication medium with an indefinite number of other users, and there is no support in case of problems. The bandwidth varies constantly, and a good access is site-dependent. VPN can provide a decent data security, but no reliability. Additionally, many operational systems may not be connected directly to a network with Internet access (due to many reasons, like security or above-mentioned availability). For many years, it was generally not recommended to use the internet as a transport technology for real-time TM/TC or other critical applications where reliability, security and data integrity play a role. It is still a viable solution for offline data transfer and for the connection between MCC and the manufacturer site, for simulations and tests with the central checkout system (CCS). This may change now, but still the decision has to be made individually, depending on tradeoff between cost, security and general usability aspects.

Communication for space missions is typically installed with respective redundancy (example shown on Fig. 10.3). Depending on availability, cost factors and considerations like security or physical access, the redundancy may be set up as a combination of previously mentioned technologies (MPLS and VSAT or MPLS and VPN over internet).

10.3.2 Data Transfer Methods

The data types that are to be exchanged via the communication lines must be considered. Telemetry, telecommand and voice interfaces require a good, reliable real-time connection with medium bandwidth, while the transfer of management information, planning, tracking and pointing can be organized more cost-effectively.

Nowadays, practically all traffic is based on the TCP/IP protocol, whereas in the past some proprietary transport protocols were used (Wikipedia DECnet 2021; Wikipedia X25 2021).

At the application level, the CCSDS Space Link Extension (SLE) standard is widely used for real-time communication, while FTP is the main file transfer mechanism. Here, some changes are expected in the future, especially for file transfer and exchange of management information, the corresponding standardization work is currently being carried out. Further online services and mission operational services will be provided. The use of cloud-based systems, centralized databases and message-based middleware enables a wide range of alternatives to the old-fashioned file transfer via FTP.

When considering the operational aspects of the GSN, it should be remembered that all connections must be tested before LEOP. The arrangement and integration of communication lines and ground stations must be done at the right time. Staff must be trained to operate the network. Procedures for different operational scenarios need to be prepared and validated.

10.3.3 GSN Examples

Now that we know all aspects of the GSN, we can take a look at the example shown in Fig. 10.4. The GSN and the communication infrastructure for the LEOP of a geostationary satellite are shown. The MCC (in this case the German Space Operations Center—GSOC) is located at the bottom. Solid lines represent voice communication, while dashed lines represent data connections (in the form of real-time TM/TC).

The MCC features a voice communications link, either via a dedicated voice system or telephone, with the launch site, the Weilheim ground station and the respective network management centers of external partners (like PrioraNet, CNES and ISRO). The data connections are implemented completely redundantly via different routes. Weilheim is integrated with two SDH 2 Mbps connections, while the connections to the PrioraNet stations are available via two levels of network management centers (NMCs) with VSAT terminal and ISDN. The CNES and ISRO networks were also integrated with one NMC each.

It can be clearly seen that the expansion of the GSN requires the partnerships with external suppliers and agencies, resulting in a complex network to support spacecraft and increasing overall reliability to a very high level. On the other hand, a such complex network needs to be managed, planned, and maintained. In many cases, it is not the technical aspects but rather the contractual and financial dimensions that are the biggest effort, with questions like: Do we always get the highest priority at the respective station? How much does it cost in the long run? Is it possible to get some discounts if we consolidate our requirements to only one provider? What are the compromises if we decided to exclude a particular station? (Fig. 10.4).

10.3.4 Cloud Based Services

In recent years, many capabilities around cloud computing and cloud services have emerged. This has significantly supported the changes in many areas related to space

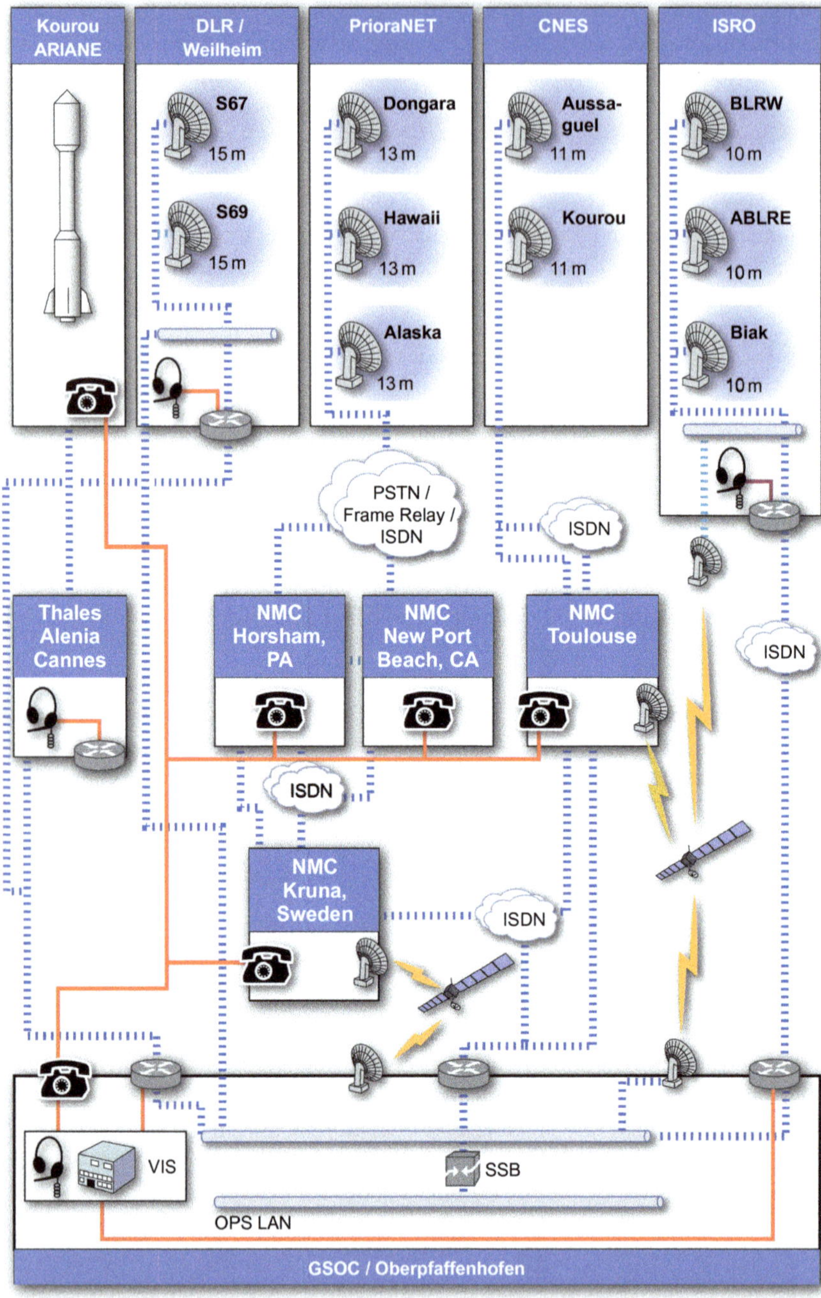

Fig. 10.4 Communication within GSN for the LEOP of a typical GEO mission

operations but was so far a rather specific part of the general infrastructure puzzle. This has changed recently with the emergence of cloud-based ground station services (especially Amazon Web Service Ground Station—AWS GS, as described in Gnat et al. 2019). Providers of such services are trying to encapsulate the entire ground station environment and related tasks into one easy-to-use and easy-to-charge unified service that is offered as a commodity. This approach could be interesting, and when setting up the GSN for a mission, the possibilities of such cloud-based services should definitely be considered.

Let's take a quick look at the advantages and disadvantages of a cloud-based solution. It allows customers to easily set up the ground station infrastructure with minimal hardware investment. Due to the fact, that the used systems require very high bandwidth between software defined radio in the cloud and the antenna, ground stations are placed in close proximity to the cloud providers' existing data centers. The customer books the time on an antenna and plans a contact with the satellite. The downloaded data is immediately inserted into the customer's cloud environment for further processing. In extreme cases, this can reduce the infrastructure required to operate a satellite to a workstation with internet access. In terms of cost, cloud services often do not distinguish between uplinks and downlinks or other variants of space communications. The service itself is charged per unit of time of use (typically per minute) and the cost of the service is relatively low (in the range of $ 3–22 per minute, depending on the service level). This allows new users to get started quite easily, they are no longer tied to a specific location and the services are offered from the cloud. Users can choose the services according to their needs and, for example, apply the "pay-per-use" principle.

There are also some drawbacks of cloud solutions. The most obvious one is that the data is not stored on the operator's local infrastructure. In many cases, this is not a problem and may even be desirable (due to availability), but there may also be (technical or legal) issues with data ownership, proprietary information, confidentiality and data security. Frequently, the interfaces provided by cloud providers do not correspond exactly to the typical interfaces used so far. In other words, some money can be saved on the ground stations as such, but more investment is required to adapt the interfaces. Cloud providers tend to offer cheap services only if the missions can accept satellite contacts as proposed by the service. As soon as the mission requires its own station usage plan, the price of the service increases. This is obvious because inexpensive services can only be offered if the service provider can maximize the use of his resources, and this can only happen if the operator can independently decide on resource usage.

In the end, each mission has to define how much risk and third-party influence it wants to take in order to reduce overall costs.

10.4 LEOP and Routine Operations

Within this chapter, we will look a bit more at the operational aspects of the control center infrastructure, network and GSN. The operational work is distributed to several specialized teams, where the so-called ground data systems (GDS) team performs most of the coordination work as well as GSN operations.

The GDS team acts as an interface between the satellite operations teams and the other communication and ground station support groups. It manages the GSN for the satellite missions and is responsible for the interfaces with external partners supporting the missions. In other words, the GDS manages operational internal and external interfaces of MCC.

The GDS accompanies each satellite project within MCC from the very early phases (mission studies, phase A, see Chap. 4) until the very end (phase F, decommissioning). Within the project, the GDS is one of the project's subsystems, where it works on fulfilling the GDS part of the project requirements, and acts as an expert for GSN and control center infrastructure for the whole MCC for the missions. This construct allows project managers to bundle all communication and infrastructure questions and requests to one person (the designated project responsible from GDS). That person manages and coordinates within his department, to the network and ground station departments or with external partners. This allows high synergy, a high reuse of resources, and an optimal work distribution. Other tasks and areas of responsibility of GDS include:

- Participation in meetings and project reviews (preliminary design review—PDR, critical design review—CDR, technical acceptance review—TAR, etc.)
- Responsibility for the project's ground station network offline and on console
- Management of interfaces to the external partners, including contractual and technical agreements
- Provision of first level expertise for all network, communication, and infrastructure questions and issues
- Coordination for all operations related activities between all parties
- Preparation of work packages, work package description
- Preparation of cost calculations related to communication and infrastructure
- Preparation of relevant project documentation (requirements, design, test plans, reports, ICDs—interface control documents, DMRs—detailed mission requirements)
- Assessment and implementation of new operational solutions for communication and infrastructure
- Participation in international standardization organizations with operational communication topics.

Most of these tasks, especially the ones related to specific project, are conducted by nominated GDS manager, who in principle plays a role of sub-project manager.

In the example of GSOC, the GDS team includes three specific subgroups, which have their own tasks. Details on the tasks of these subgroups are provided in consecutive sections.

10.4.1 GDS Engineering Team (NOPE)

The GDS NOPE (network operations project engineer) team is a group of engineers taking care of the technical and organizational tasks related to specific satellite missions. Typically, each mission has a designated GDS engineer who accompanies the project from phase C (see Chap. 4), participates in project meetings, and plays the role of point of contact in the absence of the GDS manager. The most important tasks of the GDS engineers are:

- Mission preparation
- LEOP preparation (configuration, coordination)
- Active mission support during LEOP (also on shift)
- Performance of tests (data flow tests—DFT, connection tests, configuration tests)
- Preparation of the configuration for all data connections for the mission
- Configuration coordination with external partners
- Preparation of reports
- Troubleshooting and failure analysis.

10.4.2 Systems Team (Network and Systems Control)

The network and systems control (sometimes also called network management center or systems team) is, at least from the communications point of view, in charge of on-console MCC operations. The team consists of a number of operators, who work on shift to cover 24/7 operations, and support engineers who coordinate the shift team and manage the work and operational processes within the network control room (systems room). “Systems” can be compared to a central phone switch board, where all connections (operational and technical) from all MCC control rooms are routed (switched) to the ground stations worldwide. Systems also plays the role of a voice center, as it is permanently staffed and has contact (either via telephone or a special voice system) with all projects and all stations, allowing quick reaction in case of contingencies or emergencies. This function is used to coordinate extraordinary contract requests on holidays or at night, for example, when the scheduling office is not staffed. Tasks of the systems team are:

- Network control during routine operations (establishment of connections on project request and along the schedule)
- Support for NOPE during LEOP

- Monitoring of the connections and network within GSOC and to external partners
- Support of contingency and emergency scheduling and operations.

10.4.3 Scheduling Office

A dedicated scheduling office is a functionality which becomes necessary with increasing numbers of missions and available antennas of a control center. In principle, every antenna or antenna network is available for multiple missions, especially in terms of cross-support agreements. Therefore, we talk here about a *n*-to-*n* resource coordination problem. This is essential to avoid conflicts and to increase synergy between missions, and the scheduling office defines its role in high extend due to this. The scheduling office tasks can be performed by one person and needs to be operated during office hours only. The tasks of scheduling are:

- Reception of ground station support requests from projects and coordination of allocations at the organization's own and at external ground stations
- Contact planning according to mission requirements applying mission priority rules
- Publishing of the weekly contact plan (schedule) for all MCC missions and resources
- Support and provision of solutions in case of conflicts.

When we look at the operations work aligned to the mission lifetime, most of the work may be divided into three phases: preliminary preparation (design), detailed preparation (design), and mission execution, which contains specific events like a LEOP.

10.4.4 General Tasks Throughout the Project

In the preliminary preparation phase, the main work focuses mainly on the analysis of the customer requirements. This consists of checking whether the existing system fulfills the requirements or whether changes or upgrades have to be considered. This analysis is important because the latter case will increase the costs. Based on that, detailed requirements for the subsystems are defined.

Another task is to prepare the general design (concept), which includes interface specification. This part is continued in the detailed design definition phase, where also test and verification plans need to be created.

The implementation phase is typically very busy for everybody in a space mission project, which is in particular true for the control center infrastructure, network, GDS and GSN. It is necessary to implement and integrate all subsystems, which encompass hardware, software, and service procurement (like communication lines), installation, and testing. Sometimes, hard- or software has to be delivered to cooperating partners, which means, necessary export licenses have to be issued on time.

Aside from that, the radio frequencies have to be coordinated and licensed. This is typically done at national level for ground stations. The spacecraft owner needs to apply at the International Telecommunication Union (ITU) for the allocation of the communication frequencies.

A specification for external partners needs to be prepared and issued in form of detailed mission requirements (DMR) as well. This, however, can be done only as soon as the respective contracts with these partners are signed. As one can see, there is a lot of paperwork which needs to be taken care of in advance.

At the end of that phase, the complex set of technical and operational tests and validations is performed. It is based on previously prepared test plans, and include all subsystems, from data processing, through communication, and conclude with end to end tests (including all components) and simulations. These, in particular, are important for validation of previously prepared operational procedures (e.g., emergency procedures). Technical and operational staff planning and training are equally important as well.

LEOP marks the border between the preparation phase and routine operations. All systems need to be handed over to the operational team before LEOP, typically performed formally during the operational readiness review (ORR).

During operations (including LEOP), there are a number of tasks performed repeatedly, like scheduling of ground contacts, preparation and execution of passes, reporting, and accounting. At the same time, the whole GSN and MCC infrastructure needs to be monitored and controlled; maintenance needs to be performed. The interfaces to external partners, to all ground stations, and of course the internal interfaces need to be handled. In case of any anomalies and failures, actions need to be performed according to procedures, error reports need to be generated, and any anomaly or failure shall be tracked with a dedicated discrepancy report to avoid such cases in the future.

References

Böhmer K, Gregory M, Heine F, Kämpfner H, Lange R, Lutzer M, Meyer R (2012) Laser communication terminals for the European Data Relay System. In: Proceedings of SPIE—The International Society for Optical Engineering

CCSDS—Consultative Committee for Space Data Systems. https://public.ccsds.org/Publications/. Cited 14 June 2021

ECSS—European Cooperation for Space Standardization. https://ecss.nl/standards/. Cited 14 June 2021

Gnat M, Knopp MT, Philipp R, Müller T (2019) From Apollo to Amazon—ground control changing. In: 70th international astronautical congress (IAC), Washington, DC

Stampfl RA, Jones AE (1970) Tracking and data relay satellites. IEEE Trans Aerosp Electron Syst AES-6(3):276–289

Wikipedia (2021) DECnet. https://en.wikipedia.org/wiki/DECnet. Cited 14 June 2021
Wikipedia (2021) X25. https://en.wikipedia.org/wiki/X.25. Cited 14 June 2021

Marcin Gnat received his M.Sc. degree in electrical engineering in 2000 from the Technical University of Koszalin, Poland. From 2001 to 2010 he worked as a Senior Test Engineer in the semiconductor industry at Infineon Technologies and Qimonda, followed by consultant activities in the same area. Since 2010 he works for satellite missions at DLR's German Space Operations Center (GSOC). Since 2017 he is in charge of Ground Data Systems providing communications and ground station solutions to the projects.

Chapter 11
Ground Station Operation

Amanuel Geda

Abstract This chapter introduces the basic tasks and functions of a satellite ground station. The main task of a ground station is the telemetry, tracking and command operations (TT&C) of a spacecraft to support mission preparation, as well as test and operation phases. Also, devices and measurements, protocols and interfaces are shown.

11.1 Introduction

The objective of this chapter is to introduce the basic tasks and functions of a satellite ground station. The main task of a ground station is the telemetry, tracking and command operations (TT&C) of a spacecraft to support mission preparation, as well as test and operation phases. Also, devices and measurements, protocols and interfaces are shown.

The communication to a spacecraft is performed by a ground station. A space link is defined by the Consultative Committee for Space Data Systems (CCSDS) in their *TM Space Data Link Protocol* document (CCSDS 132.0-B-2 2015) as a communications link between a spacecraft and its associated ground system or between two spacecraft.

The CCSDS also defines two types of missions (CCSDS 401.0-B-29 2019): *Category A* missions are at altitudes of below 2 million km [low Earth orbit (LEO), geostationary Earth orbit (GEO), lunar missions and Lagrange point missions (L1 and L2)]; *Category B* missions are also called deep space above 2 million km going to other bodies in the solar system or even beyond.

Large ground station facilities, governmental or commercial, often have a wide range of use cases in a variety of frequency bands supporting *category A* and *category B* missions. Their portfolio may include telemetry (TM) reception and processing,

A. Geda (✉)
German Space Operations Center (GSOC), German Aerospace Center (DLR), Weilheim, Germany
e-mail: amanuel.geda@DLR.de

F. Sellmaier et al. (eds.), *Spacecraft Operations*, Springer Aerospace Technology,
https://doi.org/10.1007/978-3-030-88593-9_11

satellite telecommanding, satellite tracking (angles, range and Doppler measurements), as well as TM simulation and testing. Support types may include launch and early orbit phase (LEOP), in-orbit test (IOT) and routine phase.

The fundamentals of space communications are covered in Chap. 3 and should be read in complement.

11.2 The RF Subsystem

The RF subsystem in downlink includes the antenna, the feed system, the diplexer, the low-noise amplifier (LNA), the frequency downconverter, and the tracking system. The feed system includes a polarizer which distinguishes right-hand circular polarized (RHCP) and left-hand circular polarized (LHCP) signals (see Sect. 18.2.2) and delivers them to two different outputs in the downlink.

The radio frequency (RF) subsystem for the uplink comprises the upconverter, the high-power amplifier (HPA), the polarization selection system for uplink, a diplexer, the feed system and the antenna.

Stations offering simultaneous telemetry (TM) and telecommand (TC) operations are the standard type of ground stations. Ranging and Doppler services are also often available. A special feed design and extra components like diplexer filters are necessary to separate the low power receive signals from the high power transmit signals.

11.3 The Intermediate Frequency (IF) Baseband Subsystem

This unit is responsible for the IF level and the baseband level signal processing for telemetry, tracking and command (TT&C), and payload signals. The IF receiver units receive the two signals, RHCP and LHCP (see Sect. 18.2.2), from the downconverter. It performs demodulation of IF signals. The IF receiver is followed by the telemetry unit, which performs bit synchronization, data decoding, frame synchronization and also demodulation at baseband level. TC data modulation is performed by the TC modulator.

Ranging Unit

Ranging is done mostly for geostationary (GEO) missions to determine the distance to the satellite. This is used as a measurement in the orbit determination process. There are several different standards in use, e.g. by ESA or Inmarsat. The ranging unit is a module inside the baseband system, which performs tone generation and demodulation. It allows to compare the uplink with the downlink signal phase and frequencies, and in that way measures the distance taking the two-way signal travelling time and the radial velocity (range-rate) using the Doppler shift.

Diversity Combiner

When both RHCP and LHCP signals are available in downlink, a polarization diversity combiner unit (DCU) can be used which maximizes the signal-to-noise ratio (SNR) at the output of the combiner. The DCU is a maximum ratio combiner with a maximum SNR gain of 3 dB at the output (Proakis 2001; Sklar 1982; Cortex CRT). The maximum SNR gain occurs when the RHCP and LHCP downlink signals are of equal strength (CCSDS 132.0-B-2 2015; CCSDS 401.0-B-29 2019; Proakis 2001).

11.4 Supporting Devices

Test-Loop Translator

An important equipment for the RF subsystem is a test-loop translator (TLT). This is a testing device that picks up the uplink signal, converts it to the downlink frequency, and returns it to the receiver with a certain attenuation. This allows to check the performance and integrity of the RF chain without a satellite in the loop.

Antenna Control Unit (ACU)

The ACU is the system that directly controls the antenna motor drives. It allows to set the pointing direction and the tracking mode. It also returns the actual angle readings to the operator. For test and maintenance, it allows manual control of the antenna. It may support the setting of linear polarization.

Time and Frequency Reference System

External frequency references are needed by the IF baseband subsystem Cortex CRT (command ranging and telemetry), downconverters, upconverters, ACU, tracking receivers, spectrum analyzers, signal generators, TLTs (test loop-translators) and the Siemens PLC (programmable logic controller). GPS-based external time references are additionally required by the Cortex CRT and the ACU.

11.5 Telemetry, Tracking and Command Operations

Support operations are divided into three phases: pre-pass, pass and post-pass. The term pass is derived from the passage of a LEO satellite over a ground station. For geostationary satellites the term support is more common, but the activities are the same.

In pre-pass operation, the station is set up for a specific mission. The mission parameters include, but are not limited to these parameters: data rate, type of pulse-code modulation (PCM) signal, type of modulation, error-correcting channel codes, ground station equivalent isotropically radiated power (EIRP), polarization, downlink frequency, uplink frequency, IF frequency, mode of tracking, TM frame length, type of ranging and ranging parameters. Ranging and Doppler calibrations are performed during the pre-pass phase. It is also a common practice to perform TM/TC and line tests and a pre-pass briefing with the mission control center in the pre-pass phase.

The pass operation starts with reception of the downlink signal, known as AOS (acquisition of signal). In the case of a station handover, the previous station or the mission control center has to confirm that no other uplink is ongoing before the carrier signal is started. Then the carrier is started and uplink sweep operation commences. The sweep frequency range and rate are according to the CCSDS standard (CCSDS 401.0-B-29 2019). The uplink sweep operation is complete when the spacecraft receiver is locked. This information can be extracted from the downlink data stream at the ground station without the need for complex processing. The state of the spacecraft receiver is contained in the TM frame. It is described in CCSDS 132.0-B-2 (2015). After the sweep operation, the station is ready for TC, for two-way ranging and Doppler. At the time of AOS and when the station is ready for uplink, announcements should be made to the mission control center. In some cases, the uplink sweep is not successful and has to be repeated. In critical spacecraft situations, it may also be necessary to stand by for immediate changes in the configuration on the request of mission control.

The pass is finished when the signal is lost (loss of signal—LOS), either due to the spacecraft vanishing behind the horizon or when the service is handed over to a different ground station. After taking the carrier down, all processing can be stopped and connections may be terminated. If products were generated (e.g. ranging data or data dumps), they shall be provided to the end users, e.g. by placing them on a file pick-up point. A briefing with the MCC ends the post-pass activity.

The Weilheim ground station is designed to use standard baseband equipment for different antennas and applications. This concept has been implemented by building up a pool of TT&C baseband units connected to the different antennas by means of switch matrices (see Fig. 11.1). This solution allows flexible and cost-effective usage of equipment in conjunction with a high grade of redundancy. The scheduling of the antennas, i.e. deconflicting and prioritizing between missions, is performed by the scheduling office at GSOC in Oberpfaffenhofen (see Sect. 10.4.3).

The TT&C Baseband unit performs the following:

Telemetry Processing

- Low rate and high rate telemetry processing
- Video demodulation (phase, frequency or amplitude modulation (PM, FM or AM)) for low rate applications
- PCM demodulation (PCM/PM, PCM/FM, phase shift keying (PSK) in several variants) for high rate applications
- Carrier identification: automatic or manual acquisition
- Pre-detection and post-detection diversity combining
- Sub-carrier demodulation (PSK, PCM/PM or PCM/FM)
- Bit synchronization
- Viterbi decoding
- Frame synchronization

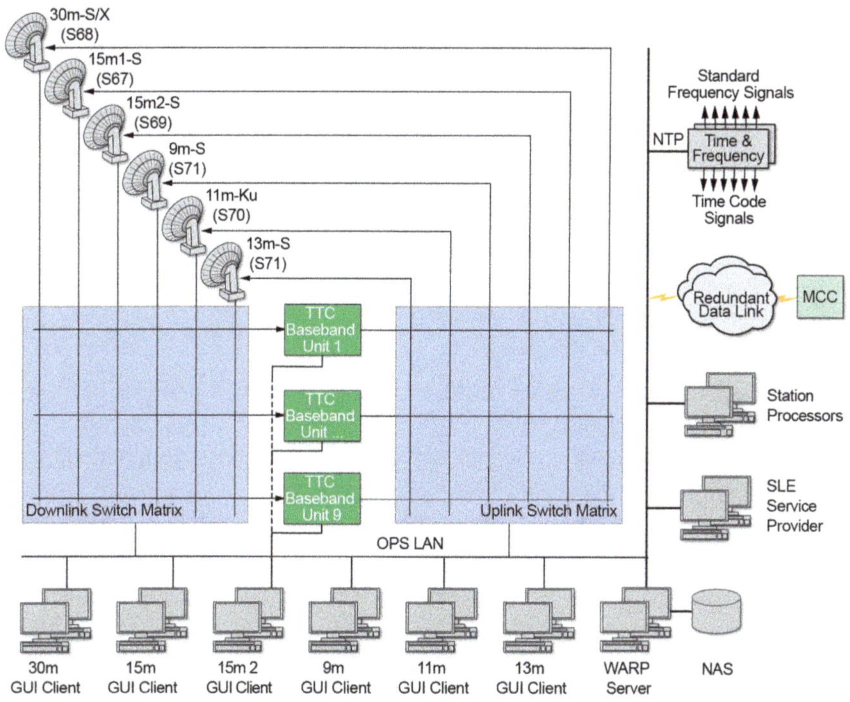

Fig. 11.1 Overview of the Weilheim ground station complex with a link to the mission control center (MCC)

- Descrambling
- Reed Solomon, Turbo and low-density parity-check (LDPC) decoding
- Telemetry storage on the hard disk and playback
- Frame (or raw data) time-tagging
- Real-time graphical display of TM Frames.

Satellite Telecommanding

- Reception and checking of telecommand messages from the telecommand clients (at the mission control center)
- PCM encoding
- IF modulation (PM or FM or suppressed carrier PSK/PSK)
- Compliance to CCSDS recommendations (command operations protocol COP-1).

Satellite Tracking (Range and Doppler Measurement)

- Reception and checking of ranging requests from the ranging clients
- Ranging tone or code generation, tone phase tracking
- IF modulation (FM or PM) and Doppler compensation
- Phase-shift measurement
- Ambiguity resolution and distance computation

- Range data time-tagging and transmission to the Ranging Clients
- Doppler measurement and time-tagging.

Simulation and Testing

Simulation capabilities for functional and performance test purposes:

- Simulation of a PSK or PCM telemetry signal from TM formats stored on the hard disk or received over data network from a remote simulation client
- IF modulation
- Automatic bit error rate (BER) measurement
- Real time IF spectrum analysis.

The functional block diagram of the baseband unit which performs the above operation services is depicted below (Fig. 11.2).

RF Compatibility testing is often performed for spacecraft to be supported by the ground station TT&C operation. These tests are conducted at the ground station to ensure compatibility between the station and the spacecraft at RF level (as described in Sect. 11.6.1).

11.5.1 Tracking the Spacecraft

In order to acquire the signal, the ground station has to know the expected track across the sky. This data is provided in the form of a time series of predicted pointing values from the flight dynamics experts at the spacecraft control center. Software at the ground station may also allow the calculation of this data from orbital elements. The antenna is oriented into a waiting direction shortly before an overflight and starts the programmed movement at the expected time or as soon as a carrier is detected (AOS). This mode is called step track.

If a tracking receiver is available at the ground station, the angular offset of the downlink signal direction against the antenna centerline can be measured. Three different methods are in use: four-horn static split system, higher order modes system and conical scan systems. Their output are delta signals for azimuth and elevation. These values can be used to set the antenna control unit into auto-track mode, which means that the system minimizes the angular offset and the antenna points more directly to the satellite. The delta values or the actual antenna pointing values can be stored into a file as a new time series. Depending on the project, this data can then be used to update and improve the orbit determination. This method was dominant for low Earth orbiting spacecraft but is usually replaced by GPS measurements in modern spacecraft. For the first orbits after launch and during contingencies it remains important.

If the system loses the signal in the middle of a track, the step tracking needs to be re-enabled in order to find the spacecraft again.

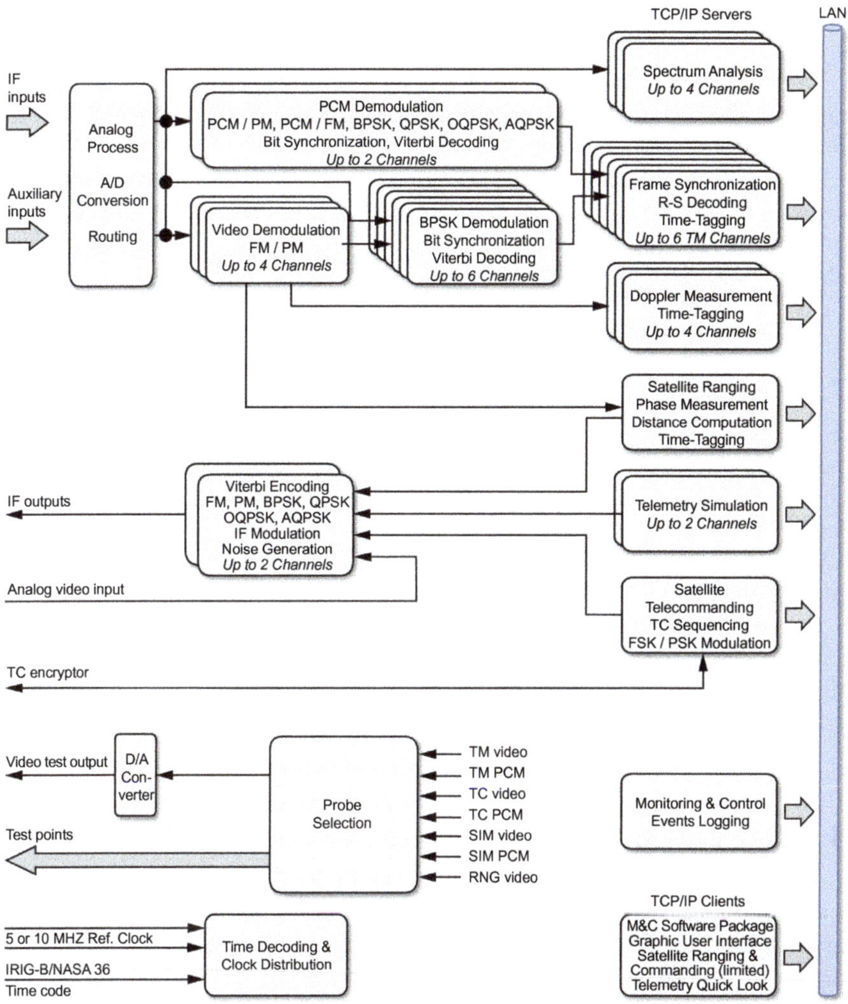

Fig. 11.2 Functional block diagram of the baseband unit Cortex CRT. Adapted from Cortex CRT Quantum user manual

11.5.2 Ranging

Geostationary satellites are not permanently stable at their position over a ground station. Their behavior needs to be monitored closely. The dominant method is to measure the distance to the spacecraft and its radial velocity. On ground this is done with the ranging unit described in Sect. 11.3, sending special tones and analyze the return signal phase. The satellite in most cases also needs equipment to support this process. For good results, it shall be avoided to have telecommanding activities during ranging. Therefore, it is usually organized in ranging sessions of five-minute duration every 30–60 min. A permanent ranging also has no advantage for the orbit determination over interval ranging. The ranging has to be coordinated with the activities of the mission control center. During LEOP campaigns, it may be difficult to find suitable gaps in flight operation. During routine operations, this is usually no problem.

Depending on the mission, a coherency in frequency between uplink and downlink has to be achieved by the on-board transponder during the ranging session. Some transponders need ground commanding by the mission control center to start this coherent mode after a loss of uplink signal or a handover from a different station.

11.5.3 Monitoring and Control of Operations

The monitoring and control (M&C) of the different subsystems is usually done centrally from a control station. An operator has full access to the complete communication chain and all antennas and devices for all projects. Some ground stations may be controlled remotely from the spacecraft control center or from a different ground station.

The pre-pass procedure includes a mission configuration that is particular for each satellite mission. A mission configuration means setting up the station with the mission parameters. The M&C sends all these parameters to the different devices as part of the pre-pass procedure.

11.6 Measurement Campaigns

In addition to the routine TT&C support, different measurement campaigns are also performed at the ground station. These measurement campaigns include RF compatibility test for TM, TC, ranging and Doppler. This test usually takes one week to complete all the tests. A LEOP could take weeks depending on the mission, as well as an IOT campaign. An IOT campaign requires a thorough rehearsal before the test campaign begins.

11.6.1 RF Compatibility Test

The objective of the RF compatibility test is to demonstrate the design compatibility between the satellite and the ground stations at the RF levels of the telemetry, telecommand and ranging signals. The compatibility tests are performed between representative models of the satellite's RF system (e.g. an engineering model or also the flight model for small satellites) and the ground station (see Fig. 11.3). A common

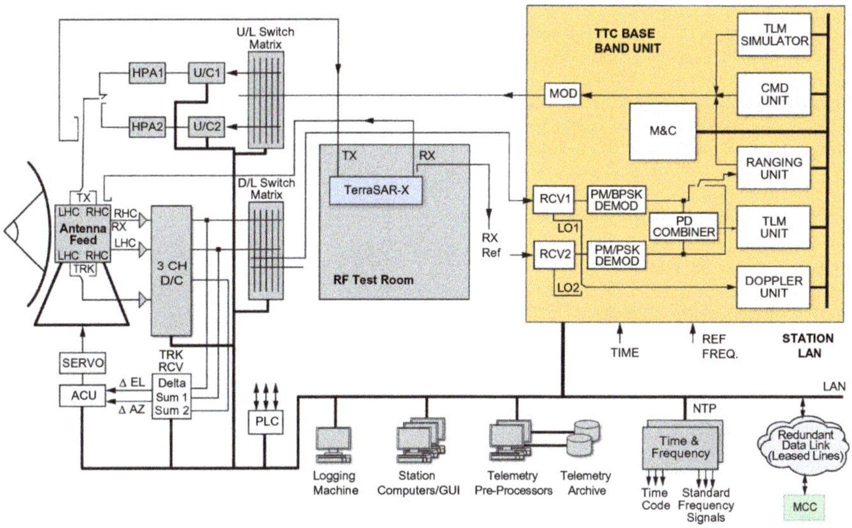

Fig. 11.3 RF compatibility test configuration Weilheim ground station S-band system with TerraSAR-X as the device under test

alternative name of this activity is "suitcase test", because in some cases only the necessary subcomponents are transported to the station in a large box.

The RF system in the test room (RF shielded measurement chamber) interfaces with the station equipment via coaxial cables (hardwired) to reduce RF interferences from outside. All measurement devices have to be calibrated before the tests are done.

A compatibility test has five main topics:

- Spacecraft radio frequency tests
- Telemetry tests
- Telecommand tests
- Ranging tests
- Earth station antenna tracking system tests.

The spacecraft radio frequency tests include the following:

- Spacecraft output power and its stability
- Spacecraft output frequency and its stability
- Spacecraft receiver signal threshold (minimum required uplink power for spacecraft receiver lock)
- Spacecraft receiver tracking bandwidth.

Telemetry (TM) tests include:

- TM carrier suppression
- Ground receiver TM threshold
- TM bit error test (BER vs. Eb/N_0)

- Uplink signal effect on TM
- TM signal spectrum plots.

Telecommand (TC) tests include:

- Spacecraft receiver TC threshold (minimum required uplink power for TC reception)
- Uplink modulation index variation.

Ranging tests include:

- Ground equipment ranging delay
- RF suitcase ranging delay
- Overall ranging delay and signal threshold
- Ranging downlink spectrum
- Ranging downlink modulation index.

The antenna tracking system test determines the minimum required downlink signal level for tracking receiver signal acquisition.

There are essentially three phases of an RF-Compatibility test:

- Test preparation (test planning, resource planning, provision of equipment and measuring equipment; this is documented in the test plan and procedure)
- Test execution [documentation of the test results in a test report and documentation of deviations via non-conformance report (NCR)]
- Test follow-up (creation of the test report and mission-specific parameter lists).

The RF compatibility test plan and procedure were developed based on previous CCSDS Green Book (CCSDS 412.0-G-1 1992), now a silver book; CCSDS historical document available at CCSDS website.

Doing an RF compatibility test is a costly effort for the customer. It may be tempting to omit it. Experience has shown, however, that in many cases incompatibilities between ground and space have been uncovered only in this test, as in other ground tests the RF equipment cannot be involved or is not fully comparable.

11.6.2 LEOP

LEOP supports vary in duration from a few days (LEO) to a few weeks (GEO). During that time TM, TC, ranging, Doppler and tracking data measurement may be requested.

Using the auto tracking system is very important in LEOP, as the actually measured elevation and azimuth angles of the antenna (tracking data) are requested by the mission control center as input for orbit determination and also because in the first orbit or following orbit maneuvers the track may differ from the prediction.

The activities of LEOP TT&C are similar to the routine TT&C, as described in Sect. 11.8, except that the LEOP is a continuous support. The orbit of the spacecraft changes during LEOP support. Hence, it requires more attention than the normal TT&C support for LEO missions. Also, because the spacecraft is new in orbit, it may take some time for the operators in the control center and in the ground station to become familiar with the peculiarities of the system. In situations where the RF contact is not stable, intensive interaction with the control center may be necessary.

11.6.3 In-Orbit Test (IOT) System

After the LEOP, there will be two phases of testing. The first is the IOT test, carried out to confirm that the spacecraft is fully functional and achieving nominal performance (platform IOT). The second is the overall system test to determine the system performance and quality of service aspects (payload IOT). The platform IOT is done by the spacecraft control center using mainly the same services as in routine operations and is not covered here. However, both test campaigns may overlap in order to reduce the necessary time.

For GEO satellite missions, the payload IOT determines the in-orbit performance of the communications payloads by direct measurement from a main anchor station. The IOTs will be started immediately after the defined IOT orbital test position has been reached.

The tests can be done with any ground station that has the necessary equipment available. It can be provided by the customer, the satellite control center or a third-party ground station that is located at a suitable latitude (Figs. 11.4 and 11.5).

Fig. 11.4 IOT system in Ka-band with 13 m diameter antenna at Weilheim ground station

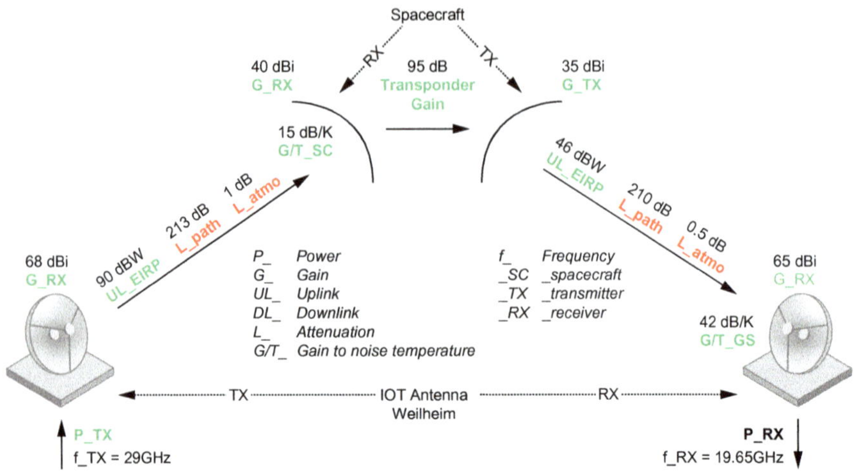

Fig. 11.5 The Weilheim IOT system and the basic budget determination steps for an IOT

The main IOT test bench elements are:

- Spectrum analyzer
- Signal generator/synthesizer
- Power sensor
- Microwave system analyzer
- Communication server.

The payload IOT measurement includes the following:

- Spacecraft EIRP measurement
- Spacecraft EIRP stability measurement
- Spacecraft G/T measurement
- Spacecraft G/T stability measurement
- Transponder group delay measurement
- Transponder frequency response measurement
- Transponder phase noise measurement
- Spacecraft transmit receive antenna pattern measurements
- Third order intermodulation distortion (IM3) caused by the non-linearity of power amplifiers.

The tests should be prepared in full detail with the customer well before the launch, as failures and repetitions or incorrect duration estimations can prolong the test duration considerably. The test conduction is done in close cooperation with the control center. The ground station staff advises the flight team in the control center about the progress and completion of the test steps.

For LEO satellites, in-orbit tests are considered part of the commissioning and are typically not an extended ground station activity.

11.7 Space Link Extension (SLE) Services

A space link is the data transfer between a ground station and the spacecraft. The SLE protocol extends this link to the control centers. It is an international standard developed by the CCSDS community and is widely supported by many space agencies worldwide. In that way, ground stations are able to cooperate with different control centers and agencies world-wide for offering real-time mission support.

11.7.1 Online SLE Service

An SLE service provider software is the interface between the baseband system and the data network. It should support the following SLE services:

- Forward command link transmission unit (CLTU) to forward TC to the Cortex for the uplink. All TCs are received in sequence and can also be configured to be sent time-tagged to the baseband equipment.
- Return all frames (RAF) to return all TM from a space link.
- Return channel frames (RCF) to return TM for specific virtual channels from a space link.

The return frame services are supported in two modes:

- Online complete: all TMs are delivered. During network congestion, TMs are queued and sent later.
- Online timely: TMs are delivered in real time. During network congestion, TM are discarded at the provider side to maintain the real-time quality of the link.

The service provider software should be able to provide concurrent support for multiple spacecrafts, and each spacecraft will have its own dedicated service sessions, identified by the service agreement id. As Fig. 11.6 shows, the service session can be seen as a container for the actual service instances. If uplink is required, a single spacecraft will then have two sessions—one "forward service session" for the uplink service instance, and one "return service session" for the downlink service instances.

By integrating the service provider into the antenna M&C system, it allows the operator to configure SLE services for a spacecraft mission via a central monitoring and control console. This greatly reduces the configuration time and makes a solid step towards the future station automation.

11.7.2 Offline SLE Service

SLE services can be provided also when the space link is not available. This can be used to fill the TM gap due to the timely limitation of the online service or to replay

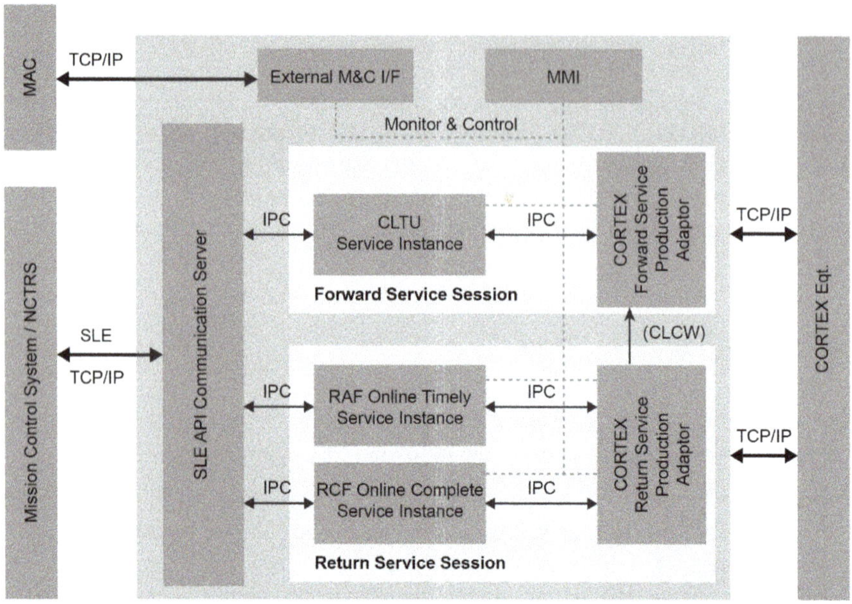

Fig. 11.6 Ground station Weilheim SLE service provider for CRT Cortex baseband unit

the TM in the case of possible data loss that might occur during online data transfer (see Fig. 11.7).

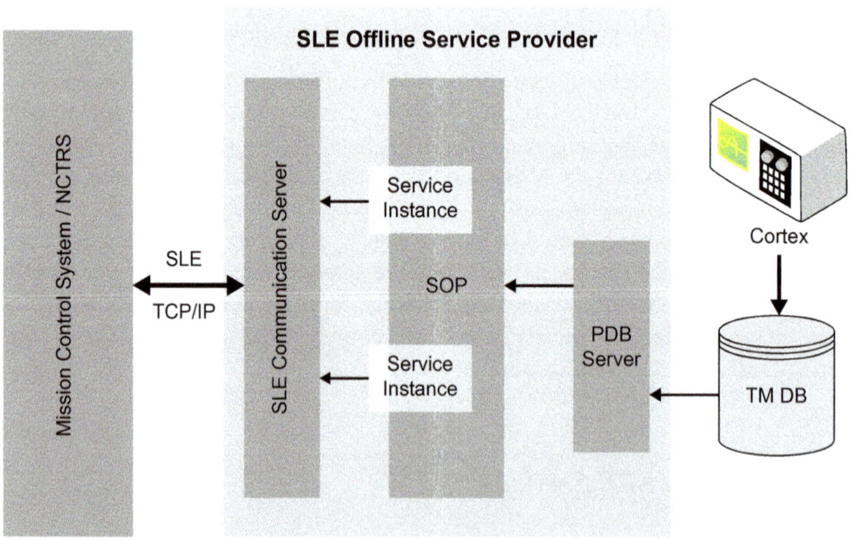

Fig. 11.7 Ground Station Weilheim SLE offline service provider

11.8 Summary

Ground stations are a central element in the access to spacecraft. Their technology and capabilities are tightly coupled to the spacecraft they can connect to. Larger, multi-purpose and multi-project oriented ground stations are profiting enormously from using standard equipment and standard protocols. Initiatives like CCSDS have established a world-wide commonality between ground stations that provide interchangeable services for spaceflight.

References

CCSDS 131.0-B-3 (2017) TM synchronization and channel coding, Issue 3

CCSDS 132.0-B-2 (2015) TM space data link protocol, Issue 2

CCSDS 231.0-B-3 (2017) TC synchronization and channel coding, Issue 3

CCSDS 401.0-B-29 (2019) Radio frequency and modulation systems. Part 1: Earth stations and spacecraft, Issue 29

CCSDS 412.0 G-1 (1992) Radio frequency and modulation systems: spacecraft-earth station compatibility test procedures. Green Book

Cortex CRT. Quantum user manual. Zodiac Data Systems, Zodiac Aerospace, Paris, France

Haykin S (1994) Communication systems, 3rd edn. Wiley, New York

Proakis JG (2001) Digital communications, 4th edn. McGraw-Hill International Edition

Sklar B (1982) Digital communications fundamentals and applications. Prentice-Hall International Editions, Englewood Cliffs, NJ

Amanuel Geda has a M.Sc. degree in Electrical Engineering (Communications) from the University of Duisburg-Essen, Duisburg, Germany. He joined DLR in 2013 and is currently working as Communications System Engineer at DLR Ground Station in Weilheim. Prior to joining DLR he was working on optical and wireless communications system with a research team at the Institute of Communications System, Department of Electrical Engineering, University of Duisburg-Essen.

Chapter 12
Software and Systems

Markus Hobsch and Michael Schmidhuber

Abstract The monitoring and control system (MCS) is the heart of a control center. In this chapter, we focus on the description of this system, since other important software (SW) systems of a control center are described in other chapters of this book. After clarifying some basic MCS terms, the data stream exchanged between the spacecraft and the MCS is explained in more detail. Using the German Space Operations Center (GSOC) as an example, it is shown which SW modules make up the MCS and which SW modules can usefully supplement the operations. An outlook on SW development and maintenance within a control center concludes the chapter.

12.1 Introduction

There are many software (SW) systems in a space mission, varying in scope and purpose. For this reason, the book devotes separate parts to mission planning (see Part V) and the flight dynamics system (see Part IV). In this chapter, we focus on the heart of a control center, without which no space mission can be carried out: the monitoring and control system (MCS). A characteristic of mission control work is that satellites are highly complex and expensive technical systems which, once in orbit, are not directly accessible. Only a thin "umbilical cord" of radio signals connects the spacecraft to its operators. Before a command can be transmitted to a spacecraft, it must be parameterized and tested by the MCS. Telemetry data sent from the satellite is processed by the MCS and made available to the engineers on ground. The engineers need to get an idea of the spacecraft's condition in space. They must anticipate and solve problems and keep the mission alive. The MCS helps these engineers to interpret the data and translate them into readable units. It can also display the data in a readable way and identify trends, for example.

M. Hobsch (✉) · M. Schmidhuber
German Space Operations Center (GSOC), German Aerospace Center (DLR), Oberpfaffenhofen, Germany
e-mail: markus.hobsch@DLR.de

F. Sellmaier et al. (eds.), *Spacecraft Operations*, Springer Aerospace Technology,
https://doi.org/10.1007/978-3-030-88593-9_12

The following section describes the software that makes up a core MCS and the functional and non-functional properties it must fulfill. In addition, other software tools are included to facilitate the operational tasks of the engineers. These are also briefly addressed and described. In order to keep up with technical innovations or increasingly complex requirements for the MCS, existing tools of the MCS are continuously improved or completely new software is developed. The last section discusses how to build a software development regime for operational systems.

12.2 Fundamentals

The subsequent explanations of terms should help you to find your way around the subject of mission control systems and to get a first idea of what is important for software in such systems (see also Chap. 19). For the position of the MCS software within the control center network, please also refer to Chap. 9.

12.2.1 Telemetry Parameters

The elements that contain information about the spacecraft to be transmitted to the ground are called telemetry parameters (sometimes also called telemetry points). They may contain status information (like ON/OFF flags), numeric data (like temperatures or counters) or binary data (unstructured). Their value or meaning must be encoded into a binary format. In most cases it is important to save bandwidth (Evans and Moschini 2013) and therefore the smallest possible coding is used: Flags can be 1-bit values, the length of the bit pattern of integer numbers depends on the value range of the corresponding parameter. Measured values either use an interpolation table or a standard real number format like IEEE 754. The used encoding is described in the spacecraft telemetry and telecommand (TM/TC) database.

12.2.2 Telecommands

Telecommands are the instructions that enable control of the spacecraft from the ground. They are also defined in the TM/TC database of the spacecraft. A telecommand has an identifier and may have a set of command parameters that modify or specify its behavior. Commands can switch devices, set values in registers or transport binary data segments.

An important part of the commands is the "address" part that describes which part of the on-board data handling (OBDH) system should receive the command. This is described in Sect. 12.3.5. The remaining part of the command is the command data.

It is composed of the previously mentioned command parameters and the command identifier.

12.2.3 *TM/TC Database*

The definitions of telemetry parameters, telecommands and their associated calibrations, limits and identifiers are stored in a TM/TC database. The relevant information needed for communication must be available for the spacecraft itself and for the counterpart, the MCS. Only if both have the identical knowledge anomaly-free communication can be established. Therefore, a change in the database can result in an update both on ground and in space. There may be much more information in the database, such as parameters that are only used on ground.

12.2.4 *Monitoring and Control System (MCS)*

Monitoring and control systems should provide users with the most up-to-date information and the ability to effectively, efficiently, and reliably operate and manage systems for which they are responsible on ground and in space. At the German Space Operations Center, which serves as an example, there is a unit responsible for the design, construction, testing, maintenance and further development of the MCS and associated software (Fig. 12.1).

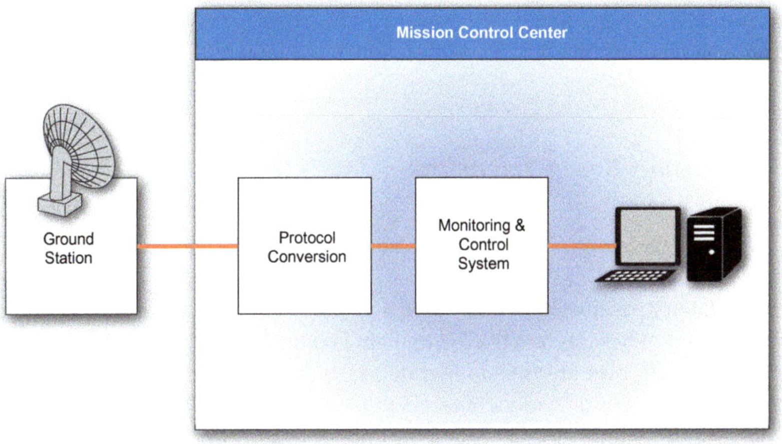

Fig. 12.1 Monitoring and control system

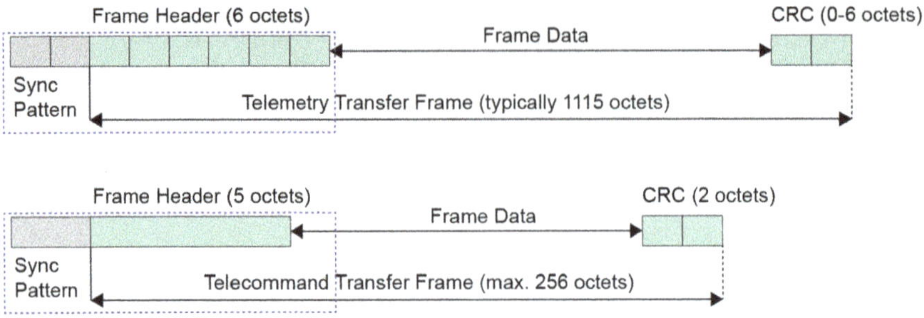

Fig. 12.2 Transfer frames for telemetry (upper part) and telecommand (lower part)

12.3 Space-to-Ground Data Streams

12.3.1 Data Transport

In both, the uplink and the downlink, the data is transported in transfer frames. We will explain the principle mainly using telemetry as an example. Telecommand transmission works in a similar way.

All data streams consist of very long rows of bit values. This stream is divided into successive pieces of equal length called transfer frames. All transfer frames start with an invariant "sync pattern" that allows the receiving station to recognize the beginning of a transfer frame even after interruptions. The transfer frames also contain header information indicating the size of the transfer frame, a frame counter, and a virtual channel identifier (Sect. 12.3.5). The frame trailer contains a cyclic redundancy check (CRC) that allows detection of transfer errors. Typical examples are shown in Fig. 12.2. Standard sizes are 1115 bytes for telemetry and 256 bytes for telecommand streams.

12.3.2 Frame-Based Telemetry

This type of telemetry is sometimes also called pulse code modulation (PCM) telemetry format.

The telemetry parameters are assigned and distributed to a set of several different "minor frames". These have a fixed length and completely fill a transfer frame. The different minor frames are transmitted one after the other and repeated cyclically. Each minor frame has a header containing the frame ID to allow identification. The complete set of minor frames is called the "major frame" or "format".

Figure 12.3 shows an example of a minor frame. Telemetry parameters are assigned to specific positions within the available data space as defined in the spacecraft database. The encoding of the parameter values depends on their use and can

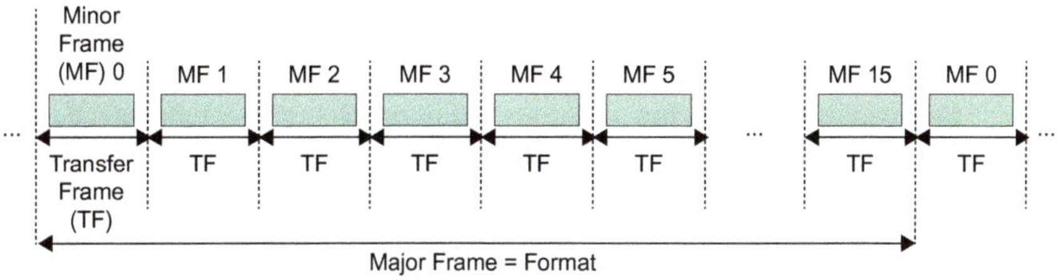

Fig. 12.3 The set of minor frames define a major frame and is repeated cyclically

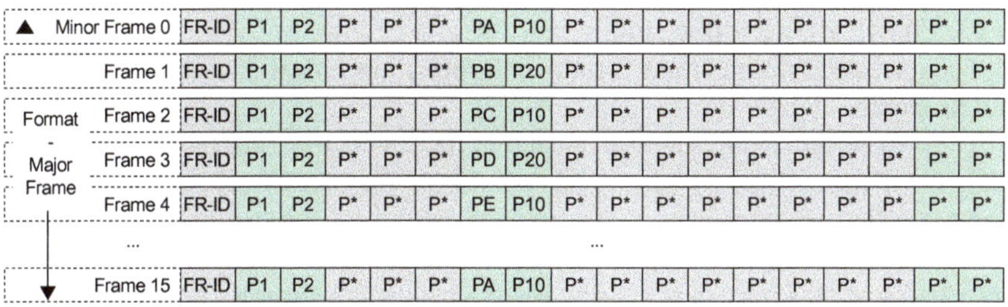

Fig. 12.4 An example for frame telemetry. P1 and P2 denote parameters that are transmitted in all 15 minor frames. Parameters PA-PP are defined only once per format (or major frame). Parameters P10 and P20 are placed in every second minor frame

occupy any number of bits. Most often, parameters are grouped and aligned in data blocks of 16 or 32 bits, referred to as words or double words.

If each telemetry parameter is assigned to only one minor frame, all values are transmitted only once per major format. However, it is possible to assign important and particular dynamic parameters to several or even all minor frames as shown in Fig. 12.4. In that way, an attempt is made to match the importance of the parameters within the available bandwidth. However, this rigid scheme may not be sufficient for all operational situations. To overcome this, it is usually possible to dynamically redefine some ranges during the flight, thus changing the selection and sampling rate of telemetered values, resulting in various improved methods and concepts such as dwell, dump, pages, oversampling and subsampling.

Frame telemetry is a basic method that allows data to be transmitted in a simple way. It has been used since the early days of spaceflight, but has been superseded by packet telemetry (ECSS 50-04C 2008) and, since the late 1990s and early twenty-first century, by packet services (ECSS 70-41A 2003), which are discussed in the following sections.

12.3.3 *Packet Data Structures*

The increasing demand for higher data rates and more telemetry parameters, as well as the general tendency to include more software features in spacecraft design, led to the need for more flexible and efficient data transport methods. A prominent example is the Consultative Committee for Space Data Systems (CCSDS) packet telemetry and command standard (CCSDS 133.0-B-1). This standard implements many modern mechanisms of data transfer.

In this concept, parameters and commands are grouped into logical packets. Any number of packet definitions can be stored in memory. Their size can vary up to 216 = 65,536 bytes (octets) (Fig. 12.5).

The packet header contains information that allows identification of the packet and its length.

The data stream itself is still organized in frames of fixed length which basically act as containers for the packets. As shown in Fig. 12.6, each frame contains a small element called segment, which can be considered as a management layer for the packets. It allows multiple packets (or packet parts) of any length to be multiplexed into the frame structure, thus distributing the bandwidth capacity to multiple target devices. The telecommand segment layer may include a sublayer for authentication. In this case, an authentication trailer is added after the segment, reducing the packet size. Authentication is performed by the TM/TC board of the spacecraft. It protects against illegal commanding of the spacecraft.

Telemetry packets can be organized to be generated at a fixed rate or on demand, or when an event occurs. For example, confirmation (or rejection) messages for command execution are generated on event. The processing of packets is performed by the on-board computer and places a relatively heavy load on it. It must provide the mechanisms for buffering and organizing the telemetry stream. The frame and segmentation layers are handled at the hardware level within the TM/TC board.

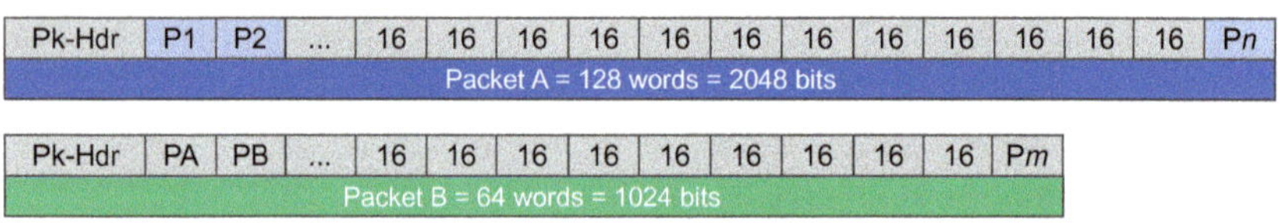

Fig. 12.5 Two example packet definitions. Typical data packets are 16 bit and are also called words. They contain telemetry parameters at defined positions

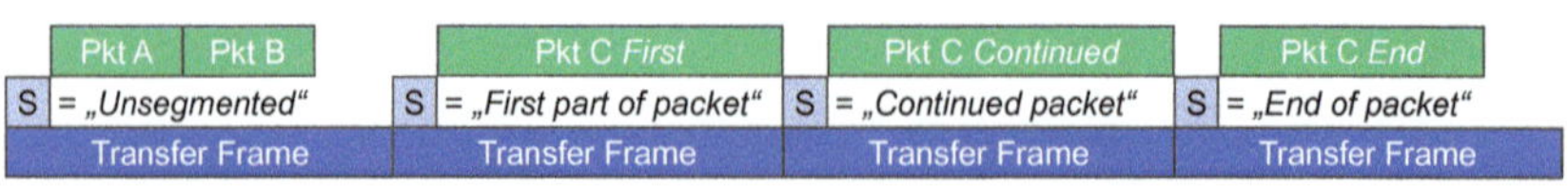

Fig. 12.6 The segmentation layer allows the transfer frames to be filled with parts of large packets or several smaller packets

Telemetry packets can be switched off or on, they can be sent at different data rates, depending on the current operational situation, which makes this telemetry concept very flexible and efficient, but also complex and not very transparent.

Again, the definition of the various packets is included in the TM/TC data base, which is also called mission information base (MIB).

12.3.4 Packet Utilization Standard

So far, we have described how information is transported, but not how it is handled on application level. To achieve a unified approach to spacecraft control, the concept of service types was defined by ECSS in the Packet Utilization Standard (PUS), ECSS 70-41A (2003) and updated to ECSS-E-ST-10-41C (2016). The idea is not to randomly write data patterns into (TC) and read others out of (TM) on-board registers and interpret the result in a mission specific manner, but to rely on standardized services.

These defined services are much more far-reaching than the classic TM/TC tasks, which may be summarized (a bit impudently) as.

- Sending a telecommand to a destination
- Loading telecommands to the time-tag buffer
- Sending telemetry to ground
- Configuring telemetry.

The services defined in the PUS now cover a vast field of data management functions, reaching from memory management to time distribution. However, not only standard services are defined in the PUS, but there is also room for mission specific definitions.

This approach ensures that the S/C manufacturer or the mission can tailor the standard for their implementation. A possible negative result is that it may happen that manufacturer-specific solutions of basic tasks are implemented in private services and are effectively undermining the standardization.

The list of defined services in Table 12.1 shows that on the one hand the tasks are now grouped into various aspects, and on the other hand, it becomes clear that very advanced concepts are also included that were previously only fragmentarily or not present at all. An example would be service 4 that allows statistical information about on-board data to be requested in a formal way. The service defines the necessary data structures and functions. In this way manufacturers and operators are "persuaded" towards thinking about advanced concepts. The hermetic art of spacecraft control now has an open language. However, as before, the implementation of advanced services can be a major development effort for the manufacturer and may therefore be avoided, or tailored. Standardization will facilitate the reuse of ground systems, and in the long run for space systems as well.

Table 12.1 The currently defined standard PUS services (ECSS 70-41C 2014)

Service type	Service name
1	Telecommand verification service
2	Device command distribution service
3	Housekeeping and diagnostic data reporting service
4	Parameter statistics reporting service
5	Event reporting service
6	Memory management service
7	Not used
8	Function management service
9	Time management service
10	Not used
11	On-board operations scheduling service
12	On-board monitoring service
13	Large data transfer service
14	Packet forwarding control service
15	On-board storage and retrieval service
16	Not used
17	Test service
18	On-board operations procedure service
19	Event-action service
20	Parameter management service
21	Position-based scheduling
22	File management service

12.3.5 TM/TC and Security Management

Flow Control Mechanisms

This function is used to assure safe operations by protecting against faulty transmission of data. The first measure is to include error control and forward correction information. This is done on low level (coding and transport level) and cannot be modified during operations. The keywords are cyclic redundancy checks (CRC) (cf. Section 12.3.1) and randomization (a deliberate coding of data to assure bit synchronization).

Operational tasks are limited to configure the MCS on ground to accept or reject faulty telemetry frames for diagnostic purposes and to monitor the uplink frame quality that is reported in low level telemetry generated by the TM/TC Board.

The second measure is the introduction of a quality of service (QOS). This is used to recognize and recover data lost during transmission. The basic means for this is counters. Frames and packets carry counters that are tracked and checked for gaps.

In the uplink channel this can be a closed-loop process by using the COP-1 protocol as described in ECSS 50-04C (2008). This mechanism is implemented on transfer frame level. The control center can choose to transmit either in AD ("acceptance") or in BD ("bypass") mode. Simplifying a little bit, in AD mode the commanding process waits until it receives the uplink confirmation of the previous command before it transmits the next one. Some MCSs allow an automatic retransmission of lost commands. Apart from synchronization tasks the command link is not necessarily slowed down compared to the BD mode if the "sliding window" mechanism is applied where a settable number of frames is uplinked before the confirmation is received. Obviously, the AD mode can only work if telemetry is available. For blind transmissions or unstable connections the BD mode must be selected.

This mechanism is called frame acceptance and reporting mechanism (FARM, ECSS 50-04C 2008) and is done on low level in the TM/TC Board. No such mechanism is standardized for telemetry. A possible retransmission must be done on MCS application level or manually initiated by the control center personnel.

Routing Mechanisms

The established standards allow to precisely address source, destination and the route of telecommand and telemetry data units. ECSS (ECSS 50-04C 2008) defines the following qualifiers:

- The spacecraft ID (SCID) is a world-wide unique number that protects against uplink signals intended for another spacecraft (SANA n.d.). It is important to note, that simulation systems, engineering models etc. usually have separate IDs. Correct selection can be a source of problems when switching from simulation to the mission.
- The virtual channel (VC) ID in the uplink path distinguishes if a signal is intended for the prime or the backup decoder (TM/TC Board). Only the addressed decoder will forward the command. Note, that this flag is set in the ground command system. Usually it is independent from the spacecraft database and should be easy to change. In the downlink, the VC allows to interleave different data channels that may be processed by separated ground systems. The low-level implementation allows separating the channels without knowledge about the spacecraft database. This is commonly used for data dumps (see Sect. 19.3.5) that are processed on ground in different ways or by different control centers.
- The multiplexing access point (MAP) ID is analyzed within the TM/TC Board and directs the command to different devices. Usual destinations are the command pulse distribution unit (CPDU), the authentication unit, and the prime and backup OBC. Most telecommands will be accepted only at a specific destination, but in the case of OBC MAP IDs this can be used for addressing the backup OBC over cross-strapped connections.
- Like for the virtual channels, this parameter can usually be set dynamically in the MCS. Alternatively, backup commands can be defined in the spacecraft database for this purpose.

- The application ID (APID) is used to specify the destination on packet level. It is evaluated by the OBC.

Authentication
Another mechanism that is located inside the TM/TC-Board is the authentication. This is used to ensure that only the legitimate control center can control the spacecraft. The uplink stream is usually not encrypted, but encrypted signatures are attached to the segmentation layer. The signature includes an encrypted counter to protect against replays. A set of secret emergency keys is implemented into the on-board authentication device, and it is possible to upload more keys for daily usage. Special telecommands are available to control the mechanism.

Flow control, routings, high-priority commands and authentication mechanisms are nicely described in MA28140 Packet Telecommand Decoder (2000).

Encryption
For military spacecraft, but also increasingly for civil applications, it has become common to encrypt either all or parts of the data transmission. There requires an encryption device and a set of keys to be on board. To enhance the protection, the keys must be changed in regular intervals. It depends on the implementation how this is exactly handled and even where the encryption device is located inside the OBDH. Two main usage cases are common:

- Encryption of the complete data stream. The en-/decryption on ground side takes place in the control center.
- Encrypting only payload data. The en-/decryption on ground side can take place in the user center. This may even be done with multiple users where each user has their own set of keys and can only extract their own data.

12.4 Monitoring and Control System (MCS) Software

In principle, satellite operations require only one tool that can process and send TM and TC, a generic mission control system (MCS) for spacecraft mission control. This does not even have to be a ready-made tool. Scripts or simple commands via a command line tool may be a sufficient approach for certain types of satellites and missions, e.g. very low-cost missions. Besides TM/TC processing, all other tools are optional at first. However, they make life easier and lead to a more comfortable and secure operation. Using the software (SW) modules used at GSOC as an example, it will be shown which tasks during operations and preparation should be covered by SW.

At GSOC, the multi-mission approach is also applied to the SW. The same binaries are used for each mission. Mission-specific properties are configured. This simplifies the maintenance immensely.

12.4.1 TM/TC Processing

TM/TC processing takes a central role within each ground segment to control communication to the vehicle in space. While the ground stations perform the physical uplinks and downlinks of communication data—telecommand (TC) data in uplink and telemetry (TM) data in downlink—the MCS processes the information within these data streams: It compiles each telecommand from the spacecraft into the appropriate bytecode, which is routed to a connected ground station and radiated to the spacecraft's TC receiver. In turn, the MCS also receives incoming telemetry data from the ground station and processes this data for monitoring or archiving purposes. For all of this, it must know the data structure and protocols necessary for communication, i.e. the Packet Utilization Standard (PUS, ECSS-E-ST-70-41C 2016). It is also essential that the satellite bus and the central MCS speak the same "language" or use the same "vocabulary". Therefore, identical TM/TC databases (see Sect. 12.2.3) must be loaded on the ground and on the satellite in space.

This central component of an MCS also handles the archiving of TM/TC data and flow control with ground stations, performs calculation, validation and the limit checking of parameter values, verifies and validates outgoing and incoming data, and presents this information to the user.

GECCOS (GSOC Enhanced Command- and Control System for Operating Spacecraft, Stangl et al. 2014) represents such a generic mission control system at GSOC. It occupies the aforementioned central role within the GSOC ground segment. GECCOS is also the appropriate tool to make all operational data accessible to other connected or offline tools within a typical mission operations system, e.g. the "ProToS" automation tool, the Satmon display tool, or to external GSOC subsystems such as flight dynamics or the mission planning system (see Fig. 12.7). All processing must be real-time, stable and reliable on a 24/7 basis. To meet the requirements of modern spacecraft control centers, GECCOS is continuously enhanced, not only on the user side (modern graphical user interface), but also "under the hood" with a modernized architecture and automated tests.

GECCOS is based on ESA's MCS development SCOS (Spacecraft Control and Operation System. The history of GECCOS started in 1999/2000 at DLR GSOC to become the leading multi-mission satellite monitoring and control system. The goal was to start an in-house development to replace various legacy systems with a single system suitable for current and future missions. Following this approach, many customizations for missions have been incorporated along with other enhancements and modernization approaches. So far, it supports as one generic "multi-mission MCS" a wide range of scientific and commercial satellite platforms (CHAMP, GRACE, Spacebus 3000, TerraSAR-X, TanDEM-X, PAZ, TET, BIROS, H36w-1, Eu:CROPIS, EDRS-A, EDRS-C).

In addition, GECCOS has the capability to act as both an MCS and a central checkout system (CCS) to support space projects from the earliest possible project phase. When the spacecraft manufacturer starts with the assembly, integration and test (AIT) phase, the spacecraft is typically controlled by the CCS. The MCS then

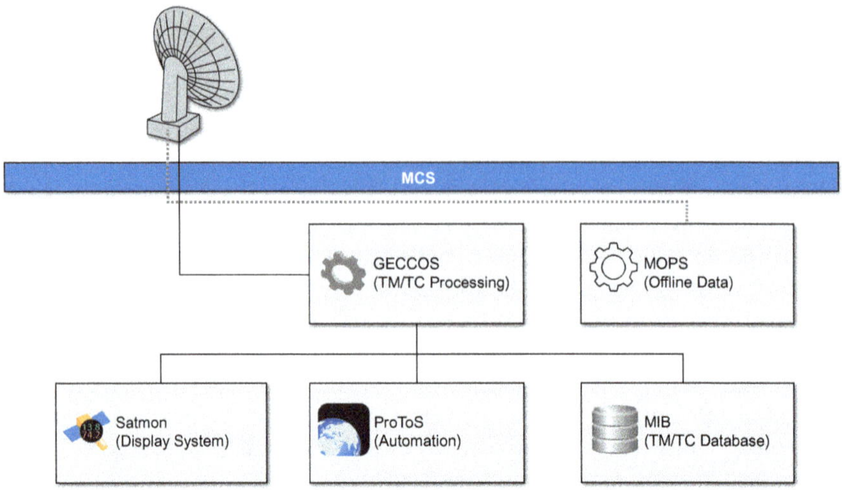

Fig. 12.7 Illustration of an MCS system at GSOC

takes over this task during the operational phase after the vehicle has been handed over to the control center. The advantage of combining CCS and MCS is not just having one system for both tasks. Rather, their combination in terms of their data handling kernels is an important paradigm that has been demonstrated in the missions TerraSAR-X, TanDEM-X, PAZ, TET, BIROS, and Eu:CROPIS. The key advantage is the inherent validation of the upcoming MCS tasks during early mission phases: It ensures the compatibility of the MCS and CCS with the spacecraft database as well as with the flight control procedures (FCP), both of which have already been validated during the early AIT and checkout phases of the spacecraft and are ready to work with the MCS in the control center (Fig. 12.8).

At ESOC (European Space Operation Centre, Darmstadt) and GCC (Galileo Control Center), also antenna control is performed by a SCOS-based MCS. This is not the case at GSOC. This is done by SW of the ground data group (see Chap. 10).

The European standard MCS, SCOS, was developed around the millennium and is therefore older than 20 years. It is still the basis for many MCS used worldwide (including GECCOS). Among other things, the impending obsolescence and lack of interoperability between the various MCS led ESA to the idea to develop a new MCS, the European Ground Systems Common Core (EGS-CC). This is a European initiative to develop a common infrastructure to support space systems monitoring and control in pre- and post-launch phases for all mission types. The goal is to harmonize between the monitoring and control system (S/C operations) and the central checkout system (S/C assembly, integration, and test—AIT). EGS-CC provides a software basis for monitoring and control operations throughout all mission phases. Its component-based and service-oriented architecture enables easily extensible functionality fitting the specific scenarios it is used in. As the EGS-CC is the result of

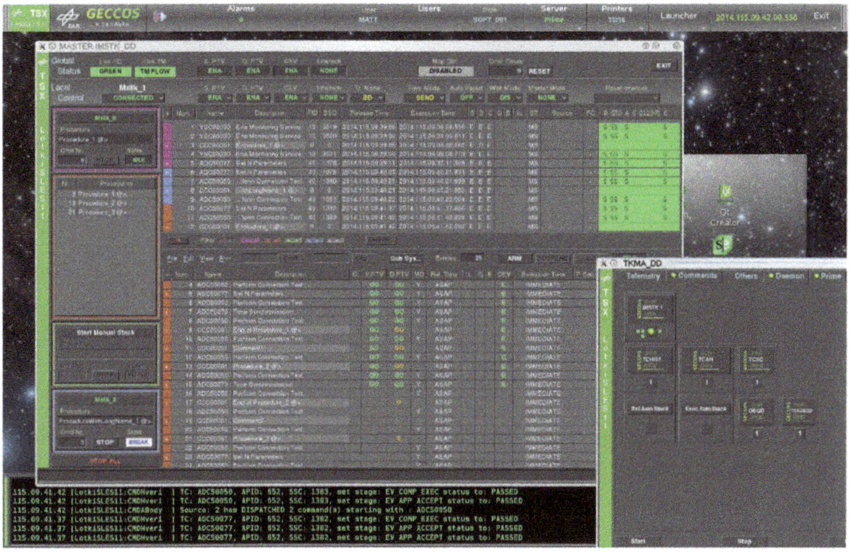

Fig. 12.8 Graphic user interface of GECCOS

collaboration between space agencies and industry across Europe, one of its goals is to increase the interoperability between partners through its standardized interfaces.

While the EGS-CC is currently under development, GSOC is taking on the role of integrator. This role involves integrating and validating the EGS-CC in scenarios that reflect realistic mission setups, including simulating data and interfacing to GSOC infrastructure and tools (EGS-CC 2013).

12.4.2 Display System

Most available TM/TC systems provide some integrated way to view telemetry data. However, a dedicated tool with the core display functionality better meets user requirements and simplifies maintenance and development.

The task of a display system is first of all to display data and prepare them adequately for the engineers. However, when we talk about data, we do not always mean measured values. Other information such as limits, minimum and maximum or even the command events can also be part of the data and are important for the engineer. Presentation is possible in various ways. For example, current values can be displayed in an alpha- numeric display (AND) page, in a time-dependent plot or in a visual graphic display, i.e. synoptics page. Decisions made by the operation engineers can only be as good and timely as the underlying information is fast, clear, unambiguous and complete. Therefore, displaying satellite telemetry must have these characteristics.

The display system receives the information and data from the TM/TC system. There should be only one source for processed telemetry: the MCS. Any coincidences must be avoided. The display system must correctly display the values from the MCS. Corrections or calibrations are made in the MCS. The display system is ignorant in this respect. Other data sources can be orbit data or pending ground station contacts.

At GSOC, Satmon is the telemetry visualization software suite for monitoring satellites developed together with Heavens-Above GmbH (Peat and Hofmann 2004). Featuring a client–server architecture, it provides users with incoming telemetry data in real time and offers fast access to archived and offline data (see Sect. 12.4.3). The client provides many options of displaying telemetry on configurable digital display pages, such as lists, aggregated parameter pages, purpose-built overview pages, procedure pages, and interactive plots. As it interfaces with the core monitoring and control system, it also allows the visualization of additional information like command history and MCS events. An integrated editor allows users to prepare project-specific display pages as well as to customize displays for personal needs. The server component features user authentication, encrypted connections, data flow control, a highly efficient telemetry database optimized for high storage density and low retrieval latency, and many admin tools for diagnostics and maintenance (Fig. 12.9).

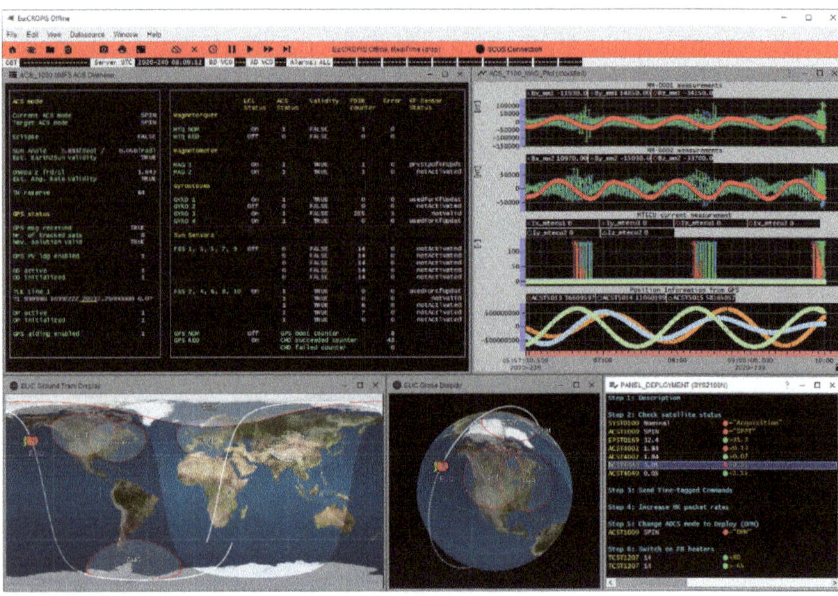

Fig. 12.9 Graphic user interface of Satmon including different types of display pages

12.4.3 Offline Data

In addition to the data transported in real time from the satellite to the MCS, there is telemetry data, especially for low-flying satellites, that is generated and accumulated between the ground contacts. This data should be inserted into the MCS and made available to the engineers just like real-time telemetry. The data packets differ from the real-time data. During a ground contact, the data can be dumped to ground in parallel while telemetry is generated and directly transmitted at the time of contact on the spacecraft. One second of generated data is loaded into the MCS in one second. The offline data can be downloaded either in this contact or in other contacts. But data generated over a long period of time have to be loaded into the MCS in a few seconds.

To avoid disturbing real-time data, many satellites can separate the real-time and offline data streams, for example by using different virtual channels (see Sect. 12.3.1). Although GECCOS can process this data, in some projects this data is processed by different processor for historical reasons (Fig. 12.10).

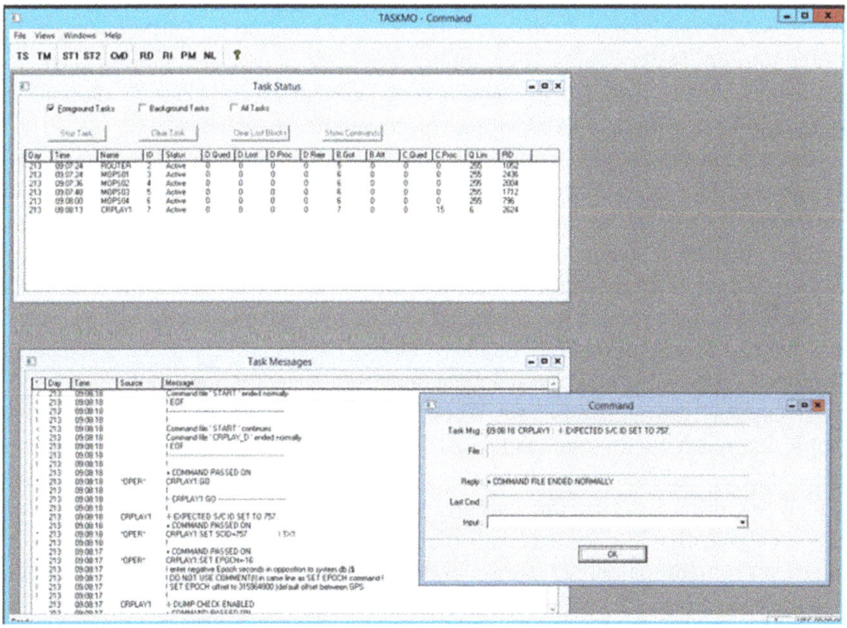

Fig. 12.10 Task monitor of MOPS

12.4.4 Automation of Mission Operations

Automation in the field of mission control is usually associated with the following needs:

- Higher speed of operational activities
- Relieving people from monotonous work
- Savings in personnel costs
- Improvement of operational process quality
- Reduction of manual errors
- Safe and efficient execution of operational processes.

The capabilities of hardware and software on ground have grown immensely compared to the pioneering days of space travel, but the complexity of on-board systems has also increased accordingly. Modern satellites are sending higher amounts of data that needs to be processed on ground. In the other direction, the increasingly complex tasks performed by spacecraft also require the transmission of longer command sequences. Automatic commanding helps to create and send complex sequences. Automatic analysis of telemetry can help engineers to detect anomalous behavior that cannot be seen through limit violation monitoring. Furthermore, an in-depth, algorithm-based analysis can detect correlations between various telemetry and events, or even predict probabilities of anomalies occurring. Predictions of degradation can serve as a basis for preventive maintenance measures.

Automation should relieve users of routine tasks without hiding processes behind the scenes. Many cases of human error can be explained by insufficient situational awareness. As a result, technical or organizational measures can be introduced that increase situational awareness or prevent the loss of situational awareness, thereby increasing overall safety in the human–machine system. Thus, every decision made automatically must remain traceable and modifiable, because responsibility cannot be delegated to machines. Only humans can react appropriately to the unexpected. The interface between man and machine must be constantly improved.

At GSOC, two approaches for automated commanding are followed. A very simple approach is to send command sequences without large if–then-else or telemetry queries. This is mainly used in S/C projects operating LEO satellites. Due to their low orbit, LEO satellites usually have contact times of around ten minutes. The main objective is to send as many commands as possible. Simpler tasks can be taken over by automation, but more complex operations cannot. The basic procedure is described in Fig. 12.11. Satellites in geostationary orbit, however, communicate continuously with the ground station. Here, it is possible to command iteratively and to react to certain conditions or to wait until they occur. The automation tool is connected to the MCS. It receives telemetry and can release telecommands based on validated command procedures (see Fig. 12.12).

The automation of operational processes in addition to commanding and monitoring are just as natural and worthwhile. The majority is provided by suitable SW

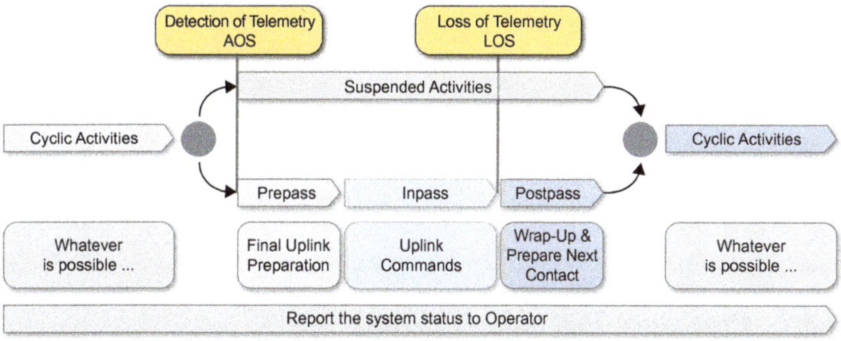

Fig. 12.11 Process for automated commanding for LEO S/C missions at GSOC

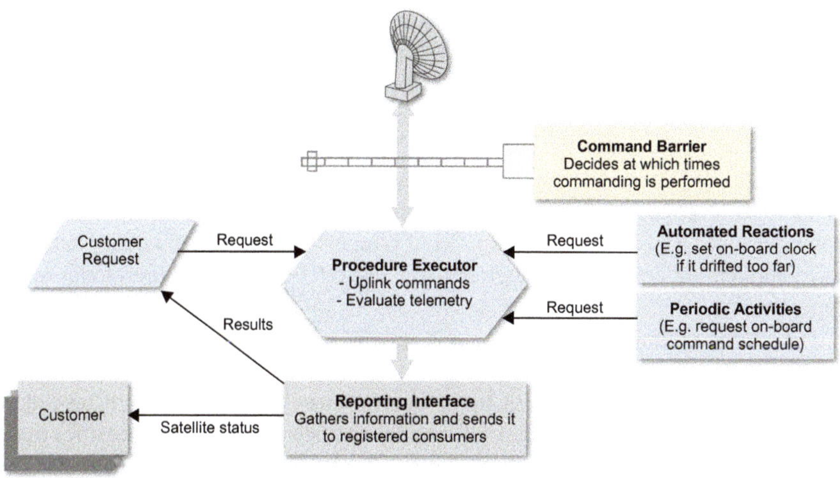

Fig. 12.12 Process for automated commanding for GEO S/C missions at GSOC

solutions, which on the one hand support the users in manual work, but on the other hand take over monotonous, recurring or manually not feasible tasks.

The automation of operational processes at ESOC is realized with the framework MATIS (Calzolari et al. 2006). EGS-CC will support this functionality, too.

12.5 Ops Support Tools

There are many tasks that have to be performed by the MCS and the engineers besides commanding and monitoring, whether in preparation of the mission or while the satellite is already in space.

For commanding a satellite, prepared, verified and validated command sequences or procedures (see Chap. 7) are used. Using ProToS these can be edited, instantiated and executed.

In operations, it is critical to share information, provide documentation and data products in a simple way. Operational workflows must be supported. The OpsWeb is a web-based platform that simplifies the exchange of information within operations at GSOC.

12.5.1 Procedure Tool Suite (ProToS)

ProToS is a multi-mission application package developed at the German Space Operations Center (Beck et al. 2016; Beck and Hamacher 2018) to support the creation, instantiation, execution, and management of software-based satellite flight control procedures (FCPs) by satellite operations engineers or ground operations procedures (GOPs) by ground engineers. It utilizes high-level procedures that encapsulate individual TM/TC checks in order to increase the efficiency of satellite and ground operations in highly complex scenarios. ProToS integrates with GSOC GECCOS or ESA SCOS-2000 mission control systems and serves as their command front end for its high-level procedures. It automates the execution of procedures and directs their control flow based on incoming TM. It is an eclipse RCP (rich client platform)-based application and can be integrated as a server/client architecture or can run in standalone mode.

The editor module enables the engineers to view and edit procedures and to validate them against the TM/TC database. Integrated version control, user and mission management, and access rights control ensure safe and secure procedure development. ProToS can be used for manual and, via various interfaces, for automatic instantiation. Parameter values can be imported and validated on-the-fly. The executor module can trigger automatic or manual execution of procedures through its direct connection to the MCS. Therefore, it is used for automatic commanding (see Sect. 12.4.4).

Procedures can be also viewed and instantiated remotely by configuring a file server location (remote procedure access). After a successful login, ProToS automatically synchronizes its database with the definitions from the file server. Procedures can be viewed and instantiated. File-based export products can then be generated and eventually be sent back to the control center (Fig. 12.13).

12.5.2 Operations Support (OpsWeb)

OpsWeb is a web-based platform for sharing information, providing documentation and up-to-date data products (e.g. timelines, procedures, calendars, intra-project messages). OpsWeb helps operations engineers comply with and optimize mission

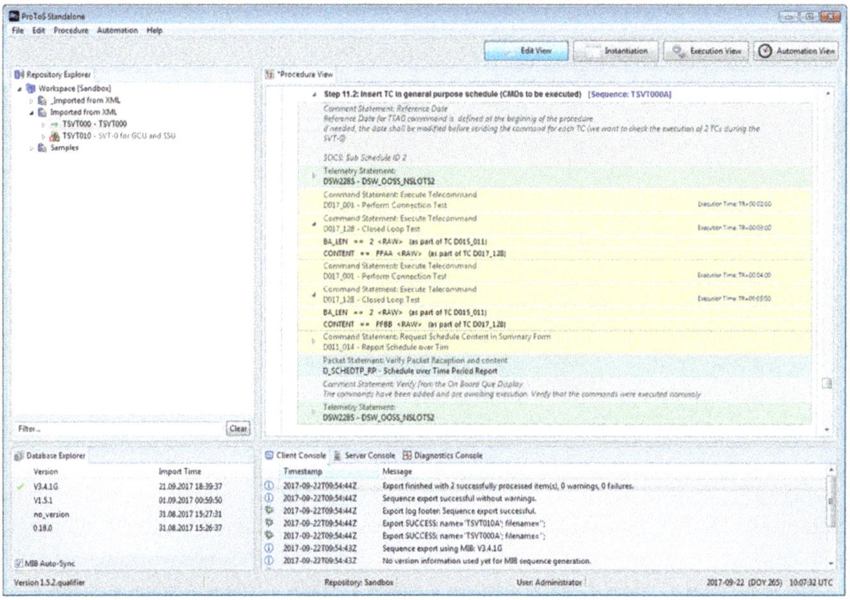

Fig. 12.13 ProToS graphic user interface

operations procedures and workflows. One of the key features of OpsWeb is a dedicated issue-tracking system. It provides the framework for controlling process flows based on digital signatures, supported by flexible user management. Among other requirements, this enables the implementation of a two-man rule, meaning that two flight directors must sign any form of ticket or report. OpsWeb's web interface is accessible from the control center's internal network, but also from the internet. Therefore, operations from home is also supported. The functionality of the OpsWeb pages displayed on the internet can be restricted to meet the security requirements of individual satellite projects. All important information can be displayed at a glance in a dashboard that each user can configure. The responsive design also supports the display of the OpsWeb on mobile devices. With all this, the OpsWeb is the central platform for collaborative operations at GSOC (Fig. 12.14).

12.5.3 Predictive Maintenance

Predictive maintenance by definition refers to a maintenance process based on the evaluation of process and machine data and is found primarily in the context of Industry 4.0. The processing of underlying data makes forecasts possible that form the basis for needs-oriented maintenance and consequently the reduction of downtimes. In addition to the interpretation of sensor data, this requires a combination of analysis technologies and in-memory database to achieve a higher access speed to the data

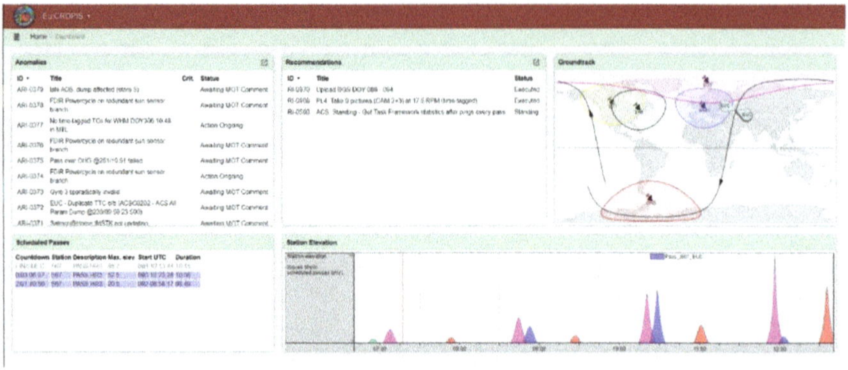

Fig. 12.14 OpsWeb dashboard

compared to hard disk drives. If everything works out, it will be possible on the operations side to assign a fix of an anomaly even before it arises. After all, economic goals can only be achieved if systems, machines and processes are working properly.

The Automated Health Monitoring System (ATHMoS) uses machine learning algorithms to find unrecognized features in data. It detects outliers by comparing the behavior of telemetry parameters to a model built on previously collected nominal data. Multi-parameter correlation is possible, where multiple parameters can be examined for known or unknown correlations (Schlag et al. 2018; O'Meara et al. 2016) (Fig. 12.15).

12.6 SW Development and Maintenance

Further advancement of mission control systems must compensate the additional workload of experts who cannot be multiplied at will due to the constantly growing satellite fleets and their complexity. The engineers on ground must have all necessary information available quickly and reliably. Ergonomics and efficiency are the key words here.

The SW for mission control is subject to constant development and dynamics. New technical standards, new operational processes and methods are created and implemented. Each mission implements operational aspects differently and software can become obsolete, which can quickly lead to obsolescence problems in long-running projects. Thus, new components or adaptations of the existing SW are always required.

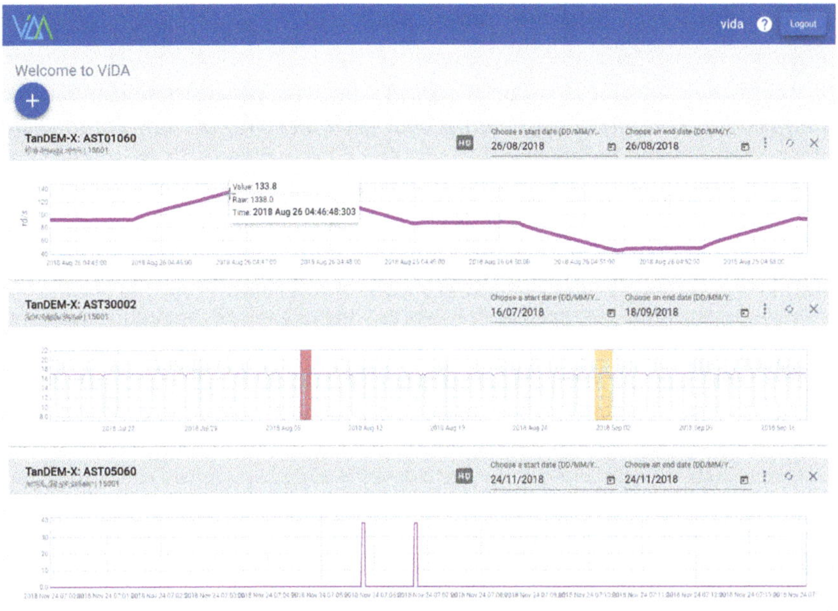

Fig. 12.15 The results from the ATHMoS analysis are stored in a database and can be displayed in plots and visualizations, on dashboards, or serve as a basis of automated alerts in the ViDA web application

12.6.1 Multi-Mission Approach

Even though satellites will remain expensive one-off products, as they always aim for the unexplored or make available the latest in communications or sensor technology, the aim on the ground is to reduce costs by reusing components as far as possible. In order to make this possible, standardization is being promoted in international committees. But also within a control center it is intended that the system engineers can fall back on a modular system of existing optimally tested and practice-proven components. This serves to reduce the time and risks involved in setting up the MCS for new missions.

At GSOC, all MCS software components are developed largely mission-independent and modular. Project specific characteristics and implementations are determined by configuration. The advantage is obvious: There is only one binary that needs to be maintained—no matter in which mission a bug is found or a change request occurs. Maintenance for a huge number of different project versions of software is not necessary any more.

12.6.2 Standards and Standardization

Whether own development within a control center or SW development for third parties and for commercial use, international space and SW development standards and processes should be applied. In most cases the standards to be followed are specified by the projects or customers themselves and adapted to the project. Self-imposed and strictly defined standards and processes help to keep the quality of the software high. International standards help in this respect and may even dictate a few things. Not always all requirements of the standards are applicable to the project or to your own organization. In this case it is worthwhile to analyze the requirements and to tailor the application of standards to the real-life situation. The adapted standard can be presented to customers. More stringent standards or conditions that deviate from the in-house standards should be discussed with the customer and reflected in the price.

The international standards continue to develop and are reviewed again and again. If necessary, they are expanded or adapted. This is done in the appropriate committees of international standardization groups. As a large space agency or space company, it is worthwhile to participate in such committees. You can look beyond your own nose and find out how other organizations develop SW and what new technologies they may be using. Standards that are applied in the European space industry are for example:

- **ECSS**: The European Cooperation for Space Standardization is an initiative established to develop a coherent, single set of user-friendly standards for use in all European space activities (ECSS n.d.). For every software that is part of a space system product tree and developed as part of a space project the standard "ECSS-E-ST-40C-Software" (ECSS-E-ST-40C 2009) shall be applied. Quote from the homepage: "This standard covers all aspects of space software engineering including requirements definition, design, production, verification and validation, transfer, operations and maintenance". They defined for example the Packet Utilization Standard (see Sect. 12.3.4).
- **ISO:** The International Organization for Standardization is an independent, non-governmental international organization that develops voluntary, consensus-based, market relevant international standards that support innovation and provide solutions to global challenges. In the context of the SW development for space systems two standards are mentioned, which are used in a control center:

  (1) ISO/IEC 27001: Information Security Management
  (2) ISO 9000 Family: Quality Management.

- **CCSDS**: Founded in 1982 by the major space agencies of the world, the Consultative Committee for Space Data Systems is a multi-national forum for the development of communications and data systems standards for spaceflight (CCSDS n.d.). They defined for example the MO Services or the standard TM frame and the space data link protocol.

12.6.3 Software Procurement

The decision which parts of a SW system in a project will be provided internally or externally is made by the project manager in close cooperation with the system managers. House policies, financial and strategic aspects shall be considered. Third-party SW, commercial off-the-shelf (COTS) SW or modifiable off-the-shelf (MOTS) SW can have benefits compared to in-house developments. For each individual SW, a trade-off must be made at least between the following:

- Availability
- Maturity
- Specialization
- Scaling
- Interfaces
- Customizing
- Documentation
- Maintenance
- Training
- Costs
- Innovation
- Feature coverage
- Flexibility
- Acceptance
- Access to source code.

Thus, rapid availability and a certain maturity are arguments in favor of COTS and MOTS, as are the pure procurement costs in most cases. However, very often the limits of specialization, customization, flexibility and maintenance are reached.

If a SW development task is awarded as a contract to an external provider, at least the following points must be considered:

- Definition of range of functions (including security functions)
- Scope of services and costs
- Delivery dates
- Scope of delivery
- Changes to former releases
- List of changed components
- (Regression) test reports
- User manual, installation guide, configuration setup, test data for acceptance tests, if needed: test tools and test pattern
- Configuration control
- Payment plan
- Scope of liability
- Description of technical modification service, non-conformance handling, engineering change handling, maintenance
- Statement to operational aspects like on-call service.

Incoming inspections shall be performed for all delivered releases and patches. If a delivery does not meet the expectations, it must be reasonably and formally rejected. Afterwards, the solution can be found in dialog with the SW provider.

12.6.4 Software Development

To steer SW development activities and to unveil delays as early as possible the entire SW development is divided into seven phases. This process is applied at GSOC and leans on the approach as in ECSS-E-ST-40 and ECSS-Q-ST-80.

1. **Feasibility Phase**

As a prerequisite, the SW requirements of the mission project have to be defined; the process is covered in the prescription requirements management. This phase unveils the feasibility of requirements implementation. The investigation is based on different SW architecture concepts to find out the most suitable solution for the SW project.

2. **Specification Phase**

This phase maps all requirements to SW functions into requirements/specification matrices (RSM) and determines the priority of each function according to the requirement specification. In parallel, a system test plan will be set up to verify testability of all requirements to be implemented. This ensures that the system architecture, describing roughly the single components and their interfaces, is reviewed critically, especially if the development team and verification team are separated. As a result of that, the SW architecture and the according system test plan are defined. For later SW lifecycles, e.g. maintenance, new versions, etc., a regression test plan is set up covering corrective and progressive regression test.

3. **System Design**

In this phase, the individual components and their interfaces are derived from the SW architecture and RSM. Parallel to the definition of components, the integration test plan must be defined in order to verify a reasonable partitioning of the system and its testability. As a result, the component definition, the component interfaces and the integration test plan are defined.

4. **Component Design**

This phase starts with the definition of component tests to guarantee the proper partitioning of components and testability of all components. Coding of components will start earliest after availability of test cases according to the requirements. The design document is prepared serving as document for component implementation. All components and their according source code management are defined.

5. **Component Coding**

The coding of the SW component can begin. From the very beginning, all parts of the SW development (source, configuration, setup, test case documents) must be managed in a version-controlled repository (development and test repository) managed by the SW developer. All parts of SW development in the repository are compilable at any time.

6. **Component Test**

The component test is performed based on the component test plan. Considering the cyclomatic complexity and lines of codes of the component, it has to be defined by the software project leader whether this task is performed by the SW engineer or by another person. Distributing this task to another person guarantees a more objective testing and adds additional inspection know-how, resulting in more robust and stable components.

7. **Integration**

This phase builds upon verified components a subsystem in the integration environment and performs the integration test based on the integration test plan defined in phase 2. Sources for integration will be managed in the integration and validation repository, managed by source code management. Anomalies will be assigned as NCRs/DRs to the component developer. Necessary changes to the system are handled in ECRs and forwarded to either system, subsystem, or component owner.

8. **System Test**

Based on verified subsystems, a system is set up in the integration environment to validate the system in real mission conditions. Sources for system tests will be managed in the integration and validation repository managed by source code management. NCRs/DRs and ECRs are handled similar to phase 4.

9. **Deployment**

The validated system is set up in the operational environment. The repository serves as documentation basis of any system deployment. In parallel, a SW maintenance plan is set up covering the way how the SW is adapted in case of faults, changes in SW environment or needed improvements or adaptations.

Validation activities may continue after phase 6, e.g. up to and including the in-orbit verification phase.

This development method, following the waterfall model, is established for extensive software projects with clear requirements and payment milestones. There are also other models: The V-model, for example, is a process model that was originally designed for software development. Similar to the waterfall model, it organizes the software development process in phases. In addition to these development phases, the V-model defines the procedure for quality assurance (testing) by comparing the individual development phases with test phases. For smaller projects and/or projects where the customer does not yet know the exact requirements on the other hand, an iterative approach has proven successful. SCRUM or Kanban as the most widely-known methods can give the development a methodical framework, if necessary.

Faster and more frequent iterations with the user or customer can be achieved. Another result are very fast release cycles. This is exactly the goal of an agile or iterative approach. Also, in the aerospace industry and especially in contact with the customer, many wishes and requirements at the beginning of a long software project cannot be foreseen or can only be formulated vaguely. Here, a pattern of constant stopping of the development, checking the results with the customer until then, and a retrospective is necessary. All described possible methods are only means to an end. It is worth considering to what extent formal constraints are absolutely necessary. After all, house processes or processes required by the customer still have to be followed.

Although the prime goal in a project is to meet the mission requirements, it will often be beneficial to start SW development based on already developed and tested SW systems. Within a multi-mission concept, this approach is essential in order to keep down costs and avoid repeating previous development errors. Additional or slightly changed requirements must be described and defined in engineering change requests (ECRs), and suitable test cases and test steps must be defined and documented in parallel. This evolutionary SW development process actually follows the same phases as described in the section above, tailored according to the needs and complexity of the software change in question. The decision whether SW development can follow an evolutionary approach heavily depends on the complexity of the planned SW and the extent of coincidence of requirements to the parent and the new SW. This decision should be made carefully. Modifications estimated as small changes to existing SW systems often influence other unconsidered parts of the system resulting in major changes, efforts and unplanned long development times with limited test coverage and multiple efforts for integration (Fig. 12.16).

Based on ECRs, existing software is adapted to new requirements or specifications. Multiple steps may be needed to fulfill requirements. ECRs have to be documented, for each iteration test definitions have to be concluded in advance to the implementation, and test execution as well as test reports have to be documented.

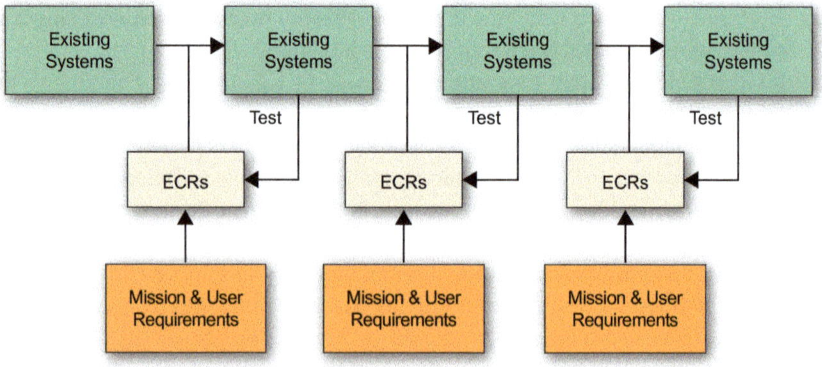

Fig. 12.16 Evolutionary approach

12.6.5 Maintenance

The handling of errors, anomalies, bugs, requests or other conspicuous features of the software running in operation must follow a defined process (ECSS n.d.). This process is ideally supported by a ticket system. A ticket opened by a user or a tester is evaluated by the SW engineers and implemented accordingly after approval. The development process described above is followed. After the appropriate tests and release of the project, the patch can be applied.

GSOC's own MCS team combines the knowledge of the self-developed MCS software modules and is responsible for the maintenance. Although the multi-mission concept has been continuously pursued for years, third party or commercial off-the-shelf (COTS) SW may complement the MCS of the S/C projects, especially if they perform project-specific tasks or tasks that are not part of the core business of a control center. It is guaranteed that the core competences remain in house and very short response times are possible. Short ways as well as direct participation in operations ensure short feedback paths, very short response times and optimal results in an iterative SW development.

A command system can be monitored and maintained in different ways. The monitoring and control system and its associated software must run smoothly. Data and information are exchanged within the system and transported to the outside via interfaces. During critical (e.g. during LEOPs) and special operations, it may be necessary to monitor the monitoring and control system in real time and respond to anomalies as fast as possible. During routine operations, a help desk in combination with oncall service may be sufficient. GSOC's DataOps team (see Fig. 12.17) makes sure that during operations the SW runs as it should, that it does what it should do

Fig. 12.17 MCS call sign

and that the data gets where it should go. In critical phases, they take over the real-time monitoring of the command system and, if required by the project, also on-call duty during routine operation. The data operations engineers blend skills of software and operation engineers. Their deep knowledge of MCS systems is combined with great flexibility, speed of response and reliability. They monitor the system, analyze problems and take care for all measures within an operational environment with all standards and processes.

12.7 Conclusion

One of the most important SW systems in a control center is the monitoring and control system. It processes TM and sends TCs to the spacecraft. Dedicated SW tools with core functionalities, such as displaying telemetry or processing the TM/TC database or editing command procedures, can usefully extend the portfolio and increase usability. It also better meets user requirements and simplifies maintenance and development.

Standards are a good way to ensure interoperability and a certain level of quality. Therefore, standardized protocols are used on ground and in ground-space segment communications. Standardized processes also lead to a reasonable level of quality in SW development. Nevertheless, it is useful to evaluate whether processes can be tailored to the daily routine and which ways are feasible. The new European monitoring and control system, EGS-CC, sets new standards and facilitates operability between control centers.

The automation of control center operations will become even more important in the future than it already is. As a result, topics such as the human–machine interface and situational awareness will become increasingly important and need to be discussed openly. Artificial intelligence and machine learning can help to manage the increasingly complex tasks of a control center.

References

Beck T, Hamacher JP (2018) ProToS: automation of flight control procedures for the European data relay system. SpaceOps Conference, 28 May–1 June 2018. Marseille, France

Beck T, Schlag L, Hamacher JP (2016) ProToS: next generation procedure tool suite for creation, execution and automation of flight control procedures. SpaceOps Conference, 16–20 May 2016. Daejeon, Korea

Calzolari GP et al (2006) Automation of ESOC mission operations. SpaceOps Conference, 19–23 June 2006. Rome, Italy

CCSDS 133.0-B-1. Space packet protocol

CCSDS. https://public.ccsds.org/, cited 19 Apr 2021

ECSS 70–41A; *Telemetry and Telecommand Packet Utilization Standard* (2003); https://ecss.nl/standard/ecss-e-70-41a-ground-systems-and-operations-telemetry-and-telecommand-packet-utilization/, cited April 19, 2021

ECSS 50-04C (2008) Space data links—telecommand protocols synchronization and channel coding (July 2008). https://ecss.nl/standard/ecss-e-st-50-04c-space-data-links-telecommand-protocols-synchronization-and-channel-coding/, cited 19 Apr 2021

ECSS (2021) European cooperation for space standardization. https://ecss.nl, cited 19 Apr 2021

ECSS-E-ST-40C (2009) Software (Mar 2009). https://ecss.nl/standard/ecss-e-st-40c-software-general-requirements/, cited 19 Apr 2021

ECSS-E-ST-70-41C (2021) Telemetry and telecommand packet utilization, 15 Apr 2016. https://ecss.nl/standard/ecss-e-st-70-41c-space-engineering-telemetry-and-telecommand-packet-utilization-15-april-2016/, cited 19 Apr 2021

EGS-CC (2021) European ground systems common core. www.egscc.esa.int, cited 19 Apr 2021

Evans D, Moschini U (2013) Ten times more information in your real-time TM. In: Schmidhuber M, Cruzen C, Kehr J (eds) Chapter 14 in Space operations: experience, mission systems and advanced concepts. AIAA 2013, ISBN 978-1-62410-207-3

MA28140 (2000) Packet telecommand decoder; Dynex Semiconductor. Publication No. DS3839-6 Issue No. 6.1 (June 2000)

O'Meara, C.; Schlag, L.; Faltenbacher, L.; Wickler, M.; *ATHMoS: Automated Telemetry Health Monitoring System at GSOC using Outlier Detection and Supervised Machine Learning*; SpaceOps Conference, 16–20 May 2016, Daejeon, Korea (2016)

Peat C, Hofmann H (2004) SATMON—a generic user interface for satellite control. SpaceOps Conference 17–20 May 2004. Montreal, Quebec, Canada

SANA (2021) Space assigned number authority. http://sanaregistry.org/, cited 19 Apr 2021

Schlag L, O'Meara C, Wickler M (2018) Numerical Analysis of Automated Anomaly Detection Algorithms for Satellite Telemetry. SpaceOps Conference, 28 May–1 June 2018, Marseille, France

Stangl C, Braun A, Geyer MP (2014) GECCOS—the new monitoring and control system at DLR-GSOC for space operations, based on SCOS-2000. SpaceOps Conference, 5–9 May 2014, Pasadena, California, USA

Markus Hobsch received his diploma in aerospace engineering in 2009 from the University of Stuttgart (Germany). In 2010 he joined Astrium GmbH, Bremen (Germany), as a Thermal Hardware and Thermal Test Engineer. He was responsible for the Multi-Layer Insulation of the ATV and the thermal testing of the RCS of the Vega rocket. In November 2011, he joined the German Space Operation Center (GSOC) in Oberpfaffenhofen (Germany) as AOCS Subsystem Engineer for the TSX, TDX, TET and BIROS satellites and as a Project System Engineer of the H36w-1 satellite. Additionally, he is head of the Mission Control and Data Systems team of the Mission Operations Technology department at GSOC since 2015.

Michael Schmidhuber works in the Department for Mission Operations at the German Space Operations Center (GSOC) where he is responsible for satellite operations training. He graduated as an aerospace engineer at the Technical University of Munich in 1994. The training activities comprise not only staff training but also courses for an external audience in the annual Spacecraft Operations Course at GSOC. Also, he is involved in the organization of the bi-annual international SpaceOps Conference.

Part IV
Flight Dynamic System

Chapter 13
Orbital Dynamics

Michael Kirschner

Abstract As all objects in space are moving all the time, it is essential to understand the dynamics of the satellite motion in the frame of satellite operations. The following chapter will introduce the reader into some theoretical background, which is necessary for the understanding of the orbital dynamics of a satellite. With this background knowledge, the reader is then guided through the major topics of flight dynamics aspects of satellite operations.

13.1 Introduction

This chapter deals with the orbital dynamics of a satellite orbiting around a central body. Other aspects of orbital dynamics are discussed in later chapters, such as the relative motion of two satellites in Chap. 25, swing-by maneuver in Chap. 26, and landing on interplanetary objects in Chap. 27.

The current chapter is divided into two main sections. The first deals with theoretical aspects in Sect. 13.2, where an overview of the description of a satellite orbit, orbit disturbances, orbital maneuvers and orbit maintenance is given.

These topics are important for the understanding and successful execution of flight dynamic operations, which is presented in the second main Sect. 13.3 flight dynamics tasks.

When reading this chapter, it is important to note that only resulting formulas of satellite orbit theories are presented in this chapter and not their derivatives. For those who are interested in such details, books on satellite orbital dynamics such as "Satellite Orbits" (Montenbruck and Gill 2000) or "Fundamentals of Astrodynamics and Applications" (Vallado 2001) should serve as a reference.

M. Kirschner (✉)
German Space Operations Center (GSOC), German Aerospace Center (DLR), Oberpfaffenhofen, Germany
e-mail: michael.kirschner@DLR.de

F. Sellmaier et al. (eds.), *Spacecraft Operations*, Springer Aerospace Technology,
https://doi.org/10.1007/978-3-030-88593-9_13

13.2 Theoretical Aspects

In the following Sects. 13.2.1–13.2.3 we consider a two-body system in which the mass of the central body is significantly greater than the mass of the satellite. The gravitational field of the central body is first approximated as an isotropic field of a point mass. Disturbances caused by deviations of an isotropic gravitational field are dealt with in Sect. 13.2.4.

13.2.1 Satellite Orbit

Each point or position in space is clearly identified by three parameters like the Cartesian coordinates x, y, and z. If we use a coordinate system, a point or position can be expressed by the radius vector r with its components r_x, r_y, and r_z.

An object in space close to a central body is moving all the time, meaning it is changing its position continuously. Therefore, we can describe an orbit as a sequence of positions, where the positions are time depending. Consequently, we need more than three parameters for the description of an orbit. As the position is changing, the time derivative of the position vector being the velocity vector can be used with its components v_x, v_y, and v_z. These six parameters are also called state vector.

For the visualization of an orbit, another set of six orbital parameters is commonly used, the so-called Keplerian elements. In the following, this set of elements is explained addressing the geometry of an orbit and its orientation in space. It has to be noted that only closed-loop orbits are addressed.

Geometry of the Orbit

Under the above assumption, the general shape of a closed orbit is an ellipse, which has two axes, the major and the minor axis. Half of the major axis is our first parameter and is called semi-major axis a (cf. Fig. 13.1). The cut of both axes (major and minor) is the geometric center of the ellipse. But it is especially important to know that the gravitational body is not located in this center, it is located in one of the two focal points of the ellipse. In Fig. 13.1 this focal point is marked as Earth center (EC).

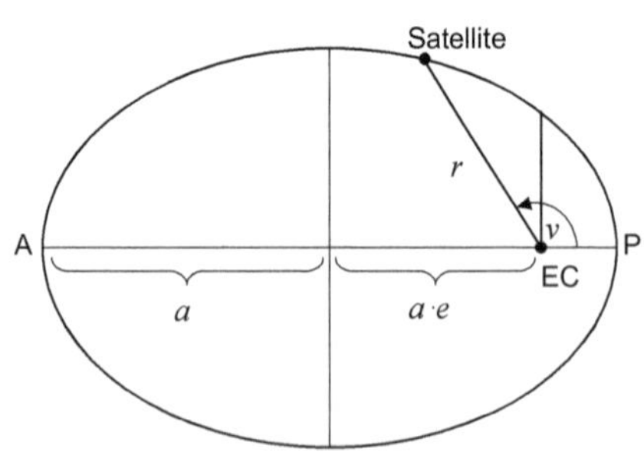

Fig. 13.1 Ellipse geometry

Therefore, an elliptical orbit has two special points, the apogee (A) and the perigee (P), where the distance is a maximum w.r.t. the EC or a minimum, respectively.

The second parameter is the numerical eccentricity e, which is a measure for the slimness of the ellipse. The smaller the eccentricity, the more circular it looks, whereas the larger the eccentricity, the smaller the relation is between the minor and major axis. The product of a and e determines the distance between the focal point and the center of the ellipse. For closed loop orbits it has a range of

$$0 \leq e < 1.$$

In the special case $e = 0$, the ellipse is changed to a circle, where the two focal points are now identical with the center of the circle.

The third parameter is the true anomaly v, which is the angle between the radius vectors to the perigee and the location of the satellite.

Orientation of the Orbit

The remaining three parameters describe the orientation of an orbit in space. We use an inertial coordinate system, which has EC as its origin. The z-axis points to the north (N) and is identical with the rotational axis of the Earth. The other two axis x and y are perpendicular to the z-axis and define the equatorial plane. The x-axis points to the so-called vernal equinox (♈), which is defined as the direction from the EC to the center of the Sun at the time, where the Sun crosses the equatorial plane from the south to the north (March 20). The x-axis is therefore identical with the cut of the equatorial plane and the ecliptic.

The point where the satellite crosses the equatorial plane from the south to the north is called ascending node (A.N.), the opposite point is called descending node. The connection between the two nodes is called nodal line. The ascending node is used to define the right ascension of the ascending node Ω (RAAN). The RAAN is the angle between the direction of the ascending node and the x axis. At the ascending node the inclination i can be defined, which is the angle between the orbital and the equatorial planes. The last of the six parameters is the argument of perigee ω, defined as the angle between the directions to A.N. and the perigee (P). For all definitions, please refer also to Fig. 13.2.

We have now found a set of six orbital parameters that are commonly referred to as Keplerian elements. They are summarized in Table 13.1 and divided into three classes. The first class defines the shape of the orbit (semi-major axis and eccentricity), the second class the orientation in space (inclination, RAAN, and argument of perigee), and the third class the position inside the orbit measured from the perigee (true or mean anomaly).

This set of orbital elements is suitable for the visualization of an orbit. But for orbit propagation this set has some disadvantages. Singularities must be treated in two cases:

1. If we have a circular orbit, there is no apogee or perigee, therefore the argument of perigee is not defined.

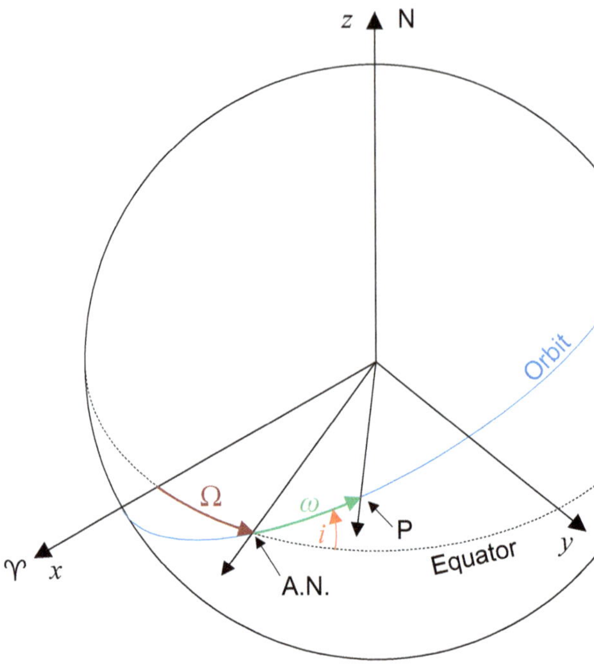

Fig. 13.2 Orientation of an orbit in space and definition of Keplerian elements

Table 13.1 Keplerian elements classification

Class	Parameter	Name
Shape	a	Semi-major axis
	e	Eccentricity
Orientation in space	i	Inclination
	Ω	Right ascension of the ascending node (RAAN)
	ω	Argument of perigee
Location	ν/M	True/mean anomaly

2. In the case that the orbit is inside the equatorial plane, as in geostationary Earth orbit (GEO) missions, we do not have a node crossing, and therefore the RAAN is not defined.

Because of these limitations, numerical orbit propagation uses a different set of elements that is free of singularities, the so-called state vector. It contains the position and velocity vectors, six parameters in total.

13.2.2 Satellite Velocity

One of the most important parameters used in many equations is the velocity of a satellite. The energy conservation law can be used to derive the equation for the satellite velocity. For closed orbits, the energy equation

$$E_{\text{tot}} = E_{\text{kin}} + E_{\text{pot}}$$

can be derived like in the following:

$$-\frac{1}{2}\frac{\mu}{a} = \frac{1}{2}v^2 - \frac{\mu}{r} \tag{13.1}$$

where μ is the gravitational parameter of the Earth ($\mu = \text{GM}_{\oplus} = 398{,}600.4415\ \text{km}^3/\text{s}^2$), a the semi-major axis, v the velocity, and r the radius from the gravitational center of Earth to the satellite.

Velocity in Elliptical Orbits
Equation (13.1) can be converted to

$$v_{ell} = \sqrt{\mu\left(\frac{2}{r} - \frac{1}{a}\right)} \tag{13.2}$$

which describes the velocity on an elliptical orbit. For example, around the perigee of the high-elliptic geostationary-transfer orbit (GTO) in an altitude of about 250 km the speed is about 10 km/s, and in the apogee in the geostationary altitude of about 36,000 km the speed is about 1.6 km/s.

Velocity in Circular Orbits
The equation for the velocity on a circular orbit can be easily derived from Eq. (13.2), because r is equal to a

$$v_{circ} = \sqrt{\frac{\mu}{a}} \tag{13.3}$$

Examples for velocities in circular Earth Orbits.

- Low Earth orbit (LEO) 7.5 km/s
- Geostationary orbit (GEO) 3 km/s
- Moon (about 380,000 km) 1 km/s.

First Cosmic Velocity
The first cosmic velocity—also referred as orbital velocity—is more an academic parameter, as it describes the circular velocity exactly above ground:

$$v_{\text{Cosmic1}} = \sqrt{\frac{\mu}{R_E}} = 7.905\,\text{km/s} \tag{13.4}$$

where R_E is the Earth radius with a size of 6,378.137 km at the equator. Of course, the atmosphere would have to be removed to orbit a satellite directly above the Earth's surface without touching the ground. And also, mountains make it impossible to orbit the Earth at that altitude. But this velocity has another important significance: it determines the absolute minimum performance of a launcher system, which is normally expressed in a velocity increment $\Delta v = v - v_O$, where v_O is zero at the moment of launch. With respect to the energy law, we would change the kinetic energy only, as we stay at the launch altitude (R_E) all the time. But in reality, the rocket has to bring a satellite to higher altitudes (above 200 km) to stay in an orbit around the Earth due to the high air drag in lower altitudes, which would bring the satellite back to ground very quickly. Going to higher altitudes increases the potential energy. In addition, the rocket also has to act against the air drag during the ascent. Due to these two effects, also called gravity and drag losses, a launcher system needs a performance between 9.2 and 9.6 km/s to bring a satellite to LEO.

Second Cosmic Velocity

The Second Cosmic Velocity is defined as the velocity needed to escape from ground into open space. The total energy for an escape trajectory in the form of a parabola is zero. Using Eq. (13.1), the second cosmic velocity can be derived for an escape from ground:

$$v_{\text{Cosmic2}} = \sqrt{\frac{2\mu}{R_E}} = \sqrt{2}v_{\text{Cosmic1}} = 11.180\,\text{km/s} \tag{13.5}$$

In the more general form

$$v_{escape} = \sqrt{\frac{2\,\mu}{a}} = \sqrt{2}\;v_{circ} \tag{13.6}$$

Equation (13.6) describes the escape velocity from any circular orbit around the Earth.

13.2.3 Orbital Period

The time for one orbit revolution, called orbital period T, can easily be derived from a circular orbit.

In general, we can use the well-known orbital equation, which depends on velocity and time

$$s = v \cdot t.$$

For one revolution s will be $2\pi \cdot a$ and using Eq. (13.3) for v we get

$$T = 2\pi\sqrt{\frac{a^3}{\mu}} \tag{13.7}$$

13.2.4 Orbit Perturbations

The orbit of a satellite is influenced by different sources of perturbations, which change the behavior of a pure Keplerian orbit. The main sources are

- Deviations from an isotropic gravitational field of the Earth,
- Drag due to residual atmosphere,
- Solar radiation,
- Influence of the Sun's and Moon's gravity and
- Thruster activity.

The first two items are addressed more in detail in the following.

Earth Gravitational Field

The shape of the Earth is not an ideal sphere; it can be approximated by an ellipsoid of rotation, where the radius to the poles is about 20 km less compared to the radius in the equatorial plane. Due to this oblateness, the nodal line performs a rotation around the polar axis. This rotation can be described by the time derivative of the RAAN Ω for circular orbits

$$\dot{\Omega} = -\frac{3}{2} J_2 \sqrt{\mu} \frac{R_E^2}{a^{3.5}} \cos i \tag{13.8}$$

where the zonal parameter J_2 of the gravitational potential for our Earth describes the form of the ellipsoid. It can be described as a function of the difference between the equatorial and polar radii of the equipotential surface of the Earth's gravity field. More accurately expressed, J_2 is a function of the difference in principal moments of inertia (for more information, please refer to the references cited in the introduction). For the Earth the value of J_2 is 1.083×10^{-3}. Applied to the orbit of the International Space Station (ISS) with an inclination of about 51° and an altitude between 300 and 400 km, the node drift would be about 3° per day to the west.

We can use this effect of node drift to describe a very special orbit, the so-called Sun-synchronous orbit. The requirement for such an orbit is that the node drift has to be equal to the mean motion of the Earth around the Sun meaning the difference between the RAAN and the right ascension of the Sun is constant. As the Earth is moving eastwards around the Sun, which is defined with the positive sign, the mean node drift has to be $+ 2\pi$ per year or + 360° per year. Using Eq. (13.8) and the constant variable C, we can now write

$$\dot{\Omega} = -C\frac{\cos i}{a^{3.5}} = +\frac{2\pi}{year} \tag{13.9}$$

The negative sign on the left side of Eq. (13.9) can be turned to plus (+), if the cos i is negative. This is possible for inclinations larger than 90°.

Figure 13.3 shows all theoretical possibilities, where the altitude is plotted versus the inclination. In reality, most of the satellite missions flying on a Sun-synchronous orbit have altitudes between 500 and 1000 km. The corresponding inclinations are between 97° and 100°.

The oblateness of the Earth has also an effect on the lines of apsides, which is the connection of perigee and apogee and therefore identical with the major axis of an ellipse. This line of apsides rotates inside the orbital plane and its drift can be approximated by

$$\dot{\omega} = -\frac{3}{4}J_2\sqrt{\mu}\frac{R_E^2}{a^{3.5}\cdot(1-e^2)^2}\left(1-5\cos^2 i\right) \tag{13.10}$$

The drift vanishes, if the expression $(1-5\cos^2 i)$ is zero, which is for an inclination of 63.43°. This effect is used amongst others by the Russian Molniya satellites, which fly on a high-elliptic orbit with an apogee placed over Russia. Due to the low speed

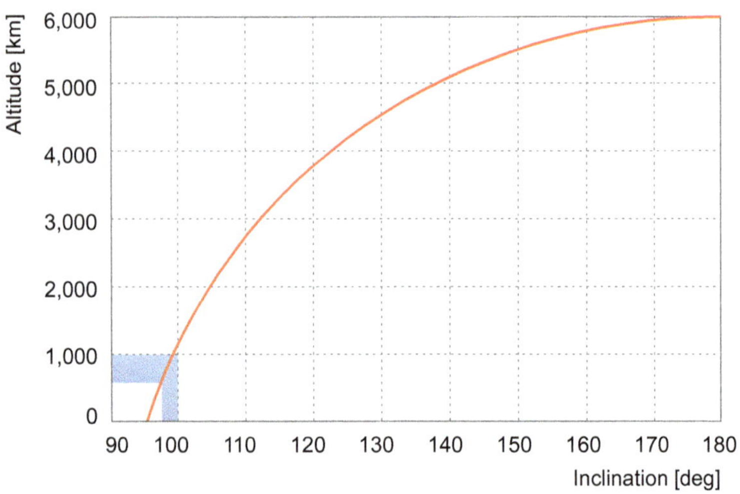

Fig. 13.3 Sun-synchronous orbits

around the apogee, the contact to the satellite is guaranteed for a long time w.r.t. the revolution time.

Drag Due to Residual Atmosphere

Even if the atmosphere of the Earth is defined up to an altitude of 2,000 km, a clear effect can be seen for circular orbits with an altitude below 800 km and for high-elliptic orbits with a perigee altitude below 250 km.

In general, the air drag reduces the orbital energy due to heating up the environment, which results in a decrease of the semi-major axis, refer also to Eq. (13.1). In the case of the high elliptic orbit with the perigee inside the "dense" atmosphere, only the kinetic energy is changed with the consequence that only the altitude of the apogee is reduced (refer also to Fig. 15.4). Once the apogee also is inside the dense atmosphere and the orbit has been circularized, the orbital height and period is decreased due to the air drag, but the satellite velocity is increased instead (Fig. 13.4).

This looks like a paradox: the air drag force acts in the opposite direction of the velocity and the velocity is increased. We have to look to the energy law: the reduction of the potential energy is larger than the reduction of the total energy. Therefore, the kinetic energy and the corresponding velocity have to be increased.

The reduction of the altitude depends on the density of the residual atmosphere which depends on the solar activity since the solar radiation can heat up and expand the upper layers of Earth's atmosphere. The solar activity has a cycle length of about 11 years (Fig. 13.5). An increased solar activity yields to a higher density of the residual atmosphere in low Earth orbit (refer also to Sect. 1.3.5).

Fig. 13.4 Influence of air drag on the orbit

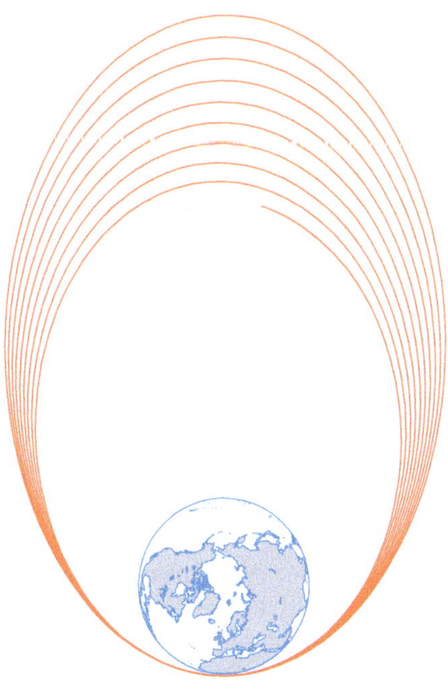

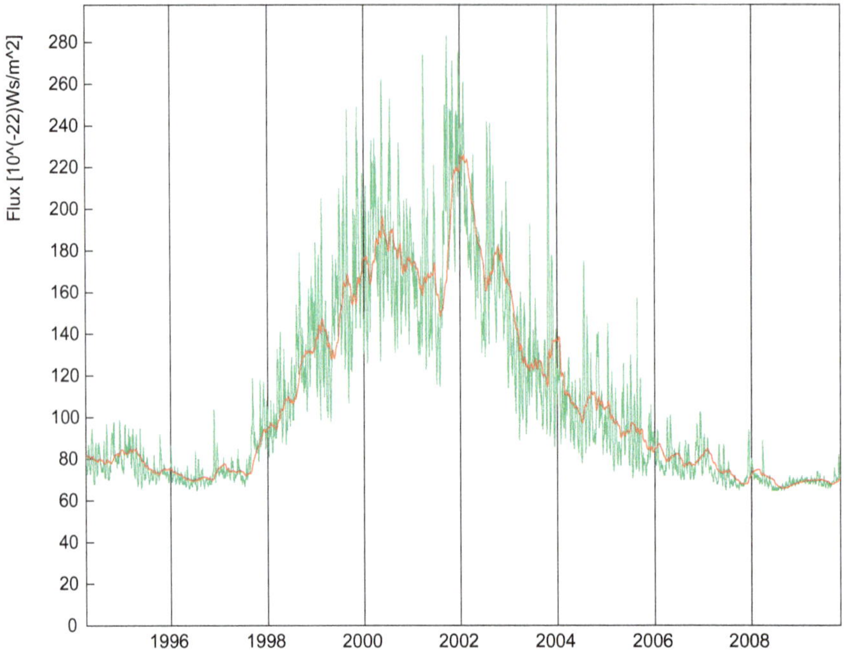

Fig. 13.5 The averaged observed and estimated profile of sunspot numbers

13.2.5 Maneuvers

If the orbit has to be changed and the natural perturbation forces are neglected, so-called orbit maneuvers have to be performed. A maneuver will always change the velocity vector only, its length and/or its orientation, and not the position, at the time of maneuver execution. In general, two types of maneuvers can be defined, the in-plane and the out-of-plane maneuvers; the in-plane maneuver changes the shape of the orbit and the out-of-plane maneuver changes the orientation of an orbit in space.

Two examples are used to explain both types of orbit maneuvers.

In-Plane Maneuver

An in-plane maneuver can be performed by a thrust in tangential or in radial direction or a combination of both. However, any in-plane maneuver will change the shape of a circular orbit into an elliptical orbit. This will create a perigee and an apogee. If we consider impulsive maneuvers only or small extended maneuvers only, where the burn duration is much smaller than the orbital period, only the kinetic energy is changed at the location of the maneuver, but the altitude is changed for all other points along the orbit. Hence, in-plane maneuvers can not only change the shape of an orbit but also the orbit height.

Usually, we change the altitude of the perigee or apogee. In the following example we want to lift the perigee of a GTO to GEO altitude (Fig. 13.6). In this example,

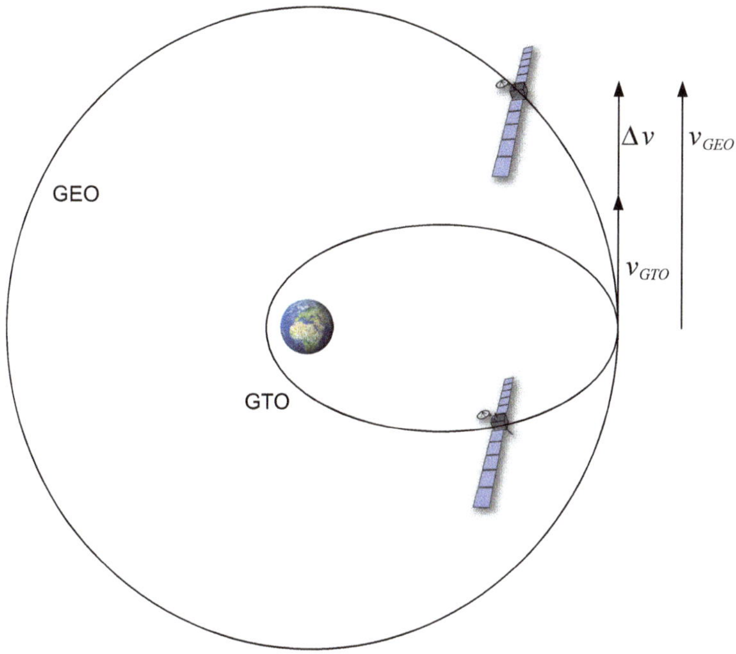

Fig. 13.6 An in-plane maneuver; lift of transfer orbit (GTO) perigee to geostationary (GEO) altitude

the satellite was injected into the high-elliptical GTO with an apogee exactly in the required geostationary altitude like for an Ariane launch. Once the satellite reaches the apogee, it has to increase its velocity. The reason is that the velocity in the apogee of an elliptical orbit inside the circular target orbit is always smaller compared to the required circular velocity.

Using the following geometric parameters.

$r_{\text{apo}} = 42{,}164$ km (GTO apogee radius).

$a_{\text{GTO}} = 24{,}400$ km (GTO semi-major axis).

$a_{GEO} = 42{,}164$ km (GEO semi-major axis).and Eqs. (13.2) and (13.3), we can calculate for the velocity in the apogee $v_{\text{GTO}} = 1.603$ km/s and for the required circular velocity $v_{\text{GEO}} = 3.075$ km/s. The necessary in-plane velocity increment is

$$\Delta v_{\text{IPL}} = v_{\text{GEO}} - v_{\text{GTO}} = 3.075 - 1.603 = 1.472\,\text{km/s}$$

Out-of-Plane Maneuver

This class of maneuver only changes the orientation of an orbit, i.e. the inclination i. Consequently, only the velocity vector is rotated and its length is kept constantly. The out-of-plane velocity increment Δv_{OPL} can be calculated by the geometry of an isosceles triangle (refer also to Fig. 13.7).

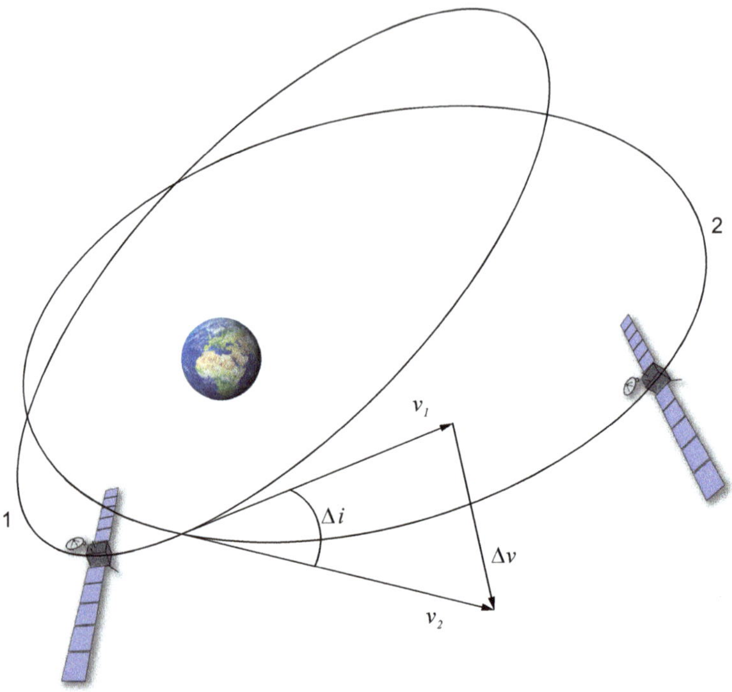

Fig. 13.7 Out-of-plane maneuver; change of the inclination

$$\Delta v_{\mathrm{OPL}} = 2 \cdot v \cdot \sin(\Delta i/2) \tag{13.11}$$

with $v = |v_1| = |v_2|$, which can be calculated with Eq. (13.2), and Δi is the required inclination change.

Performing maneuvers in space is always connected with optimization. The reason is that the fuel consumption is directly proportional to Δv. In the current case, the velocity along an elliptical orbit is different. The minimum velocity increment for a rotation of an orbit can be achieved in the apogee. On a circular orbit, of course, the velocity is constant and therefore the velocity increment cannot be minimized.

Again, using the GTO of the previous chapter as an example, and due to the location of the launch site, an initial inclination of, e.g., 7° has to be compensated. Using the value of the apogee velocity of the previous example, we can calculate a velocity increment of

$$\Delta v_{\mathrm{OPL}} = 2 \cdot \mathrm{v} \cdot \sin(\Delta i/2) = 2 \cdot 1.603 \cdot \sin\,(3.50°) = 0.195\ \mathrm{km/s}.$$

Combined Maneuvers

For a mission to GEO it is always necessary to lift the perigee to the geostationary altitude *and* rotate the initially inclined injection orbit into the equatorial plane.

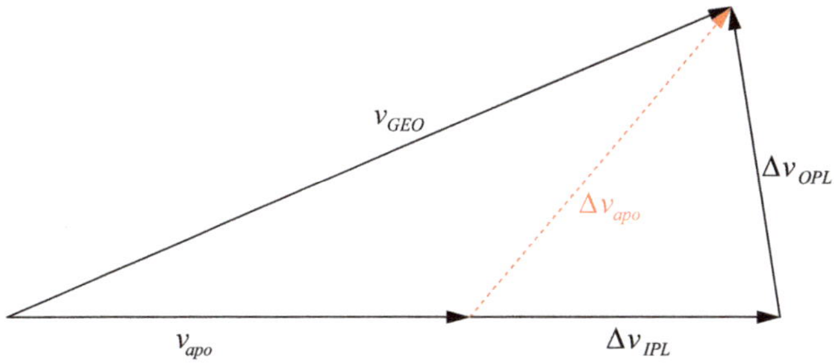

Fig. 13.8 Combined maneuvers

The two previous sections showed how we can calculate the necessary change in the velocity vector. In reality, we will not inject the satellite by two subsequent maneuvers into GEO because of the vectorial nature of the maneuver (Fig. 13.8). The advantage of vector geometry is used to optimize the resulting Δv, as:

$$|\Delta v| = |\Delta v_{\mathrm{IPL}}| + |\Delta v_{\mathrm{OPL}}| \geq |\Delta v| = |\Delta v_{\mathrm{IPL}} + \Delta v_{\mathrm{OPL}}|.$$

Therefore, both types of maneuvers are combined in the apogee. For the optimized maneuver a Δv of only 1.497 km/s is needed compared to a Δv of 1.667 km/s, which extends the mission lifetime by about three years, as about 1 m/s per week are needed for keeping the satellite inside its box.

13.3 Flight Dynamics Tasks

The involvement of the flight dynamics (FD) team in a satellite mission begins years before launch with the mission preparation. In general, the FD team supports mission operations and the ground systems network during the preparation and, if the satellite is launched successfully, the execution of a satellite mission.

13.3.1 Mission Preparation

The main tasks before launch are the performance of a mission analysis comprising the orbit selection, a ground station visibility analysis, launch window calculation, first acquisition analysis, and maneuver strategies as well as the implementation of an operational flight dynamics system (FDS). The major topics of these tasks are addressed a bit more detailed in the following.

Orbit Selection and Ground Station Visibility Analysis

At the very beginning of a mission it has to be discussed with the customer, which orbit can fulfill the mission requirements. Once the orbit is selected, a ground station's network has to be searched, which can support the mission. In general, during the launch and early orbit phase (LEOP) and possibly during the commissioning phase, more ground stations are needed compared to the routine phase. Which ground stations are usable depends on the launch site and the orbit type. As an example, Fig. 13.9 shows the ground track of a LEO satellite mission for the first day after injection. The satellite was launched from Plesetsk and injected into a close-polar orbit close to the Hawaiian Islands. The first contact over the Weilheim ground station, which has been used as the main antenna supporting all mission phases, occurred about half an orbit after the injection. Another orbital revolution later, the second contact occurred with a visibility time of about eight minutes. It is typical for locations like Weilheim that, due to the Earth rotation to the east, the next contact possibility is about 10 h later, when the Earth has performed roughly half a revolution and the descending orbits come into the visibility area of Weilheim again. This is one of the reasons why more than one ground station is used during LEOP.

Figure 13.10 shows the additional ground stations from the NASA polar network, Spitzbergen, Pokerflat, and Wallops used for the CHAMP mission. A fourth one, McMurdo, is not visible in this figure, as it is located close to the South Pole. The first visibility after injection took place over the South Pole station McMurdo (MCMUR) and the second one was a relatively long and combined contact including Weilheim (WHM), Spitzbergen (SPZBG), and Pokerflat (PKF). This triple pass had a total visibility time of about 25 min and clearly shows the advantage of the geographical

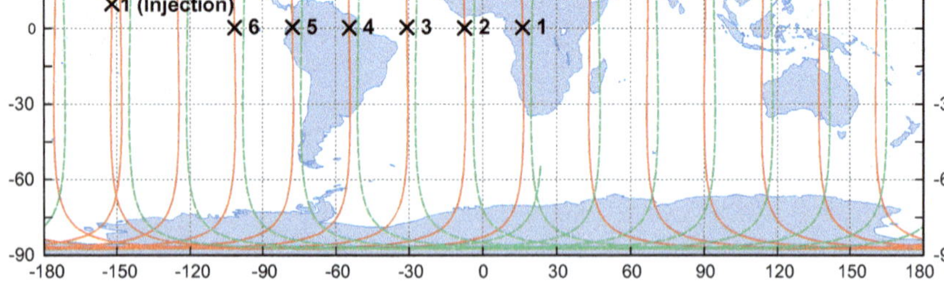

Fig. 13.9 Ground track of the first 24 h for a LEO mission (example CHAMP)

locations of the selected ground stations for the early phase of this mission. All theoretical visibilities of the involved ground stations are shown in Fig. 13.11 where the elevation is plotted over time. A contact is possible at least each half orbit over the close-polar stations, but only the contacts over 10 degree of elevation are used. In addition, it can be seen that full coverage by ground stations is not possible for LEO missions.

Launch Window

Another important task is the calculation of the necessary launch window. Two aspects have to be considered, the launch date and time as well as the length of the window. As both depend on the mission requirements, two examples are used to explain some of the constraints. In case of TerraSAR-X, an Earth observations mission in LEO, the satellite has to be operated along a reference orbit with a repeat time of 11 days and a local time of the ascending node of 18 h. In order to optimize the number of maneuvers and the fuel consumption, the injection time had to be exactly at the same time for each launch date and the length of the window was only a few seconds. For the injection of a communication satellite into GTO the launch window

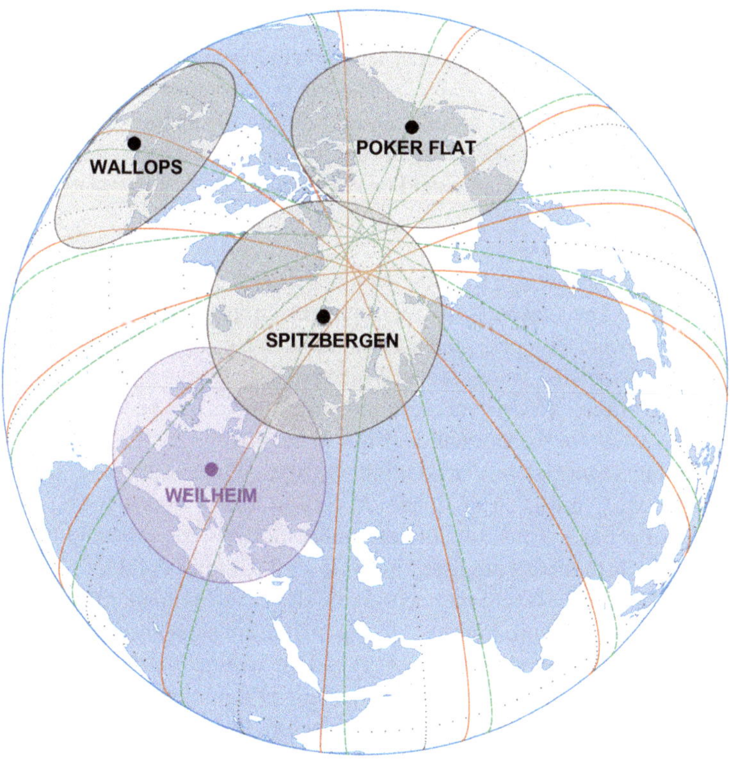

Fig. 13.10 Northern hemisphere with ground track of the first 24 h of a LEO mission (example CHAMP)

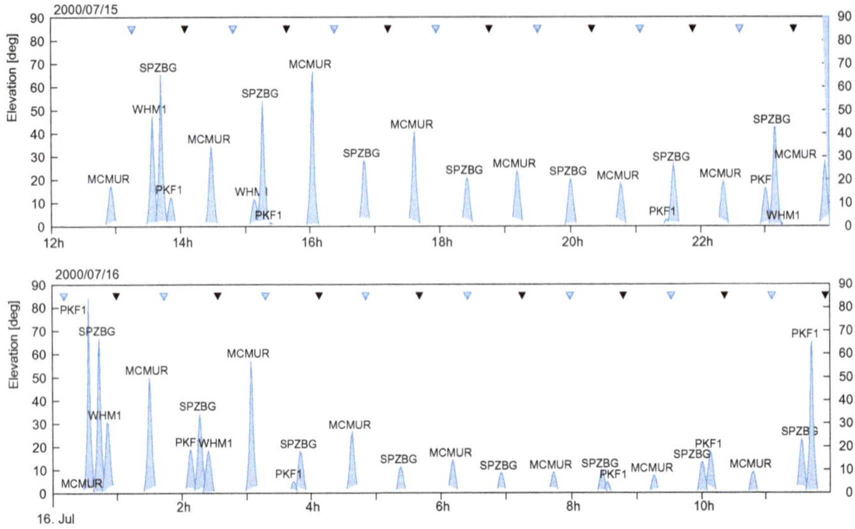

Fig. 13.11 Station visibility of the first 24 h of a LEO mission (example CHAMP)

is completely different. The start of the launch window can vary within 1 h and the length of it can be between 1 and 2 h. The main constraints are the Sun location as well as the visibility of the ground stations during the apogee boost maneuvers. In the latter case, problems occurring during the countdown can possibly be solved during the launch window. This would not be possible in the first example.

First Acquisition Analysis

During mission preparation, a so-called first acquisition analysis has to be performed for all missions. The reason is the possible injection dispersion of the launcher rocket. The launcher company releases nominal injection elements a few weeks before launch; dispersion values are also given mainly for the altitude and the inclination. In case of the CHAMP mission, the uncertainties had been $\pm$ 10 km for the altitude and $\pm\,0.02°$ for the inclination. Both values change the geometric conditions of a ground station contact, but, in addition, the first influences the orbital period and therefore also the acquisition of signal (AOS) and loss of signal (LOS) times of a ground station contact. Thus, two offsets can be derived, the time and the azimuth offsets at AOS. Both offsets are listed in Table 13.2 (time) and Table 13.3 (azimuth) for the first six contacts, where the second row shows the nominal (N) values for the AOS times and the azimuth. In rows 3–6 the offsets are shown for low (L) and high (H) dispersions for the semi-major axis (a) and the inclination (i), respectively.

Some of the cells are highlighted indicating possible problems. In general, a ground station points to the nominal azimuth before the nominal AOS time. In case it receives no signal at the nominal AOS time, it will wait a maximum time of 10 s before it starts searching the satellite. Besides the time offset due to altitude dispersion, the maximum azimuth offset at which a signal can be received, is defined

Table 13.2 Required time offsets for the listed ground stations due to injection dispersion (semi-major axis *a* and inclination *i* for a LEO mission (example CHAMP)

a	*i*	MCMUR	WHM	SPZBG	PKF	MCMUR	WHM
0 km (N)	0° (N)	12:52:02	13:29:49	13:37:36	13:47:58	14:24:40	15:05:15
-10 km (L)	0° (N)	+2 s	−7 s	−9 s	−9 s	−18 s	−26 s
+10 km (H)	0° (N)	2 s	+7 s	+9 s	+9 s	+18 s	+27 s
0 km (N)	-0.02° (L)	+1 s	0 s	0 s	+1 s	0 s	−1 s
0 km (N)	+0.02° (H)	1 s	0 s	0 s	−1 s	0 s	+1 s

Table 13.3 Required azimuth offsets for the listed ground stations due to injection dispersion (semi-major axis *a* and inclination *i* for a LEO mission (example CHAMP)

a	*i*	MCMUR	WHM	SPZBG	PKF	MCMUR	WHM
N	N	70.71°	166.00°	177.17°	336.73°	31.20°	240.95°
L	N	+0.83°	−0.39°	−0.20°	−1.12°	+0.53°	+0.75°
H	N	−0.79°	+0.37°	+0.19°	+1.03°	−0.51°	−0.68°
N	L	0.22°	−0.11°	−0.18°	−0.29°	+0.18°	−0.19°
N	H	−0.23°	+0.11°	+0.18°	+0.28°	−0.19°	+0.18°

by the beam width of the antenna. In the current example, a beam width of $\pm$ 0.3° is used for acquisition with an S-band antenna. If the azimuth offset is greater than the beam width, the ground station can never contact the satellite. The ground station will wait for the given time offset before it starts searching for the satellite. Therefore, it is important to provide the ground stations with both offset tables so that they know what to expect and when to start searching.

Implementation of an Operational Flight Dynamics System
In the case, mission specific software has to be developed or existing software has to be modified depending on the analysis of specific FD mission requirements; this software has to be tested and validated. Together with the software packages in the FD multi-mission environment, which have to be used to perform all tasks required for the mission, the whole FDS has to be tested and validated, together with all involved subsystems of the Satellite Control Center, in particular the interfaces and the input and output data of the FDS.

13.3.2 Mission Execution

Once the satellite has been successfully launched and separated from the upper stage, the mission execution phase begins with FD operations, which includes tasks such as orbit determination including Δv estimation, orbit prediction, production and delivery of orbit-related products, maneuver planning if thrusters are on board, monitoring and handling of critical conjunction events with other space objects, as well as lifetime and reentry predictions. Some of these tasks are discussed in more detailed below.

Orbit Determination and Prediction
Once the satellite has been separated from the upper stage, it is very important to refine the nominal injection elements due to possible injection dispersions (see also Sect. 13.3.1). If the ground stations are able to lock to the signal from the satellite, they can generate tracking data that can be used in orbit determination (OD) software to calculate a new set of orbital elements. As described in Sect. 13.2.1, a satellite orbit is uniquely defined by six parameters, so a set of six equations must be solved. In theory, six measured parameters would be sufficient to solve the equations. In reality, however, many more are needed, since the measurements are not exactly on the real orbit, but more or less describe a cloud around the orbit (see also Fig. 13.12).

Measurement data generated by ground stations are called tracking data (see also Fig. 13.13) and can be angular data (azimuth and elevation), range data (distance between spacecraft and antenna), and Doppler data (relative velocity of the spacecraft towards the ground station). A fourth type of measurement is the GPS navigation data, which is generated and stored on board a satellite and transmitted to the control center via a ground station contact.

Orbit determination software searches an orbit characterized by r and v, which minimizes the sum of the differences between the measured and calculated values

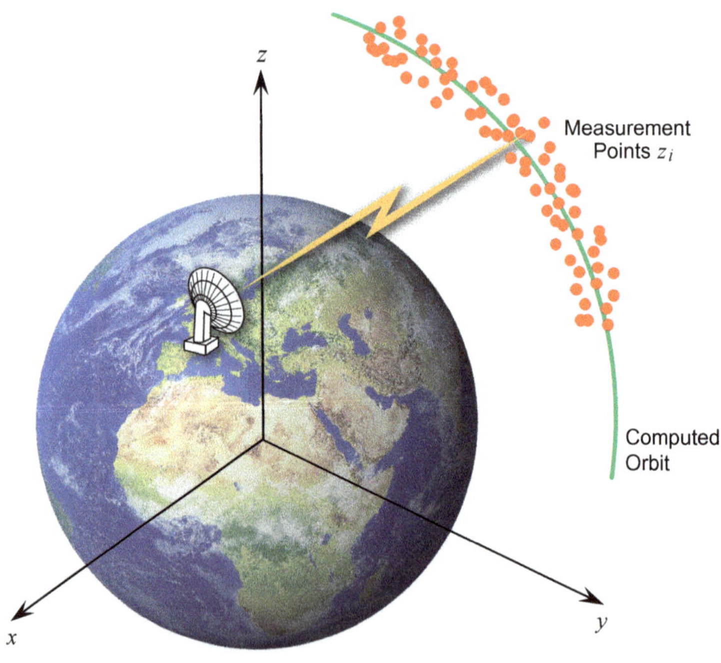

Fig. 13.12 Cloud of measurement points around the real orbit

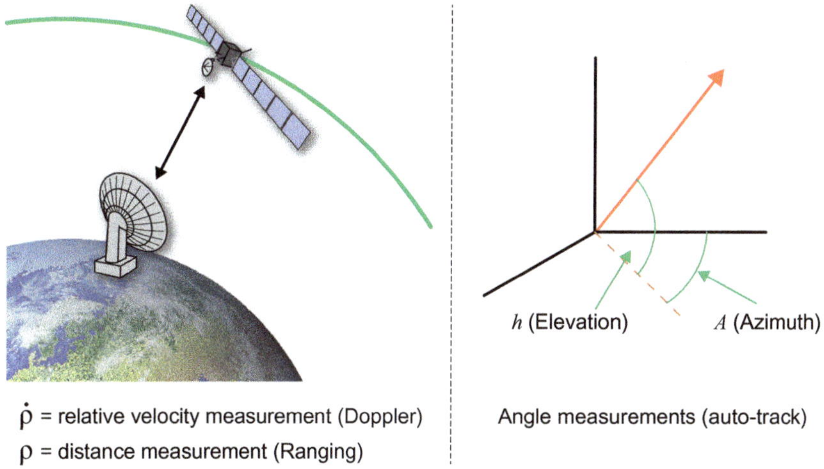

Fig. 13.13 Tracking data types

according to Eq. (13.12).

$$\sum_i [z_i - f_i(r, v)]^2 = \min \tag{13.12}$$

where z_i is the measured parameter and $f_i(r,v)$ is the computed value of the measured parameter based on the current orbital elements (r,v). Two main techniques are used for orbit determination, the batch least squares fit and the Kalman filtering. If the OD process is successful and calculates a set of refined orbital elements, these orbit parameter sets can be used for orbit prediction and to generate required orbital products such as the AOS and LOS times for the ground station network.

13.3.3 Maneuver Planning

If orbit thrusters are on board a spacecraft, maneuvers have to be planned according to the mission requirements. In the following, three examples are presented for maneuver planning.

Orbit Maintenance in LEO

Maneuvers are necessary to maintain an orbit of a satellite. As the requirements for orbit maintenance are very mission specific, the TerraSAR-X mission is used as an example. Here a short list of general requirements is given, which had to be fulfilled for the TerraSAR-X mission:

1. Compensate the orbit decay

2. Keep the ground track stable
3. Keep the time relation of the orbit stable
4. Keep the orbit form stable
5. Achieve the mission target orbit.

These general requirements were derived from mission requirements for TerraSAR-X like:

- Sun-synchronous orbit (1), to keep the illumination conditions constant—both for solar panels and the radar instrument
- Repeat cycle of 11 days (1, 2, 3)
- Frozen eccentricity (4), to keep the altitude profile constant w.r.t. the latitude
- Control of the satellite along a reference orbit within a tube with a diameter of 500 m (1, 2, 3, 4), to support interference methods.

A node crossing requirement of $|\Delta s| = \Delta\lambda \cdot R_E \leq 200$ m could be derived to fulfill the overall and very challenging tube requirement, where $\Delta\lambda$ is the longitude difference of two subsequent ascending node crossings. This has the advantage that the control algorithm for the node-crossing requirement is easier to implement. When we start in the reference altitude, the air drag is responsible for the decrease of the altitude. Consequently, the orbital period is a little bit smaller compared to the reference one and we reach the next node crossing a bit earlier. Therefore, the longitude of the next node crossing is in the eastern direction w.r.t. the reference node. We now let the satellite drift, until it reaches the upper limit of + 200 m, where we have to perform an altitude correction maneuver. This maneuver overshoots the reference altitude in order to maximize the time between two orbit maintenance maneuvers. Due to the higher altitude, the satellite has a higher orbital period and reaches the next node crossing a bit later, which is the opposite behavior compared to lower altitudes. The node crossing is now drifting to the west. We calculate the overshoot so that the reference altitude is reached at the western limit of − 200 m w.r.t. the reference node crossing. This strategy was simulated for the first 70 days of the TerraSAR-X mission and is shown in Fig. 13.14. With this strategy the time between two orbit maintenance maneuvers could be maximized to about two weeks for the start of the mission.

As for each mission an estimation of the fuel consumption for orbit maintenance has to be done, a simulation of the whole mission lifetime was performed a few months before launch. Figure 13.15 shows the expected altitude changes at that time, which can be converted into Δv by the following equation derived from Eq. (13.13):

$$\Delta v_t = -\frac{1}{2} \cdot \Delta a \cdot \frac{v}{a} \tag{13.13}$$

At the start of the mission the altitude had to be corrected by about 40 m, which had to be increased by a factor of 4 to the end of the mission correlated with orbit raise maneuvers each 1–2 days. The reason is that for the end of the mission a solar activity maximum had to be expected. But the amplitude of the solar maximum was finally much smaller compared to the prediction in 2007, where the simulation was

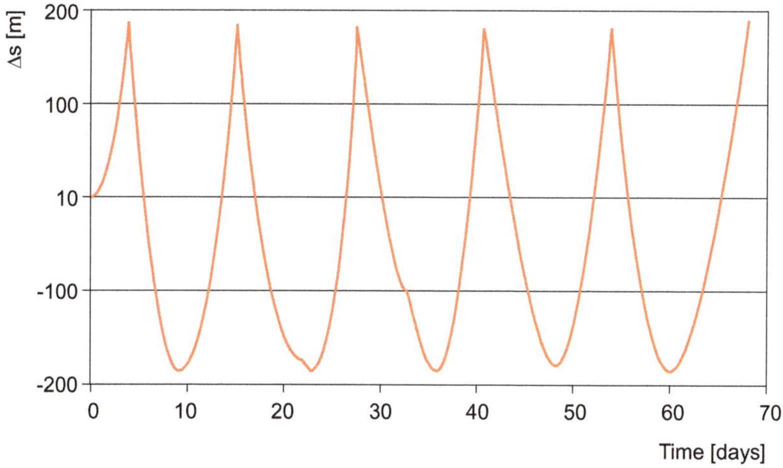

Fig. 13.14 A node crossing simulation of the first 70 days of the TerraSAR-X mission

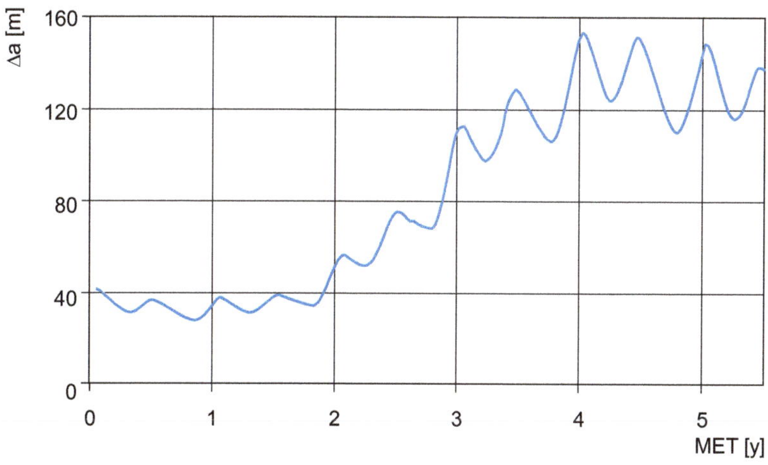

Fig. 13.15 Simulation of altitude change (Δa) maneuvers over mission elapsed time (MET) for TerraSAR-X

performed. A consequence is that fuel could be saved, which can be used to extend the mission lifetime of TerraSAR-X.

Maneuver Planning for the Transfer from GTO to GEO

As already shown in Sect. 13.2.5, in-plane and out-of-plane maneuvers have to be performed to bring the spacecraft into the required target box in the geostationary ring. There are at least two main reasons why such an injection into GEO cannot be performed with only one apogee maneuver:

1. The geometry of the high-elliptical transfer orbit is directly linked to the launch site, so all subsequent apogees have their own longitude. There is a high probability that these apogee longitudes will never meet the longitude of the target box. Consequently, the perigee of the GTO has to be increased step by step in order to reach a drift orbit to the target box.
2. The performance of the thrusters on board has typically an initial uncertainty of about 2%. In case of an over-performance of a single injection burn with a corresponding Δv of, e.g., 30 m/s, the apogee will be exceeded, which has to be reduced again with the same amount of Δv. Such a waste of fuel would reduce the lifetime of the satellite by about 60 weeks (station keeping in GEO requires about 1 m/s per week).

In general, at least three apogee maneuvers are performed to raise the perigee to the geostationary altitude. With the first one, the thruster performance can be calibrated, and the uncertainty reduced to less than 1%, and with a relatively small final maneuver the absolute over-performance can be reduced to a few centimeters per second.

For the positioning of a satellite in GEO, so-called intermediate orbits (IO) are therefore used (see also Fig. 13.16). How many apogee maneuvers have to be used for the positioning depends on the mission requirements, such as the number of ground stations involved, the visibility of dual-stations during an apogee maneuver, and the backup strategies for the nominal maneuver sequence.

Station-Keeping in GEO

Figure 13.17 shows the "final station acquisition" of a typical communications satellite in GEO, where the altitude offset w.r.t. the GEO altitude is plotted versus the east longitude. After the last apogee maneuver as well as the complete solar array and antenna deployments, which changes the orbit due to attitude thruster activation, the

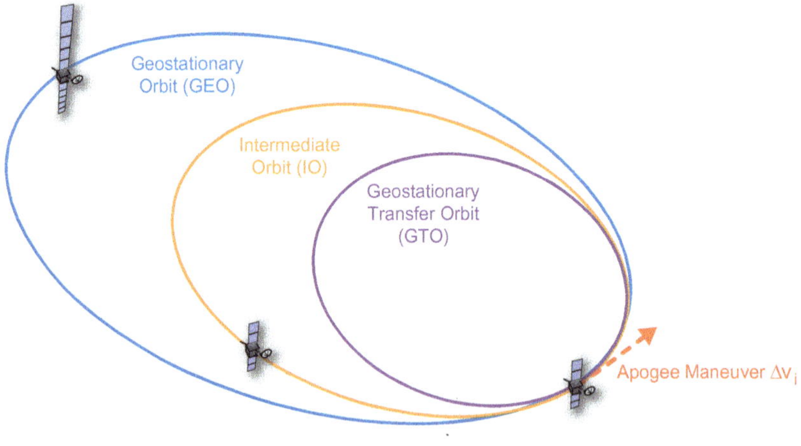

Fig. 13.16 GTO mission profile

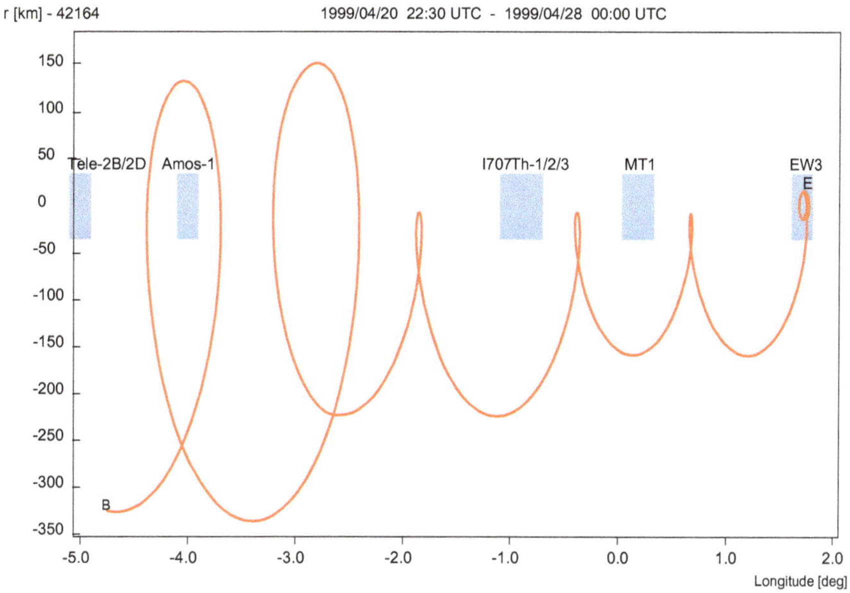

Fig. 13.17 Fine station acquisition of a GEO mission (example Eutelsat EW3 in 1999)

apogee shows a higher altitude as required. However, this situation could be used to perform a safe flyby of the boxes of other satellites, which are located between the begin of the drift and the target box.

Once the satellite is placed in its target box, its orbit is also affected by disturbances in the gravitational fields of the Earth, Sun and Moon. As a result, the satellite begins to drift inside the box in north/south and east/west direction (see also Fig. 13.18). Before the satellite leaves its box, the drift has to be reversed or corrected. Consequently, the routine control center has to perform an orbit determination every day to monitor the drift and to plan and execute so-called station-keeping maneuvers depending on the requirements.

As Fig. 13.18 also shows, more than one satellite can be placed and controlled in one box. In such a case not only the natural drift must be monitored, but the control center must also prevent a collision between the satellites in a box.

Collision Avoidance Operations

In January 2009, a collision between two satellites, IRIDIUM 33 and COSMOS 2251, resulted in about 1500 tracked pieces of debris and other small fragments still orbiting in the wide range of LEO. Since this event, the monitoring and handling of critical conjunction events between active satellites and space debris has become an important part of flight dynamics operations. Currently, more than 50,000 objects are orbiting the Earth, with orbital data publicly available (see also Fig. 13.19).

The Joint Space Operations Center (JSpOC) in the USA tracks objects larger than 10 cm in size through radar antennas for objects in LEO and through optical sensors

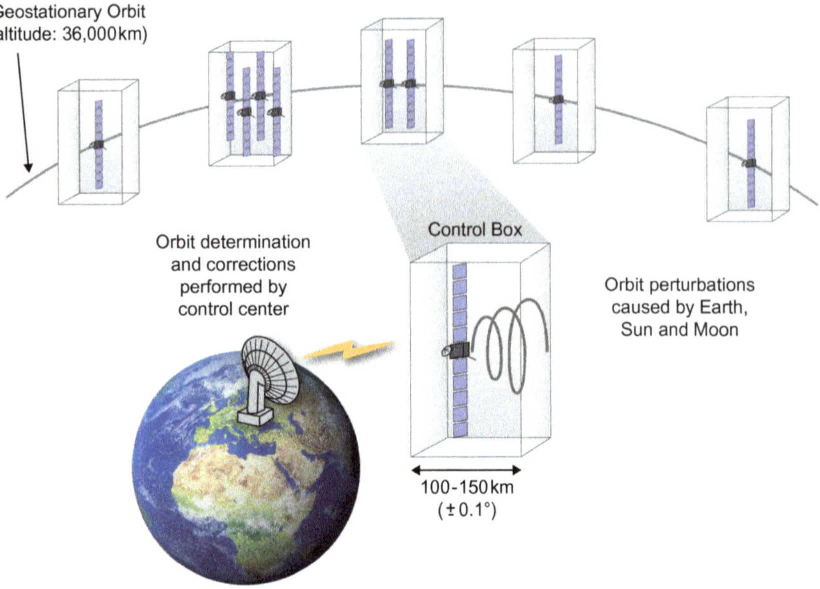

Fig. 13.18 Station-keeping in GEO

for objects in higher orbits like GEO. Based on the tracking data of these sensors and an orbit determination for each object in space, JSpOC performs critical conjunction calculations for all spacecraft in operation. As soon as they detect a close approach, they contact the Control Center and provide data about the event, including precise orbit data of both objects. The precise orbit data allows the FD team to recalculate the conjunction event based on the information of JSpOC and the precise orbit data of its own satellite. This is especially necessary when a maneuver was just executed or planned. Depending on the event geometry, the collision probability, and the error ellipsoids around the space objects, the FD team must plan a collision avoidance maneuver, taking into account the specific mission requirements for the satellite concerned.

As an example of such an event, Fig. 13.20 shows a critical conjunction between a space debris particle and a close formation of two satellites in LEO. The conjunction occurred near the North Pole, where the formation had its maximum radial separation of a few 100 m. The space debris passed through the formation between the two satellites at a critical distance from one of them. As a result, the FD team had to perform a collision avoidance maneuver a half-orbit before the event, and a formation reacquisition maneuver a half-orbit after the event.

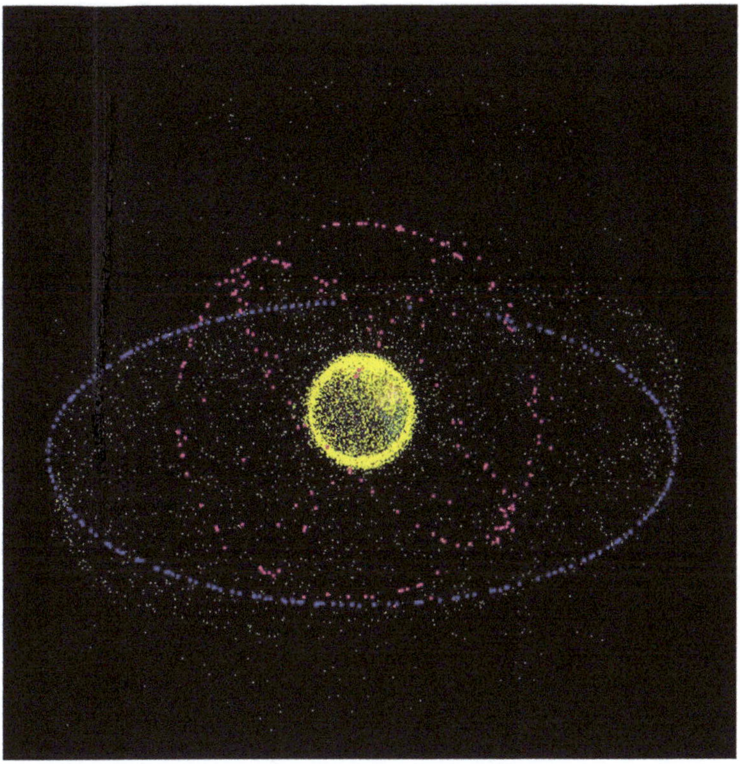

Fig. 13.19 Man-made space objects orbiting the Earth (*Source* DLR)

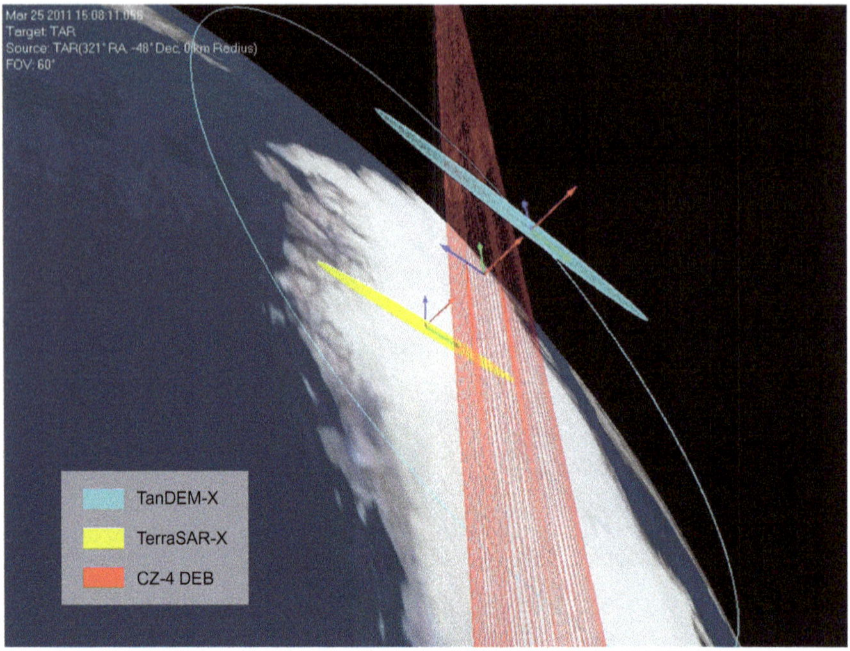

Fig. 13.20 Close approach between a CZ-4 debris (part of the error ellipsoid in red) and the TerraSAR-X (yellow ellipsoid)/TanDEM-X (cyan ellipsoid) formation (*Source* DLR)

References

Montenbruck O, Gill E (2000) Satellite orbits—models, methods, and applications. Springer, Heidelberg

Vallado DA (2001) Fundamentals of astrodynamics and applications, 2nd edn. Kluwer Academic Publishers, Dordrecht, Space Technology Library

Michael Kirschner received the Dipl.-Ing. degree in aerospace engineering from the University of Stuttgart (Germany) in 1982, and the Dr.-Ing. degree from the Technical University of Munich (Germany) in 1990. In 1990, he joined DLR/GSOC as a Flight Dynamics (FD) Engineer. He was and is involved in all satellite mission operation at GSOC. His field of work as senior engineer at the flight dynamics group comprises, besides space mission analysis, FD software development and execution of Launch and Early Orbit Phase, routine FD operations as well as the teaching and training of new colleagues.

Chapter 14
Attitude Dynamics

Jacobus Herman, Ralph Kahle, and Sofya Spiridonova

Abstract The theory of attitude control for satellites is presented. The definition of "attitude" is followed by a description of the several disturbances and of the methods to determine the current status of rotational motion. An attitude prediction into the near future allows for active control, either in one or in three axes, either done autonomously on board or by commanding. The principles of attitude propagation and control are described, as well as the possible types of control mechanism. Comparisons between theory and practice are made and several examples are given from real missions.

14.1 Introduction

The attitude of a satellite is always defined in relation to an external reference system. A few of the more obvious examples of such reference systems are a coordinate system based upon the trajectory, one with the Earth, the Sun, or another celestial body at the center, or an inertial coordinate system defined by the "fixed" stars (see also Sect. 13.2). The three angles between the satellite's body-fixed axes (for example denoted by a unit vector x_{sat}, y_{sat}, z_{sat}) and the chosen reference coordinate system (e.g., the unit vector x_{ref}, y_{ref}, z_{ref}) uniquely define the attitude. The transformation matrix from this reference frame into spacecraft body coordinates is often referred to as the attitude or direction cosine matrix:

$$\boldsymbol{A} = \begin{pmatrix} \boldsymbol{x}_{sat} \cdot \boldsymbol{x}_{ref} & \boldsymbol{x}_{sat} \cdot \boldsymbol{y}_{ref} & \boldsymbol{x}_{sat} \cdot \boldsymbol{z}_{ref} \\ \boldsymbol{y}_{sat} \cdot \boldsymbol{x}_{ref} & \boldsymbol{y}_{sat} \cdot \boldsymbol{y}_{ref} & \boldsymbol{y}_{sat} \cdot \boldsymbol{z}_{ref} \\ \boldsymbol{z}_{sat} \cdot \boldsymbol{x}_{ref} & \boldsymbol{z}_{sat} \cdot \boldsymbol{y}_{ref} & \boldsymbol{z}_{sat} \cdot \boldsymbol{z}_{ref} \end{pmatrix} \tag{14.1}$$

whereby each dot product is the cosine of the angle between the two axes referred to (see e.g., Wertz 1978).

J. Herman · R. Kahle (✉) · S. Spiridonova
German Space Operations Center (GSOC), German Aerospace Center (DLR), Oberpfaffenhofen, Germany
e-mail: ralph.kahle@DLR.de

F. Sellmaier et al. (eds.), *Spacecraft Operations*, Springer Aerospace Technology,
https://doi.org/10.1007/978-3-030-88593-9_14

Easiest to grasp intuitively are the roll, pitch, and yaw angles which are also widely used in the navigation of ships and airplanes (see Fig. 14.1 and also Arbinger and Luebke-Ossenbeck 2006). Roll is the movement around the direction of motion (the length axis in case of ships and planes). Pitch denotes the angle around the axis perpendicular to the orbital plane—the stamping motion of a ship—and yaw the angle around the local vertical axis.

It is always possible to choose an orthogonal reference system denoted by r_x, r_y, r_z, for example, such that the instantaneous attitude of the satellite is defined by

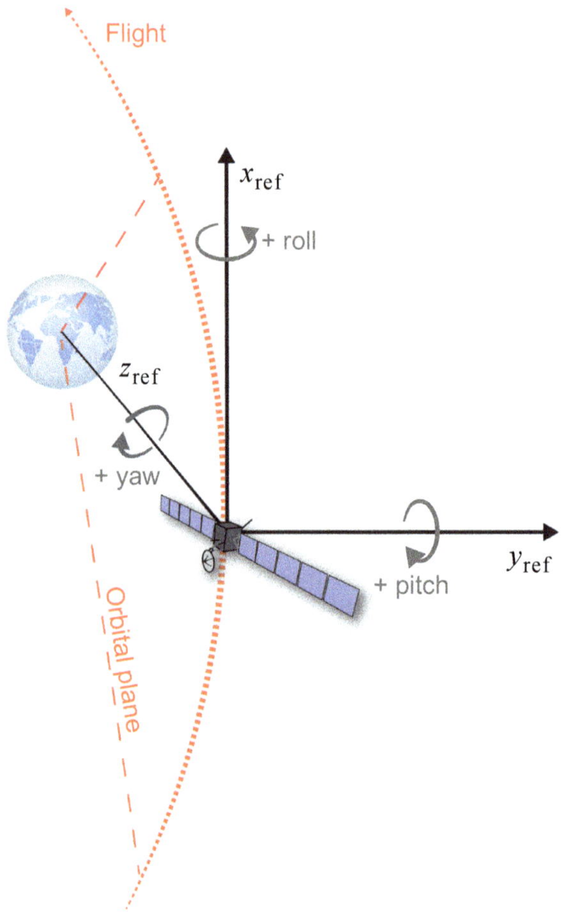

Fig. 14.1 Example of the roll, pitch and yaw angles. Reference is normally the orbit frame. Roll is the movement around the direction of flight, pitch around the direction perpendicular to the orbital plane and yaw the movement around the local vertical axis and pitch around the direction perpendicular to the orbital plane

the rotation around a single axis and thus by a single angle θ.[1] This leads to the definition of quaternions, or so-called Euler symmetric parameters. More than one representation is possible, but a possible choice is (see e.g., Wertz 1978):

$$\begin{aligned} \boldsymbol{q}_1 &= \boldsymbol{r}_x \sin\frac{1}{2}\theta \\ \boldsymbol{q}_2 &= \boldsymbol{r}_y \sin\frac{1}{2}\theta \\ \boldsymbol{q}_3 &= \boldsymbol{r}_z \sin\frac{1}{2}\theta \\ \boldsymbol{q}_4 &= \cos\frac{1}{2}\theta \end{aligned} \tag{14.2}$$

whereby $q_1{}^2 + q_2{}^2 + q_3{}^2 + q_4{}^2 = 1$ and q_4 is the scalar component of the quaternion. Quaternions were especially designed for three-dimensional space and thus offer, from a computational point of view, the most efficient way to handle all necessary attitude operations, such as propagation, vector and matrix multiplications, rotations, transformations, etc. Note that an attitude quaternion will normally be time dependent and its interpretation not straightforward anymore. Therefore, results are normally re-translated into more easily understood quantities such as the deviations from a default attitude or into roll, pitch and yaw angles.

In general, the attitude will not be constant in time. Changes are introduced by internal disturbances, e.g., caused by the firing of a thruster or a change in the speed of a reaction wheel (see Larson and Wertz 1992 and references cited therein for descriptions of all sensors and actuators that are commonly used), but also by external influences such as radiation pressure or gravity gradients. The deviations from a prescribed attitude which are still acceptable depend upon design and environment but most of all, on the task at hand. Clearly, providing sunlight on the panels or pointing the nadir-antenna towards the Earth has less stringent requirements than imaging a star or mapping part of the Earth's surface with a resolution of one meter. Most missions therefore have three distinct regimes of attitude control:

1. Rate damping
2. Coarse control in a basic safe mode (e.g., designed to survive a situation with low power or with thermal problems)
3. Fine control used for payload operations and orbit control.

The exact definitions vary from mission to mission, but a rough indication is provided by the following numbers:

1. Rates > 0.2–0.3°/s must be damped before operations can start; high rates can, for example, occur shortly after launch when the satellite is separated from the upper rocket stage.

---

[1] The single rotation axis is an eigenvector of the direction cosine matrix A; for example, $A \cdot r_x = r_x$, when r_x is the rotation axis.

2. Coarse attitude control points the satellite with a typical accuracy of a few degrees, which is good enough to guarantee power on the solar panels or to establish a link to a ground station.
3. Fine pointing generally is in the order of arcminutes or better.

Determination of the spacecraft's attitude and its variability depends upon the desired accuracy and the available environment. It might be feasible to use the magnetic field or radiation for attitude determination when flying close to a central body, but at some distance the signal strength might become too weak to be usable. Widely implemented is a configuration like the one discussed below, but each mission may and will have individual solutions.

1. Large rates are often measured with an optical gyroscope (IMU = inertial measurement unit which can handle up to ~ 15°/s; Larson and Wertz 1992). Smaller rates can also be obtained by determining the time derivative of subsequent attitude measurements.
2. A simple (and hence robust) bolometer can yield the directions of the Earth (or another central body) and the Sun with an accuracy of a couple of degrees. Close to the Earth a magnetometer (Larson and Wertz 1992) may deliver similar results. Reaction times of such devices may be of the order of several seconds (CESS = coarse Earth Sun sensor. Bolometric Sun sensors with higher precision have also been developed but are less frequently used).
3. A star tracker (Larson and Wertz 1992) typically provides information with an accuracy < 1 arcmin in all three axes. Computations can be done at least once per second, thus also enabling rates to be determined accurately.

Several general principles must be taken into account before using such measurements for attitude control. First of all, the on-board computer normally calculates the desired corrections at a fixed frequency, so that the measurements have to be propagated to the next grid point. Secondly, all measurements exhibit noise which is normally suppressed by applying filters. Spurious or missing points must be flagged as unusable and some kind of damping mechanism must be included in order to avoid "bang-bang" control.

Actuators are then used to actively control the attitude. Once again, the exact choice will depend upon environment and mission goals. Commonly used are thrusters, reaction wheels, and, when a sufficiently strong magnetic field is present, magnetic torque rods (Larson and Wertz 1992). Other, more exotic solutions have also been tried, but will not be covered here. A more exhaustive enumeration of equipment used for attitude control can be found in Chap. 22 on AOCS (attitude and orbit control system) operations.

14.2 Disturbances

The attitude of each satellite is influenced by several intrinsic and external factors, the relative importance of which depends once more upon design and environment. Some of these influencing factors may be actively used to control the attitude.

14.2.1 Satellite Intrinsic

Reaction wheels provide a powerful and accurate way to control the attitude in a fine pointing mode. Disturbances which are several orders of magnitude larger than the allowed tolerances may occur in case of malfunctioning, or when one of the wheels is switched off completely. Turning off all wheels simultaneously will also lead to severe attitude deviations due to the unavoidable asymmetrical run-down.

Thrusters used for orbit correction maneuvers are designed to act through the center of gravity. However, the center of gravity (CoG) may shift during the mission due to the emptying of the fuel tanks, for example. Originally equal thrusters may develop dissimilarities with time, which also leads to attitude disturbances if maneuvers are made with several thrusters at the time.

Movable or flexible parts of the satellite may influence its attitude. A camera, antenna, or other instrument might be movable and lock into a prescribed slot at the end of a slew. A boom or another flexible appendix leads to disturbances as does the sloshing of fuel in an emptying tank.

14.2.2 External Influences

In the vicinity of the Earth, or another massive central body, the major disturbance of the attitude stems from the differential gravitational force acting on the satellite.

The gravity gradient tends to point the length axis, or more precisely the axis with the smallest moment of inertia, towards the central body.

The influence of solar radiation pressure and to a far lesser extent the pressure of the reradiated infrared (IR) radiation of a central body depends upon distance, the area-to-mass ratio, and properties of the satellite such as reflectivity. The geometry of the orbit also plays a role. The disturbance of radiation pressure will be more or less constant when flying in Sun-synchronous orbit but will be variable in other orbits.

Ram pressure or air drag can become important for an orbit close to a central body. It is a differential effect depending on asymmetric properties of the spacecraft. In general, this effect is several orders of magnitude smaller than the effects mentioned above. Satellites flying around the Earth at altitudes ≤ 400 km will be affected, whereby the height of the remnant atmosphere depends upon solar activity.

The smallest effect of all is caused by the interaction with the magnetic field. Satellites which use magnetic torque rods for attitude control are confined to low Earth orbits $\leq$ 500 km where the field strength is of the order of 40,000 nT. Disturbances are typically an order of magnitude smaller than the effects of air drag. However, if the satellite possesses a significant magnetic dipole of its own the torques can be as large as the disturbance caused by the gravity gradient. That occurs, for example, if the satellite has a relatively large residual magnetic moment due to currents on the solar panels (see the example below), or in "SAR" missions that use a synthetic aperture radar which sends out a strong electromagnetic signal.

A quantitative example is included here for a satellite with a mass of 600 kg and a cross section in flight direction of 1 m^2, flying in a polar orbit at 495 km altitude.[2] The residual magnetic moment was measured to reach up to 2 Am^2 in the roll- and pitch-axes leading to disturbance torques $\leq 10^{-4}$ Nm. The gravity gradient is in the order of a few 10^{-5} Nm, and the disturbances by the solar radiation ~ 5×10^{-6} Nm (see also d'Amico 2002). The aerodynamic drag at this altitude is only ~ 10^{-7} Nm during years with low solar activity. This can be up to two orders of magnitude larger at lower altitudes during solar maximum, though (Fig. 14.2).

14.3 Attitude Determination

In principle, two external reference points suffice to determine the attitude. The practical implementation is considerably more complicated, though. This will be illustrated by looking at three different methods. The first delivers a coarse attitude by finding the direction to the Sun and/or Earth with temperature sensors. The second, much more precise method is based upon the measurement of stars. These two computations are carried out on board. It is also possible to do the attitude determination on ground, either as a desired profile to be sent beforehand to the satellite, or as a refinement of the on-board calculation. This constitutes the third method.

14.3.1 Coarse Attitude

A rough idea of the satellite's attitude was all that was possible in the early days of space flight, mostly due to limitations in on-board computing power (see Wertz 1978). Star cameras were available to the military only and it was not before the last decade of the twentieth century that these and arrays of photocells became generally available. Sensors for a coarse determination (see Chap. 22 for descriptions and accuracies)

[2] Properties of the two GRACE (Gravity Recovery And Climate Experiment) follow-on satellites launched in 2018. Peculiarities of this mission are the microwave and laser links between their frontends over 220 ± 50 km. This implies that both fly with a ~ -1° pitch bias with respect to the flight direction.

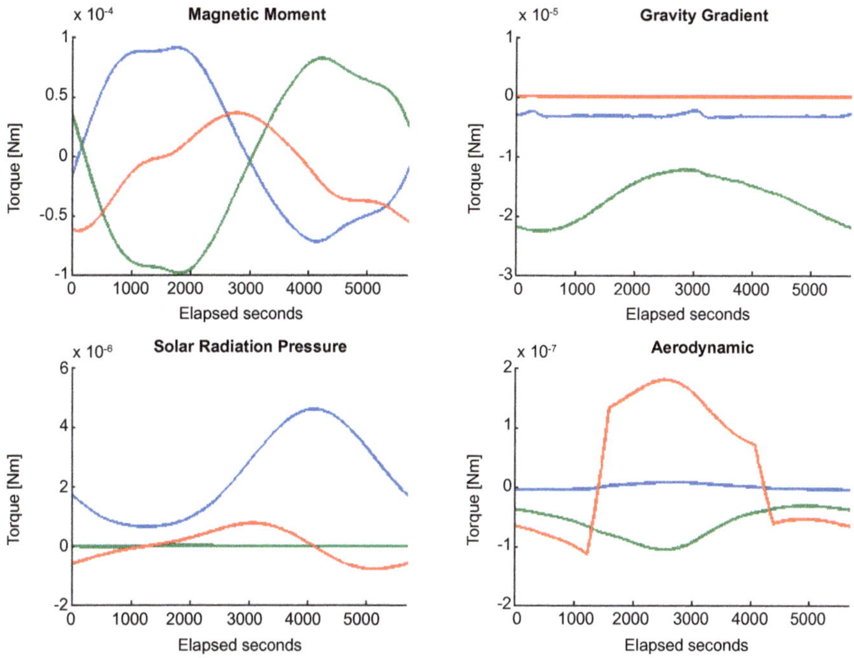

Fig. 14.2 The torques that are caused by the four relevant disturbances for low-flying satellites, are shown in all three axes over one orbit. The blue, green and red colors refer to the roll-, pitch- and yaw-axes, respectively. The properties are those of the GRACE follow-on satellites (see text and footnote[2]) and the environmental conditions represent those of January 2019. The theory is compared with the situation on-board in the next panels

continue to play an important role and are still used to guarantee a reasonable attitude for thermal and power control in case the more precise method on board fails.

One Axis, or Spin Stabilization

A stable attitude is achieved easiest for a spinning satellite with one axis pointing towards an external reference (e.g., Earth or Sun). Denote the reference vector in body coordinates as

$$r_{ref} = \begin{pmatrix} 0 \\ 0 \\ 1 \end{pmatrix} \quad (14.3)$$

in this case implying that the z-axis should be pointed towards the reference. The actual direction with respect to the satellite can be computed from the temperature measurements.

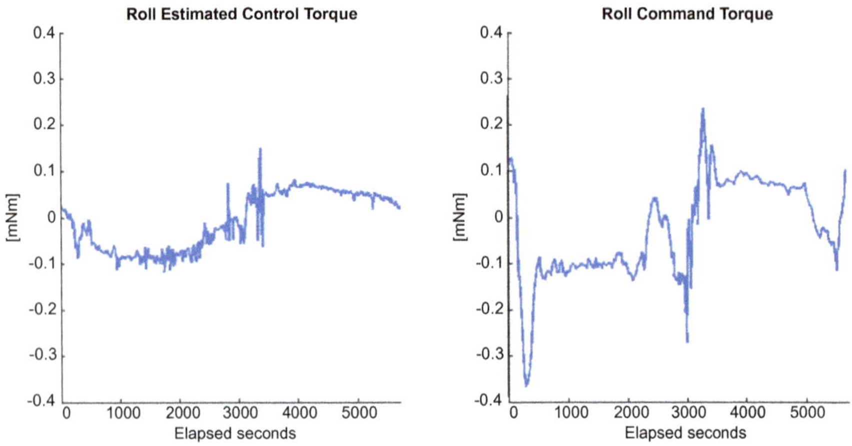

Fig. 14.3 The total disturbance in roll is the sum of the four components as shown in Fig. 14.2. The autonomous attitude control system has to counteract the disturbance torques and must also correct attitude and rate errors to stay within prescribed limits. The estimated control torque (left side panel) includes the reaction to the theoretical disturbances as well as the contribution to correct the observed attitude deviations. It is compared with the actually commanded torque which is shown in the right-hand panel. The one orbit shown uses an average of the measurements over one full day

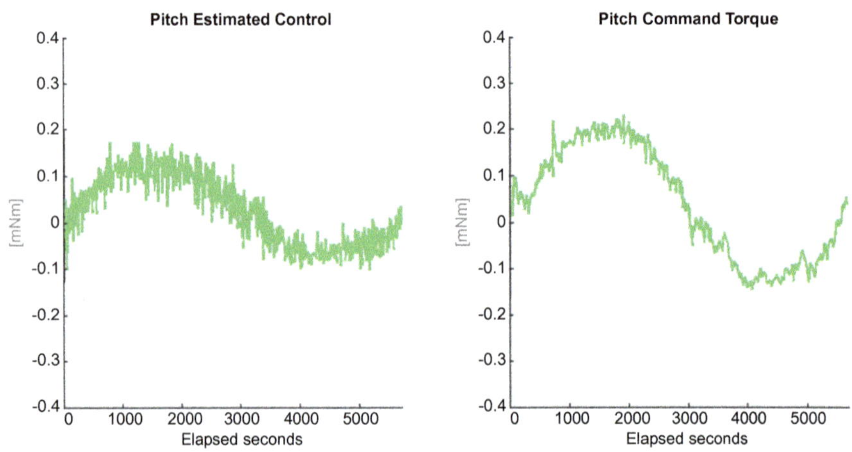

Fig. 14.4 The same comparison as in Fig. 14.3 is made for pitch. Note that the contribution of the gravity gradient to the disturbance torque is more pronounced due to the permanent $-1°$ pitch bias[2]

$$r_{meas} = \begin{pmatrix} x_{meas} \\ y_{meas} \\ z_{meas} \end{pmatrix} \tag{14.4}$$

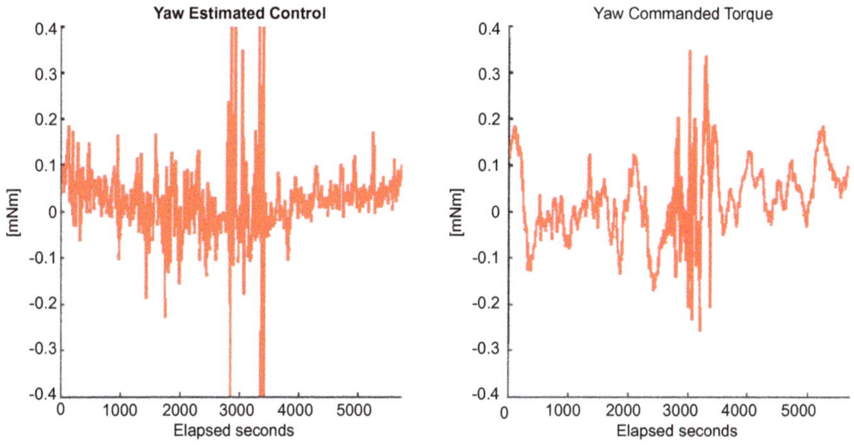

Fig. 14.5 Comparison for the yaw axis (see descriptions above). The disturbances are relatively small (see Fig. 14.2) and most of the corrective torque is applied near the nodes where magnetic authority for this axis is largest

and the angle between the measured and desired attitude follows from the cross-product.

$$\sin\varphi = \frac{|r_{ref} \times r_{meas}|}{|r_{ref}||r_{meas}|} \tag{14.5}$$

The angle φ can then be minimized by on-board control (see Sect. 14.5). The instantaneous rotation axis is given by

$$e = r_{ref} \times r_{meas} \tag{14.6}$$

No model is required for this but the expected temperature of the reference. The measurements must be converted using an a priori calibration.

Three Axes Stabilization

Slightly more difficult is stabilization in three axes. The references are taken to be the Earth and the Sun and the angles that are to be minimized are computed as in Eq. (14.5). The reference vectors require modelling because Sun and Earth directions are seldom perpendicular. A possible way is to let one satellite axis point towards the center of the Earth and then optimize the angle between the Sun direction and the plane defined by this axis and a second one. That requires a simple model of the Sun's direction over an orbit and during a year.

Note that the attitude thus determined is ambiguous, which is normally irrelevant or resolved by taking other factors into account. No three axes solution is possible if the reference directions are too close together, i.e. when Sun, satellite and Earth are aligned.

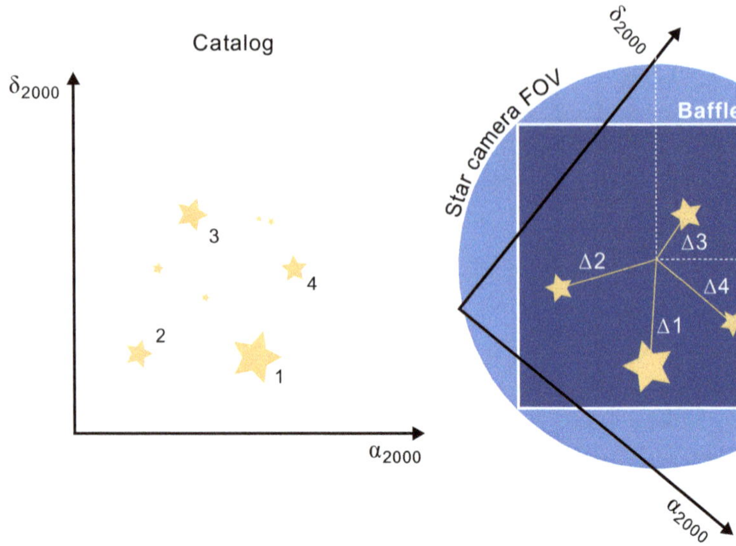

Fig. 14.6 The bore sight and orientation of the star tracker is determined in Y2000 coordinates by comparison with a catalog. Magnitude information is also used to guard against spurious detections. The high redundancy allows an estimate of the quality of the solution to be made

14.3.2 Precise Attitude

A detailed description is given in this section of the steps to get a precise determination of the attitude in all three axes from star camera measurements.

1. A CCD (charge-coupled device) image of a portion of the sky is made each duty cycle of the attitude control system. The field-of-view is typically ~ 100 deg^2 comprising some 20–100 bright stars (Fig. 14.6). The limiting magnitude and minimum signal-to-noise ratio might be autonomously adjusted or be set per ground-command. A baffle provides protection against stray light from, e.g., the Sun or the Moon.
2. Comparison with a star catalogue, which is stored on board, yields the pointing direction of the star tracker's bore sight and the orientation of the field of view (FOV). The apparent stellar positions must be corrected for the effect of aberration.[3] Very precise measurements require the inclusion of a correction for the proper motion of the stars. In principle it is enough to have two identified stars (position on the sky and magnitude) to fix the camera's looking direction

[3] Aberration is the apparent displacement of a star due to the motion of the observer; for a satellite moving at a velocity v with respect to the fixed stars the displacement $\approx v/c \sin \varphi$, where c is the velocity of light and φ denotes the angle between motion and star direction.

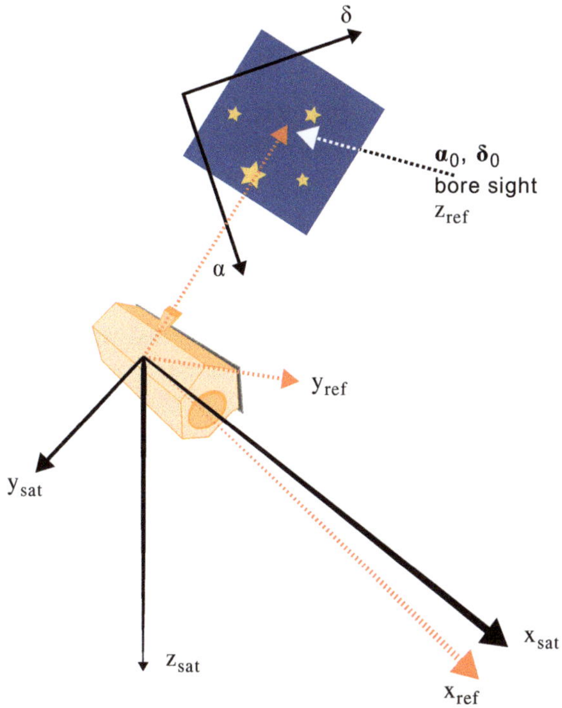

Fig. 14.7 Transformation of the measured star tracker's attitude (here indicated by the system with suffix "ref" which has for illustration purposes one of its axes aligned with the bore sight) into the body frame yields the attitude of the satellite in Y2000 coordinates (suffix "sat"). The relation with the measured star field is indicated

and orientation. The surplus information from 20 to 100 is used to get an estimate of the accuracy of the determination and will also guard against spurious identifications.[4]

3. The attitude of the star tracker is now known in the coordinate system of the catalogue, usually right ascension and declination (α, δ) at epoch 2000.0.[5]
4. The location and built-in angles of the star trackers on the satellite are known from prelaunch measurements, so that it is easy to transform the result into the attitude for the three body axes (see Fig. 14.7). This might already be sufficient if an instrument has to be pointed towards a galaxy or a star. Note that the built-in angle of the star tracker and the motion of the satellite along its orbit imply that the accuracy of the attitude determination will in general not be identical in the three body axes.

---

[4] Another satellite, a comet, an asteroid or meteoroid could transitorily be identified as a star. Blinding could lead to afterglow on one or more pixels with the same result.

[5] Right ascension is measured along the Earth's equator towards the east; zero point is the direction of the vernal equinox at the specified epoch. Declination is measured perpendicular to the equator and is positive towards the north.

5. If, however, the purpose is to point an instrument at a certain location on the Earth's surface, further transformations are necessary. The first step is to insert the precession and nutation from the epoch of the catalogue until the exact moment of the determination. Use of the instantaneous direction of the vernal equinox and the current value for the obliquity of the ecliptic (the angle between the equator and the plane of the ecliptic) leads to the denomination "true-of-date" (Fig. 14.8).
6. Additional information on the current position of the satellite in its orbit allows the transformation from inertial to Earth-fixed coordinates. Position information can be sent to the spacecraft in the form of an ephemeris, or as a single state vector at a given epoch which is then propagated with an on-board model. The current generation of satellites often use GPS receivers to determine this information autonomously on board (only feasible in orbits around the Earth

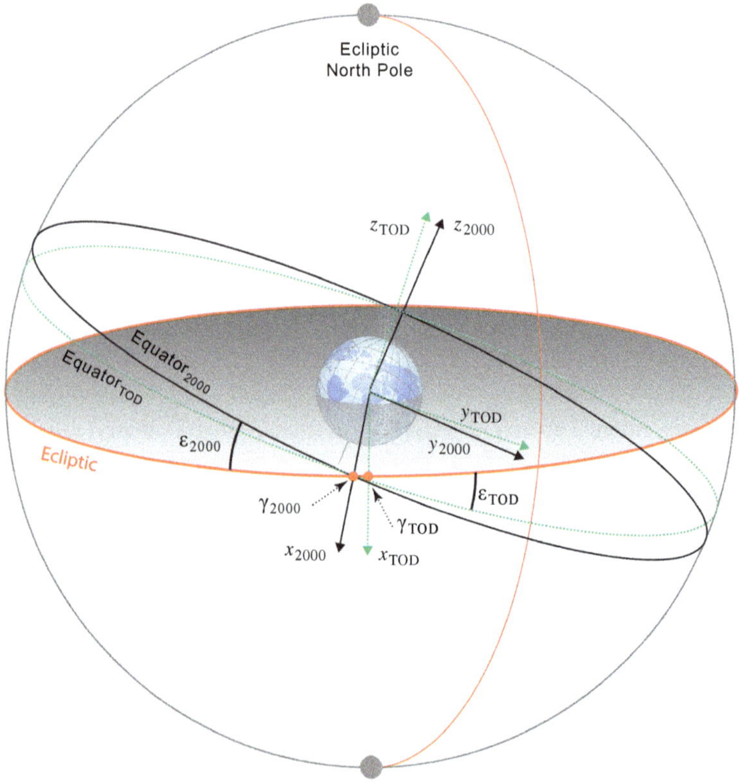

Fig. 14.8 The relation is shown between the inertial Earth-centered coordinate systems Y2000 and "true-of-date" (TOD; see text for further details). Targeting of a point on the surface requires time information and knowledge of the current position in orbit also. Thereafter the attitude can be expressed in the familiar roll, pitch and yaw angles. High precision applications might require additional corrections such as for the UT1-UTC time difference or for the oblateness of the Earth

with an altitude < 10,000 km). After this step the attitude can be expressed in the intuitively understood roll, pitch and yaw angles.

7. Further corrections in order to relate the attitude to a target on the surface might need to take the oblateness of the Earth or the random polar motion into account. The latter can be implemented by correcting the time from UTC to UT1 (Seidelmann 1992).
8. Payload operations require a final transformation into an instrument frame.

The above is by no means unique, but different missions, different payloads, different sensors, or a different central body might all lead to adaptations of this scheme. Several operational aspects always have to be considered though.

- The sensor used for attitude determination in fine pointing mode will normally have at least one backup. The mutual alignments of prime and backup(s) have to be matched to such a degree that payload operations are not disturbed if a temporary switch-over is necessary. The same holds true for redundant sensors in other modes.
- Such mounting matrices are of course measured when building the satellite but might change during the mission, especially at launch. Also, characteristics in orbit might differ from those in the laboratory. Note that in case of misalignments detected in flight one sensor has to be chosen as reference and the orientation of all others will be related to this one. A good choice would be a star tracker which serves as cold backup, i.e., it is not operationally used at the time.
- Similarly, results from different types of sensors have to match within their respective accuracies.
- Computations might use measurements from different types of sensors which requires their relative weights to be set. The direction to the center of the Earth can, for example, be derived from CESS and magnetometer data with a relative weight of 2:1 (numbers are arbitrarily chosen in this example).
- Additional information is usually required before the sensor information can be evaluated. The below would be exemplary for a satellite in low Earth orbit.

  (a) Star tracker:

  Catalogue (e.g., in Y2000 coordinates) containing all stars brighter than a given magnitude
  Current epoch
  Precession and nutation matrices as a function of time
  Mounting matrix with respect to the satellite
  Orbit position.

  (b) GPS receiver:

  Mounting matrix
  GPS SV (or corresponding PRN) usability
  GPS SV ephemerides
  GPS clock correction
  Model for orbit propagation.

(c) CESS:

Position of heads and thermistors
Calibration of temperatures
Model of terrestrial (or central body) albedo
Position of the Sun as a function of time (a simple sinus function normally suffices).

(d) Magnetometer:

Mounting matrix
Model of the terrestrial magnetic field Orbit position.

(e) IMU:

Mounting matrix

- Propagation to next grid point (model and or rates)
- Rates from time differentiation or from separate measurement
- Noise may require use of filters; obviously wrong measurements should be discarded. This requires quality control, e.g., by monitoring the point-to-point variation or signal-to-noise ratio and flagging of "invalid" points
- Gaps require special handling.

(f) Expected outages:

E.g., the star tracker used will be blinded by the Sun (provided the orbit continues as is)

(g) Expected outages:
(h) Unexpected outages:

Star tracker blinded by another satellite
Long stretch of invalid measurements e.g., due to sensor malfunctioning or wrong parameter adjustment
Software problem or any other problem causing sensor outage.

Mitigation measures are designed to minimize the effect on payload operations. A switch to the redundant sensor, either autonomously done on board, or by a priori commanding, is the easiest solution.

It might be possible to wait until the sensor delivers valid data again and, in the meantime, to bridge the gap by propagating the last valid attitude measurement. This usually requires a model of the disturbances acting on the satellite and/or the availability of independent rate measurements. However, even with sophisticated means a propagation of attitude information usually does not remain within the limits necessary for payload operations for a long time.

Cruising through the outage, i.e., disable active attitude control, is another possibility, but obviously deviations will grow even quicker than in case of propagation.

Finally, there could be an option to switch to another mode that uses different sensors which are not affected.

14.3.3 Ground-Based Attitude

Attitude control runs in real time and is only feasible as an autonomous process on board. There are a few exceptions though that will be discussed briefly here.

Reconstruction
The orbit can be determined with much higher precision on ground than on board. Long intervals of GPS measurements can be used as well as tracking data from ground stations (e.g., range, range-rate, or angle measurements). The reconstruction of the attitude is used to improve the coordinates of single images, or to facilitate the combination of images taken at different times.

Initialization
Lasers are used more and more frequently on satellites. They may serve as communication link between them, transfer data to a ground station or to a relay satellite, or they may be employed as a ranging instrument. The requirements for pointing are μrad or better, whereas the best that can be done by on-board attitude control is in the order of 100 mrad.

Precise orbit prediction allows accurate target information to be sent to the satellite in order to initialize the link. Movable mounting or mirrors can maintain the link once it is established and optimize the pointing of the laser.

Predictions become inaccurate in relatively short time so that fresh information has to be sent to the satellite several times per day.

Guidance
Sometimes specific operations, such as orbit maneuvers or image acquisitions, require computation on ground and an attitude profile is subsequently sent to the satellite as reference. This can take the form of a single target point or a guidance list over a certain interval. Time corrections may be included to account for small along-track deviations detected shortly before execution.

Such scenario might be applicable for a satellite with defunct components or with limited on-board autonomy or computing capacity.

14.4 Attitude Propagation

A prediction over a certain time interval in the future can be made as soon as the current attitude and angular rates have been determined. The simplest propagation would be a linear extension of the measurements, a so-called kinematic model. The combination of an attitude matrix A [e.g., the one from Eq. (14.1)] with the skew matrix for rotation Ω leads to the former's time derivative:

$$\frac{d\boldsymbol{A}}{dt} = \boldsymbol{\Omega} \cdot \boldsymbol{A} \tag{14.7}$$

Ω is defined in terms of the angular rates around the three principal axes of the reference coordinate system ω_x, ω_y, and ω_z (denoted by ω in the following):

$$\boldsymbol{\Omega} = \begin{pmatrix} 0 & \omega_z & -\omega_y \\ -\omega_z & 0 & \omega_x \\ \omega_y & -\omega_x & 0 \end{pmatrix} \tag{14.8}$$

The equivalent when Euler symmetric parameters are used, see Eq. (14.2), becomes $dq/dt = ½\Omega \cdot q$ with the skew matrix defined in this instance as

$$\boldsymbol{\Omega} = \begin{pmatrix} 0 & \omega_z & -\omega_y & \omega_x \\ -\omega_z & 0 & \omega_x & \omega_y \\ \omega_y & -\omega_x & 0 & \omega_z \\ -\omega_x & -\omega_y & -\omega_z & 0 \end{pmatrix} \tag{14.9}$$

A kinematic model can, in practice, only be used in the absence of disturbance torques (e.g., interplanetary flight without active control) or in order to bridge short time intervals (normally less than one second for a propagation to the next grid point in the on-board computer).

A more elaborate and better prediction can be attained by adding a dynamic model, which takes all internal and external disturbances into account (see Sect. 14.2 for an overview). The complexity increases with the accuracy to be obtained. Each of the disturbance torques has to be modeled and projected onto the chosen reference system, for example, body-fixed coordinates. All terms then have to be included in the right-hand side of Eq. (14.7). The number of disturbance torques taken into account will depend upon the actuators used, the design of the satellite, and the accuracy that has to be reached. Euler's equation of dynamic motion is mostly written in the form of

$$\boldsymbol{I}\frac{d\boldsymbol{\omega}}{dt} = \boldsymbol{T}_d + \boldsymbol{T}_c - \boldsymbol{\omega} \times (\boldsymbol{I} \cdot \omega) \tag{14.10}$$

where T_d and T_c denote the overall disturbance and control torques, respectively, I is the tensor for the satellite's moments of inertia, and the vector $\omega = (\omega_x, \omega_y, \omega_z)$. T_c will normally include the torques by the wheels, and/or the thrusters and maybe also the magnetic torque rods when present, whereas T_d will comprise the torques caused by the gravity gradient, by the air drag and possibly by the solar radiation pressure and by the magnetic field also.

The vector product $I \cdot \omega$ is the total angular momentum of the satellite, often denoted as L. Note that the last term in Eq. (14.16) is a vector cross-product. It can be seen from Eq. (14.16) that the relative influence of disturbance torques diminishes if either ω or I become large. The attitude of fast rotating satellites or of large structures such as the international space station is not that sensitive to disturbances.

Finally note that Eq. (14.16) is valid for a rigid body only. Inclusion of the dynamics of flexible or irregularly moving structures is normally too complicated to be considered. A more elaborate and detailed treatment falls outside the scope of this chapter, but examples can be found in Wertz (1978), Sidi (1997) and Wiesel (1997).

14.5 Attitude Control

Attitude control can either be active or passive. Passive control can be achieved by stabilization due to gravity gradients, by the alignment of an on-board magnet with an external field, or by a high spin rate sustained over a long time which, for example, could be applicable during the cruise phase in interplanetary flight. Active control is far more common though.

Once the satellite's attitude and rates are determined and a model for all disturbing influences implemented, future trends can be predicted and thus counteracted by applying one or more of its actuators. Commonly used are

- **Thrusters**: A pressurized propellant in a central reservoir is led to several thrusters that singly, or in combination, can control one of the axes. The torque is achieved by expelling a small amount of mass and is described well by the Tsiolkovsky rocket equation in the low-mass approximation (see, e.g., Wiesel 1997). Thrusters can be used in any space environment but have the obvious disadvantage that the amount of propellant is limited.
- **Reaction wheels**: A change in the rotation velocity of one (in case of a three-wheel configuration) or several (in case of the more often used four-wheel configuration) wheels leads due to conservation of angular momentum to an opposite rate change in one of the satellite's axes. The transfer of momentum will in practice be less than 100% due to friction. Reaction wheels can also be used universally, but require a significant amount of power which must either be provided by an internal source (e.g., a battery) or by the Sun. Also, the amount of lubricant carried is limited meaning that at some time the friction will become too large for further operation.
- **Magnetic torque rods**: These are mounted parallel to each of the satellite's axes. A current sent through the rod will produce a Lorentz force in the presence of a magnetic field. This requires a power source and a field that is strong enough. The torque is perpendicular to the current and to the field, which implies that an axis aligned with the field can't be controlled at that particular position in orbit.

Attitude control is normally implemented as a fully autonomous process on the satellite (see Chap. 22, but also Sect. 14.3.3). Manual commanding is impracticable due to the short duty cycle (≤ 1 s in general), but the possibility to intervene in the control loop in case of problems is still present on most satellites. The way to correct the measured deviations from a desired attitude by autonomous on-board control is described at the end of this section. Such algorithms can be used for any deviation, but are normally limited to a certain range, that might be $\leq 30°$ for a coarse mode and $\leq 0.5°$ for a fine pointing mode (the numbers are exemplary only). A slew, i.e., a planned change of the default attitude, can in principle also be carried out by just specifying a target and let the controller do its work. This method is used on several satellites, but it has some drawbacks. The acceleration is governed by the moments of inertia and thus different for each axis. The speed of the slew depends upon the maximum torque delivered by the actuators, which might be dependent upon the position in orbit (e.g., there is no magnetic authority in yaw near the poles). Thus, the maximum angular rate, the duration and even the direction of the slew (180° could be made in either direction depending upon the accidental attitude error at the start) are not well predictable. The maximum torque must be throttled in order to prevent that slews become too fast, whereby different limits need to be applied for the three axes and perhaps also for several ranges (e.g., angles $< 0.5°$, $< 3°$, or $\leq$ 180°; examples and number of ranges are again exemplary only).

Therefore, a better way is to handle a change of the default attitude manually. The commanding of a 90° yaw slew will be treated in detail below as an example, but it might also apply to the pointing of the instrument towards a target or preventing irradiation of certain parts of the satellite by turning it away from the Sun.

Consider a satellite flying around the Earth with its orbit control thrusters all located on the backside.[6] Then, in order to change the inclination of the orbit an "out-of-plane" maneuver is necessary, preferably near one of the nodes. This requires a ±90° yaw slew with the offset reached at a very precise moment, namely just before the start of the burn. It is implemented by defining a profile connecting start and target attitude (ψ_0 and ψ_t respectively) with a user-defined rate $\dot{\psi}_0$ and acceleration $\ddot{\psi}$.

The acceleration and deceleration phases are equally long and determined by the time it takes to reach the cruising rate.

$$t_1 - t_0 = t_3 - t_2 = \frac{\dot{\psi}_0}{\ddot{\psi}} \tag{14.11}$$

The angle traveled in the acceleration phase is

$$\Delta\psi = \frac{1}{2}\ddot{\psi}(t_1 - t_0)^2 \tag{14.12}$$

so that $(\psi_t - \psi_0) - 2\Delta\psi$ is the angle remaining to be traversed in the cruise phase (Fig. 14.9). This of course takes

[6] TerraSAR-X, TanDEM-X, and PAZ are all build like that.

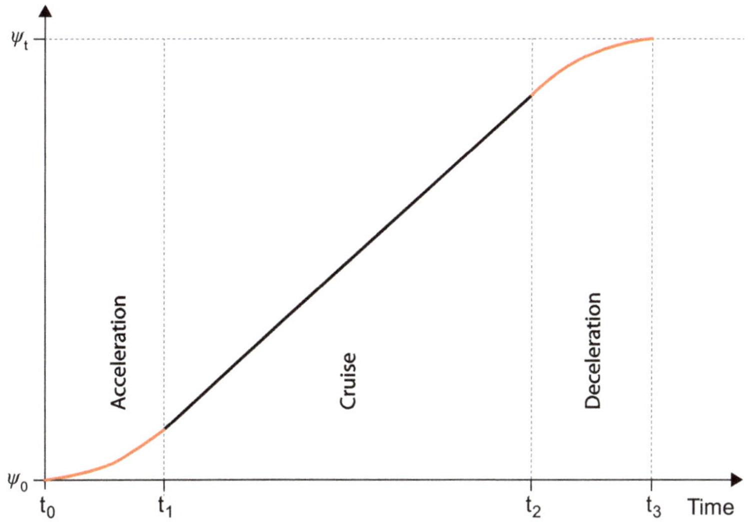

Fig. 14.9 Profile for a yaw slew from ψ_0 (default) to ψ_t (target, e.g., 90°). Acceleration starts at t_0 and ends at t_1, where the cruise phase at constant rate starts. The deceleration phase starting at t_2 lasts just as long as the acceleration and ends at t_3

$$t_2 - t_1 = \frac{(\psi_t - \psi_0) - 2\Delta\psi}{\dot{\psi}_0} \tag{14.13}$$

The maximum cruise rate and acceleration that can be chosen of course depend upon the power of the actuators, but more critically on the limits, set by the sensors, that are required to deliver attitude and rate information during the slew. Most star trackers for example will not function above a rate of ~ 0.5°/s. Finally, note that the dwelling time at a non nominal attitude might be limited e.g. by power constraints.

A practical example can be found in Fig. 14.10, where the measured yaw angles and rates are shown for a – 90° yaw slew. The prescribed profile is followed closely, but some slight deviations occur. For $t < 700$ s and for $t > 1800$ s the yaw angle is not 0°; this is due to the fact that the TerraSAR-X satellite performs so-called yaw-steering, a roughly sinusoidal motion superimposed on the orbit with an amplitude of ~ 3° to compensate for the Earth's rotation during data takes. Yaw-steering is switched off in preparation of the slew ($\psi = 0$ at $t = 750$ s, rates > 0 from $t = 700$ to 750 s). Acceleration starts at $t = 750$ s and ends at ~ 920 s, see Eq. (14.12). The choice of parameters in this case is such that the cruise phase [Eq. (14.13)] that now begins is very short and ends at $t = 925$ s already with the onset of the deceleration phase. The yaw angle shows a small overshoot at – 90° and some rate disturbances are visible around $t = 1200$ s due to the maneuver. The return slew to $\psi = 0°$ starts at $t =$ ~ 1250 s. The operation is almost symmetrical, the only difference being that the re-enabling of the yaw-steering has a delay of 1–2 min (allows different maneuver sizes to be made with the same procedure).

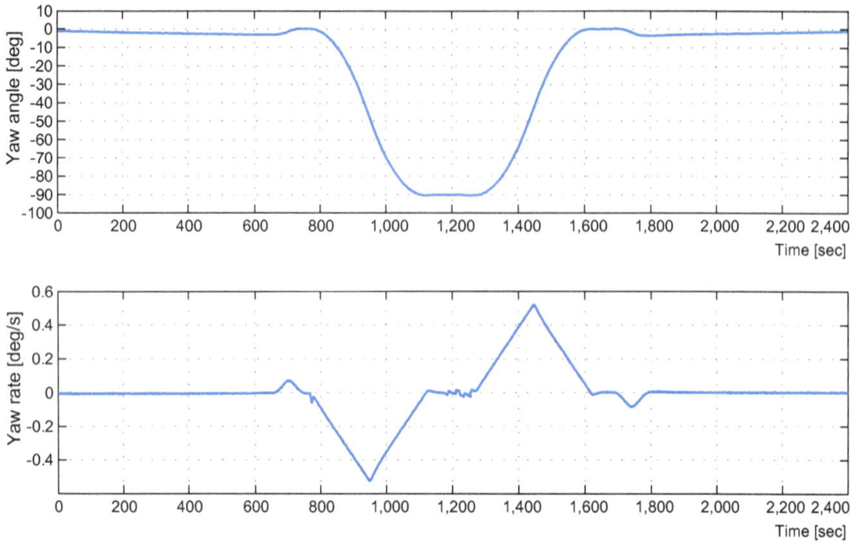

Fig. 14.10 The top panel shows the actual achieved profile for a yaw slew from ψ_0 (default, here 0°) to ψ_t (target, here − 90°) preparatory to an inclination maneuver on TerraSAR-X. The corresponding rate measurements are shown in the bottom panel

The elaborate method described above is not feasible for the real time correction of small attitude errors which is rather done by autonomous control on board. This will be based upon the equations for attitude determination [Eq. (14.7)] and propagation [Eq. (14.8)]. The latter will normally include several disturbances (see Sect. 14.2). The deviation from the desired attitude, which can either be a default attitude or a profile as defined above, and the rate of change in this deviation yield the direction and magnitude of the torque to be applied to the spacecraft in order to correct these errors.

The desired torque might then be distributed over several types of actuators if that is wanted. One could, in order to save fuel for example, send as much to the magnetic torque rods as possible and only if that is not sufficient, invoke the thrusters for the remainder. The computed torques must finally be transformed into the coordinate system of the actuators; in case of a four-wheel configuration this will be a 3×4 matrix.

The calculation of the desired torques from the measurements can take several layers of complexity. The principle will be illustrated here for a one-dimensional controller. Denote the predicted deviation in roll by δ_{roll}. The time of the computation could, for example, be the next grid point of the on-board processor. The prediction itself might be complex too (see Sect. 14.2). The easiest controller to build is a proportional one that reacts directly on the predicted deviation. The desired torque T_{des} can be written as:

$$T_{des} \sim -G\delta_{roll} \tag{14.14}$$

The gain parameter G determines the strength of the reaction and is normally configurable. The result will be a rather crude bouncing between the upper- and lower- dead bands. Therefore, a better way is to include the rate of change as well as the predicted deviation. The result is a so-called PD (proportional differential) controller which takes the form

$$T_{des} \sim -G_1\delta_{roll} - G_2\frac{d\delta_{roll}}{dt} \tag{14.15}$$

The gain parameters G_1 and G_2 are once more configurable and can themselves be complex functions of actuator properties. A final refinement not only takes the predicted values, but also includes the time interval before. The length of the interval $(t_1 - t_0)$ can be chosen and it is also possible to, of course, include a function that gives less weight to older values. This yields a so-called PID (proportional integral differential) controller which is widely used nowadays.

$$T_{des} \sim -G_1\delta_{roll} - G_2\frac{d\delta_{roll}}{dt} - G_3\int_{t_0}^{t_1}\delta_{roll}\mathrm{dt} \tag{14.16}$$

The performance of the attitude control system can and normally will be evaluated by high precision attitude determination a posteriori. The on-board restriction that requires a full new computational cycle for each grid point (≤ 1 s) vanishes and long stretches of data can be used. Such a high precision attitude reconstruction might also be required by the users of the payload, for example, for very accurate image processing or in order to combine images taken at different times.

14.6 Tasks of AOCS

Each mission normally has one or more engineers dedicated to each particular subsystem. It normally comprises such areas as instrument or payload operations, power/thermal, data and on-board computer, as well as AOCS. Tasks of AOCS are manifold and generally it is one of the busiest subsystems during LEOP (launch and early orbit phase) and also during the further mission.

- In LEOP the performance and alignments of all sensors and actuators used for attitude determination and control must be verified (including redundant ones). A reconfiguration might be required in case of defunct components (more likely to occur later in the mission).
- AOCS has a close interaction with the users of the instrument. Control parameters must be adjusted to optimize payload output; sometimes this can continue throughout the mission [prime examples are the GRACE and GRACE follow-on missions where the satellites themselves are the instrument (d'Amico 2002).
- There is also a close interaction with FD (flight dynamics). The performance of maneuvers must be evaluated and taken into account in future planning. The

direction of thrust might be different from the a priori calibration and might shift during the mission. Single thrusters can under- or over-perform and might even become defunct.
- Non-standard attitude maneuvers can be required, especially during LEOP, which need manual computation and commanding (also with FD).
- Precise attitude determination might be required for payload checkout and instrument calibration (again in co-operation with FD).
- Close interaction with P/T (power/thermal) subsystem; the temperature and power range of each sensor and actuator must be monitored, especially when switched on or off. The performance in terms of current (cost function) can become important when the batteries start degrading in the later stages (Herman and Steinhoff 2012).
- The performance of all actuators must be monitored during the entire mission (e.g., to detect increasing friction or leakage).
- Resources must be administered (fuel expenditure, book-keeping of the number of thruster cycles).
- Sensor gaps (e.g., intrusions) must be avoided by switching to a redundant sensor or by a configuration change.
- Special wishes (orbit correction maneuvers, imaging at special attitude, new targets) must be accommodated.
- A complete re-design might be required in case of serious changes in the mission's design. Examples of these are the new magnetorquer safe mode on the TerraSAR-X and TanDEM-X satellites that is not using the thrusters anymore (became necessary to safely fly in a formation with mutual distances < 150 m; Schulze et al. 2012), or the magnetic yaw-steering for GRACE, which was designed to align the satellite with the magnetic field (necessitated by the failure of several CESS thermistors).

The above also requires offline support from the flight dynamics group and for most missions also from the users of payload instruments. An in-depth overview of the AOCS subsystem is given in Chap. 22. Here only two examples will be given of the interaction of AOCS and FD with the on-board attitude control system.

14.6.1 Example 1

Interaction with the satellite is normally not per single command but with prepared procedures which not only contain the several commands required for a certain action, but also the timing, pre- and post-conditions that must be met, possibly the setting of the correct attitude (which for example in case of a 90° yaw offset will take a certain amount of time to achieve and might need to be carried out at a certain orbit position), possible changes in fault detection, isolation and recovery (FDIR) settings, timely interruption of payload operations, mode changes, switch on of additional telemetry packets, and—in case the action is done in real time—telemetry checks to be performed at the console.

The procedure shown in Fig. 14.11 can only be used time-tagged, i.e., the complete sequence is prepared beforehand, uploaded to the satellite and then stored in the on-board buffer. It will be carried out autonomously at the specified time.

In case a flight procedure can be carried out in real time, i.e., during a contact, every single command will be accompanied by several telemetry checks. Parameters are checked before sending the command and after on-board processing the effects become visible. Expected values, parameter names, and pages are embedded in the procedure. An example is shown in Fig. 14.12.

The flight procedures must be developed well in advance and be tested either on the satellite itself, on a real-time test bed, or on a software simulator.

14.6.2 Example 2

The second example deals with part of the evaluation of star tracker data as collected during the LEOP of GRACE (d'Amico 2002, but see also footnote[2] in Sect. 14.2). The measured attitude quaternions showed erratic behavior soon after the star tracker had been switched on with a one-orbit periodicity and each time lasting about 5 min (see Fig. 14.13).

The activity of the attitude thrusters and hence fuel expenditure increased considerably during and immediately after these disturbance periods. The cause could soon be identified. The star tracker started to deliver invalid quaternions as soon as the Moon was in, or close to its FOV (see Fig. 14.14).

The conclusion was twofold: In the first place it was seen that it is necessary to switch to the redundant star tracker as soon as the Moon enters the FOV of the prime and in the second place it was clear that the sensitivity for stray light extended over an area ~ 30% larger than specified (see Fig. 14.15).

This strategy was implemented and then used for the entire mission. In the late phases of the mission the FOV had to be enlarged even further, because the sensitivity of the star trackers to stray light worsened over the years. Finally, it should be noted that a switch to the redundant star tracker is only possible if that is not blinded by the Sun at that moment.

Procedure for an inclination maneuver with +90° yaw slew on TerraSAR-X (Note that the only parameter input is the start time of the maneuver and its duration)		
1	Pre-conditions	Satellite must be in the correct AOCS mode. The thruster branch must be available and active. Time-tagged commanding only Payload and other AOCS operations are not possible from 9 m 40 s before until 14 m 50 s after the start of the maneuver. The maximum burn time is 180 s.
2		The reference time, **t_0**, is the start time of the maneuver.
3	**t_0-00:59:52**	Additional heaters near the thrusters are switched on.
4	**t_0-00:11:00**	Additional telemetry packets containing parameters pertaining to the maneuver are switched on.
5	**t_0-00:09:50**	The autonomous reaction which triggers if deviations larger than a specified limit occur on the Sun-vector are disabled.
6	**t_0-00:09:50**	The table showing the settings of the on-board surveillance is dumped.
7	**t_0-00:09:45**	Payload operations on TerraSAR-X require continuous correction of the yaw angle in order to image the desired swash on the Earth's surface. This so-called "yaw-steering" is disabled.
8	**t_0-00:09:44**	The autonomous recovery from "attitude hold mode" into the mode for payload operations is disabled.
9	**t_0-00:09:40**	The satellite is commanded to attitude hold mode; this mode uses the same sensors and actuators as the mode for payload operations, but has larger tolerances on attitude and rate deviations. It also allows one to command any orientation with respect to the orbit frame. A margin of three minutes is included to let the spacecraft stabilize.
10	**t_0-00:09:35**	The commands in step #10, …, #14 set and activate an on-board
11	**t_0-00:09:34**	safety mechanism that would stop the thrusters in case they
12	**t_0-00:09:31**	would still be active after the specified end of the burn. Terra-
13	**t_0-00:09:28**	SAR-X uses off-modulation to control the attitude during the
14	**t_0-00:09:26**	burn, so a 10 % margin of the maneuver duration is added.
15	**t_0-00:06:40**	A yaw slew to +1.574 mrad is commanded. Note that the 90° maneuver is not made with exactly 1.571 mrad, but with 1.574 to account for the difference in performance of the thrusters. Slew and stabilization last exactly 400 s.
16	**t_0**	Start of the maneuver with the input burn time. As soon as the accumulated thruster on-time reaches the specified duration the thrusters close automatically.
17	**t_0+1.1*(burn time)**	Additional, normally superfluous, maneuver stop command; safe- guards e.g., against a zero to many in the input duration
18	**t_0+00:00:01**	The yaw-steering (see #7) is re-enabled. Commanded here, but takes effect only once the satellite is back in attitude hold mode
19		Autonomous start of the slew back to the nominal yaw attitude. Slew and stabilization last 400 s
20		Autonomous transition into attitude hold mode at the end of the slew
21	**t_0+00:13:00**	Well after the end of maneuver and slew the additional heaters switched on in step #3 are turned off again.
22	**t_0+00:13:51**	The on-board surveillance of deviations of the Sun-vector is re-enabled.
23	**t_0+00:13:52**	The table showing the settings of the on-board surveillance is dumped.
24	**t_0+00:14:00**	The autonomous transition from attitude hold mode into the mode where payload operations are performed is re-enabled. The transition will occur as soon as attitude deviations are small enough; statistics over a large number of maneuvers shows this normally occurs between four to eight minutes.
25	**t_0+00:14:54**	The additional telemetry switched on in step #4 is switched off again.
26	Post-condition	Next maneuver can't be made within 60 minutes (unless step #21 is deleted and the timing adapted).

Fig. 14.11 Example of a flight procedure for an "out-of-plane" orbit maneuver, i.e., a burn with a 90° yaw offset

…	Previous part	see Fig. 14.11
		Verify telemetry ASD00367 PSI_FROM_QUAT = 0.0 rad Page = AOC5505A
15	**If TLM OK then**	Slew to -90° is commanded. Slew and stabilization last exactly 400 s
	Wait 400 s then	Verify telemetry ASD00367 PSI_FROM_QUAT = -1.574 rad Page = AOC5505A
…	Following part	see Fig. 14.11

Fig. 14.12 Example of telemetry verification during a station contact. Such checks may be part of a procedure as shown in Fig. 14.11, when it is completely or partly carried out in real time

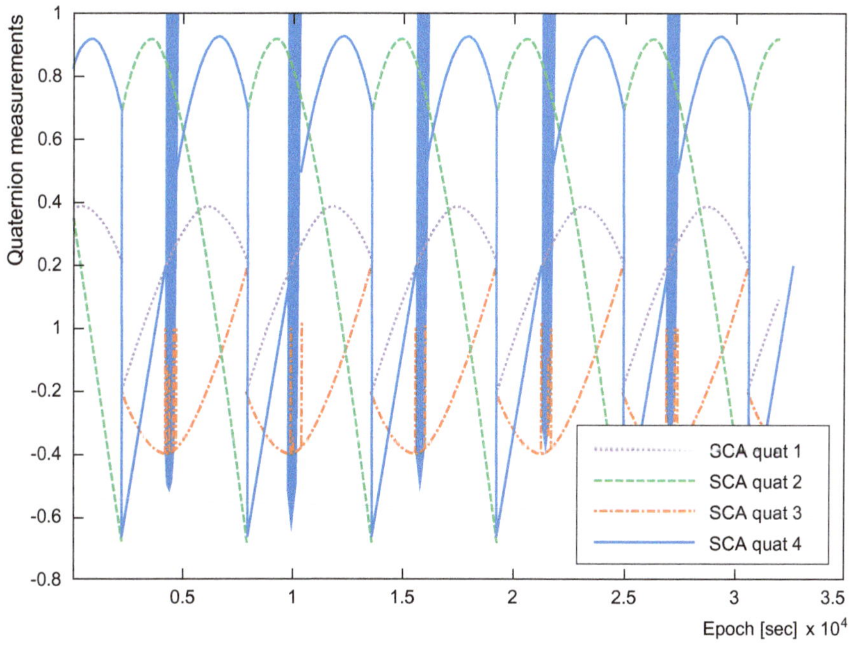

Fig. 14.13 [© (d'Amico 2002)] The measured star tracker quaternions on GRACE displayed strong disturbances during the first days of LEOP. These recurred each orbit and lasted about 5 min each. Note that the discontinuities (e.g., around epoch 0.25, 0.75 etc.) are no error but are caused by a parity switch; changing the sign of all four components does not change a quaternion

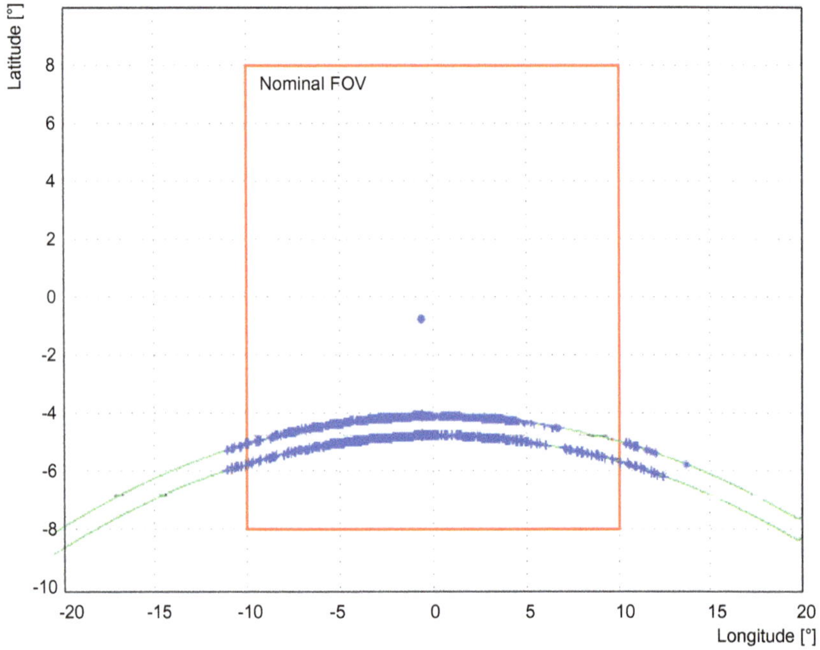

Fig. 14.14 [© (d'Amico 2002)] The path of the Moon through the star tracker's FOV is shown for two consecutive orbits (green line). The blue marks denote the points where the solution became invalid. The nominal FOV is shown by the red lines; it is rectangular due to the baffle in front of the camera

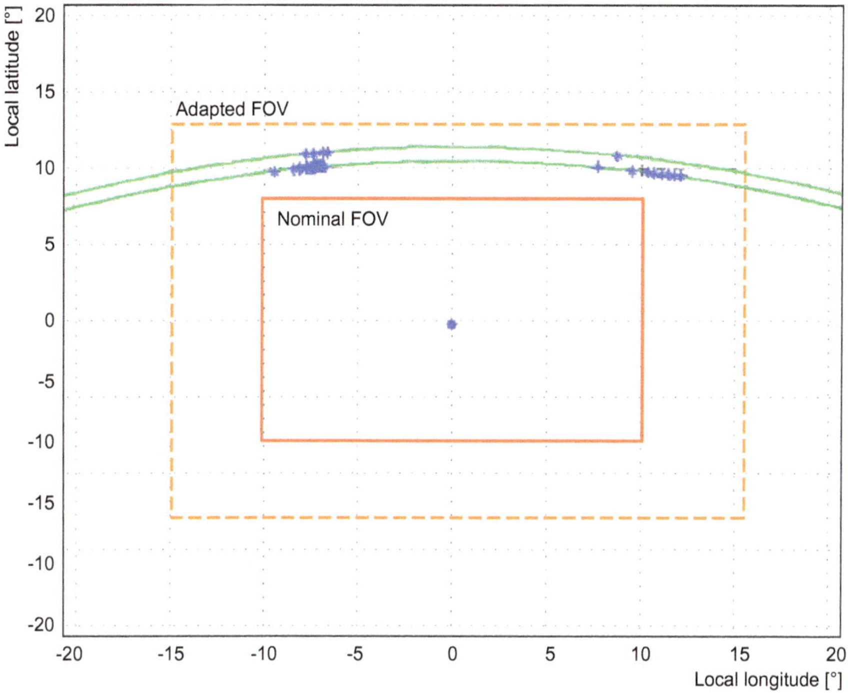

Fig. 14.15 [© (d'Amico, 2002)] Two orbits are plotted a day later than the ones shown in Fig. 14.14 Symbols are the same as in Fig. 14.14. A switch to the redundant star tracker is commanded from this moment on as soon as the Moon enters the enlarged FOV of 30° × 26°. Note that on the right-hand side, where the Moon leaves the FOV, an additional safety margin of ~ 1° is implemented

References

Arbinger C, Luebke-Ossenbeck B (2006) Chapter 4.5 Lageregelung. In: Ley W, Wittmann K, Hallmann W (eds) Handbuch der Raumfahrt-technik. Carl Hanser Verlag GmbH & Co. KG

d'Amico S (2002) Attitude and orbital simulation in support of space mission operations: the GRACE formation flying; Master Thesis, Politecnico di Milano

Herman J, Steinhoff M (2012) Balancing, turning, saving special AOCS operations to extend the GRACE mission. In: SpaceOps conference in Stockholm (Sweden)

Larson WJ, Wertz JR (eds) (1992) Space mission analysis and design. Kluwer Academic Publishers

Schulze D, Herman J, Löw S (2012) Formation flight in low-earth-orbit at 200 m. In: SpaceOps conference in Stockholm (Sweden)

Seidelmann PK (ed) (1992) Explanatory supplement to the astronomical almanac. University Science Books, New York

Sidi MJ (1997) Spacecraft dynamics and control. Cambridge University Press

Wertz JR (ed) (1978) Spacecraft attitude determination and control. Kluwer Academic Publishers

Wiesel WE (1997) Spaceflight dynamics. The McGraw-Hill Companies, Inc.

Jacobus Herman is leading the AOCS group at GSOC since 2001. He joined the DLR in 1989, at first in the flight dynamics department, and wrote a study on space debris commissioned by the ESA in 2000. After getting his Ph.D. in astrophysics in 1983 in Leiden (the Netherlands), he worked for several years at ESTEC (Noordwijk, the Netherlands) and the Max Planck institute for extraterrestrial physics in Garching (Germany).

Ralph Kahle received the Dipl.-Ing. degree in aerospace engineering from the Technical University of Dresden (Germany) in 2001, and the Dr.-Ing. degree from the Technical University of Berlin (Germany) in 2005. In 2004, he joined DLR's German Space Operations Center (GSOC) as a Flight Dynamics (FD) Engineer. He is responsible for FD system engineering and operation of various satellite projects. His field of work also comprises space mission analysis, FD software development, and execution of Launch and Early Orbit Phase and routine FD operations. Since 2016 Dr. Kahle is head of the Flight Dynamics Services group at the German Space Operations Center (GSOC).

Sofya Spiridonova joined GSOC/DLR in 2012 as a flight dynamics engineer. Since then she has taken part in FD operations of several low-earth orbit missions, geostationary satellite missions, as well in FD software development and mission analysis activities.

Part V
Mission Planning System

Chapter 15
The Planning Problem

Christoph Lenzen and Sven Prüfer

Abstract In this chapter, we discuss key concepts for designing a mission planning system for a satellite mission. This includes in particular the GSOC modelling language for planning problems which will be illustrated by various example use cases from current missions at the German Space Operations Center (GSOC).

15.1 Introduction

Mission planning plays a key role in the operations of a spacecraft, as it ensures that all resources are available and used to an optimal level and the goals of the mission are achieved. There are various ways how to define the exact meaning of mission planning, and different operations centers throughout the world use different definitions.

Within this book, we will consider mission planning as the task of preparing, organizing, and planning all relevant activities that happen during the mission, on board as well as on ground. It is therefore distinct from mission analysis or mission preparation tasks, which serve to analyze and define a mission beforehand and provide the necessary means to execute the mission.

As such, the responsibility of mission planning is to deliver the plan in form of a timeline or so-called sequence of events (SoE) right in time for the relevant activities (e.g., before uplink to the spacecraft). This might happen only once (e.g., in case of an asteroid lander) or very frequently (e.g., an Earth observation satellite fed with customer orders). The plan has to be conflict free, i.e., under the given on-board and ground constraints it can be executed on the spacecraft and on ground without any errors. In addition, the timeline should assure that the available resources are used efficiently in order to meet as many mission goals possible.

The overall scope of the activities for which mission planning is responsible, varies a lot from mission to mission. It can range from very limited duties, e.g.,

C. Lenzen (✉) · S. Prüfer
German Space Operations Center (GSOC), German Aerospace Center (DLR), Oberpfaffenhofen, Germany
e-mail: christoph.lenzen@DLR.de

F. Sellmaier et al. (eds.), *Spacecraft Operations*, Springer Aerospace Technology,
https://doi.org/10.1007/978-3-030-88593-9_15

planning only simple transmitter switching for ground station contacts, to almost full responsibility for all spacecraft commands including ground station scheduling.

In this chapter, we will introduce the key concepts for designing and operating a mission planning system and present some examples of planning systems for current active missions at the German Space Operations Center (GSOC).

15.2 General Overview of a Mission Planning System

Since the requirements for a mission planning subsystem (MPS) and the tasks that it shall perform vary considerably between different missions, there is no general planning system that can cover all use cases. In practice, depending on the nature of the planning problem at hand, one will encounter a wide variety of planning systems, ranging from very large software systems to small mission planning components within other systems.

They can be categorized by their degree of automatism, ranging from fully automated planning software without any necessary user interaction to a completely manual planning process. In most cases the system is somewhere between these two extremes: A human mission planner may be supported by GUI-based software tools, which algorithmically create some aspects of the desired plan and check the overall consistency in the end. This is usually the most economical solution.

An example for a fully automated concept is the combined TerraSAR-X/TanDEM-X (Krieger et al. 2007) mission planning system (Maurer et al. 2010). It creates a timeline including the required telecommands for every payload-related activity on the two satellites and distributes the generated information to the affected parts of the ground segment.

On the other hand, the mission planning system for the GRACE mission (Braun 2002) consisted only of a small algorithm that was triggered by human mission planners and created only the sequence of commands necessary for the downlink transmitter switches. It is therefore an example of a software-assisted planning concept.

As a third example, the ISS mission planning (see Chap. 17) depends entirely on manual creation of the timeline, which is only fed into software tools to display it and to allow further processing.

All the mentioned examples can be planned quite accurately using a time-based planning system. This concept may fail when the planning entity cannot foresee the consequences of executing activities with sufficient accuracy. An Earth-based planning system for Mars rovers, for example, can plan an operation "*Drive to target A*", but it can't estimate when the rover will have performed this task, due to obstacles which may block the direct path. Such planning problems therefore require different techniques, such as event-based or goal-based planning, which we will not discuss further in this book.

Another aspect by which mission planning systems can be characterized is the periodicity of the planning process. Some missions have limited duration and a final timeline might be created before the execution of the mission, e.g., for a flyby such as New Horizon's Pluto flyby. For others, a predefined timeline might be adjusted to the current situation in regular intervals, e.g., once per week. Some Earth observation satellites are examples for which the amount of planning required and frequent customer requests make it necessary to limit the planning horizon and repeat the process very often, e.g., once before every commanding possibility of the spacecraft or even every time new planning relevant information is available.

The periodicity of the planning process defines the way the timeline is generated which in turn has profound implications on both operations and development of the mission planning system. In general, we distinguish between three timeline generation modes:

Fixed Plan

The mission is planned during the preparation phase from launch until end of mission. There is either no need or no possibility to update the plan during execution. For this approach to work, the mission objective and the approach to reach it need to be known exactly in advance. It usually requires a quite conservative estimation of available resources.

On the other hand, the plan can be heavily optimized by using complex algorithms with large runtime durations and by iterating the results with the customer.

Repeated Replanning

Replanning is done on a regular basis, e.g., each time before a low Earth orbit (LEO) satellite is visible by an uplink station. It allows ingesting updated information about resources, orbit events etc. and new tasks, which need to be performed, in particular new planning requests of the customer.

Major modifications of the timeline are very common, especially when trying to optimize the result. Optimization, however, is usually limited to fast running algorithms. Another drawback is that the user has to rely on the decisions of the planning process: when ingesting a new order, the user will not know about modifications in the timeline until the succeeding replanning has been performed. He might also not have time to overrule the decisions of the planner, because replanning is usually performed just in time.

Reactive Planning

The operational concept of this type of mission planning system resembles a booking system. Starting point is an empty timeline, which is filled with each input it gets. In particular, when a planning request is sent to the system, the algorithm checks what other requests might need to be (re)moved and can act on this situation. As another application, it may send this information back to the user who then decides whether to accept these modifications or to discard the current request.

Of course, there are a lot of variations on automatic decisions; nevertheless, the reactive planning concept allows keeping the modifications of the existing timeline as little as possible. This type is chosen in case the timeline shall remain stable or in case the user requires full control over the timeline. Algorithmic optimization, however, is even more restricted due to the performance requirements of the system. Besides, it impedes user control and stability of the timeline.

Most mission planning systems have in common that they form a central part of the ground segment, usually having interfaces to many other ground systems, both inside and outside mission operations. Typical interfaces of a mission planning system are depicted in Fig. 15.1. They comprise input interfaces for incoming planning requests, updated orbit and maneuver information, payload configuration updates, etc. Often, the availabilities of external resources have to be fed into the planning system, like ground station availabilities and booking times or astronaut availabilities. Sometimes, additional external information will influence the planning result and needs to be fed into the system. For example, the cloud forecast might be used for an optimized selection of image acquisition targets or evaluating optical downlink opportunities. A mission planning system might create spacecraft commands by itself, so interfaces to the spacecraft control systems are needed to transfer the created commands. In this case, the system may also receive feedback whether these commands were uplinked successfully to the spacecraft and whether they have been executed on board according to telemetry. Finally, the output of the planning process needs to be distributed to various customers, e.g., the timeline is published for the operations personnel, ground stations are informed which downlinks to expect, and archive systems are fed with the planned activities. Some mission planning systems offer immediate feedback to customers, who can then control the feasibility of their requests, see their planning status, and decide on possible alternatives.

Since different missions impose different requirements, the composition and the size of a mission planning system greatly varies from mission to mission. For completely monolithic specialized mission planning applications, this would mean, that they would need to be re-implemented from scratch for every new mission. Hence, it is beneficial to maintain a set of loosely coupled generic tools as building blocks that can be extended and tailored to support the mission at hand. This provides greater flexibility and reusability while allowing easier development in a team. A prerequisite for this is to maintain the MPS-internal interfaces as strict and stable as possible to prevent tedious adaptations for different missions. A general way of expressing planning problems is needed that can be used to specify most mission requirements. For this purpose, a generic planning language is required; one example, the GSOC Planning Language, is described in more detail in Sect. 15.3.2.

One of the components that may be used in such a modular MPS is a central database that holds the planning model, i.e., the objects which are to be planned. This database also stores the latest timeline version, when generated. All other components can then interact with the planning model and perform the necessary manipulations. Ingestion modules will import new data into the model, like planning requests, input from flight dynamics, external resource states, etc. A scheduling engine can be used to generate the plan algorithmically by fulfilling as many requests as possible while considering all constraints and available resources. Various export components can access the final plan and produce the desired outputs, like files containing a plan overview, command files, or downlink information files for the various ground stations. GUI (graphical user interface) components can visualize timelines and the

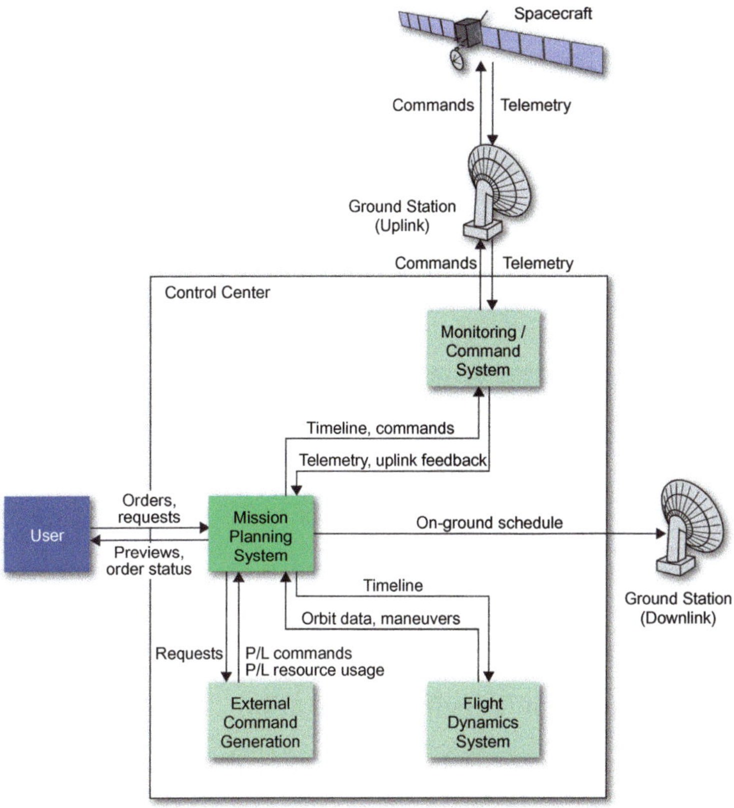

Fig. 15.1 Typical MPS interfaces

current state of planning for users and allow manual modification of the plan or starting of other automatic planning tools. Automation components might be included which trigger the planning process upon certain events or at specified times. As an example, the components of the combined TanDEM-X/TerraSAR-X MPS are depicted in Fig. 15.2. Almost all the above-mentioned components are realized in this system. For the reactive planning concept, however, a message driven approach simplifies communication between the various components and prevents threading issues which can occur when multiple components access a common database. Here the central database is replaced by private databases of certain components, which merely serve for recovery purposes.

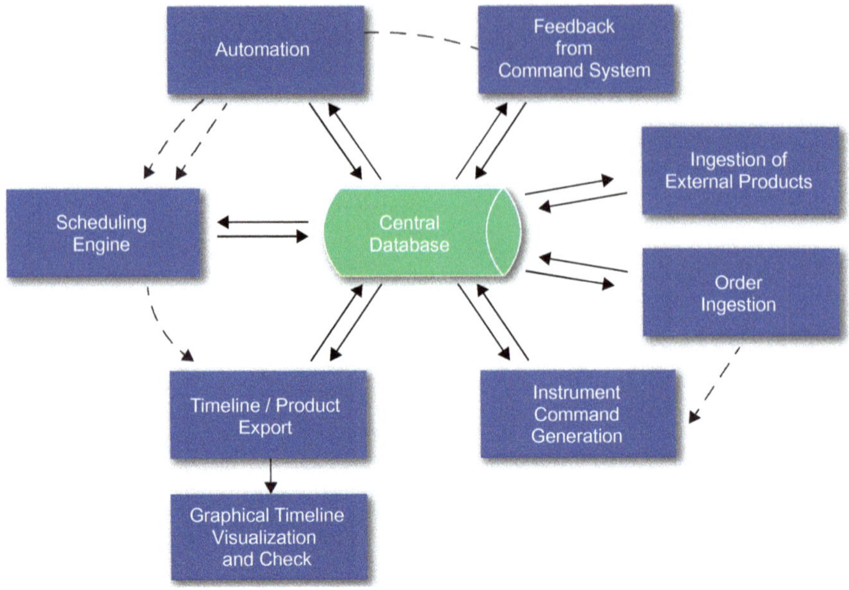

Fig. 15.2 Components of the TerraSAR-X/TanDEM-X MPS

15.3 Techniques for Timeline Generation

15.3.1 General Considerations

As mentioned above, there are several characterizations of a mission planning system, which all drive the design of the planning system and its integration into its environment. When looking at its core, however, the objectives of the timeline generation process are the same:

- **Feasibility and Safety**: Each subsystem must be able to execute the timeline (feasibility) without being exposed to inacceptable high risk (safety). In particular, the timeline shall not rely on on-board safety mechanisms of a spacecraft. For example, it must be avoided to generate a timeline which leaves a high energy consumer switched on at the end of the commanding horizon, even if a later uplink or an on-board safety feature is expected to switch it off at the proper time.
- **Benefit**: The timeline shall yield the biggest possible benefit according to criteria defined by the mission.
- **Traceability**: In most missions, the user wants to understand why the timeline looks like it does. Since the decision on which planning request is included in the timeline is made during the planning run, evidence for this choice must be supplied by the mission planning system for later analysis.
- **Performance**: The timeline generation process must be sufficiently fast.

Whereas feasibility and safety should be given the highest priority for all missions, the emphasis on the other three objectives may vary significantly between different spacecraft and missions. Even during the lifetime of a spacecraft, the goals of the mission may change, e.g., because an instrument of the spacecraft degrades or breaks down, because an additional satellite is added to the existing system (as happened with the TerraSAR-X/TanDEM-X satellite twin) or because the user of the satellite changes [as happened for TET (Axmann et al. 2010), which became part of the FireBird mission (Reile et al. 2013)]. In all these cases, the timeline generation process needs to be adapted and, in some cases, this has to be accomplished fast. Thus, another objective may be added for the timeline generation process:

- **Flexibility**: It shall be possible to adapt the timeline generation process to meet any modified mission requirements.

Whereas the first four objectives might be served best using an individually implemented system, possibly based on a small set of core components, the need of flexibility requires a full generic tool suite, which can be adapted to each mission's needs. A most welcome side effect of such a generic tool suite is that in the long term it can save a lot of manpower.

15.3.2 GSOC Modeling Language

The GSOC modeling language is one example of a system which allows modeling the planning problem of a typical spacecraft mission. When modeled properly, a conflict-free timeline will meet the main objective of feasibility and safety. Therefore, a modeling language must, on one hand, allow sufficient and accurate means to represent the real world and, on the other hand, must remain descriptive and user-friendly in order to avoid modeling mistakes. Of course, it must also be manageable by a planning engine. It turns out that the following basic elements and features supply a good trade-off between these three goals.

Activity

An activity represents something that may be executed at some time, e.g., one telecommand or a procedure, i.e., a set of telecommands, which are executed in a fixed order. In order to be able to store all required information, an activity may be given numeric, temporal or string parameters.

Timeline Entry

A timeline entry describes when an activity takes place. Consequently, the properties of a timeline entry are its *startTime* and its *duration*, resp. its *endTime,* and its activity. There may exist multiple timeline entries for one activity in case the activity is repeated.

Slider and Offset

Often a timeline entry needs to be mapped to a point in time, e.g. in order to specify when a resource modification shall start and when it shall end. For this, two numbers, the *slider* and the *offset* may be defined. They transform the start and the end time of a timeline entry into a reference time of the timeline entry as follows:

$$referenceTime = startTime + slider * (endTime - startTime) + offset$$

Sliders and offsets are illustrated in Fig. 15.3.

Minimum and Maximum Duration of an Activity

These values specify the allowed durations of timeline entries for this activity, as illustrated in Fig. 15.4.

Time Dependency with Other Activity

This specifies a minimum mandatory temporal separation between the timeline entries of two activities, as shown in Fig. 15.5. For both activities, a slider can be defined transforming the activity's timeline entry to a reference time. Note that in this case, we don't need to specify offsets together with the sliders, because each offset may be incorporated into the minimum separation of the time dependency.

Also, a maximum allowed separation may be modeled that way by swapping the activities and using a negative minimum separation.

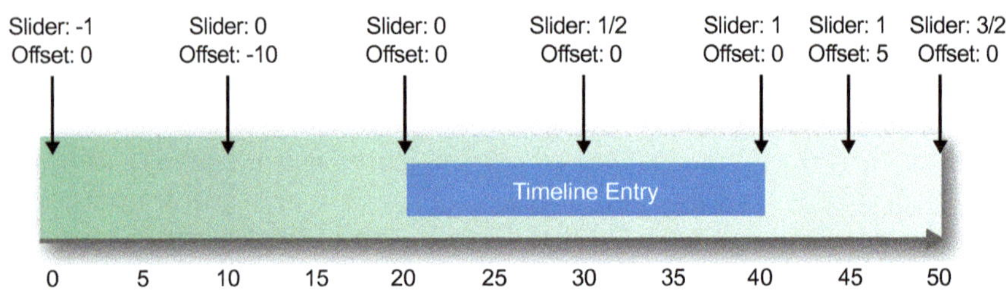

Fig. 15.3 Examples for slider and *offset*

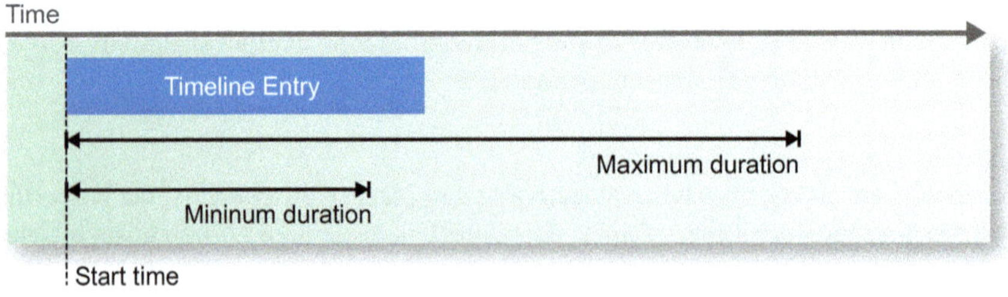

Fig. 15.4 A timeline entry consists of a start time and a duration. The duration can be restricted via a minimum and a maximum duration value

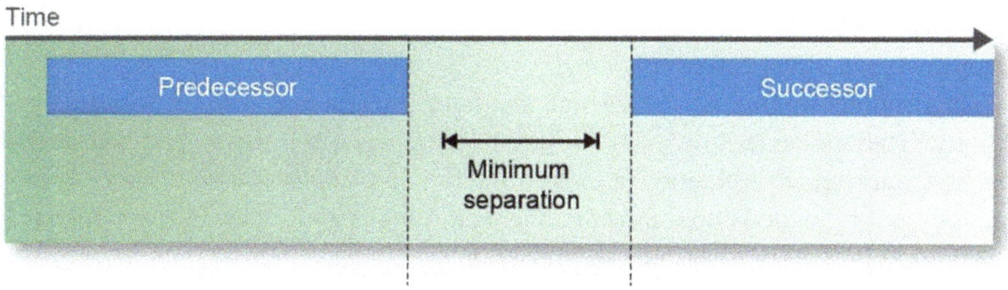

Fig. 15.5 Two timeline entries, separated by a minimum offset between the end time of the predecessor (*slider* = 1) and the start time of the successor (*slider* = 0)

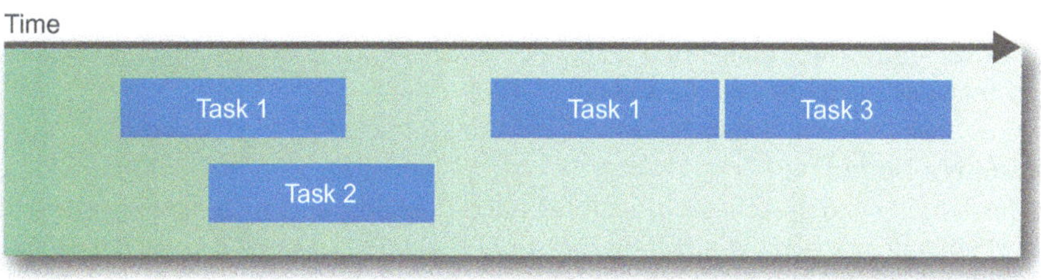

Fig. 15.6 A timeline consists of timeline entries corresponding to activities

Timeline

A timeline is a set of timeline entries, as shown in Fig. 15.6.

Child Relations of Activities

Often some low-level activities need to be executed in order to achieve a common goal, e.g., in order to create an image of a large region one may have to execute multiple image acquisitions. In order to express such links, we use parent–child relations between activities. In our example, we create a parent activity, which we identify with the request to map the total region, and for all individual image acquisition activities, we define a parent–child relation indicating that the image acquisition belongs to the request.

The benefit of this approach is not only to allow easier implementation of algorithms exploiting this information. Such a hierarchical structure also supports the common approach of long-term and short-term planning. In this scheme, a rough plan is created by assigning the parent groups a timeline entry, indicating roughly when this request shall be considered. Only when the short-term plan is generated, the child activities are assigned their respective timeline entries.

In order to distinguish them, we call activities with children *groups* and activities without children *tasks*. Note that a *group* can have *tasks* as well as *groups* as children.

Planned When, Minimum to Plan and Maximum to Plan

Usually an activity is considered to be planned in case there is at least one corresponding timeline entry for it. However, if the activity has children, the property *plannedWhen* indicates the number of child activities, which need to be planned in order to consider this activity planned.

The properties *minimumToPlan* and *maximumToPlan* may also be defined for an activity. They indicate how many children of the activity must be planned at least and how many may be planned at most. Note that if the activity itself is not planned, the number of planned children is not checked, because only planned activities may have conflicts. This is illustrated in Fig. 15.7.

Resources

A *resource* consists of a scalar time profile called *fill level*. Optionally, capacity limits can be attributed to a resource: whenever these limits are exceeded, the surplus is cut off (*lost values*), as shown in Fig. 15.8.

In the following, constraints and operations on resources as well as between resources and activities are described.

Resource Bound

A *resource bound* specifies a globally defined time profile, which the resource's fill level must not exceed, as shown in Fig. 15.9.

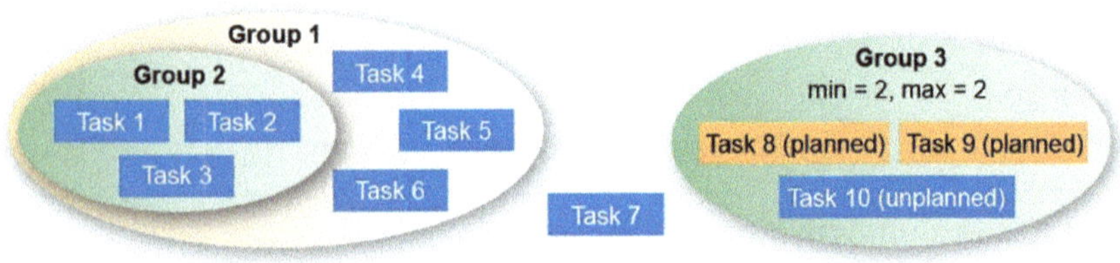

Fig. 15.7 Group 1, Task 7, and Group 3 are top level. Group 3 has both *minimumToPlan* and *maximumToPlan* set to 2 and thus requires exactly two tasks to be planned

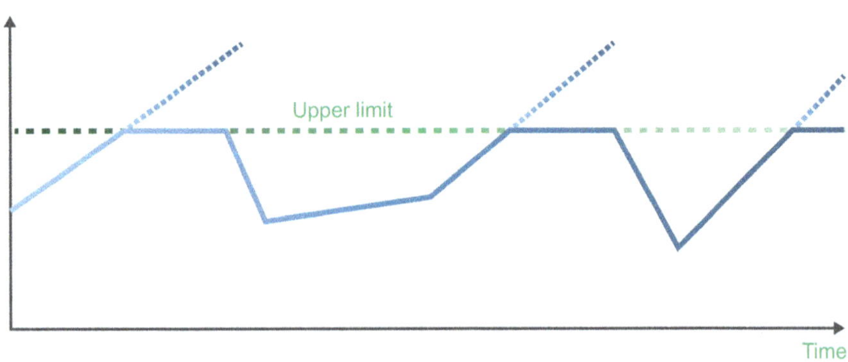

Fig. 15.8 This fill level profile might represent a simplified battery model: when the capacity is reached, the surplus supply gets lost

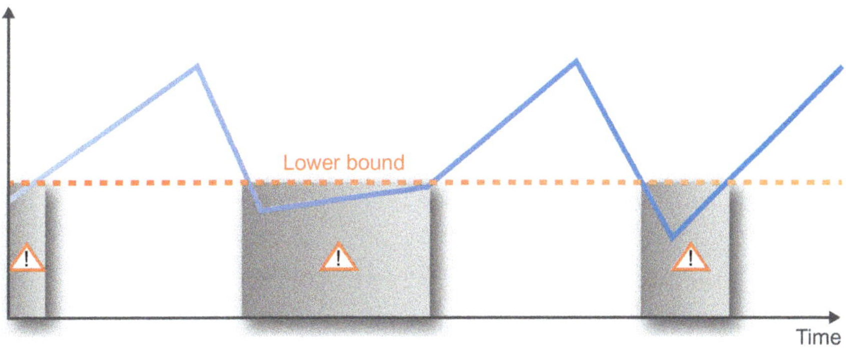

Fig. 15.9 The fill level causes a conflict, when it falls below the lower bound

Resource Comparison

A *resource comparison* is a resource bound applicable to a time interval depending on an activity's timeline entry. The resource bound does not necessarily have to coincide with the duration of the timeline entry. Instead, for both start and end time of the constraint, a slider and an offset is defined, which transforms the timeline entry interval into the constraint interval, as depicted in Fig. 15.10.

Resource Modification

A *resource modification* specifies how an activity's timeline entry modifies a resource. The modification is defined by an active change profile, which is mapped to the time axis via two pairs of sliders and offsets, one specifying the profile start (*startSlider,*

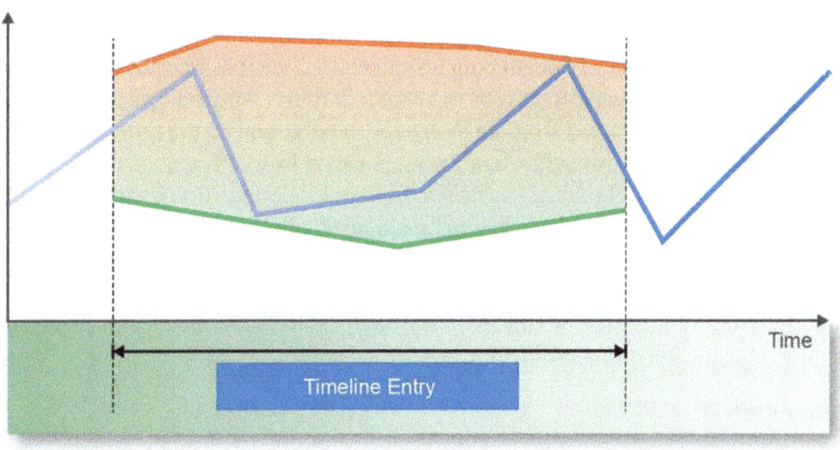

Fig. 15.10 The resource (*blue*) is constrained by upper and lower boundaries (*red, green*). The interval where these boundaries apply (the region between the two black vertical lines) is derived from the timeline entry via the constraint's sliders and offsets

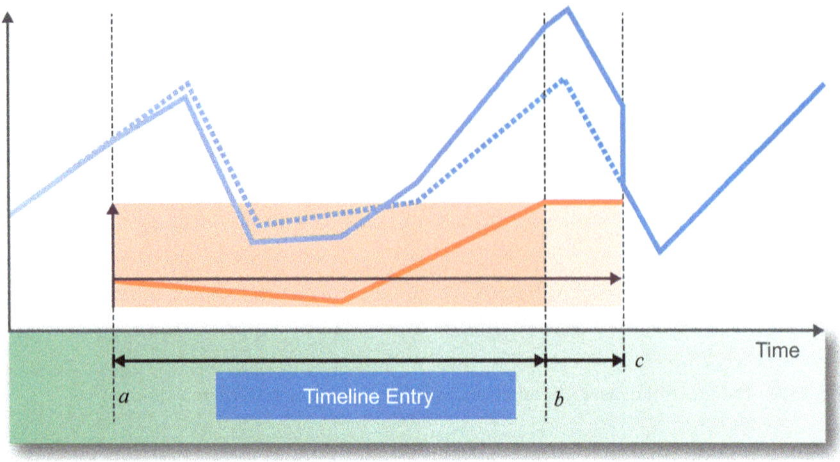

Fig. 15.11 The timeline entry has an active profile (*red*) starting at *a* and ending at *b*, which influences the resource (*blue*) not only during the time frame defined by the duration of the active profile, but also during the extended validity period ending at *c*

startOffset) and the other specifying the profile end (*activeProfileEndSlider, activeProfileEndOffset*). A typical example for this is a data downlink activity, which reduces the available bandwidth in the timeframe, in which it takes place. However, the modification of the resource can also last longer than the end of the active change profile. In that case it is required to distinguish between the end of the active change profile and the end of the resource modification. We therefore also introduce the pair *dependencyEndSlider* and *dependencyEndOffset*: In case the dependency end is later than the active profile end, the value of the profile at active profile end is extended until dependency end, as shown in Fig. 15.11.

For the special case of accumulating resource modifications like power consumptions the *dependencyEndOffset* needs to be set to infinity: After being consumed, the power which was extracted from a battery for the duration of the timeline entry is never available again. Instead it could be replenished by a different activity.

For the case of a finite *dependencyEndOffset* and a modification profile which has non-negative values, we may add another parameter called *releaseSlope*. If specified, it must have a negative value: Instead of cutting off the resulting resource modification profile at the dependency end, we append a segment with $slope = releaseSlope$ starting with the resource modification profile's value at *dependencyEndTime* and ending when the resulting profile reaches the value 0.

Suitabilities

A suitability is a special utilization of the resource concept. It considers the resource as a measure for the benefit to schedule an activity at a given time. This way, it supplies information which times to prefer for the execution of an activity. A suitability therefore cannot cause a conflict. The benefit of a timeline entry is derived from a resource's fill level around the considered timeline entry by taking the integral,

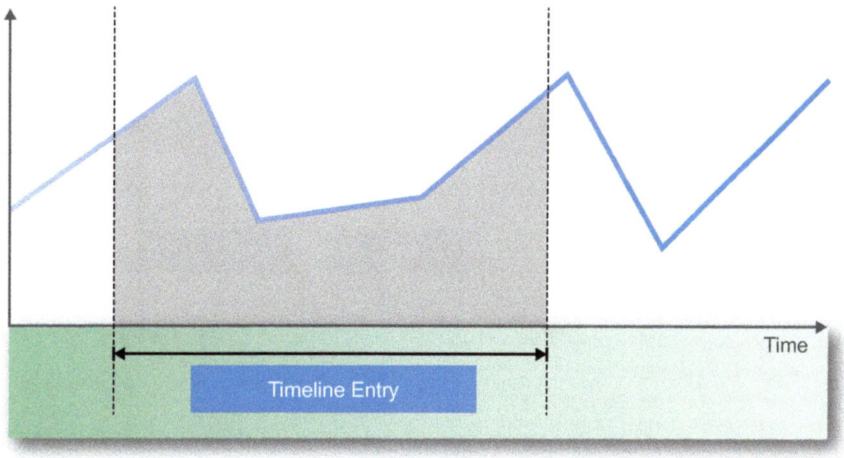

Fig. 15.12 The benefit of scheduling a given timeline entry at a given time can be quantified by a suitable mathematical operation (e.g., the maximum or the integral) applied to the corresponding suitability profile during the time interval defined via two offsets and sliders

maximum, or minimum of the fill level, as depicted in Fig. 15.12. The interval around the timeline entry is specified again by two pairs of offsets and sliders for start and end, respectively. An algorithm may prefer timeline entries with maximum benefit or it may even try to optimize the sum of all benefits of the whole timeline.

15.3.3 Application Examples of the Modeling Language

In this section, we present a few examples how to apply the above-presented basic building blocks in order to illustrate the flexibility of the GSOC modeling language.

Opportunity Resources, e.g., Ground Station Visibilities
In order to avoid scheduling downlinks outside ground station visibility, we define a resource with fill level equal to one wherever the ground station can receive data from the satellite, and zero otherwise, as shown in Fig. 15.13. Such a resource may be referred to as an *opportunity resource.*

Each downlink activity on the other hand is given a lower bound comparison, which checks that this opportunity resource has value greater than or equal to one.

Equipment Resources, e.g., Downlink Antennas of a Ground Station

In order to model the antenna availability of a specific ground station, we define a resource whose integer values represent the number of antennas in use at this ground station. Furthermore, we define

- an initial fill level of zero and
- an upper bound corresponding to the maximum number of available antennas.

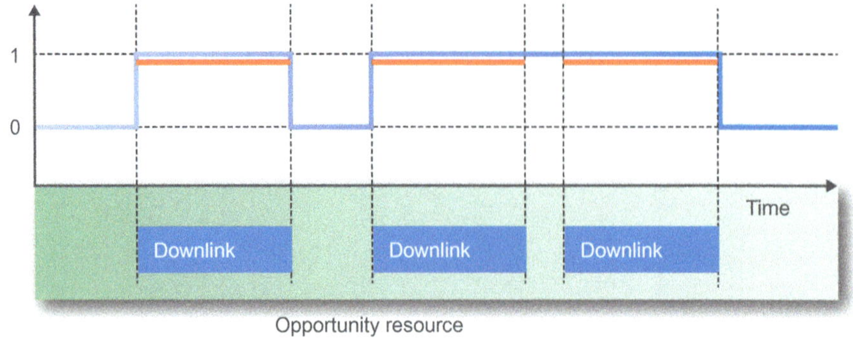

Fig. 15.13 The fill level of the opportunity resource indicates where downlinks may be scheduled

Whenever a downlink for one satellite shall be planned, we also need to allocate one antenna of the respective ground station. The downlink therefore increases the fill level of the antenna availability resource. Wherever the number of available antennas is reached (i.e., the upper bound), no further downlink may be planned anymore, as depicted in Fig. 15.14.

Renewable Resources, e.g., Battery Discharge Level

In order to model the battery of a spacecraft, we may define a continuous resource with

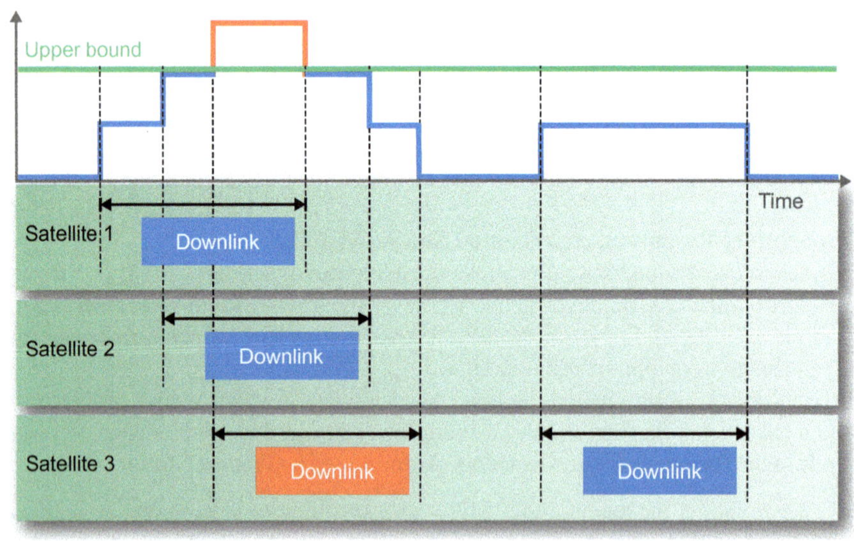

Fig. 15.14 Planning a downlink increases the number of antennas in use, including an offset for preparation and cleanup

- an initial fill level of zero,
- an upper limit corresponding to the battery capacity and
- a lower bound of zero, as you cannot discharge an empty battery any further.

In order to model consumption and recharging of a battery, we define resource modifications:

- For consumer tasks, we define a resource modification with
  - change profile defined in the interval $[0, \infty]$, usually with constant negative slope,
  - *startSlider* $=$ *0*, *startOffset* $=$ *0*, *activeProfileEndSlider* $=$ *1* and *activeProfileEndOffset* $=$ *0*, which means that the change profile is applied during the time interval of the timeline entry and
  - *dependencyEndOffset* $= \infty$, which means that the final consumption at the active profile end applies to the whole future. This resource modification describes therefore an accumulating resource consumption.
- As a supplier task, we define e.g. a solar panel task with a similar constraint, but with a positive slope on the change profile. This task is scheduled all the time when the solar panels collect sunlight.

Modelling the battery charge in this way, we see that as soon as the battery is full, the surplus power supply is lost, as shown in Fig. 15.15.

However, the more consumer tasks are planned, the less power is lost, as illustrated in Fig. 15.16.

Gliding Windows for e.g. Thermal Constraints

Thermal constraints are usually too complex to be modeled directly. However, they may be described as "Don't turn on the instrument for more than 10 min per orbit." In order to model such a constraint, we define a resource with

- initial fill level of zero and

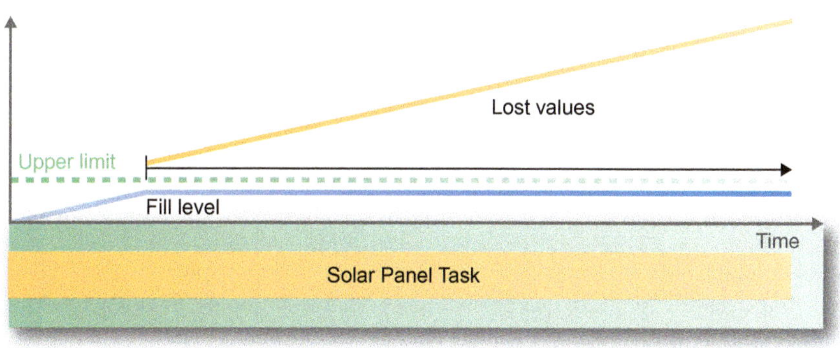

Fig. 15.15 Without satellite activities, much energy is lost, because it cannot be stored

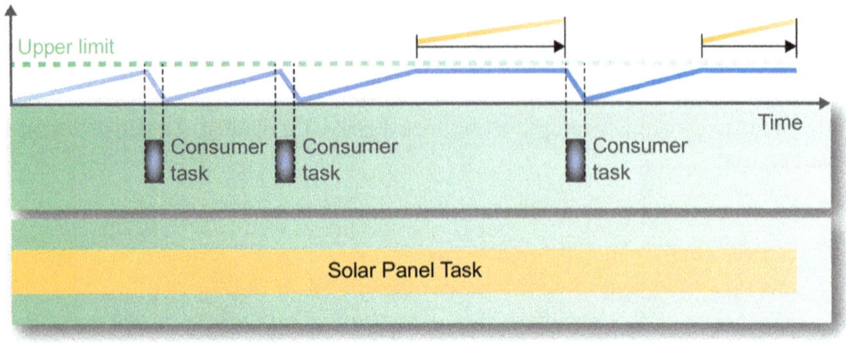

Fig. 15.16 With planned consumers, only little energy is lost

- an upper bound of e.g. 600 [s].

For each task, that needs the instrument to be switched on, we define a resource modification with

- a linear change profile with slope 1 [per s],
- activeProfileEndSlider = 1 and activeProfileEndOffset = 0,
- dependencyProfileEndOffset = one orbit duration and
- releaseSlope = −1.

The effect of this modeling is visible in Fig. 15.17. Each scheduled task increases during its duration continually the "clock resource," and this modification is kept for the duration of one orbit, before it is reset to zero. The "clock resource" has an upper bound of 600 s, which corresponds to the requirement to keep the overall "clock count per orbit" below 10 min.

In order to make this model precise, the release slope is defined which replaces the ending step of the modification profile by a ramp. This detail is relevant for two consumers, whose timeline entries are separated by less than 1 orbit, but which are not included within a 1 orbit interval.

Combining Resource Types

In the examples above, we have presented different resource types such as, e.g., opportunity, equipment or renewable resources. Although the resources may differ in their defined upper and lower limits, the difference of the resource types originates mainly from the constraints defined on them. Given the fact that the GSOC modeling language does not explicitly distinguish these resource types, they can be combined arbitrarily.

For example, one may introduce a resource comparison on a renewable resource modelling a battery's state of charge in order to assure that some on-board experiment is executed only if the battery's voltage exceeds a certain level.

Another use case is the unavailability of equipment: One may schedule an unavailability task with a resource comparison of an upper bound of zero. This effectively

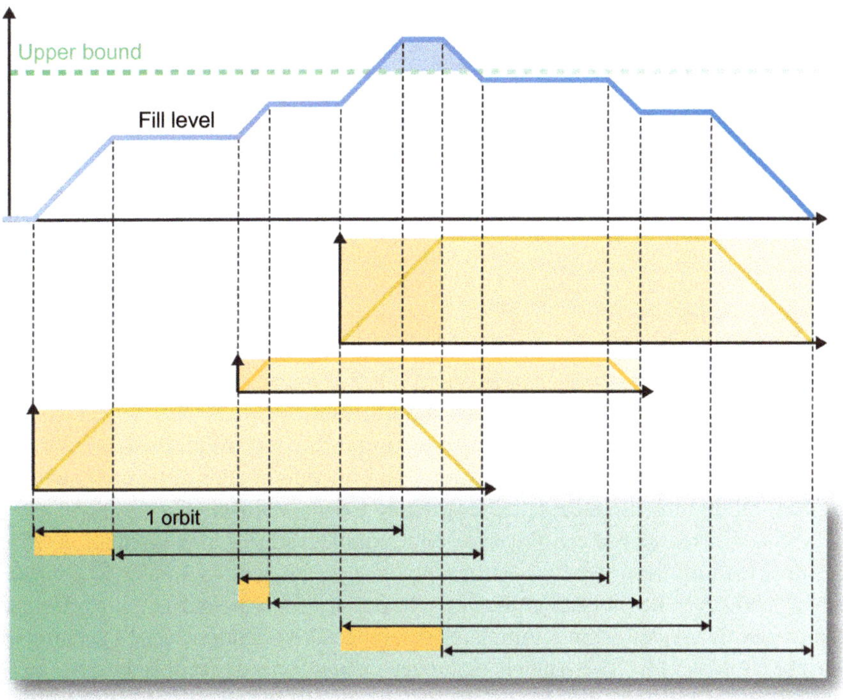

Fig. 15.17 The "one-orbit gliding window" assures that only a limited amount of on-board activities may be scheduled during any time interval of length one orbit. In the depicted example, the three planned activities are in conflict, because the resulting fill level exceeds the upper bound

adds a local upper bound of zero on the equipment resource during the time of unavailability. This prevents any task needing (i.e., increasing) this resource from being scheduled there.

Besides, it turns out to be useful to define all initial fill levels of resources to have a constant value zero and to avoid modeling any upper or lower bounds on resources. Instead, it is recommended to introduce setup tasks, which are planned during the whole timeline horizon. A non-zero initial fill level of a resource can be supplied by a resource modification of one setup task together with a comparing resource dependency which will replace the upper or lower bound of the resource. This way one may prepare different configurations by specifying different sets of setup tasks and easily switch in between these configurations just by planning the desired set of setup tasks. In fact, resource bounds may be implemented using such setup tasks.

15.3.4 Templates

A typical example for an activity onboard a low Earth orbiting, Earth observing satellite is the acquisition of an image. To model this, we create a root activity representing the request itself and holding all information provided by the customer. Due to the low Earth orbit, we need to calculate when the target of the request is visible to the satellite. For each of these visibilities, we create an opportunity activity as a child of the root activity. Each opportunity holds all information necessary for taking the respective image. More sophisticated instruments require preparation before the image acquisition, e.g., the instrument may have to be heated before it can provide the desired performance. This preparation may depend on the size of the gap to the preceding image acquisition—if two image acquisitions are executed immediately one after another, no heating is necessary between the two acquisitions. We therefore add two activities to each opportunity: one which models both pre-heating and image acquisition and one which models image acquisition without pre-heating. Figure 15.18 depicts the structure for one request.

A standardized way of creating such opportunities reduces implementation effort. For this, we introduce the concept of *templates*. A template is similar to a normal activity; however, no timeline entries may be defined for template activities. Instead, copies can be created from a template including all its children which themselves must be templates too. These newly generated instances need to reflect the customer's input, e.g., for an image request, at least the location of the target needs to be stored. In order to instantiate templates with correct parameters, variables may be referenced within a template. On instantiation, all referenced variables are evaluated and used as parameters of the instance. Variables may refer to

- project parameters,
- arguments provided by the customer,
- constant values or
- simple operations on two other variables.

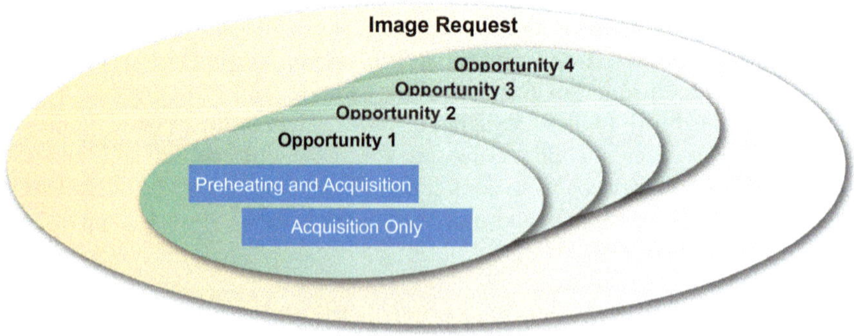

Fig. 15.18 Model of one image request containing multiple opportunities, each of which contains activities for two possibilities, either including pre-heating or not

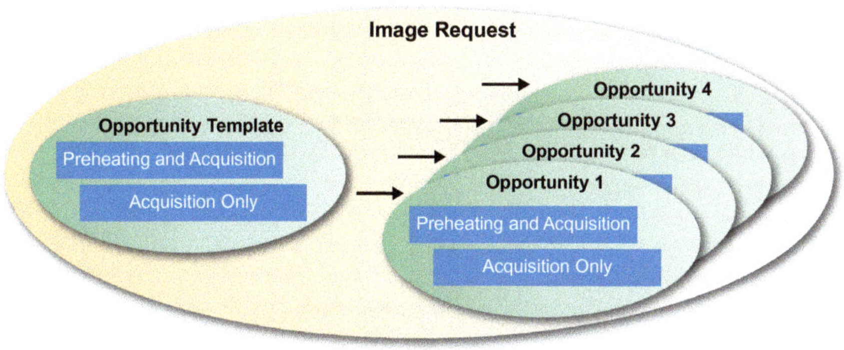

Fig. 15.19 The opportunities belonging to an image request are generated as instances from a template

Thus, our model looks like in Fig. 15.19, where all opportunities have been generated from the opportunity template.

As image requests are received on a regular basis, image requests themselves should be generated in a similar way. We therefore define a template for the whole image request. However, on instantiation, we might not yet know what opportunities exist. We therefore need to instantiate the whole request in two steps: We first instantiate the image request template by copying the included opportunity template, which therefore remains a template. Only during the second step of instantiation—when we calculate the image opportunities—we use the opportunity template of the new image request instance, to create the new opportunity instances with their children, see Fig. 15.20.

The benefit of this approach is obvious in case of manual operations: If the operator needs to update the model using a GUI, instantiating a template is as easy as providing all arguments required by those variables which are referenced by the template. But also when running a fully automated system, this approach provides some benefit: Due to the fact that templates are part of the model, they can be modified by a mission planning operator during runtime, supported by our graphical planning applications. This means that all properties and constraints of templates are part of the configuration of the mission planning system.

15.3.5 Planning Algorithms

Having translated the mission into the modeling language, we can now focus on generating timeline entries, i.e., the execution times for the different activities. This can be done by a human operator using interactive software such as GSOC's planning tool PINTA (Nibler et al. 2017). When using such software, operators are shown a visual presentation of all non-conflicting timeline entries for each activity they want

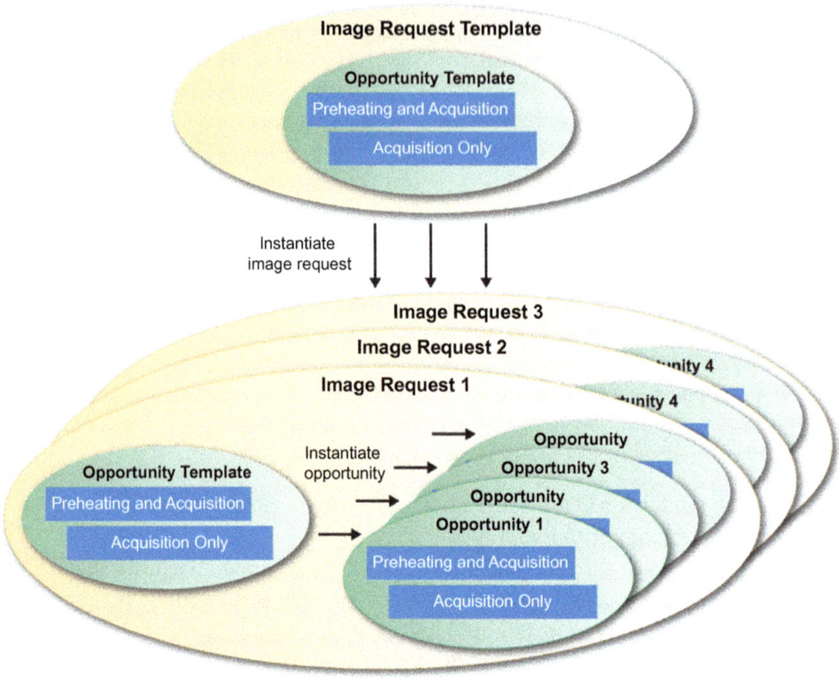

Fig. 15.20 Instantiation of an image request in two steps, first, the image request instance and, second, the opportunities of the individual image request instances at some later time

to plan and they will be warned about existing conflicts in the timeline. However, this manual approach may be quite work-intensive; therefore, it may be a good idea to support the process by supplying algorithms, which the operator can apply to certain activities. This approach can be enhanced by more and more complex algorithms, until the only remaining task of the operator is to start the algorithm and check its result. This kind of planning has been implemented, e.g., within the TET mission (Axmann et al. 2010).

When the same type of scheduling is applied repeatedly or when the dependencies are too complex for a human operator to survey, the operator may be excluded completely from this process, ending up in a fully automated scheduling system. An example for such a fully automated planning system is described in Maurer et al. (2010). In this example, a priority-based greedy algorithm (Korte and Vygen 2005) is used, i.e., the existing planning requests are considered one by one in descending order of their priorities and included into the timeline if this is possible without causing a conflict. However, more sophisticated algorithms exist, which perform global optimization on a given quality criterion. If such an alternative algorithm was implemented and proved to be superior to the existing one, exchanging the algorithm would be quite easy as long as both algorithms use the same model.

The question of how much automation should be implemented and how much should be left to a human operator, depends heavily on the type of the mission. As mentioned earlier, we may also distinguish between fixed planning, repeated replanning or reactive planning approaches which are suitable for different types of mission. All in all, these design decisions have considerable influence on the involved algorithms.

When planning manually, the timeline generation may be supported by using generic off-the-shelf algorithms. In this case, a simple setup may be sufficient to generate a first version of the timeline, which thereafter is checked and improved by the operator.

On the other hand, complex missions may require a highly specialized optimization algorithm. There exist many solvers providing optimization algorithms for their specific planning domain and if one manages to map the problem at hand to such a domain, one should give the solver a try. It may very well be the best solution one can retrieve. However, the more diverse the constraints are, the less likely it is that one finds such a mapping. In this case, modeling the problem in a standardized way has advantages, such as:

- Conflict check of the final timeline using generic tools
- Display of the final timeline using generic tools
- Simple integration of the specialized algorithm with existing scheduling software.

Repeated replanning and reactive planning support ingestion of new information during the mission execution phase. This is necessary if, e.g., new requirements are added to the planning process during execution time. In this case, it may be of great advantage to compose the mission's planning algorithm from generic algorithm snippets. For example, switching transmitters on, respectively off, before and after a ground station passage, is nothing complex, so a generic sub-algorithm may be used. This saves implementation effort and assures that the sub-algorithm is well configurable, allowing to adapt its settings during the mission. A combinable and extensible set of sub-algorithms has been described in Lenzen et al. (2012). However, combining such sub-algorithms via configuration files seems not to imply any benefit.

Unfortunately, most satellite missions include very specific requirements, which cannot be covered by a fully generic sub-algorithm. Nevertheless, modelling the problem in a standardized way allows using generic functions for, e.g., finding conflict free timeline entries or for analyzing why conflicts exist. With these functions, implementing a mission specific algorithm becomes much cheaper and the resulting algorithms are much more reliable compared to writing a new mission specific system from scratch.

15.4 Summary

GSOC's modeling language constitutes the core of its planning software. All libraries and applications are built on top of it, which allows combining and reusing them according to the missions' requirements.

In the past, GSOC's planning library has been driven by heuristic algorithms, which generate good results in a rather short time and which are designed to keep the timeline stable and allow for tracing reasons when a request cannot be served. For such heuristic algorithms, sophisticated analysis- and preview-functions have been implemented, simplifying the implementation of new mission algorithms. Using this approach, it is possible to freely combine all features of the modeling language.

Missions which don't require timeline stability and traceability may benefit from mapping the model to the domain of an optimization problem for which a dedicated solver exits, and which might calculate better solutions. This, however, puts severe restrictions on the planning model. Still, one may find approximations of the model in the domain covered by an optimizer. The obtained optimal solution of this restricted model may then be used as a starting point of a heuristic algorithm for the full model.

References

Axmann, R., Mühlbauer, P., Spörl, A., Turk, M., Föckersperger, S., Schmolke, J., Operations Concept and Challenges for 11 different Payloads on the TET-1 Mini-Satellite; 4S Symposium – Small Satellite Systems and Services, Funchal, Madeira, Portugal (2010)

Braun, A., GRACE Mission Planning: Application of the Linear Programing Method to On-Board Memory Management; Proceedings of SpaceOps (2002)

DLR GSOC; Planning Modelling Language; https://www.dlr.de/rb/en/Portaldata/38/Resources/dokumente/GSOC_dokumente/RB-MIB/GSOC_Modelling_Language.pdf (2010), cited October 25, 2019

Krieger G, Moreira A, Fiedler H, Haynsek I, Werner M, Younis M, Zink M (2007) TanDEM-X: A Satellite Formation For High-Resolution SAR Interferometry. IEEE Trans Geosci Remote Sens 45(11):317–334

Korte, B., Vygen, J., Combinatorial Optimization; Issue 3. Springer, ISBN 3–540–25684–9 (2005)

Lenzen, C., Wörle, M., Mrowka, F., Spörl, A., Klaehn, R., The Algorithm Assembly Set of Plato; SpaceOps (2012)

Maurer E, Mrowka F, Braun A, Geyer MP, Lenzen C, Wasser Y, Wickler M (2010) TerraSAR-X Mission Planning System: Automated Command Generation for Spacecraft Operations. IEEE Trans Geosci Remote Sens 48(2):642–648

Nibler, R., Mrowka, F., Wörle, M., Hartung, J., Lenzen, C. and Peat, C., PINTA and TimOnWeb – (more than) generic user interfaces for various planning problems; IWPSS 2017 – 10th International Workshop on Planning and Scheduling for Space (2017)

Reile H, Lorenz E, Terzibaschian T (2013) The FireBird Mission – A Scientific Mission for Earth Observation and Hot SpotDetection; Small Satellites for Earth Observation, Digest of the 9th International Symposium of the International Academy of Astronautics, Wissenschaft und Technik Verlag, ISBN 978-3-89685-574-9

Christoph Lenzen is member of the mission planning group of the Department for Mission Technology at the German Space Operations Center (GSOC). His special subject is the automated timeline generation process, where he significantly contributed to the Mission Planning System of the TerraSAR-X/TanDEM-X missions. Current work includes contributing to design and development of the Reactive Planning System and its adaptation for the EnMAP mission and participating in the CCSDS working group for establishing an international standard for Mission Planning services. He received his mathematics diploma from the University of Munich (LMU) in 2001. After one year as scientific assistant at the LMU, he started work at GSOC in 2002.

Sven Prüfer is a Mission Planning Engineer at the German Space Operations Center (GSOC) since 2017. He received his Ph.D. in mathematics from the University of Augsburg in 2017. After joining the DLR, he worked on the mission planning system for the European Data Relay System (EDRS), the Environmental Mapping and Analysis Program (EnMAP) and the Columbus Ka-Band Antenna (ColKa). Furthermore, he is a developer of GSOC's modern planning library Plains and the Reactive Planning Framework.

Chapter 16
Mission Planning for Unmanned Systems

Tobias Göttfert, Sven Prüfer, and Falk Mrowka

Abstract This chapter illustrates the various manners in which the overall design of an unmanned satellite mission can influence its mission planning system. We build on the basic principles from Chap. 15 and discuss the mission planning system for the TerraSAR-X/TanDEM-X missions in more detail. Furthermore, we will have a look at the numerous timescales at which mission planning operates and that need to be considered together.

16.1 Introduction

One central aspect of a space mission is the planning of activities in a way that ensures the safety of the spacecraft and makes efficient use of available resources. The basic principles, which have been introduced in Chap. 15, will now be applied to unmanned missions. Space missions without humans on board are typically better suited for automated planning systems than manned missions. One reason is that humans may react spontaneously to events and act less deterministically than computers. It is also more common for unmanned missions that the overall mission objective and duration is already fixed and known beforehand. Another reason that unmanned space missions are more amenable to automated planning is that the complexity of an unmanned spacecraft's subsystems is typically lower than the one of, e.g., a manned space station. There is usually a quite limited list of tasks that can be planned, depending on the features of the spacecraft, and only few resources need to be modelled. Still, the requirements and operations scenario of unmanned spacecraft missions can vary a lot, and no single mission planning system (MPS) fits all missions that are flown in one control center. In general, the more complex and more agile the planning problem gets, the more work should be invested into automation and a sophisticated software system. Nevertheless, for simpler planning problems that have to be performed less often, regular manual creation of the mission timeline is

T. Göttfert (✉) · S. Prüfer · F. Mrowka
German Space Operations Center (GSOC), German Aerospace Center (DLR), Oberpfaffenhofen, Germany
e-mail: tobias.goettfert@DLR.de

F. Sellmaier et al. (eds.), *Spacecraft Operations*, Springer Aerospace Technology,
https://doi.org/10.1007/978-3-030-88593-9_16

often still the more economical solution (see also Chap. 17). However, in both cases, a set of common software tools for planning and scheduling is highly beneficial as it reduces implementation effort on the long term as well as allows for more reliability due to the reuse of well-tested and validated components.

16.2 Mission Planning System Example

As an example for applying the GSOC scheduling language from Chap. 15, we will describe the combined MPS for the TerraSAR-X and TanDEM-X (TSTD) missions at the German Space Operations Center in the following (Lenzen et al. 2011). At the moment, this system is the most evolved and feature-rich MPS at GSOC, having grown from a single-satellite, single-mission automated planning system into a dual-satellite, dual-mission planning system that takes various correlations between spacecraft and missions into account (Mrowka et al. 2011). The TerraSAR-X and TanDEM-X spacecraft are described in more detail in Krieger et al. (2007). For our purposes, it is only necessary to know that both carry a synthetic aperture radar (SAR) instrument that is able to capture either two-dimensional imagery via one satellite or three-dimensional radar-based digital elevation models by using both spacecraft simultaneously in an interferometric mode. Therefore, the two satellite missions need to be considered as tightly linked in terms of planning.

Additionally, the mission is operated in a public–private partnership, where several parties, such as researchers, commercial customers, and emergency services, can place planning requests for radar imagery, which all have to be processed by the MPS.

In total, the TSTD-MPS has to model more than 300 resources and their constraints, which are associated with the two satellites and is responsible for the scheduling of all payload-related activities, both on board and on ground. This includes not only the radar instrument activities, but additionally mass memory and bus-related activities like X-band downlink of the radar data. In addition, an on-ground schedule is created that allows the participating ground stations to control their antennas accordingly, as well as the data processing centers to properly file the incoming payload data. The GSOC generic scheduling language (see Chap. 15) provides all building blocks that are needed to describe the planning problem for this mission and the implementation of a priority-based scheduling algorithm. For example, the simple addition of one 2D radar image is modelled via the addition of a timeline entry of a task on the timeline. This task corresponds to the 2D radar image acquisition request and contains all information necessary for planning and processing. When scheduled, the timeline entry affects several resources, e.g., the power or time usage of the radar instrument. Of course, scheduling the timeline entry is only possible if none of the resource limits are violated. In addition, scheduling the data take is only possible if there is no danger that the partner spacecraft is illuminated by the SAR instrument during that data take, which is modelled via a separate resource. Furthermore, the data take requires various file handling tasks including corresponding timeline entries, e.g., creation and deletion of data files for the image. These tasks in turn affect the mass memory resource usage which has a natural upper bound determined by the on-board memory. On top of that, data downlink tasks have to be scheduled during a following ground station contact, which in turn requires the scheduling of antenna switch-on and switch-off tasks. This way, the scheduling of a simple data take task has a profound influence on the timeline with multiple subtasks that each come with their own constraints that need to be satisfied. The TSTD planning algorithm therefore needs to be tailored to this problem.

During the scheduling process, the algorithm looks at all planning requests in descending order by priority and decides if all conditions necessary for scheduling can be met. Only then, the data take is planned together with all additional tasks, and the resources are modified accordingly. In the end, an export tool creates telecommands for the spacecraft from the tasks on the timeline, which are then uplinked and executed. Further data files are generated for the various shareholders on ground, such as ground stations, operators, or data processing. Figure 16.1 shows a simplified example of a TerraSAR-X/TanDEM-X timeline, taken from the GSOC GUI tool for timeline analysis, PINTA (see Nibler et al. 2017), in which some important tasks and resources can be seen in a graphical view.

The planning process within the TSTD-MPS is performed regularly and is strictly coupled to the ground station contacts for command uplink, see section about repeated replanning in Sect. 15.2. On average, this results in a planning run every 12 h, when a new version of the timeline with a planning horizon of three days into the future is generated. However, each upload contains time-tagged spacecraft commands only for the next 24 h, still leaving 12 h of overlap from each command upload to the next

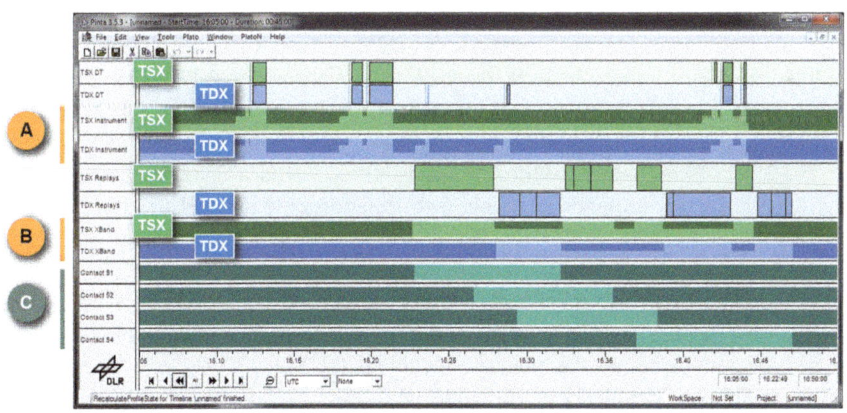

Fig. 16.1 Graphical view of parts of a TerraSAR-X/TanDEM-X timeline within PINTA. A short-time period (approx. 45 min) and a limited set of activities are displayed on a horizontal time axis. Tasks and resources of the TerraSAR-X satellite (*TSX*) are shown in green, and for the TanDEM-X satellite (*TDX*) in blue. Note how the algorithm schedules instrument activation for the data takes (third and fourth row, marked with *A*) and the switchover of X-band downlinks (seventh and eight row, marked with *B*) when passing over several ground stations (last four rows, marked with *C*). Three-dimensional data takes are scheduled in parallel on both satellites

one. This ensures that one failed uplink contact will not stop the mission execution while requiring not too many deletions of uploaded commands in case of a timeline change more than 24 h into the future.

16.3 Considerations on Designing a Mission Planning System

Mission planning systems for unmanned missions are software systems, for which the degree of automation and human interaction depend heavily on the precise mission requirements. Therefore, creating an MPS is strongly coupled to software engineering. The design of such a system is influenced by the following key factors:

- How to map mission objectives and capabilities to the planning model and planning algorithms
- How to interact with the rest of the ground segment and serve its interfaces
- How to design the system as robust, failure tolerant, and automated as needed
- How to maximize the amount of reuse of existing components.

During the mission development phases A to C (cf. Chap. 4), the baseline and complexity of the MPS are usually defined already. Small changes in the mission design or requirements can thus influence the needed mission planning effort considerably. In the following, we will describe a couple of aspects that need to be considered as early as possible during the ground design process.

If a mission consists of several spacecraft, one needs to determine whether the spacecraft can be planned independently or if they are inherently coupled to each other. For example, the TerraSAR-X and TanDEM-X spacecraft are inherently coupled in their formation flight because of synchronized data taking for three-dimensional images, and therefore one common MPS is needed because it needs to consider both satellites and both missions at the same time.

The level of intelligence of the on-board software needs to be taken into account, e.g., whether telecommands available to MPS address high-level or rather detailed low-level functionalities. For the TerraSAR-X spacecraft, the MPS also has to command instrument activation and instrument standby in several levels, whereas other spacecraft perform these tasks on their own if data takes are commanded. In some cases, there may even be some on-board planning system available that needs to be taken into account. An example for this is the Mars 2020 Perseverance rover which is able to plan and execute additional tasks on its own after some limited support from on-ground planning (Rabideau and Benowitz 2017).

Another important question is whether the command uplink scheme is very rigid or rather flexible. For example, for geostationary satellites a permanent uplink is possible which gives the MPS the capability of exporting commands whenever there is a change in the timeline. As an example, for less-flexible uplink schemes the MPS for the GRACE mission (Braun 2002; Herman et al. 2012; Tapley et al. 2004) is built using weekly planning runs and incremental uplinks of the resulting commands every day. The TSTD-MPS, on the other hand, is based on specific ground stations contacts, as explained in Sect. 16.2.

The flexibility and optimization requirements from the user's side need to be analyzed and taken into consideration. For example, when multiple users can order data takes (as, e.g., in the TSTD mission), it is quite common to require that the plan should be stable. This is because otherwise the users cannot rely on any status update on their requests as it might change shortly thereafter. In other cases, however, one may want to optimize the total mission output and thus optimizes the timeline during every planning run.

It is crucial to clearly define which resources have to be modeled to what extent. For some resources, a detailed model is required; for other approximations are sufficient. For the TerraSAR-X spacecraft, for example, memory usage is modeled exactly; the battery level is modeled in a linear approximation; and thermal resources are modeled by a simplified heuristic using a predefined time window approximation. However, other missions may not be able to model, e.g., the memory usage exactly because they compress data files on-board. Sometimes, additional resources are not needed for operating the spacecraft, but for improving customer feedback, e.g., by giving more detailed reasons why a request was not planned. In general, the development of the resources and their constraints which precisely capture the requirements of the mission tends to form the bulk of the design work for the MPS.

Another important aspect is the modeling of the ground segment. It could be very complex, but often a simplified representation is good enough. For example, the MPS may have a choice between various ground stations for data downlinks which have varying network bandwidths. In this case, one may schedule a particular ground station contact which offers a high downlink capacity but can affect the overall timeline profoundly. Another example is the mechanism of ordering a ground station contact. Some ground stations use a complex request-response workflow for their contact booking whereas others may allow the mission to use any available slot on short notice. Of course, the complexity of this workflow is reflected in the MPS.

Furthermore, the complexity and safety of the commanding concept needs to be taken into account. For, e.g., the TerraSAR-X spacecraft, every 12 h commands for the next 24 h are uplinked to the spacecraft for safety and autonomy reasons. Of course, to allow this, the MPS needs sophisticated delta-commanding capabilities. In contrast, for a geostationary satellite mission with continuous contacts, the MPS may be informed at short notice about any command upload failures and can try to remedy the error in a live fashion. This allows the MPS to keep the spacecraft timeline in sync with its own at all times.

There is one more quite common and important issue at the intersection of resource modeling and the commanding concept, namely the treatment of time-tagged telecommands (TTTC). If the satellite bus has a very limited amount of slots available for saving TTTC, the MPS will have to take this additional constraint into account. It is also common that the satellite bus is able to execute only a small number of TTTC at the same time, very often just one. In this case, the MPS needs to be aware of any other systems that might make use of TTTC slots, potentially requiring additional interfaces.

Finally, the mission operations concept has a large influence on the MPS design regarding interactivity and operator interaction. If it is foreseen that operators may need to look at the timeline or even resolve certain conflicts manually, it is necessary to include elaborate interfaces to GUI tools into the MPS. Of course, algorithm-supported semi-manual MPS, such as the MPS for the Firebird mission (Wörle et al. 2016), require GUI tools to allow the operator to create a timeline. But even in automated MPS, such as the TSTD-MPS, a mission may require a display of the timeline for the near future or a way of manually modifying the timeline in case of an anomaly. This interaction of automated algorithms and, potentially arbitrary, manual modifications might cause considerable implementation effort for the MPS to ensure safe operations.

This list of issues influencing the MPS design is by no means complete. However, most of these items have come up in one way or another during the MPS design of any mission operated at GSOC so far and should thus be kept in mind for any upcoming mission.

16.4 Mission Planning at Various Time Scales

Mission Planning is a challenge taking place on a wide variety of timescales, from years down to sub-seconds. This section will give an overview about these different timescales and the different types of tasks that are associated with them.

Mission planning is involved during all phases of the mission (cf. Fig. 4.1), starting with the analysis phase A, i.e., *years* before the launch. The mission planning team provides consulting and participates in the analysis and mission definition process. It collects mission and user requirements and ideas about the goals of the mission and prepares concepts for planning and scheduling. During phases B and C, it defines the design of the mission planning system and the necessary components, which either can be taken from existing components or have to be newly developed. In the end, a concept for the MPS and its operation and the mission planning requirements are created and the MPS is developed and integrated during phase D, together with all necessary tests.

Activities that happen in the timescale of *months* are usually considerable changes in the mission, e.g., major orbit-configuration adjustments. Other examples include changes in the operational concept, e.g., modifications of the uplink or downlink scheme, different usage of the payload, and introduction of new requirements. The mission planning team consults the mission management and ensures the feasibility of the change.

For planning systems that handle stable and predictable scheduling problems, a long-term preview of the timeline may be given, which covers the timescale of months. This preview of course is subject to changes, depending on the input of planning requests, but can serve as a valuable tool to enable the coordination between different parties that may influence the timeline, such as users or mission management. For example, flexibility in the execution of orbit maneuvers can be used to schedule them at a time reducing disruption of the nominal mission or different users can coordinate their requests to allow for more mission output. Also, a coarse preplanning of activities is sometimes performed months ahead.

Normally, the planning horizon of an unmanned mission is in the order of *days*. Within this time frame, all relevant nominal input needs to arrive at the MPS, well ahead of execution time. Those inputs are, e.g., planning requests, the ground station availabilities, orbit and maneuver information, or other flight dynamics input. Planned maintenance also falls into this category.

The timeline is then usually generated some *hours* before uplink, depending on the mission planning concept, degree of automation, and requirements on the reactivity of the MPS. This short-term planning can take the latest status of flight dynamics data for products with a high requirement on accuracy and a limited time span of validity into account. The moment of timeline creation also determines the order deadline, after which no late planning requests can be considered anymore.

Sometimes, activities can be planned *minutes* before their execution time. This is usually only the case during LEOP (launch and early orbit phase) because extended real-time operations with human personnel in the form of mission planning and flight operators are carried out only in this phase. However, automated MPS for geostationary satellite missions may be able to do such last-minute planning, too. The activities that have to be scheduled also happen on different timescales, depending on their nature. Hereby, both duration and required timing accuracy can vary a lot, from days and hours down to the second and sub-second range. A few examples are maintenance phases, which can last hours and days and usually start at manually chosen times, or ground station contacts, which have durations in the range of minutes for low Earth orbiting satellites and need to be scheduled with a few seconds precision, or maneuvers, which usually last a few seconds and also require precision in the range of seconds.

Most satellite buses provide an accuracy of execution time for time-tagged commands in the range of one second, which is normally sufficient for most activities on board.

For applications that require a more precise control over their execution time, special provisions have to be made: First, the on-board hardware has to support this, usually via GPS-supported time correction features in the affected subcomponent of the satellite. Second, the MPS has to be able to fine-tune the start time and duration of the activity when the latest orbit information with the required accuracy is available. A prominent example is the scheduling of imaging data takes for low Earth orbiting satellites, which require a timing accuracy of less than 1 ms for a ground track accuracy in the range of meters.

Figure 16.2 shows various timescales of activities encountered by the MPS in a timeline of a fictitious satellite mission.

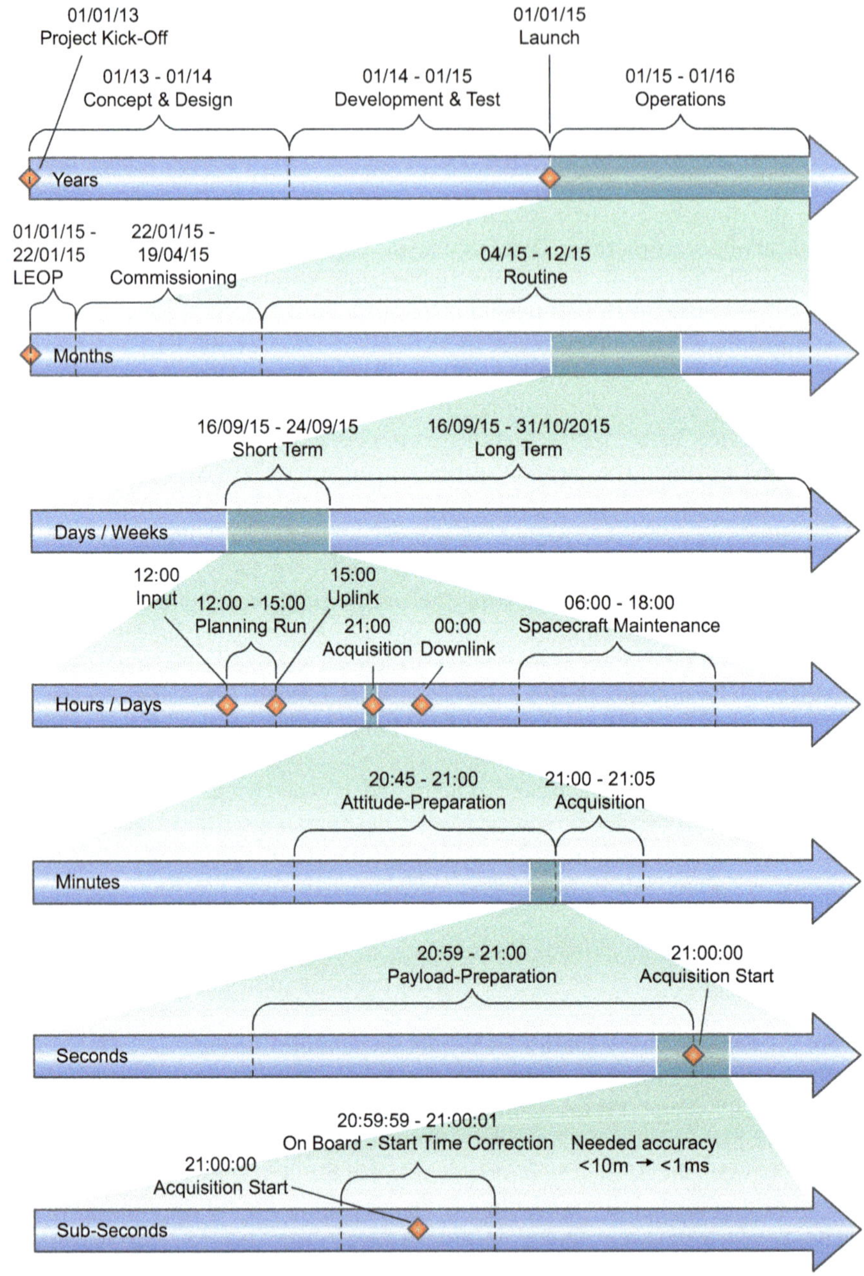

Fig. 16.2 Timeline with example activities of a fictitious satellite mission. The various levels of zoom exemplify the different timescales on which the mission planning processes, and scheduled activities take place

16.5 Conclusions and Outlook

Mission planning for unmanned missions has to cope with many different types of spacecraft and operational concepts, leading to a variety of MPS designs. However, the well-known set of plannable activities and their potential frequent repetition usually make some sort of automatic scheduling feasible. The tools that compose the MPS may then be crafted to the automation and operational needs of the mission. For this, a proven set of planning tools that supports all relevant scheduling functionalities, as well as easy inspection of the resulting timeline, is a crucial prerequisite. The mission specific requirements define the precise planning model that is used by the MPS to ensure safe operations.

Following the development of highly automated mission planning systems and the increasing computational capabilities of on-board hardware, new approaches increase the flexibility of mission planning systems even further to cope with the growing demands of reducing overall reaction times of the satellite system. One example is the so-called on-board planning, which moves parts of the planning process onto the spacecraft, making it possible to invoke last-minute changes to the timeline without contact to ground (Wille et al. 2013; Wörle and Lenzen 2013).

References

Braun A (2002) Application of linear programming to on-board memory management. In: SpaceOps 2002, proceedings. AIAA. https://doi.org/10.2514/6.2002-T3-10

Herman J, Davis A, Chin KB, Kinzler M, Scholz S, Steinhoff M (2012) Life with a weak heart; prolonging the grace mission despite degraded batteries. In: 12th international conference on space operations (SpaceOps 2012), 11–15 June, Stockholm, Sweden

Krieger G, Moreira A, Fiedler H, Haynsek I, Werner M, Younis M, Zink M (2007) TanDEM-X: a satellite formation for high-resolution SAR interferometry. IEEE Trans Geosci Remote Sens 45(11):3317–3341

Lenzen C, Wörle MT, Mrowka F, Geyer M, Michael P, Klaehn R (2011) Automated scheduling for TerraSAR-X/TanDEM-X. In: IWPSS 2011, 8–10 June, Darmstadt, Germany

Mrowka F, Geyer M, Lenzen C, Spörl A, Göttfert T, Maurer E, Wickler M, Schättler B (2011) The joint TerraSAR-X/TanDEM-X mission planning system. In: IGARSS 2011, 24–29 July, Vancouver, Canada, symposium proceedings, pp 3971–3974. ISBN 978-1-4577-1004-9

Nibler R, Mrowka F, Wörle M, Hartung J, Lenzen C, Peat C (2017) PINTA and TimOnWeb—(more than) generic user interfaces for various planning problems. In: 10th international workshop on planning and scheduling for space (IWPSS 2017), 15–16 June, Pittsburgh, PA, USA

Rabideau G, Benowitz E (2017) Prototyping an onboard scheduler for the Mars 2020 Rover. In: international workshop on planning and scheduling for space (IWPSS 2017), 15–16 June, Pittsburgh, PA, USA

Tapley BD, Bettadpur S, Watkins M, Reigber C (2004) The gravity recovery and climate experiment: mission overview and early results. Geophys Res Lett 31:L09607. https://doi.org/10.1029/2004GL019920

Wille B, Wörle MT, Lenzen C (2013) VAMOS—verification of autonomous mission planning onboard a spacecraft. In: 19th symposium on automatic control in aerospace (IFAC 2013), 2–6 Sept, Würzburg, Germany

Wörle MT, Lenzen C (2013) Ground assisted onboard planning autonomy with VAMOS. In: 8th international workshop for planning and scheduling in space (IWPSS 2013), 25–26 March, Moffett Field, CA, USA
Wörle MT, Spörl A, Hartung J, Lenzen C, Mrowka F (2016) The mission planning system for the firebird spacecraft constellation. In: 14th international conference on space operations (SpaceOps 2016), 16–20 May, Daejeon, Republic of Korea

Tobias Göttfert is deputy head of the Mission Operations Technology group in the Mission Technology Department at the German Space Operations Center (GSOC). Starting after his physics Ph.D. in 2010 in the GSOC mission planning group, he was responsible for several existing and newly developed automated mission planning systems, namely for the TerraSAR-X/TanDEM-X, EDRS and ColKa missions at GSOC. Currently his work concentrates on software engineering for data analysis, machine learning and operator assistance systems in the recently founded Mission Operations Technology group.

Sven Prüfer is a Mission Planning Engineer at the German Space Operations Center (GSOC) since 2017. He received his Ph.D. in mathematics from the University of Augsburg in 2017. After joining the DLR, he worked on the mission planning system for the European Data Relay System (EDRS), the Environmental Mapping and Analysis Program (EnMAP) and the Columbus Ka-Band Antenna (ColKa). Furthermore, he is a developer of GSOC's modern planning library Plains and the Reactive Planning Framework.

Falk Mrowka is head of the mission planning group of the Department for Mission Technology at the German Space Operations Center (GSOC). He received his physics diploma and his Ph.D. while working at the University of Leipzig on the subject of low temperature physics. While working at the GSOC since 2002, he was working for various projects as mission planning system engineer.

Chapter 17
Mission Planning for Human Spaceflight Missions

Thomas Uhlig, Dennis Herrmann, and Jérôme Campan

Abstract Planning is an essential part of human space flight. Therefore, it is crucial to understand its development during the last years and the current planning concepts. This chapter gives an insight into crew and ground activity planning, based on the International Space Station (ISS) as example.

17.1 Introduction

Planning is a key part of human spaceflight operations, was already part of the first Mercury flights, and has significantly developed ever since. While the first flight plans were text-based tables on paper, today's plans are displayed graphically with dedicated applications, have sophisticated possibilities to include or extract information, and are available via internet. In the meantime, the terminology "timeline" is commonly used to describe the schedule of the astronauts and the related ground activities.

This chapter describes the planning of human spaceflight operations using the International Space Station (ISS) as an example. This limited focus might sound like a loss of generality. However, all previous human spaceflight efforts cumulate in the ISS project. In this sense, the ISS planning processes, phases, and tools can be considered as the heritage of the planning processes of American Mercury, Gemini, Apollo, Skylab, Space Shuttle, and the Russian Восток (Vostok), Восход (Voskhod), Салют (Salyut), and Мир (Mir). All that experience and lessons learned were used to design the mission planning for the ISS. It should be emphasized that there are a number of planning and scheduling processes and products for the ISS in various areas: For example, planning is done for station attitude, for robotic tasks, for extravehicular activities (EVAs), for consumables, and for critical resources like water, as shown in Fig. 17.1. This article focuses only on the crew and ground activity planning—

T. Uhlig · D. Herrmann · J. Campan (✉)
German Space Operations Center (GSOC), German Aerospace Center (DLR), Oberpfaffenhofen
82234 Wessling, Germany
e-mail: Jerome.Campan@DLR.de

F. Sellmaier et al. (eds.), *Spacecraft Operations*, Springer Aerospace Technology,
https://doi.org/10.1007/978-3-030-88593-9_17

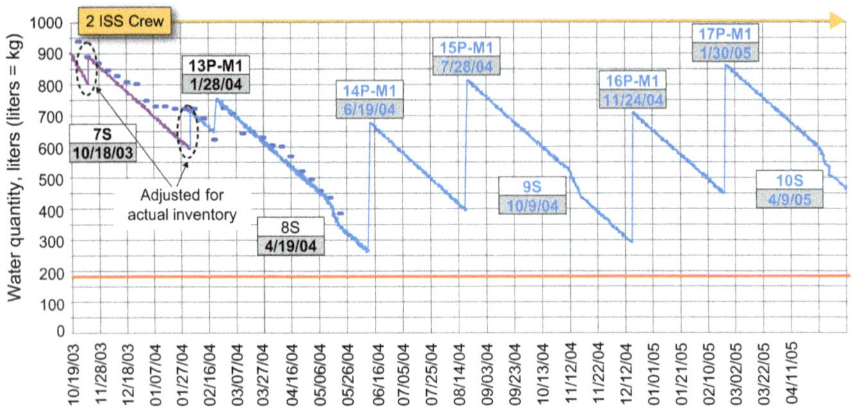

Fig. 17.1 Planning for human spaceflight missions takes various forms. Here, you see an example of the planning of water availability on board (reprinted from Kitmacher et al. (2005) with permission from Elsevier)

the process which is commonly referred to as "mission planning." Other aspects of mission planning are covered in Chaps. 15 and 16.

Although the terminology and the planning theory are similar to mission planning for unmanned missions, the automatization of planning for the ISS is marginal. In principle, it would be possible to model activities, to exactly define the constraints, relationship to other activities, and dependencies of resources, as described in Chap. 15. However, since the mission goals for ISS operations cannot be broken down to a few science objectives, like it is possible for a satellite (i.e. "Earth observation"), the number of resources and conditions to be modeled would be immense. If the goal of a mission would be Earth observation only, the conditions could be limited to some parameters concerning the position and attitude of the spacecraft, the orbital constellation with respect to the Sun, if sunlight is required for optical pictures, etc. Resources would also have to be modeled, such as power availability, the data storage situation on board, and downlink paths.

But ISS mission goals do not only include Earth observation, but also experiments of almost all scientific disciplines. The experiments might have requirements for special microgravity conditions, temporal constraints on a variety of different time scales, interdependencies to "private" crew tasks (like fasting or blood sampling), priority orders, and even international or political dimensions. Figure 17.2 gives an overview about the most important resources and constraints.

Although it might be theoretically possible to generate a complete list of all thinkable resources and conditions, and a state-of-the-art computer system with a corresponding algorithm could predict these parameters continually, the effort to link each activity to its related resources and conditions, and thus to model it correctly, would outweigh the benefit of an automated planning system. This, in particular, is true, since a large fraction of the activities on the timeline occur only once.

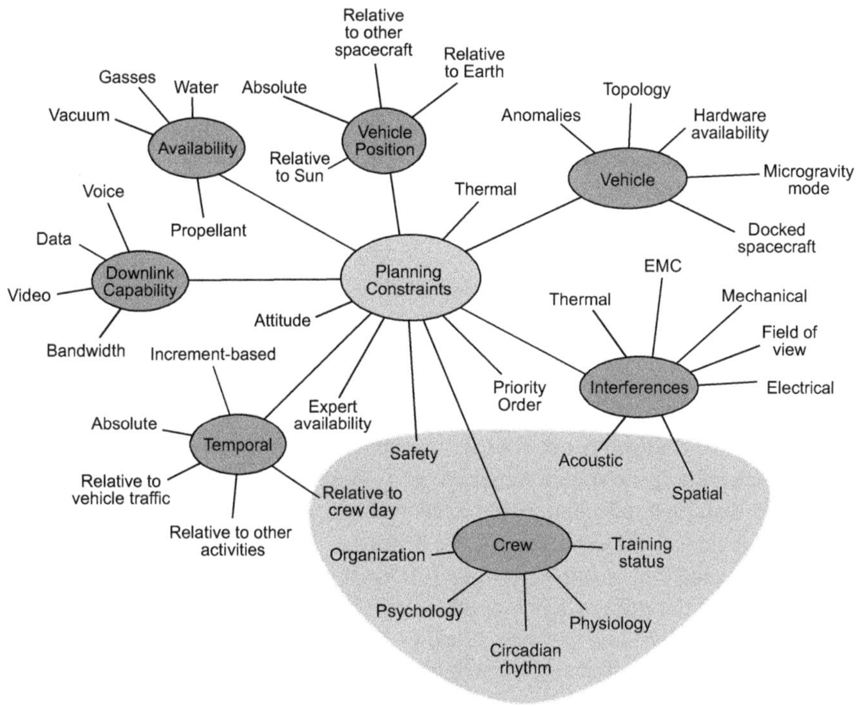

Fig. 17.2 Planning for human spaceflight can be constrained by a variety of factors. The most important ones are shown here; the ones directly linked to human presence are summarized in the shaded area. A mathematical modeling of them to support automatic planning algorithms is not feasible

The human factor also has to be considered. It is difficult to put into a mathematical model. Astronauts have personal preferences, which need to be respected and incorporated into the planning, if possible. There are medical and psychological aspects, which only appear if the integrated schedule is checked by an expert, e.g., there might be concerns to do more than one blood draw per day.

Due to the internationality of the ISS and multiple participants, the planning also has a political dimension, which might influence the final planning product, but definitely needs to be reflected in the corresponding processes.

For those reasons, human spaceflight planning is still a manual process and involves different teams around the world to accomplish it. The multitude of influencing factors, which can easily change in near real time, also makes the mission planning a highly dynamic endeavor.

17.2 Basic Considerations

Planning for projects like Apollo or the early Space Shuttle flights is fundamentally different than planning for the Mir or the International Space Station. If the planning is done for short duration flights, the preparation and the execution phase are clearly separated. The optimization of such plans has to be very high to make best usage of the limited flight time, and due to the abovementioned separation, this goal can also be accomplished (Popov 2002).

For long-term missions like the ISS, which is continuously manned for several years, the plan preparation and the plan execution processes need to be run in parallel, which adds complexity to the planning, but also decreases the need for high optimization, since overall crew time is no longer a limiting factor (Korth and LeBlanc 2002).

However, to be able to better cope with continuous ISS operations, it was decided to break down the space station timescale into smaller pieces, which then can better be worked with methods comparable to project management. Those timescales are called "increments" or "expeditions", are driven by the ISS crew rotation, and have numbers assigned. For example, in March 2020 the ISS was in increment 62. An ISS increment has a duration of a few months up to half a year; each Soyuz undocking event triggers the start of a new one. Since two Soyuz crews alternate, this translates into a crew stay on board of approximately six months, which is compatible with the time a Soyuz spacecraft is designed for and approved to remain in space under the very harsh environmental conditions, which are described in more detail in Chap. 1.

The planning processes are therefore adapted to the increment concept, as described in more detail in Sect. 17.5. For historical reasons, two increments are most of the time treated together in one combined planning effort—so all the considerations below are applicable for double increments, e.g. increment 37/38. For the future, some adaptations to the concept of increments can be assumed, since significant changes in crew transportation are currently ongoing with SpaceX already flying and Boeing in preparation of its first crewed flight (see Sect. 24.3.3).

The elementary planning items in human spaceflight planning are activities (sometimes also referred to as planning requests (PR)), which are either executed by the ground team or by the astronauts on board. The heart piece of each activity is normally a procedure (or parts of a procedure), which are called operations data files (ODFs). They contain a detailed description of what has to be done using well-defined formats and standards.

These procedures also define the required resources to conduct the activities. To complete the set of planning data required for the scheduling and a proper execution of the activity, some additional info is attached to each planning request, partially detailing the data already contained in the ODF, and partially adding new planning information like the resource utilization. The relationship between the ODF and the planning request is depicted also in Fig. 17.3.

The planning requests are maintained in a database at each participating control center (see Sect. 17.3) and are then fed into the common planning system.

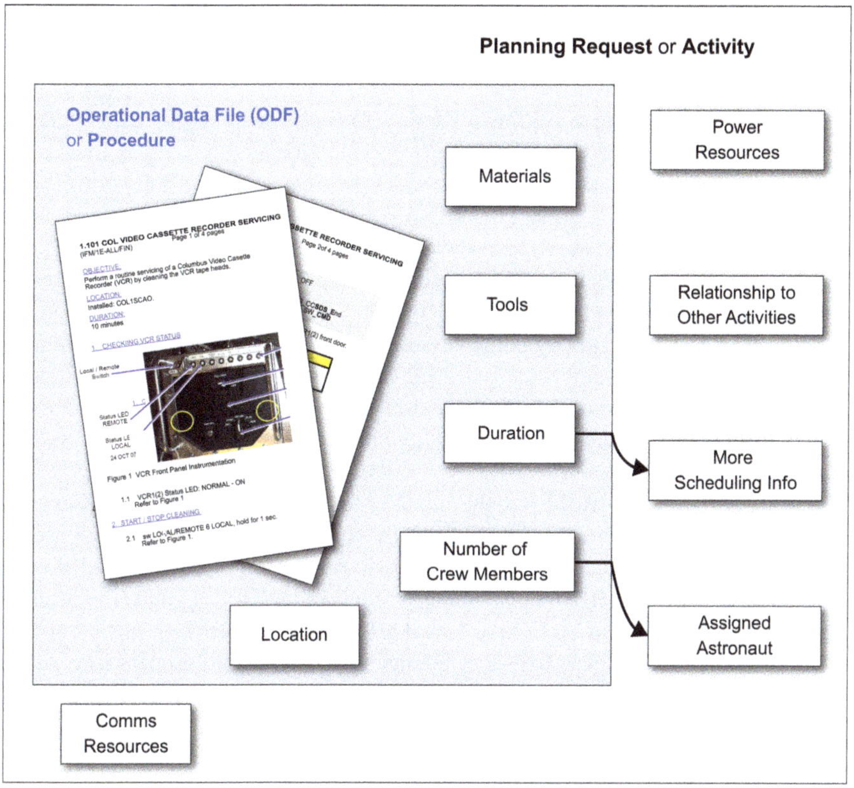

Fig. 17.3 An activity definition or planning request as elementary planning item contains a procedure called operational data file (ODF), which defines the actual task to be performed. It also contains various pieces of information about the task, like the materials or tools required for execution. The procedure is supplemented by some additional information, like the assigned astronaut—often referred to as planning attributes—which complete the set of data required for proper planning of the planning request.

During the planning process, the limited ISS resources like crew time or data bandwidth need to be shared between the different international partners. It was agreed that the ratio every partner is eligible to use is directly related to its overall contribution to the ISS program. This results, for example, in an ESA utilization share of 8.3% of all common ISS resources, including crew time.

17.3 Planning Teams

The increment concept also allows better assignment of key personnel and ensures a rotation of tasks and responsibilities with a frequency of about half a year.

The first steps of defining the content of the increment are taken at management level by each international partner during the strategic and tactical planning phases described below.

Each ISS control center (see also Chap. 24) then facilitates its own planning department, whose experts are assigned for the increments. For a given increment, a number of people are usually involved in the preparation phase. The terminology long-range planner (LRP) is used by NASA, ESA calls them Columbus lead increment planner (CLIP), and the Japanese supply the 3PO (**p**re-increment **p**roducts and WLP **o**fficer). We will see later that the preparation phase does not end with the start of the increment, but the plans have to be finetuned again as the actual execution day approaches due to the highly dynamic environment of human spaceflight operations. In all ISS control centers, this near-real-time processing of the timeline is performed by the same people who prepared the first draft of the timelines in order to minimize the handovers and to facilitate the knowledge built up by the planners who worked the preparation phase.

One week before the execution, the timeline for a specific day is handed over by the planning staff in the backrooms to the flight control team (FCT) on the consoles. In addition, a planning function is implemented within each FCT via a special position staffed by an expert from the planning group (like the OPSPLAN at the Johnson Space Center in Houston or COMET at the Columbus Control Center (Col-CC) in Oberpfaffenhofen). These real-time planners coordinate and process late formal changes to the timeline and can assist the flight controllers in case real-time replanning of activities "on the fly" is required.

17.4 Concept of Crew Flexibility

It is human nature that an astronaut prefers to have certain freedom in his working day. Thus, the timeline is generally considered only as a suggestion from the ground teams, and the crew is generally allowed to deviate from the plan or to adapt the schedule to their individual preferences in consultation with the relevant ground team.

Within the ISS community, some effort is currently made to group activities into different categories. Activities marked as flexible in the timeline can be performed by the crew at their discretion within the day—these tasks are not dependent on anything else. The daily exercise of an astronaut is a typical example.

Team activities are tasks with constraints that are expected to be executed as scheduled. However, they may be performed differently than scheduled, but in this case a prior ground and crew coordination is required. Such activities might be dependent on special resources or conditions or are requiring support from the ground teams.

Time-critical activities need to be executed exactly at the time when they are scheduled. If this is not possible, the activity is canceled for that day. In the timeline viewer those activities appear with a blue frame. Typical examples are live public affairs events.

In case some activities are related to each other and have to be conducted in a certain sequence, those activities are marked with the same color in the timeline viewer. Thus, the astronaut still has some degree of freedom in the timeline execution, as long as the given sequence of events is respected.

Weekends and pre-agreed holidays are kept free of work as far as possible. However, on crew request a concept of voluntary science/maintenance was established in the recent years. In case the crew expresses their interest prior to their mission, the ground team puts together tasks which can be autonomously executed by the crew on weekends or holidays. For each weekend, the crew then receives a pick list approximately two weeks in advance, from which they can select their preferred activities. These are then put into the timeline of the crew-off day. Nevertheless, the crew always reserves the right to withdraw from the tasks and protect the days off.

All activities, which appear scheduled in the timeline, are called "hard scheduled" and the time required for their execution is budgeted to the corresponding space agency. However, there is an alternate concept in use, which is called job jar or task list. The task list is populated each week with possible candidate activities which can be performed by the crew at any time—similar to flexible activities. The idea behind the task list is to provide the crew a repository of activities which need to be executed sometimes, but are (at least for the moment) not critical. Typical examples are stowage audits or recording of public affair-related messages. In case the crew has some "gray space" (i.e. spare time) on a dedicated day, they might consider to check the task list for further work and get ahead with the increment goals, if they wish.

17.5 Planning Phases Overview

For the ISS project, different planning phases have to be distinguished. The phases range from very generic planning of vehicle traffic and crew rotation well in advance to the very detailed daily planning of activities. As shown in Fig. 17.4, three phases can be distinguished. The strategic planning phase reaches far into the future and becomes tangible about five years before the execution phase. It goes directly to the tactical planning phase, which is initiated about 1.5–2 years before the start of the double increment.

The phases of strategic and tactical planning follow an annual rather than an incremental schedule. Therefore, the time relationship to the start of the increment is, by nature, only an approximation.

One year before the increment, the execution planning processes are entered, which can be split into two subphases: the pre-increment and the increment execution planning. Since the planning products (see below) of the increment execution have a lead time of 17 days, this phase is started about two weeks before the increment begins. This phase covers both increments. After the increments, the planning

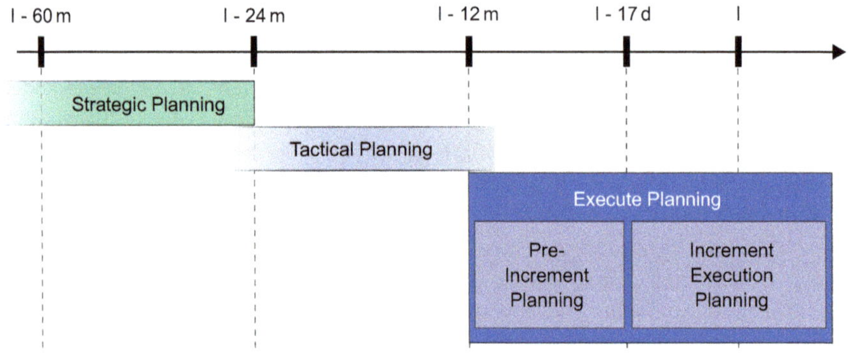

Fig. 17.4 The planning process for a dedicated increment can be divided into different phases. These phases use the beginning of the increment (referred as "I") as a time reference. The strategic planning phase takes shape about 60 months before the start of the increment, the tactical planning about 24 months before. The execute planning phase starts at I-9 months and can be divided into the pre-increment and the increment execution time frame. The latter begins 17 days before the increment with the creation of the first execute planning product (see below) and continues throughout the increment

community has to generate corresponding reports, which will not be discussed further here.

17.6 Planning Products and Processes

17.6.1 Strategic Planning

The strategic planning phase is also called multi-increment planning because it covers a longer time frame. Its planning horizon is five years in advance (Leuttgens and Volpp 1998). Like all planning phases, it is a multilateral process and defines the ISS assembly sequence, all visiting vehicle traffic, the crew rotation on ISS and ensures that the corresponding resources and consumables are available to achieve the high-level operational and scientific objectives also defined in that planning phase.

The most important documents produced during this planning phase are shown in Fig. 17.5.

The operations summary document is used for the generation of the composite operations plan (COP), which contains projections of the utilization capabilities allocated to each international partner, and for the development of the composite utilization plan (CUP), which contains the utilization plans of all ISS partners. These two documents are combined and harmonized in the consolidated operations and utilization plan (COUP), which is also published once a year. The COUP covers five consecutive years of ISS operations, of which the first is the just completed one, for which a utilization report is provided, and the second is the one in the tactical/execute

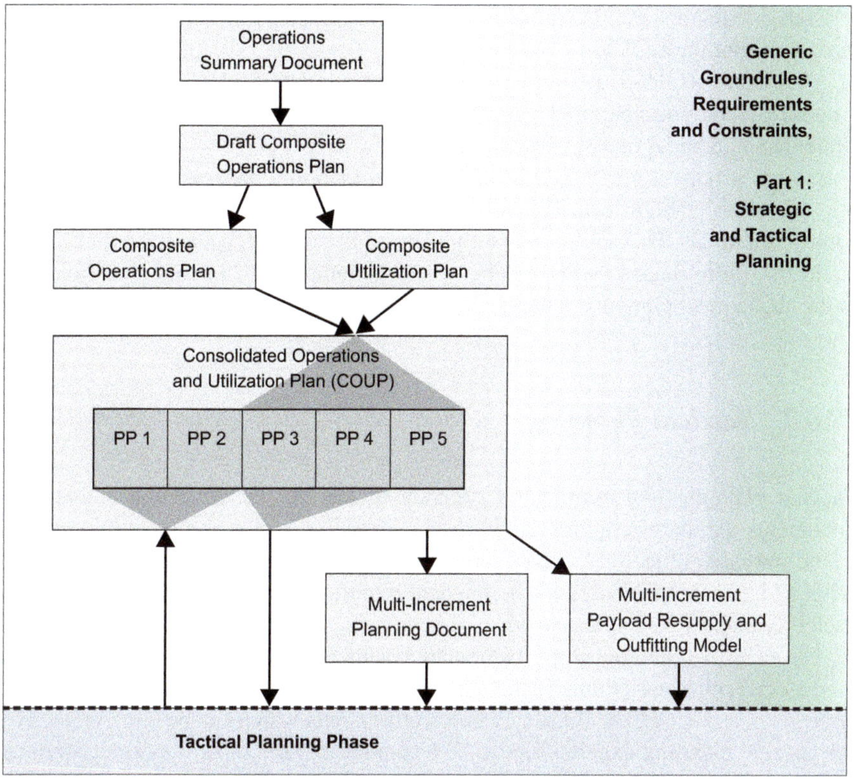

Fig. 17.5 The strategic planning process involves the creation of several documents, which are iterated annually before final publication. The multi-increment planning Document, the multi-increment payload resupply and outfitting model and the COUP document feed the tactical planning phase. The results of the tactical planning are fed back into the COUP. The generic ground rules, requirements and constraints, part 1, defines the rules and standards for the planning process (simplified schematic)

planning phase. The subsequent years are in strategic planning: Years three and four were already contained in the last issue of the COUP, while year five is the newly added outlook for the next five years.

The multi-increment planning document (MIPD) and the multi-increment payload resupply and outfitting model (MIPROM) are derived from the COUP and other source documents.

The MIPD defines tactical content of the ISS program and flight definition required to allow consistent planning and resource control. In addition, it identifies the projected ISS resources, accommodations, and supporting services available to the operations and utilization communities.

The MIPROM provides long-range facility-class payload launch, disposal plans and the topology of the external payloads installed on ISS.

Both documents are also published annually and have the same time horizon as the corresponding COUP.

All documents described above are governed by a set of rules called the "generic groundrules, requirements and constraints" (GGR&C), which accompanies the entire strategic and tactical planning process.

Several groups are involved in the strategic planning process and the products are approved by high-level ISS program panels such as the program integration control board (PICB), the Space Station Control Board (SSCB), and the multilateral payloads control board (MPCB). The strategic planning documents give directions to the subsequent planning process, the tactical planning.

17.6.2 Tactical Planning

Tactical planning is a multilateral function of the ISS Program that defines and documents the main program requirements, priorities, resource allocations, vehicle traffic, research objectives, the system-related assembly, logistics and maintenance work to be accomplished, and the payload manifest down to a sub-rack level, for each increment 1.5–2 years ahead.

The tactical planning phase begins approximately 1.5–2 years before the start of the corresponding double increment. The high-level program documents of the strategic phase need to be translated into requirements which can be implemented by the execute planning experts. Figure 17.6 summarizes the most important products of this planning phase.

The main document being developed in that phase is the increment definition and requirement document (IDRD), including its annexes, which are, in fact, also self-standing documents.

However, before starting to define the increment, the strategic considerations based on an annual timescale must be broken down to double increments. All transportation systems capabilities and schedules, the ISS capabilities and available resources and the operations and utilization requirements are now assigned to a specific increment and the associated flights.

IDRD Annex 5, also known as payload tactical plan (PTP), marks the transition from the multi increment based planning to a double increment driven approach. It also details the payload planning down to a sub-rack level, and provides the corresponding up- and download manifest information, including mass, volume and additional resource requirements of each item during the transport phase. The research objectives for the increment time frame are clearly defined and the resources required for each experiment are compared to the available quantities. This includes the use of common station items such as tools, disposable gloves, wipes and tapes. The topologies of the experiment hardware on board are defined, which means the configuration of the entire space station is described down to a locker or rack insert level, and changes to that configuration during the increment are defined. A table lists all requirements for cold stowage of items. It is already documented what time has to

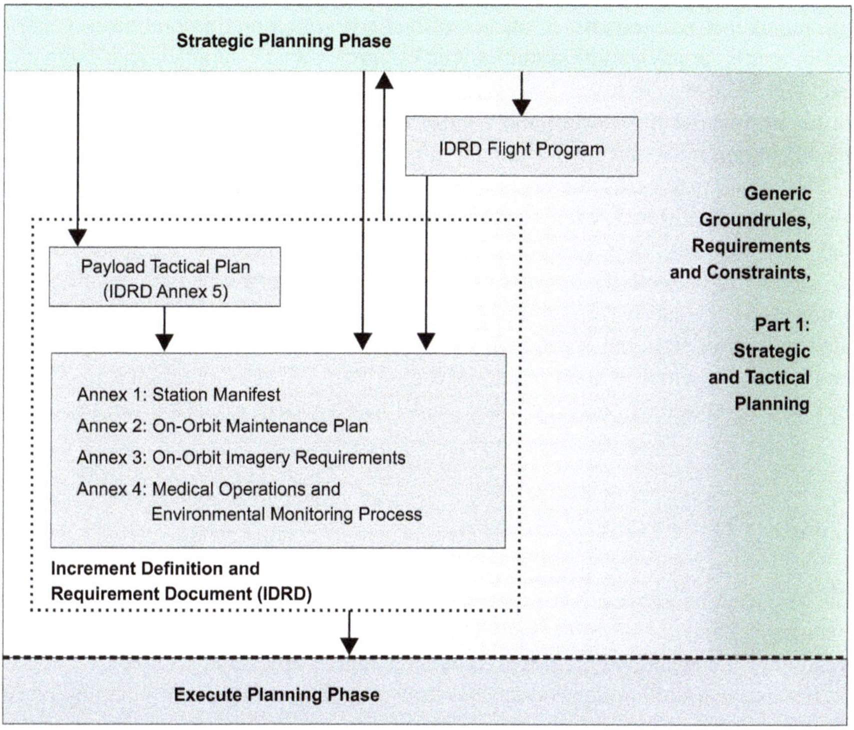

Fig. 17.6 During the tactical planning phase the outcomes of the strategic processes are translated into the increment definition and requirements document (IDRD) and its annexes, which then serves as input for the execute planning

be allocated for each activity of the research crew. The details of the training to be given to the astronauts on ground are documented. Reference is also made to any special agreements between the international partners that apply to the increment.

The document generation process cumulates in its baselined version approximately one year prior to the start of the corresponding double increment.

Annex 5 of the IDRD is an essential driver for further planning, which is described in great detail in the other annexes of the IDRD. These annexes follow their individual development schedules and are under the control of different working groups and decision boards. Annex 1 (station manifest) is essentially a detailed inventory list of all planned vehicle "flights" to the space station and back to Earth. This includes not only payload items, which have been identified in Annex 5, but also system or assembly parts, resupply, propellant, cryogenics, water, or crew items. Annex 2 (on-orbit maintenance plan) of the IDRD lists all maintenance activities which are required in the double increment time frame to ensure that the station remains in a good shape. In general, it is distinguished between corrective and preventive maintenance and the corresponding requirements are derived from the various design documents of all station hardware. Annex 3 (on-orbit imagery requirements) documents the requirements of on-board imagery, which might come from generic requirements, or payload or system driven requirements, or other sources. Both still and video imagery are listed. Finally, all medical requirements, guidelines and rules for environmental monitoring and control of the ISS are summarized in Annex 4 (medical operations and environmental monitoring).

The IDRD serves as the main input for the next planning phases, which is commonly referred to as execute planning. Since the IDRD also contains a priority ranking for all activities, it serves as an important document in that regard during the increment as well. There might be the need for changes of the content which is driven by the latest real-time developments. Therefore, the document is translated into an operational document, the so-called current stage requirement document (CSRD), which is then maintained by a joint effort of the management and operations community and serves as a guideline document for task priorities.

17.6.3 Pre-increment Planning

The transition into the execute planning phase also marks the point in time when the planning is handed over to the actual planning teams, which then continue with the creation of increasingly detailed executable planning products.

The two main planning products developed in the pre-increment planning phase are the on-orbit operations summary (OOS) as a rough outline of the future timeline (see Fig. 17.7) and the increment-specific ground rules and constraints (GR&C), which comprises a set of planning rules that apply for the activities of the increment. Both products are created by the execute planning teams in a common attempt, which cumulates in two technical interface meetings (TIM), the so-called OOS TIMs. Here, the planning for the corresponding double increments is consolidated among all international partners.

The OOS is the first product which has a clear planning character in its literal meaning: It assigns dedicated activities to a dedicated crew member and already puts them into a temporal context. The duration of each activity is already known. Thus, the planning can be accomplished to a granularity of one day.

The planning teams do not only consider the requirement documents of the tactical planning phase, but also receive the actual activity definitions from the corresponding activity owners (i.e. the payload expert centers), which already contain all information necessary for the planning process, as shown in Fig. 17.3. This technical information includes the duration of the activity, the assigned and trained crew members, any requirement for resources (e.g. power) and conditions (e.g. Ku-band coverage for video or data transfer), the associated procedure, as well as correlations or interferences with other activities.

The international ISS partners work their contributions for the OOS and the GR&Cs independently, but in close contact with each other. Eight weeks before the increment start (I–8w), the payload parts of OOS and GR&Cs are collected by the Payload Operations and Integrations Center (POIC) in Huntsville/Alabama and

Mo	GMT 98						
09-Mai-11	ISS CDR	FE-1	FE-2	FE-3	FE-4	FE-5	FE-6
EHS-ALM-START						00:10	
EHS-ALM-STOP						00:50	
EHS-TEPC-RELOCATE							00:20
PAO-HD-SETUP							00:10
PAO-PREP						00:10	00:10
PAO-HD-EVENT						00:20	00:20
HMS-AED-INSPECT						00:05	
CMS-T2-DIS-PWRDN						00:05	
WHC-UR/IF-CHGOUT						01:30	
WRS-WATER-BALNCE-PLC							00:30
WRS-WATER-BALNCE-PLC							00:30
ICV-USND2-PREP						00:25	
ICV-R/ECHO-SCAN_FE5						01:00	
ICV-ECHO-SCAN_OPR50							00:50
USND2-H/W-PWROFF						00:05	
USND2_DATA-XFER						00:10	
USND2-H/W-SU/PWRON						00:30	
JRNL-NOM-ENTRY_FE6							00:15
BASS-FAN-CALIBRATION							00:20
BASS-VIDEO TAPE-EXCH							00:10
BASS-EXP-OPS							01:00
MSG-FACILITY-ACT							00:10
MSG-FACILITY-PWR DN							00:10
VI-SCAN-OPS_FE5						00:45	
BLB-GAS SPLY-OPN						00:05	
BLB-GAS SPLY-CLS						00:05	
GHF-SCAM DOOR-OPEN							00:15
GHF-SCAM DOOR-CLOSE							00:15
GHF-HICARI-CTRG-INST							00:25
БВС-СУБА-R&R-PREP							
СОЖ-MNT							
Д33-12-EXE							
TEX-39-DNLD-D/L							
TEX-39-RSE-LCS-ON							
IMS-EDIT							
ТГК-47P-IMS-XFER							
Subtotals	06:40					05:30	05:50
Minus	00:00					00:00	00:00
TOTALS	06:40					05:30	05:50

TOTAL	TIME
MEDICAL	
OBT	
ROUTINE	
US MAINTENANCE	
RS MAINTENANCE	
ESA MAINTENANCE	
JAXA MAINTENANCE	
EVA	
US UTILIZATION	
RS UTILIZATION	
ESA UTILIZATION	
JAXA UTILIZATION	
XFERSTOW	
RS SYSTEMS	
US SYSTEMS	
ESA SYSTEMS	
JAXA SYSTEMS	
TOTALS	
WEEKLY CREW TOTAL	

Fig. 17.7 The on-orbit operations summary contains already a full list of all payload and system activities, assigns them to a dedicated day of the increment and to a crew member

are integrated by the NASA experts within five weeks. Thereafter, the integrated payload OOS and GR&Cs are delivered to Houston three weeks before the start (I–3w) and combined by their planners with the system inputs, which will also be provided to all international partners at that time.

The integrated, preliminary OOS and GR&C documents are thus ready four months before the start of the double increment and are now being discussed in the aforementioned TIM between all planning teams in order to identify and resolve any conflicts.

This is followed by a second iteration of the OOS and GR&C integration: All payload inputs are collected again by the POIC at I–11w and forwarded to Houston two months before the start of the increment, where the final OOS/GR&C document is created. A final planning TIM is held, resulting in the approval of the final planning products for the upcoming double increment.

The OOS plans for each day of the two increments do not yet contain the actual scheduling times of the activities. The standard elements of a crew day (e.g. exercises, meals) are not included either. This level of detail is only reached in the next planning step.

The increment-specific GR&C contain special planning rules that must be followed by the planning teams to ensure that all activities are planned in a proper way compliant with all scientific, technical, or medical constraints and can be executed on a non-interference base.

The GR&Cs are used throughout the increment for the subsequent planning processes and planning product developments.

17.6.4 *Increment Planning*

During the increment planning phase, the OOS is successively translated in more detailed planning products until the final on-orbit short-term plan (OSTP) is reached, which serves as the final product and is executed by the operations teams. The time relationship between the different products is shown in Fig. 17.8.

Each week of the increments, a seven-day time period (Monday to Sunday) is extracted from the OOS and converted by the long-range planners (LRP) into the

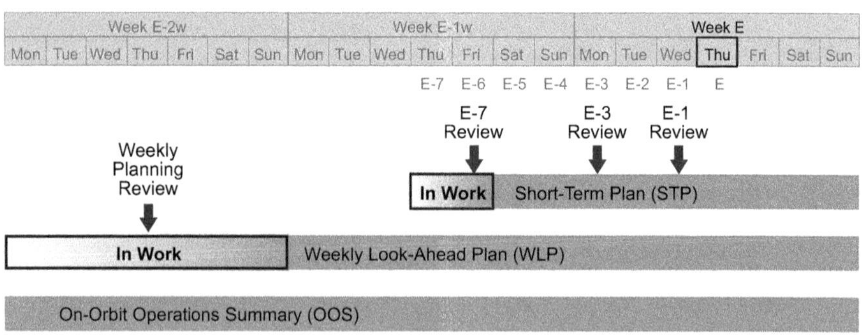

Fig. 17.8 The execute planning phase is illustrated for a given increment day, the execution day E. A first version of the planned activities for that day is already included in the OOS, which is prepared and approved before the increment starts. Two weeks before day E, the weekly look-ahead plan (WLP) for that week is prepared, using the OOS as baseline, which already contains a more detailed schedule for E. From the WLP the STP is derived seven days in advance (E-7) and finalized at E-6. This product is reviewed by the console teams three more times (“E-7”, “E-3” and “E-1”) before it is finally executed

weekly look-ahead plan (WLP). This timeline usually already contains the exact timing for all activities, including the daily tasks of the crew. A full working week is planned for the creation of each WLP.

For the WLP development, which in turn is decentralized by all international partners and finally integrated in Houston, POIC derives a payload planning template from the OOS, adds the availability of communications resources and ensures that the crew's sleep cycle is adapted to the latest agreements. This template is used by all control centers to fine-tune the appropriate payload part of the week under consideration. The info is then sent back to POIC who creates a preliminary WLP for the payload.

The preliminary payload WLP will then be further integrated by the Houston planning team with the system inputs from all partners, leading to an integrated preliminary WLP that will be available in the middle of the generation week. During the remaining week, the preliminary product will be discussed to resolve any conflicts within the WLP and create a final WLP that will be available on Friday. Formal approval of the planning product is due on Monday of the following week, during which the translation of the WLP into the next planning product, the short-term plan (STP), already begins.

The WLP development phase is supported each week by regular teleconferences (international execute planning telecon, IEPT) between all international planning teams on Monday (approval of last week's final WLP, discussion of first issues), Wednesday (discussion of preliminary WLP), Friday (discussion of final WLP) and the weekly plan review (WPR), which is a high-level meeting involving management, various increment lead functions and the planning community.

The approved WLP is used to create the STP. Each day of the week, a new development cycle is initiated for the day one week in advance: A separate day is extracted from the WLP one week in advance (E-7) to execution; each partner generates its input for the STP and sends it to Huntsville and Houston. At the POIC in Huntsville, the payload part of the STP is integrated and then sent to Houston where the final integration takes place. The preliminary, integrated STP is then compiled into the OSTP format, which can be considered the executable version of the STP. Using a special software, the OPTIMIS viewer, this is displayed on the consoles and can be checked by the flight control teams.

This review, also called "E-7", is performed by all three console shifts and all change requests for the STP/OSTP are integrated by each partner planner. The Houston planning team then integrates all changes made by the flight controllers, resulting in a final, integrated STP for the execution day (E) on E-6. This final plan is also converted into an OSTP and published in OPTIMIS. At that time, the STP is considered consolidated and approved, the actual planning work is completed for day E and the final product, the OSTP, has been handed over to the flight control teams and the astronauts for execution.

The STPs are also subject of discussion in the IEPTs mentioned above, which are held every second working day (Mon, Wed, Fri). The STPs for the weekends are usually easy to create, since Saturday and Sunday are usually "crew off days". They are therefore developed together with the STP of the corresponding Friday.

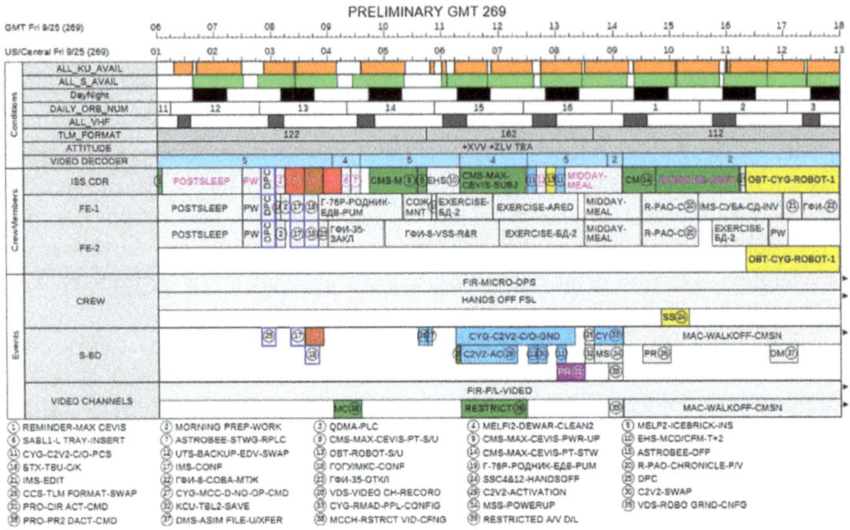

Fig. 17.9 The level of detail of the weekly look-ahead plan (WLP) and the short-term plan (STP) are similar. All plans are published in a web-based online viewer, called OPTIMIS Viewer

For the time of a visiting vehicle launch and subsequent days, the planners may decide to develop a special slip STP in addition to the nominal STP, which will come into effect when the launch and thus the docking with the station is delayed. This usually only happens when a launch slip is likely and when a slipped launch has significant impacts on the crew's timeline.

The format of the WLP and the STP is identical (Fig. 17.9), only the timespan differs (week vs. day).

17.6.5 *Real-Time Planning*

For the flight control teams, the first time they get to see the so-called timeline on console is the publication of the preliminary STP seven days in advance. The publication is done via the OPTIMIS viewer and appears as shown in Fig. 17.10. As described earlier, the team conducts a joint "E-7" review of this STP and passes all input to update the timeline to the respective planning position which then implements it in collaboration with its counterpart. The subsystem experts thus help to consolidate the final STP, which is then published the next day in a corrected version. The responsibility for the timeline is at this time transferred from the long-range planners to the planning function within the flight control team.

As the execution of a particular day approaches, the flight control team performs two additional reviews of this timeline: The "E-3" review three days in advance and the "E-1" review the day before. The agreement for these timeline reviews is that

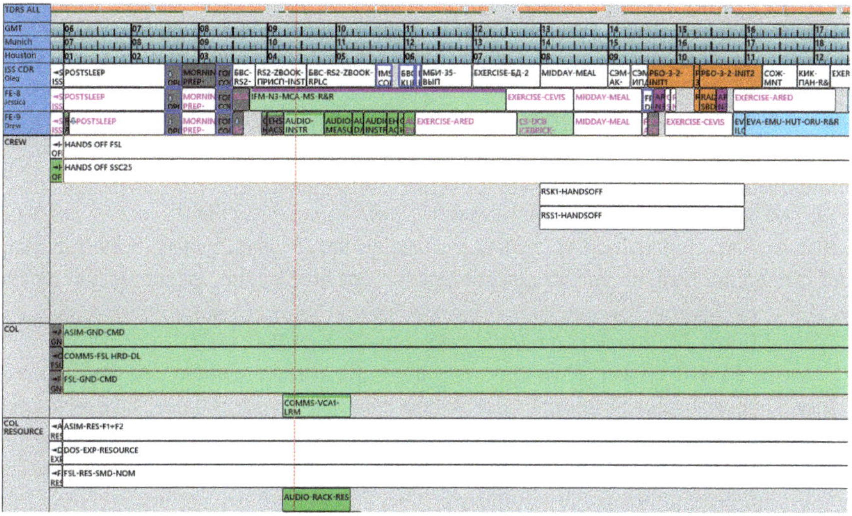

Fig. 17.10 The OPTIMIS viewer displays the timeline both on board and in the control rooms

only significant errors with potentially large impact, if not corrected, may be subject of change. Should a timeline change be required, a formal change request process is triggered that requires approval from the Houston planning team as the primary timeline integrator, as well as all international partners impacted by the change.

Changes to the timeline can be requested through this process outside of the formal reviews, if, for example, the latest developments on board or any issues on ground would require an adjustment to the final, approved planning products.

Although the timeline has been developed and optimized over weeks or even months and has passed various reviews, the astronauts still have some degree of freedom to deviate from the plans, as explained in Sect. 17.4.

The relevant control centers monitor the crew's activity execution and can take appropriate action with regard to timing in real time, if the crew is unable to work through the prepared timeline as planned. In this case, the control centers' planning functions work in the background to suggest how to reschedule the remaining crew day, to identify activities that can be postponed to another day or shifted to another crew member without impact, or—with the help of the corresponding flight director—to cancel lower-priority activities.

17.7 Planning Tools

Being an international and distributed effort, the execute planning is highly dependent on computer and web-based tools used by the concerned parties.

Some tools are used to collect the planning data needed to create the appropriate plans and schedules. All ISS planning partners use the same planning tool suite called OPTIMIS. This unifies the efforts of all centers and minimizes discrepancies and wasted time during integration. Planners create a timeline using software called SCORE. The tool is connected to a plan repository that allows live collaboration by users who have the right to modify a plan in "real-time".

When the OSTP is ready for publication, the web-based OPTIMIS viewer software is utilized, which provides an easy-to-use and effective visualization of the schedule and is used not only by all ISS control centers, but also by the astronauts aboard the ISS (at least the non-Russian crew—Russian cosmonauts are required to execute the so-called Form 24, which is a tabular representation of the STP).

The OPTIMIS viewer allows not only a graphical representation of the timeline, but also access to the additional information associated with each activity, as shown in Fig. 17.2. The associated procedures are also linked and can be opened using additional software, the international procedure viewer (IPV). The associated stowage notes, which list all the tools and materials needed to perform a crew procedure and their stowage location, are also linked.

The process for changing one of the released planning products (WLP, STP) is also supported by a web-based tool, the planning product change request (PPCR) software. This interface grants all partners access to a "from-to" form that provides all the information needed to process the change request. The tool also provides a function to obtain formal approval from all international partners for the requested change.

All international partners (JAXA, ESA, RSA and NASA) currently use the same suite of tools, with different subscriptions depending on the center. While RSA still maintains its own internal process for developing the WLP and STP (Form 24), the other partners use SCORE to enter their own activities into a timeline.

This standardization of tool use has improved the efficiency of communication between partners so that each center knows what the other partners are doing in their respective timelines in a collaborative manner.

17.8 Conclusion

Planning is a key element for mission success. Planning for human spaceflight missions is similar in many aspects to the corresponding processes of unmanned missions, but there are also areas of significant differences. It is a manual and thus more process-oriented endeavor, whereas unmanned missions can also have a high degree of automation, which requires good mathematical modeling of the planning problem. The fact that humans, rather than on-board computers, are the subject of planning, introduces an element that also requires "men in the loop" for the various planning processes.

Acknowledgements The authors gratefully acknowledge the kind support of Hilde Stenuit and Eric Istasse, the former members of the ESA Mission Science Office, who contributed to this chapter with their knowledge of the strategic and tactical ISS planning phases.

References

Korth D, LeBlanc TP (2002) International space station alpha operations planning. In: Proceedings of the international workshop on planning and scheduling for space, Houston, TX

Kitmacher GH, Gerstenmaier WH, Bartoe JF, Mustachio N (2005) The international space station: a pathway to the future. Acta Astro 57:594–603

Leuttgens R, Volpp J (1998) Operations planning for the international space station. ESA, Bulletin 94

Popov A (2002) Mission planning on the international space station program. Concepts and systems. IEEEAC paper #1439

Thomas Uhlig received his physics diploma and his Ph.D. from the University of Regensburg/Germany, where he worked on electromagnetic imaging of nanostructures. After working as scientist at the Argonne National Laboratory in Chicago he changed into the spaceflight business and to the Columbus Control Center in Oberpfaffenhofen, where he first worked as Payload expert on the Space Shuttle mission STS-122, then for several years as Columbus Flight Director. He then took over the training for the European ISS flight controllers, before he was assigned as team lead for the geostationary satellite team at GSOC. In between, he was working as expert for functional safety, first at BMW, then at Heidenhain GmbH before returning to DLR in 2022. He was awarded with the Helmholtz price 2005 of the Physikalisch-Technische Bundesanstalt. He is author of various scientific articles.

Dennis Herrmann studied aerospace engineering and received his diploma in 2007. He later joined DLR and worked as part of the Columbus team as Columbus Operations Planner (COP) and Columbus Operations Coordinator (COL OC). In 2009, he assumed the group leader role for this position before being promoted to the Columbus Flight Director group. In 2012, he acted as Lead Increment Flight Director and received a project management certification before leaving the space sector to join the automotive safety industry, where he still works for BMW today.

Jérôme Campan is the head of the Columbus Flight Director Team at the Columbus Control Center in Oberpfaffenhofen. After receiving his engineering diploma from the ICAM school (Institut Catholique des Arts et Métiers) in Toulouse, he graduated from the ISAE school (Institut supérieure de l'aéronautique et de l'espace) in the "Space System engineering" field. He started his professional career as a Columbus Operations Coordinator for the first three years, then became the team lead of the European Planning and Increment Coordination team before moving to his current position.

Part VI
Spacecraft Subsystems

Chapter 18
Telemetry, Commanding and Ranging System

Michael Schmidhuber and Tarsicio Lopez-Delgado

Abstract This chapter describes operations of the telemetry, commanding and ranging components of a satellite. Using radio frequency transmissions, they allow remote monitoring and controlling of the spacecraft.

18.1 Definition of Subsystem

Operations of spacecraft components enable radio frequency (RF) transmission of remote spacecraft monitoring and control information and the RF reception of commands from the ground station.

Although in many projects and system descriptions this function is handled as a sub- component of the data handling subsystem (see Chap. 19), we decided to treat it here as a self-standing subsystem.

As it is described below, it provides the radio link and not any processing of or insight into the transferred data.

The subsystem can also be used for orbit determination, a function that is commonly referred to as "ranging" or "tracking".

Common names used for this are telemetry/telecommand (TM/TC) or telemetry, commanding and ranging (TCR).

The future prospect of optical communication is not covered here.

18.2 Signal Characteristics

A good and fundamental description of the physical properties of the communication is contained in Chap. 3; only special aspects that are of interest for the TCR operations task are highlighted here.

M. Schmidhuber (✉) · T. Lopez-Delgado
DLR, Munich, Germany
e-mail: michael.schmidhuber@DLR.de

F. Sellmaier et al. (eds.), *Spacecraft Operations*, Springer Aerospace Technology,
https://doi.org/10.1007/978-3-030-88593-9_18

18.2.1 Frequencies

Electromagnetic signals are typically divided by their frequency into ranges called "bands". Refer to Table 3.1 in Chap. 3 for an overview of ranges and frequencies.

The frequencies used for a particular spacecraft are selected by its corresponding communications requirements.

S-band (~2 GHz) is currently used mostly for low Earth orbiting spacecraft that transmit actively only over selected ground stations, or for all space missions during their launch and early orbit phase (LEOP). This comparably low frequency band is suited for a relatively easy design of antenna systems with an omnidirectional radiation characteristic. Obviously, this is a useful feature in case the spacecraft orientation and the direction towards the ground station (antenna aspect angle, see Sect. 18.3) is continuously changing.

It is also beneficial for contingency cases with uncertain spacecraft attitude which does prevent an accurate antenna pointing. S-band antennas are currently widely available around the Earth at many ground stations which makes cooperations during LEOPs and emergencies easier.

X-band (~8 GHz), Ku-band (~12 GHz) and Ka-band (~20–30 GHz) are using higher frequencies which result in higher possible data rates. These bands are mainly used during the routine phase of the spacecraft lifetime. Their bundled signal characteristic is used to avoid interferences and to save energy, but it requires exact pointing.

There is currently a tendency to shift operations to higher frequencies. This is partly due to pressure from other spectrum users (mobile ground communications) who want to use these bands for other applications and the fear of spacecraft operators of signal interference with increasingly crowded frequency regions. But mainly it is caused by the increased demand in telemetry downlink bandwidth.

A disadvantage of the usage of high frequency ranges like Ka-band is that the transmission of signals is highly dependent to atmospheric conditions and is thus affected by moisture and rain.

All frequency uses have to be coordinated with and approved by the International Telecommunication Union (ITU).

18.2.2 Polarization

Like visible light and any other type of electromagnetic radiation, radio signals can be polarized, which means that the direction of the oscillation is not randomly distributed but follows a defined behavior.

Two main types of polarization are distinguished: linear polarization and circular/elliptical polarization, as depicted in Fig. 18.1.

Both types of the phenomenon allow two independent ways of transmission on the same frequency: two signal waves with perpendicular linear polarization can be

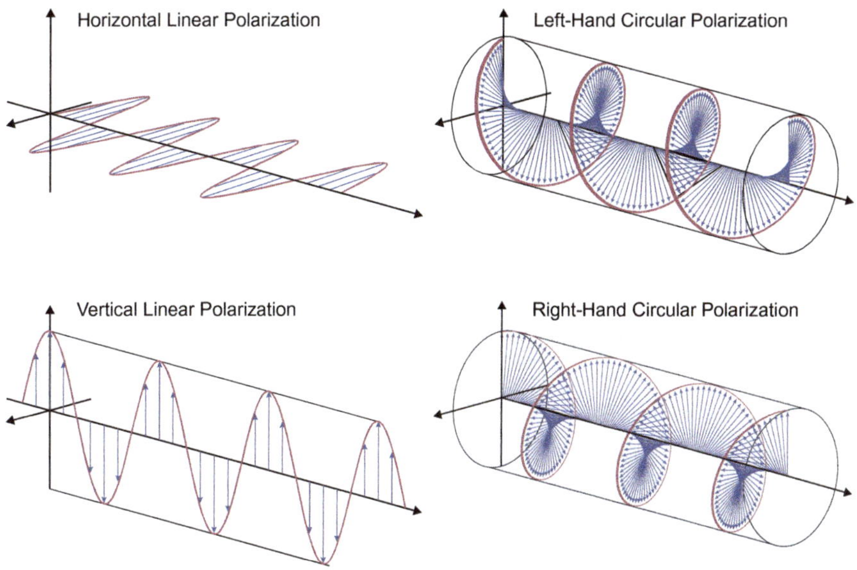

Fig. 18.1 The different main types of polarization of electromagnetic waves. The oscillations of the electric field are used to describe the polarization type

considered as independent channels as well circular/elliptical polarized waves with different senses of rotation. This can be used intentionally in communications to make double use of a frequency. The receiving antenna then has to be designed to filter out a selected polarization.

For 3-axis stabilized geostationary satellites, the linear type is easiest to realize. E.g. the polarization patterns of satellite television signals are either X- or Y-polarized. If receiver and transmitter antennas are likely to be rotated (or even rotating) around the signal direction, the circular polarization is used. Left-hand (LHCP) and right-hand circular polarized (RHCP) can then be used to distinguish the two independent channels.

18.2.3 Side Bands and Side Lobes

Every antenna produces a signal not only in its main direction, but also side lobes, as depicted in Fig. 18.2.

Depending on the antenna design, these side lobes are weaker in strength and can normally not be used for signal transmissions. However, they may result in false receiver lock conditions during the phase of antenna alignment, as discussed below. See Sect. 3.4 for further details.

Not only in the spatial domain, also in the frequency domain inadvertent side effects can occur.

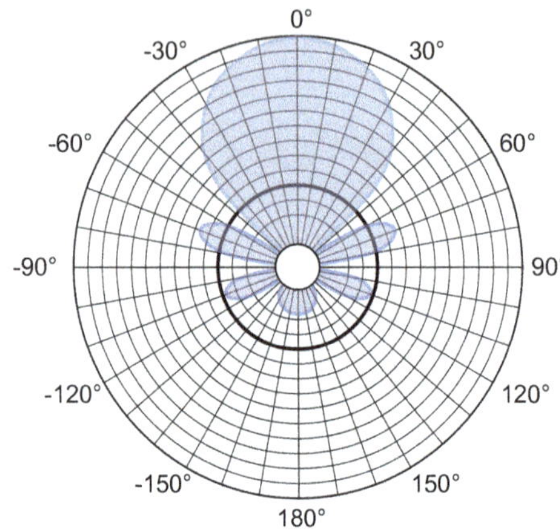

Fig. 18.2 The antenna pattern including the main and the side lobes

If a carrier frequency is modulated by a signal, side frequencies are generated. They contain the signal information. The characteristic of this frequency pattern is dependent on the transmitter design (filtering). In the initial phase of frequency alignment between satellite and ground station, strong side frequencies can lead to an incorrect receiver lock (misinterpreting the sideband as carrier frequency). The resulting demodulated signal is normally not usable as it is considerably weaker in strength.

18.3 Design

18.3.1 Subsystem Elements

The design of the telemetry, commanding and ranging subsystem is largely driven by the mission requirements, hence system details will rarely be identical on two spacecraft. But the main building blocks are mostly recurring. They are:

- Antenna
- Receiver
- Transmitter.

In the block diagram in Fig. 18.3, an uplink signal from the ground station is picked up by one of the antennas and guided to the receiver. There it is demodulated from the carrier signal. The resulting output is still an analogue waveform. It is routed to the TM/TC board of the on-board data handling subsystem. This board is considered part of the data management system and is described in Sect. 19.2.1. Reversely, the transmitter gets its input for the downlink from the on-board computer. It modulates it onto a carrier frequency and routes it to the transmitting antenna.

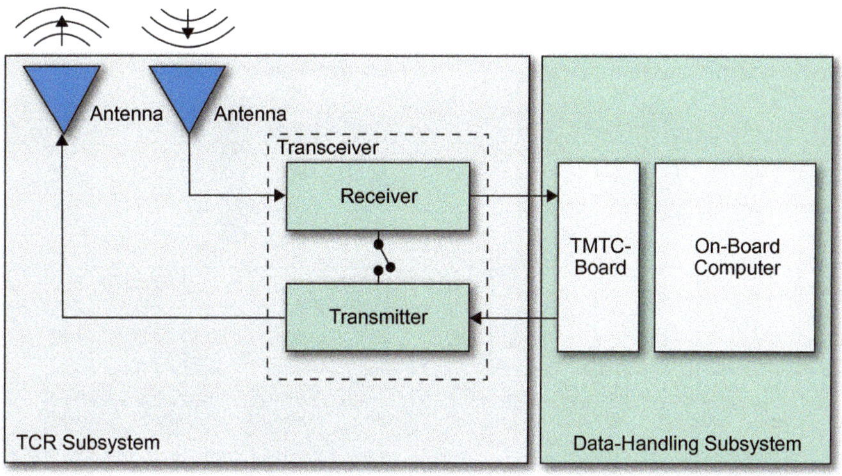

Fig. 18.3 Block layout of the TCR subsystem. It shows the fundamental components and the differentiation to the data handling subsystem

Some systems are designed to allow a direct signal connection between receiver and transmitter for the ranging function. This technique is rarely used in low Earth orbit (LEO), where angle tracking and GPS-tracking is dominant, but is standard in orbit altitudes where GPS cannot be used (geostationary Earth orbit (GEO)/interplanetary missions). Ranging is explained in more detail in Sect. 18.5.2.

Antennas can be used for reception and transmission at the same time if interference of signals is avoided. This is done by using an uplink frequency that is different from the downlink frequency and by using filters in the signal paths.

The functional unit of receiver/transmitter is called a transceiver (transmitter–receiver).

For payload data, a dedicated communications system operating in a different frequency band is often used (see Table 18.1 and Fig. 18.5). In case of scientific

Table 18.1 Examples of frequency usages for different types of satellite missions

	Mission type	Orbit	S-band	X-band	Ku-band
GRACE	Scientific Earth observation	LEO	TM/TC and payload	–	–
TerraSAR-X	Scientific/commercial Earth observation	LEO sun-synchronous	TM/TC	Payload	–
Eutelsat W24	Geostationary communications	GEO	TM/TC during Launch and Early Orbit phase and emergencies	-	TM/TC during routine phase Payload channels

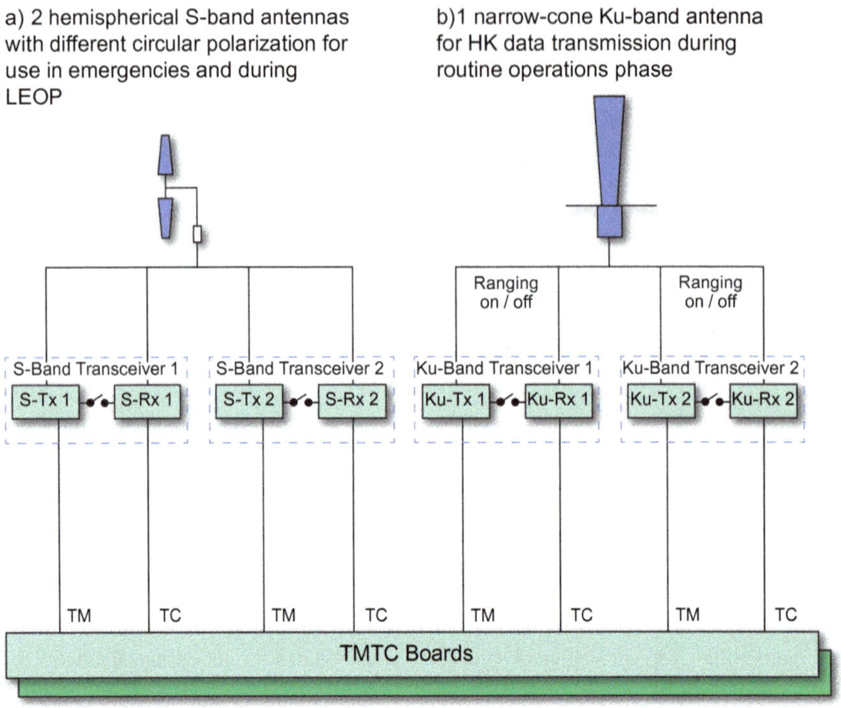

Fig. 18.4 Example TM/TC-System layout of a geostationary communications satellite

satellites, this follows the same principles as described here. For communications satellites, refer to Chap. 23.

Figure 18.4 shows an example of a geostationary communications satellite TCR subsystem. The same data can be received and transmitted through any of the four shown transceivers. The S-band systems are used as a redundant pair during LEOP and during emergencies: their omnidirectional antenna patterns make transmission independent of the spacecraft orientation. The S-band receivers cannot be switched off to ensure they are available during emergency situations.

The Ku-band systems are used during the routine phase when the spacecraft is fixedly oriented to the Earth. The antenna has a bundled characteristic that needs less energy to operate and does not as easily interfere with other RF signals. The Ku-band transceivers offer more functions like selection of different frequencies and some cross-coupling that is not shown in the figure. All four transceivers allow ranging, as this is the most precise orbit determination method available for geostationary satellites.

In comparison, Fig. 18.5 shows the TCR system of a scientific LEO satellite, based on the TerraSAR-X design (Grafmüller et al. 2005). A ranging function is not used, as the spacecraft can use GPS data for orbit determination. An S-band system is used for real-time and stored data. The transmitter output power can be adapted

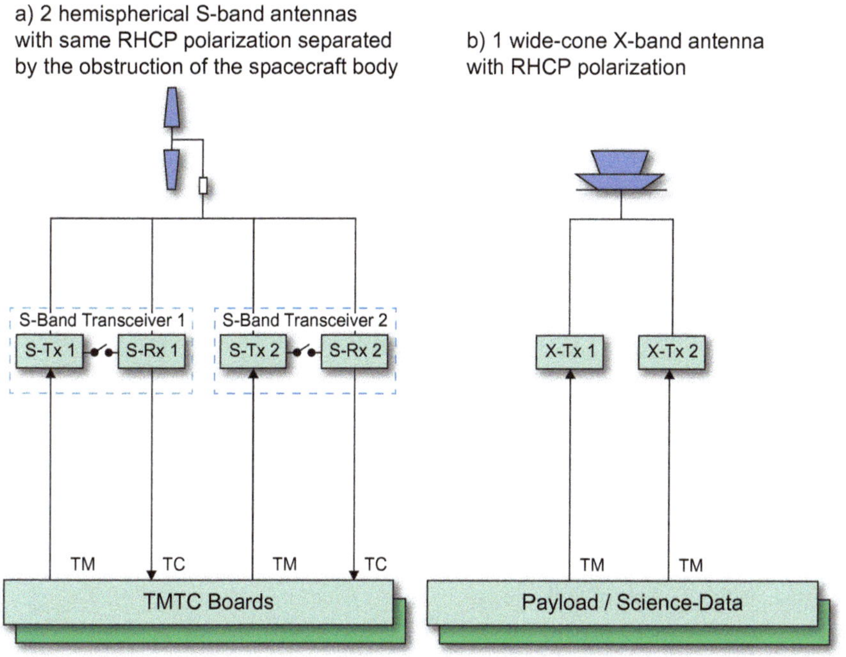

Fig. 18.5 Example layout of the TMTC system of a scientific satellite in low Earth orbit

to allow transmission with high data rates requiring more power. The payload data system is separated and uses X-band, as an even higher data-rate is needed for the payload data. A wide-angle pattern is needed on the payload antenna because the spacecraft is oriented along its orbital track and its attitude relative to the ground stations therefore changes during passes. The payload antenna is nadir pointing.

18.3.2 Spacecraft Antenna Layout

Antennas can be distinguished by their transmission and reception characteristics, the so-called antenna pattern. It describes the sensitivity of the antenna dependent on the direction the radiation is coming from. It can either be directional or non-directional (omnidirectional). The higher the frequency the more directional is the characteristic. Normally the pattern is the same for reception and transmission. Typically, directional antennas provide a higher antenna gain and are used for routine tasks with high data volume. Non-directional antennas are used for tasks that need to be robust against spacecraft attitude changes like in emergency situations or when no high rate data transmission is needed.

Fig. 18.6 A typical S-band antenna for LEO application with hemispherical antenna pattern and either right-hand or left-hand circular polarization. The shown model has a height of 100 mm. Image used with friendly permission of STT-SystemTechnik GmbH

A real spherical, omnidirectional characteristic cannot be achieved with one antenna, as the spacecraft body and the antenna structure will always block a significant part of the wave propagation. Therefore, two antennas on opposite sides of the spacecraft body are used to form a spherical coverage. This is in particular the case on LEO missions. Two S-band antennas with hemispherical antenna patterns are used here (Fig. 18.6).

When two antennas radiate the same signal on the same frequency and polarization in the same direction, the signal quality will be significantly degraded due to interference effects. One way to avoid this is to design the patterns of the two antennas with a gap between them, causing a belt with no transmission, as shown in Fig. 18.7 (left). This case is easy to implement and does not put additional requirements on the ground stations. However, if the satellite orientation causes the belt pointing in the direction of the ground station, the contact to the spacecraft will likely be lost or severely limited. Attitude and antenna aspect angle have then to be carefully monitored and considered for operational impact.

An alternative design is shown in Fig. 18.7 (right). The two antennas have wider angle of reception, causing a significant overlap and thus transmission in all directions, but to avoid interference they emit and receive radiation with different polarization. This design produces no gaps, but needs a more sophisticated ground segment, being able to receive and transmit with two polarizations.

The two polarizations chosen for this design are typically left-hand circular polarized (LHCP) and right-hand circular polarized (RHCP), as those are not dependent on the axial orientation of the antennas with respect to each other.

In some cases, a complex overall antenna pattern is required. This can lead to designs with antenna arrays of more than 20 elements. Figure 18.8 shows the example of the Meteosat second generation spacecraft (Van't Klooster et al. 2000). The S-band

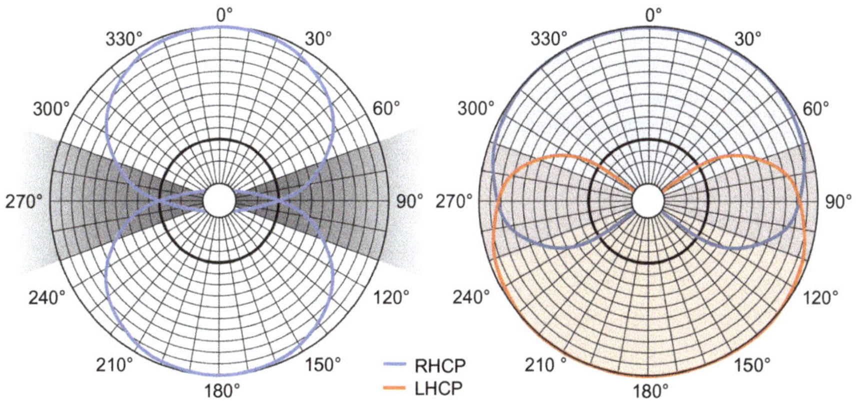

Fig. 18.7 Left: The two signals coming from the two antennas have the same polarization. In the overlap region (darkly shaded belt) the signals are too weak to result in a receiver lock. Right: The two signals coming from the two antennas have different polarizations: Right-hand circular polarized (RHCP) and left-hand circular polarized (LHCP). In the overlap region either signal can be used

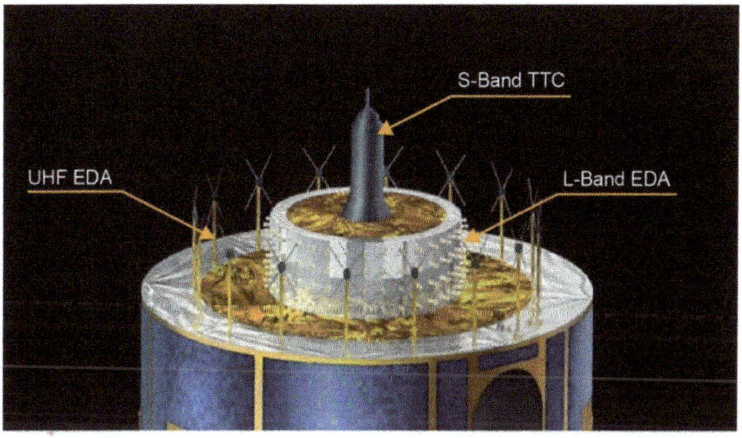

Fig. 18.8 The antenna section of a geostationary meteorological satellite. The spacecraft is spin-stabilized and rotates around its symmetry axis (vertical direction in image). *Photo* ESA

antenna is located on top at the spin axis and is of the type shown in Fig. 18.6. The L-band antenna (~1.5 GHz) and the UHF (~40 MHz) antenna are composed of multiple elements pointing in radial direction. To avoid interference and unnecessary power consumption only the element pointing towards Earth is activated (electronically despun antenna—EDA).

18.3.3 Redundancies

To guarantee the function of the subsystem even after failures of components, more than one unit is usually supplied. Full redundancy is only reached when the backup component can take over the complete functionality. Complex electrical components like receivers and transmitters are therefore provided redundantly and independently, typically by providing a second system in parallel. Mechanical components like antennas are very robust and, in many cases, only a single item is implemented. A real redundancy is therefore not given. In these cases, a limited workaround can be reached by using other antenna systems, as in the example case of one of GSOC's GEO missions where a cabling problem during LEOP caused the complete loss of the Ku-band system. The satellite was then operated in S-band for the complete mission duration, without any loss of functionality, but with the negative effect of permanent omnidirectional S-band radiation that caused interferences with other spacecraft.

Another example was the Galileo to Jupiter mission. Interplanetary missions typically have an omnidirectional low-gain antenna in addition to a high-gain main antenna that is folded for launch. In the case of Galileo, the main antenna reflector failed to deploy (Jansma 2011). The complete mission was then handled with the low-gain antenna (Marr 1994), causing higher efforts in on-board preselection and preprocessing and accepting a reduced data return.

18.4 Monitoring and Commanding

18.4.1 Automatic Gain Control (AGC)

A central role in operations is the monitoring of the uplink automatic gain control (AGC) level. The AGC is a circuit in the receiver that controls the signal amplification (or attenuation) to keep the signal strength in a defined range for subsequent components (Bullock 1995). The characteristic AGC parameter is the ratio of the signal level to an internal reference level. The term is used synonymously to the received uplink signal strength. The units are decibel, scale is logarithmic and negative. Typical values are between −50 dB (strong signal–low amplification) and −110 dB (low signal–high amplification) depending very much on the receiver type and the antenna characteristics. Values that are either far too high or too low may result in difficulties to demodulate the signal. If no signal (i.e. only noise) is received, very low values e.g. −150 dB will be indicated. Any remaining indicated signal level is caused by internal and external radio noise.

Operations shall be done only when the signal is within the range given by the manufacturer. It shall be planned ahead to either pause operations or switch over to another ground station with a better transmission situation.

The monitoring of the AGC level is important for successful operations. However, the received signal level cannot easily be influenced. Therefore, it is important to monitor the AGC evolution and make predictions about imminent development.

For example, it is not advisable to perform commanding for an orbit maneuver over a station whose signal will become weaker or will be out of sight. An early switch to a different ground station should be considered. In some cases, like an unusual spacecraft attitude, a boost of uplink power can be advisable.

As the signal strength is dependent on many aspects, a lot can be learned about the status of the mission when closely looking at the AGC behavior over time. For more details, Sect. 6.4 shall be consulted.

18.4.2 Loop Stress

The signal is demodulated by a phase-locked loop (PLL) circuit in the receiver (Bullock 1995). This circuit generates a reference oscillation at the nominal frequency and compares it to the actually received signal. If the received signal is of exactly the same frequency, the loop is "in sync" or "locked". If a difference is detected, then the loop circuit is able to adapt the reference frequency automatically within a certain range to eliminate the difference and establish the lock status. This offset to the nominal frequency is called loop stress. It is provided by the PLL circuit as a voltage and can be converted into kHz. It has limits beyond which the PLL cannot compensate the difference. In this case, the lock status is lost and the signal cannot be decoded. Note that the PLL circuit cannot follow very fast changes in frequency as it has a certain inertness to adapt the frequency. Nor can it detect and adapt to frequencies too far away from the reference frequency, even if the signal frequency is within the allowed loop stress range. To still achieve a lock the ground station has then to adapt the frequencies of its transmitter.

The difference in frequency causing the loop stress can come from imperfect adjustment in manufacturing or shifts in oscillator properties, both on the sender and the receiver side (temperature variations, aging). The largest part, however, is caused by the Doppler effect due to radial velocities between sender and receiver. This part has a typical value between plus and minus 50 kHz for a LEO spacecraft transmitting in S-band when it appears at the ground station horizon. The variation of loop stress during a ground station pass is depicted in Fig. 18.9. A possible compensation for this effect is task of the ground station operations. Either a constant offset or a variable adaptation of frequencies can be applied to keep the loop stress within the limits and enable a successful receiver lock.

For interplanetary missions this effect can be much higher and the receivers have to be designed for this task. The interplanetary probe Voyager 1 is heading away from the Sun at around 17 km per second (slightly higher than the escape velocity of the solar system, which actually makes it an interstellar probe by now). This causes a S-band Doppler shift of approx. -100 kHz. Finally, a ± 185 kHz due to the Earth orbital velocity over the course of a year has to be considered. However, this Doppler

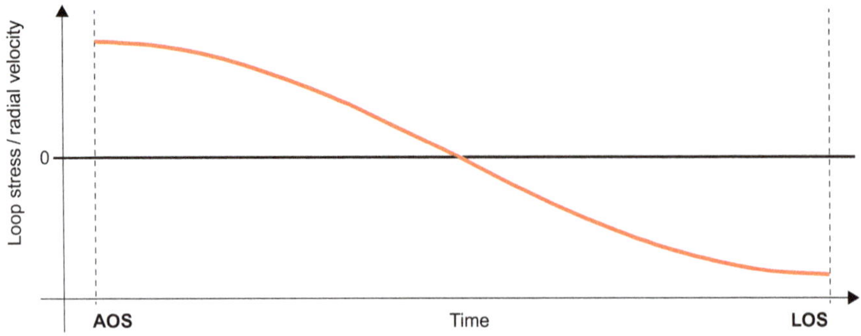

Fig. 18.9 The expected frequency shift caused by the Doppler effect during the ground station pass of a LEO satellite. AOS (acquisition of signal) and LOS (loss of signal) are the times when the satellite appears and disappears at the ground station horizon. The frequency shift is proportional to the radial velocity between satellite and ground station

shift remains largely constant over short periods. Only a small shift of ±3 kHz shift variation caused by the Earth rotation is observed during ground station passes of a few hours.

18.4.3 Lock Status

As described above, a lock status is necessary for the correct processing of the signal. Most receivers will indicate the lock status in telemetry. In other cases, the status can be deduced from the AGC level and loop stress readings. It is worth noting that according to the ECSS-104 standard (ECSS 50-04C 2008) a summary indication of the reception and lock status of all receivers is shown in the so-called command link control word (CLCW) telemetry. This flag indicates if at least one receiver is in lock. Since the CLCW has a fixed position in the telemetry transfer frames, it can be extracted from the telemetry stream by ground station equipment without complicated telemetry processing. Station personnel can use this to check successful uplink acquisition after an uplink sweep.

18.5 Operational Situations

18.5.1 Acquisition and Loss of Signal (AOS/LOS)

An acquisition of signal (AOS) has to be performed each time a change in uplink is done, e.g. caused by a station handover. It is a task done by ground station personnel and under normal conditions no action is required from the TCR engineer. To ensure a lock of the spacecraft receiver, the uplink signal frequency is shifted up and down in small steps in the vicinity of the expected frequency. This is called a “sweep”. At one of these steps the receiver will be able to lock onto the signal and can follow further variations. However, as described above, the station personnel checks only if any receiver goes into lock and does not analyze the signal quality. It may happen that the satellite receiver goes into lock on a side band either in direction (side-lobe) or in frequency (side-band). This may produce a lock status, although the signal is not usable. It will result in a low AGC level (low signal–high amplification) and in case of a frequency side band a high loop stress. Uplinked telecommands can fail. A low AGC will also be seen in the ground received AGC and can therefore be detected by observant station personnel. If such a situation remains undetected by the ground station, it is the task of the TCR engineer to inform the ground station and suggest a re-sweep.

The signal reception quality and strength are also affected by the operational situation. A major factor is the antenna aspect angle (Fig. 18.10). This angle changes throughout a pass for all satellites that are not Earth-oriented or do not have steerable antennas. Depending on the antenna pattern the received and transmitted signal strengths are weakening when the angle becomes larger.

All transmissions are affected by atmospheric attenuation and refraction. This effect becomes very large and unpredictable when the signal has to travel large distances through the atmosphere near the ground station horizon. This is the reason why the time of loss of signal (LOS) cannot be determined exactly. Operations planning shall take this into account and place critical operations into periods of stable reception.

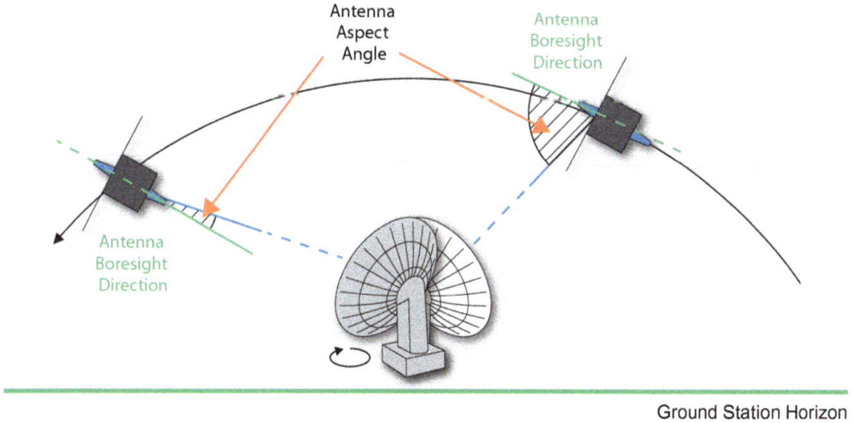

Fig. 18.10 The antenna aspect angle is the angle between the main receiving axis of the spacecraft antenna (boresight direction) and the direction to the ground station

18.5.2 Ranging

Ranging is a function that allows the determination of the spacecraft orbit. Unlike vehicles in lower Earth orbits, which nowadays simply use GPS to track their positions, satellites in higher Earth orbits or on interplanetary cruise trajectories cannot locate their own position. Usually this is done using the ranging technique. Here the uplink signal is routed from the receiver directly to the transmitter and is radiated back to the ground stations. The on-board delay between reception and transmission time is exactly known and was measured on ground before launch. The ground station measures the signal round trip time, from which the distance between spacecraft and station can be calculated.

Taking and processing the measurements are the tasks of the ground station and the flight dynamics team. The impact on the flight operations team is that during the ranging period, roughly 5 min every 30 min for a GEO mission, no commanding shall take place in order to not distort the ranging signal. And also, depending on the transceiver model, the ranging function has to be switched on and off. Some receivers require to reset the uplink coherency function (see next chapter) after each interruption of the uplink.

For interplanetary missions, the impact is larger because the ranging tones takes away power from the telemetry signal, which might be critical due to the weak signals coming from large distances. Also, the operations pauses needed for ranging are considerably longer than for Earth orbits (Bryant and Berner 2002).

18.5.3 Doppler and Coherency

Along with ranging, usually Doppler measurements are taken to improve the accuracy. The radial velocity component between satellite and ground station can be measured by determination of the frequency shift due to the Doppler Effect. For this method, the downlink signal frequency has to be known precisely. As spacecrafts usually have no highly stable oscillator on board, the self-generated downlink signal cannot directly be used for this purpose. In that case, the uplink signal is used as frequency reference. Within the receiver, the downlink frequency is generated in a defined ratio (221/240 for S-band) from the received uplink signal and thus becomes as stable as the ground station signal. This is called coherency mode.

Depending on the transponder model, the coherency status may have to be commanded manually each time the uplink signal starts (or each time after the uplink was lost). In times without uplink the transmitter generates the frequency based on its own (imprecise) oscillator. A frequency jump will likely happen when coherency is started or stopped. This jump can be too fast or too far for the PLL circuit in the ground station receiver and can result in a loss of lock at the ground station. In that case, telemetry will be lost and has to be reacquired.

Doppler shift, coherency and ranging are normally used in combination. Ranging without coherency is possible and has been used in interplanetary missions (Reynolds et al. 2002). Also, some geostationary satellites use this.

Interplanetary missions may actually use a highly-stable oscillator at least during some mission phases like asteroid fly-bys in order to use Doppler measurements without the need for coherency.

18.5.4 Antenna/Transponder Selection

It depends very much on the design of the spacecraft if it is necessary to select the antennas used for transmission and reception. The less switching is involved the more robust the design and the simpler the operations. Mechanical devices like waveguide switches should be reconfigured only if necessary and only without radiation load.

Typically, receivers that are also used for emergencies are never switched off and are always able to receive over some antenna. Transmitters are usually only switched on if they are in use. This saves energy and avoids unnecessary frequency blocking. Only in contingencies these settings are commanded manually. Usually, the commands are generated as part of the operational timeline and are automatically included.

18.5.5 Polarization Prediction

Spacecraft that are not oriented to the Earth (but e.g. to the Sun) have large changes in the antenna aspect angle (see Figs. 18.10 and 18.11) during a ground station pass.

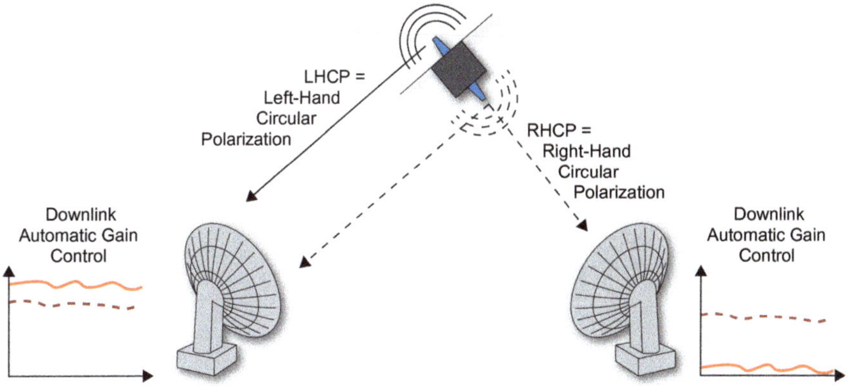

Fig. 18.11 The two ground stations can determine the reception condition at the satellite by looking at both received signals and use the stronger signal's polarization also for uplink

The antenna aspect angle is the angle between the ground station and the antenna boresight direction as seen from the spacecraft.

Section 18.3.2 showed two typical layouts of antenna systems. It stated that using two different polarizations has significant advantages. However, although most ground stations can receive the downlink in more than one polarization (diversity), all stations uplink only in one polarization. A change of this polarization can take several minutes, during which the contact to the spacecraft is lost and a new uplink sweep has to be performed. This situation can be anticipated and mitigated if the TCR engineer makes a prediction about the expected polarization during an uplink period. The necessary calculation can be made easier if the antenna aspect angle is routinely predicted by the flight dynamics system for each ground station (see Sect. 18.5).

18.5.6 Interferences with Another Spacecraft

In the tendency to move away from the increasingly congested S-band frequencies, some communication satellites are using Ku-band frequencies for TM/TC during positioning activities. Traditionally this band was used only in routine phase for payload data and for TM/TC. Ku-band allows higher data rates, a fact that is convenient for modern spacecraft.

A problem that arises then is that during positioning, the spacecraft is moving in front of other communication satellites as seen from the ground stations (see Fig. 18.12). On the one hand, the uplink signal might then interfere with the stationary spacecraft and, on the other hand, the downlink signal of the moving satellite could

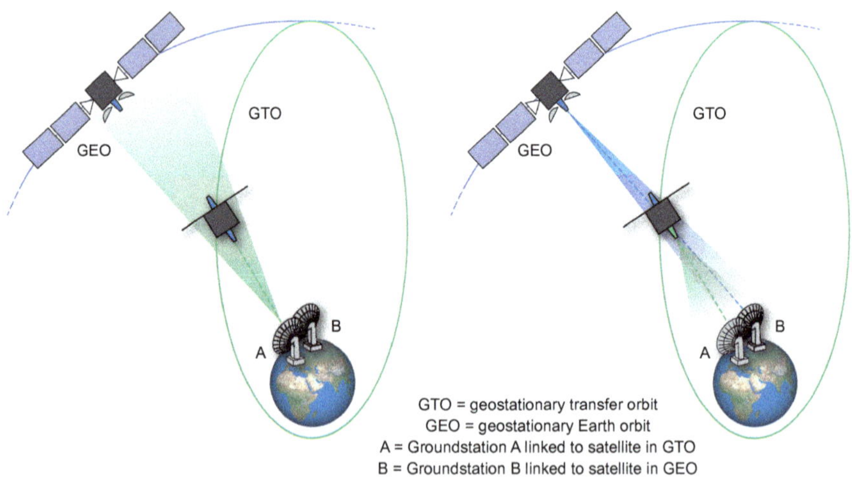

Fig. 18.12 Interferences with another spacecraft, left: uplink hinders, right: downlink hinders

interfere with the monitoring signal of the satellite being passed: the moving satellite is experiencing a hindrance or "hinder" (zone of possible interference).

Experience has shown that most hinders affect the TC link, i.e. the uplink of the LEOP control center ground station interfering with the payload or the TM/TC link of those spacecrafts in the line of sight of the uplink antenna. This is especially true if the hinder happens during orbit raising maneuvers. There, the received signal strength at the spacecraft is at its lowest value due to a typically unfavorable alignment of the spacecraft antennas tangentially to the orbit. This often forces the ground station to uplink at its maximum power, which will increase the risk of interference with other spacecraft in the line of sight of the ground station.

For a hinder, a cone with a two-degree aperture shall be respected by default. During a LEOP, the complete geostationary belt will be passed.

ITU rules require that interference has to be avoided as far as possible. As a good practice the control center of the moving spacecraft has to check the hindrances or "hinders" (possible interferences) and negotiate with the corresponding stationary satellite's operator a zone inside which countermeasures needs to be taken.

Interference can be mitigated by choosing a different frequency, by using a different ground station or to stop transmission completely.

For the first option, the design of the spacecraft needs to include receivers and transmitters that allow to select at least two different frequencies each for down- and uplink.

Most of the time, one hinder affects only one of the two available frequencies, hence the hinder can be mitigated by using the other available frequency.

For the telemetry link, the control center needs to configure the spacecraft to work in the allowed frequency during hindrances. This is done via time tag commands based on the output of the prediction software. For the telecommand link, this means to coordinate with the affected ground stations to use the frequency with no hinder for uplink.

The second option of having a second ground station available does not help much the closer the spacecraft transfer's orbit comes to the GEO belt, since the line of sight of both ground station with respect to the potential interfered spacecraft is practically the same.

The third option is the last resort: if all available frequencies are affected by hinders, no carrier uplink is allowed during the hinder's duration.

The task of hinder avoidance is in the responsibility of the mission control center of the moving spacecraft and all involved ground stations have to be considered.

Before the launch it is advisable to coordinate with all affected satellite operators: ITU provides a list of all transponder frequencies of a given spacecraft, which does not necessarily mean they will be in use. Permissions might be obtained to transmit within a defined hinder because it does not affect the services. Often, the hinder's definition can be relaxed: the default two-degree cone is a very conservative approach. One degree has shown to generate no noticeable interference and to provide some operational freedom.

In critical anomalous situations, where the loss of mission might be a consequence, the control center needs to inform the ITU and officially declare an emergency. From that time on until the control over the spacecraft is regained, a non-stop telecommand uplink may be used ignoring all possible hindrances.

Due to the high number of spacecrafts located in the geostationary belt, it is recommended to include the hinderance calculations in the event prediction software which is provided by the flight dynamics experts.

A complete precalculation and evaluation of the expected situation must be made while setting up the mission operations plan to find out if critical operations are affected and how the reduced contact times and the interruptions caused by hinders might affect the sequence of events. It has to be noted, that there may be unannounced spacecraft relocations between the evaluation and launch time, so the situation can be different. Calculations have also to be repeated after each replanning of the launch date or changes in the mission sequence in flight (e.g., orbit maneuver postponement).

18.6 Outlook to Future Developments

The future of the TCR subsystem and operations is predicted to lie in higher frequency domains and optical transmissions.

The proposed use of relay satellite systems may result in different antenna designs with steerable antennas that allow tracking of the relay spacecraft. The TCR subsystem may be supplemented by an optical communications system that also allows receiving telecommands and transmitting telemetry.

References

Bryant S, Berner J (2002) Operations comparison of deeps space ranging types: sequential tone vs. pseudo-noise. IEEE Aerospace Conference

Bullock S (1995) Transceiver system design for digital communications. Noble Publishing, 1-884932-46–0

ECSS 50-04C (2008) Space data links—telecommand protocols synchronization and channel coding, July 2008

Grafmüller B, Herschlein A, Fischer C (2005) The TerraSAR-X antenna system. Radar conference, IEEE International

Jansma PA (2011) Open! Open! Open! Galileo high gain antenna anomaly workarounds. Aerospace Conference, Big Sky, MT, USA

NASA JPL (2021) Voyager. https://voyager.jpl.nasa.gov/. cited June 14

Marr JC (1994) Performing the Galileo mission using the S-band low-gain antenna. In: Proceedings of 1994 IEEE aerospace applications conference proceedings, Vail, CO, USA

Reynolds MK, Reinhart MJ, Bokulic RS, Bryant S (2002) A two-way noncoherent ranging technique for deep space missions. In: IEEE aerospace conference

Van't Klooster K, Di Fausto M, Florio I, Rosa A, Robert B (2000) Antenna sub-system for Meteosat second generation satellites: modeling tools and needs. In: MMET 2000, Kharkov

Michael Schmidhuber works in the Department for Mission Operations at the German Space Operations Center (GSOC) where he is responsible for satellite operations training. He graduated as an aerospace engineer at the Technical University of Munich in 1994. The training activities comprise not only staff training but also courses for an external audience in the annual Spacecraft Operations Course at GSOC. Also, he is involved in the organization of the bi-annual international SpaceOps Conference.

Tarsicio Lopez-Delgado works for Airbus Defense and Space in Ottobrunn near Munich (Germany). Until 2021 he worked in the mission operations department at GSOC and was involved in the maintenance and operations of both the ground and space segments of geostationary missions. He graduated as a telecommunications engineer at the Technical University of Málaga, Spain in 2008.

Chapter 19
On-board Data Handling

Michael Schmidhuber

Abstract This chapter describes the operations of the spacecraft components that handle the distribution and processing of on-board data. This subsystem is referred to as on-board data handling (OBDH). Modern spacecraft are built around a powerful processing module that uses software to perform many functions. Complex data structures that need to be handled are defined to communicate with the ground stations, but also to handle on-board components. Using classic communications satellites and scientific satellites in low Earth orbit as examples, we present the basic components of OBDH in more detail, including typical operational tasks and considerations.

19.1 Definition of Subsystem

This chapter describes the operations of the spacecraft components that handle the distribution and processing of on-board data. This subsystem is referred to here as on-board data handling (OBDH), although other abbreviations are sometimes used, like the terms on-board computer (OBC) or satellite board computer (SBC). Nevertheless, these terms only describe the core computing component.

Often, the telemetry, commanding and ranging (TCR) signal transfer components are included in the definition of the OBDH subsystem, but in this book, they are described separately in Chap. 18.

In our definition, the subsystem includes the telemetry and telecommand (TM/TC) board, the on-board computer units (OBCs), and the on-board data distribution components. Although the design of the subsystem is extremely different between spacecraft, it is also a showcase example on how the use of international standards can communitize the use of spacecraft.

M. Schmidhuber (✉)
German Space Operations Center (GSOC), German Aerospace Center (DLR), Oberpfaffenhofen, Germany
e-mail: michael.schmidhuber@DLR.de

F. Sellmaier et al. (eds.), *Spacecraft Operations*, Springer Aerospace Technology,
https://doi.org/10.1007/978-3-030-88593-9_19

19.2 Fundamentals

In the following, we present the basic components of the OBDH in more detail. The examples given here are from a classic communications satellite series (Spacebus 3000 from Thales) and from low Earth orbit (LEO) scientific spacecraft (TerraSAR-X from Astrium and TET from Kayser-Threde). They allow demonstration of the basic principles found in some form on all unmanned spacecraft.

19.2.1 Subsystem Elements

Nowadays, spacecraft in LEO have a Global Positioning System (GPS) receiver for time synchronization and orbit measurements, as well as a mass memory module for storing large amounts of payload data. In addition, in many cases the attitude and orbit control system (AOCS) software can be integrated into the main processor.

Figure 19.1 shows an example of an OBDH system. It contains a fictitious collection of components from a classical satellite. Actual implementations will use subsets

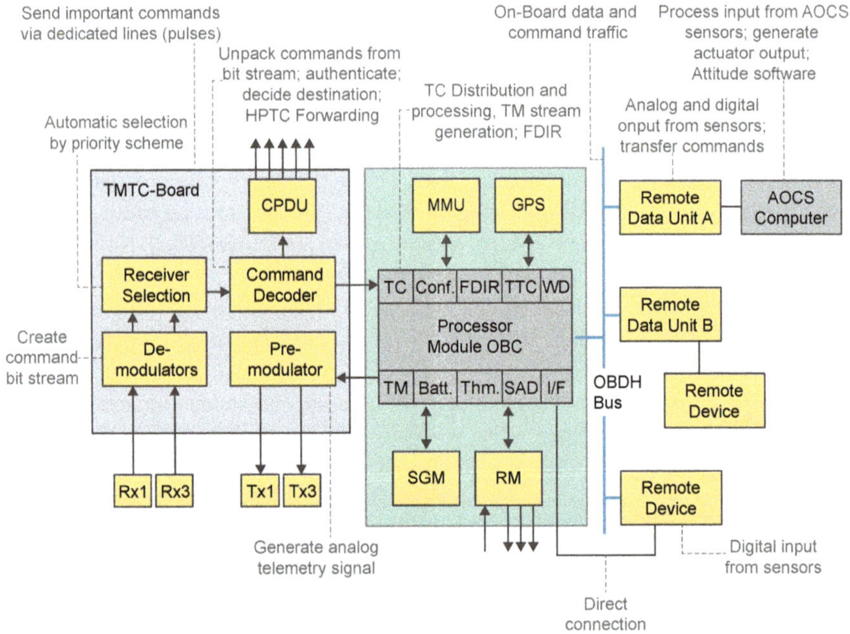

Fig. 19.1 Fictitious block layout of the components found on typical satellites. They are shown without redundancies. Abbreviations not explained in the text: receiver (Rx), transmitter (Tx), memory management unit (MMU), telecommand (TC), fault detection isolation recovery (FDIR), telemetry, tracking, and command (TTC), telemetry (TM), interface (I/F), watchdog (WD), safeguard memory (SGM), reconfiguration module (RM), solar array drive (SAD)

and variations according to the mission requirements and dependent on the design decisions of the manufacturer. The components are explained below.

The TM/TC board has the task of decoding telecommand signals into logical structures, transporting them to their destinations and, in the opposite direction, to generate the telemetry signal from the bits and bytes. It allows isolation of dedicated command streams such as high-priority commands (see next paragraph), and it includes authentication functions (see Sect. 12.3.4). Information about the result of telecommand execution is merged into the telemetry stream along with the standard telemetry coming from the on-board computers (OBC). The board is typically implemented as firmware running on robust and fast hardware. The generic design is standardized and largely independent of the spacecraft. The main standards for telecommand and telemetry are coming from organizations like the Consultative Committee for Space Data Systems (CCSDS), International Organization for Standardization (ISO) and European Cooperation for Space Standardization (ECSS) formats.

Not all telecommands are processed by the OBC. Low-level commands are already processed by the TM/TC-board. They serve very basic functionalities such as switching from OBC to backup, which must also function in emergency situations. The TM/TC-board contains logic that allows a limited number of these commands. This is called command pulse distribution unit (CPDU). The pulses are routed to their destinations via dedicated cables. This mechanism is also used by devices that require a higher electrical current than can be provided by normal data lines. They are therefore also referred to as high-power telecommands (HPTC). A common example is the ignition of pyrotechnic devices that can release a folded antenna structure. The corresponding command pulse requires a certain duration. This is usually an adjustable command parameter. If an HPTC command is not successful, it may be helpful to increase the pulse duration. It is not possible to time-tag HPTCs.

On-board Computer (OBC)

Sometimes referred to as the processor module (PM), the OBC is a programmable computer that contains the satellite-specific software used to manage complex spacecraft and payload functions. Some functions may be housed in dedicated, separated units: In Fig. 19.1, the attitude control functions are implemented in a separate computer module (AOCS Computer).

All software can potentially contain programming errors or require updates, e.g. to adapt to deteriorating hardware. Therefore, mechanisms are implemented to monitor the activity of the OBCs (see Sect. 19.3.2) and to allow the upload of new software (see Sect. 19.3.3).

Additional modules such as a safeguard memory, mass memory units and reconfiguration modules (RM) are located in the immediate vicinity of the processor modules. They are described in Sect. 19.3.2.

Mass Memory Unit

Only spacecraft with permanent, reliable ground contact can be designed without extensive data storage capability. Interplanetary probes and LEO satellites must store data until it can be transmitted to ground. Today, solid-state memory units have replaced devices, such as tape recorders that were used in the past. These memory units may be attached to or integrated with the OBC, or they may be located adjacent to the payload instruments. A mass memory device is not shown in Fig. 19.1 because a geostationary satellite was used as an example. See Sect. 19.3.5 for more details.

Data Bus and Data Cable Harness

Most spacecraft use a data bus system for communication between the OBC and remote devices. Standards such as MIL-1553 (U.S. Department of Defense 2018) or Spacewire (ECSS-E-ST-50-12C 2003) are used as the protocol. Selected data may be transported to the OBC via dedicated cabling if it is of particular importance in terms of safety or robustness, or if the remote unit does not support the bus protocol. Remote data units (RDU) are used to connect sensors, actuators to the bus.

With telecommands it should be noted that some commands are converted into electrical pulses that are transmitted via dedicated cabling. These are the high priority commands that are generated by the TM/TC board as described above.

Additional Components

Spacecraft in LEO contain a GPS receiver for time synchronization and orbit measurements. Please note, that GPS reception is conventionally only possible at orbital altitudes up to 7,000–8,000 km above the Earth. Nevertheless, modern geostationary satellites also include GPS receivers for use in transfer orbit and, with limitations, also in geostationary orbit.

For classical geostationary communication satellites, the AOCS software usually has its own processor located in a remote device.

Other remote devices can be used for functions such as power distribution and data bus interfacing.

19.2.2 Redundancy

Due to its key function, the OBDH system is usually designed with redundancy. This means that all components are duplicated to protect against failures. Depending on the design of the spacecraft, redundant equipment may be powered or may be placed in a standby mode. In any case, at least the redundant TM/TC Board is constantly powered, since it must be able to immediately take over telecommand reception from the active board.

The redundancy concept requires careful consideration of costs and risks. It is quite common that not every function is fully and independently redundant. The latter means that, for example, a particular redundant fuel valve can only be controlled by the redundant on-board computer (including telemetry), but not by the primary one. This significantly reduces the complexity of the system and thus the costs, but also reduces the degree of redundancy: If the redundant fuel valve has to be used, the on-board computer also has to be switched over and thus also loses its redundancy.

These constraints must then be considered in the in-orbit test (IOT) campaign and also in the contingency procedures.

In technology demonstrations, projects have tested more complex and advanced layouts. The TET satellite, for example, has a layout with two pairs of computing units that can monitor each other and swap the roles of "worker" and "monitor".

Typically, redundancy is for safety purposes. However, in some cases the redundant units can be used for additional activities such as generating a second, independent telemetry stream.

19.2.3 Telemetry Parameters

The elements that contain information about the spacecraft to be transmitted to Earth are called telemetry parameters (or sometimes telemetry points). They may contain status information (such as ON/OFF flags), numeric data (such as temperatures or counters), or more complex binary data. Their value or meaning has to be coded into a binary format. In most cases, it is important to save bandwidth and therefore the smallest possible coding is used: Flags can be 1-bit values, the length of the bit pattern of integer numbers depends on the value range of the corresponding parameter. Measurement values use either an interpolation table or use a standard real number format like IEEE 754. The used encoding is described in the spacecraft database.

The inner structure and meaning of complex binary data chunks are not defined in the mission information (data) base (MIB). This data is simply dumped to the ground system where it will be further down in the processing chain. An example may be payload image data intended to be evaluated at a payload data center or data files as described in Sect. 19.3.5.

19.2.4 Telecommands

Telecommands are the instructions that allow controlling the spacecraft from ground. They are also defined in the spacecraft database. A telecommand has an identifier and may have a set of command parameters that modify or specify its behavior. Commands can switch devices, set values in registers, or transport binary data segments.

An important part of the commands is the "address" part which describes which part of the OBDH should receive the command. This is described in Sect. 12.3.5. The remaining part of the command is the command data. It is composed of the previously mentioned command parameters and the command identifier.

19.2.5 Spacecraft Database

All information about how commands are packed into the uplink stream and how they are decoded from the downlink stream is defined in the spacecraft database. Its elements include the complete set of available telecommands, the complete telemetry item set, the definition of packets or frames, or the scripts running on board. In most cases, it also contains so-called derived parameters. These items are treated like telemetry, but their values are basically the result of ground system calculations that use telemetry or ground processing information as input. The formulas behind the calculations are usually developed by the flight team according to the manufacturer's specifications and added to the database.

For telemetry parameters, limit values can be defined in the database. Violations of these limits can be indicated in the telemetry display system with a visual or audible alarm message, but it can also be subsequently checked in an offline system (e.g. for dumped telemetry that has not been processed by the real-time MCS). Alarm staging is possible if additional limit sets are provided that trigger only pre-warnings. Multiple sets of limits can be provided, allowing limits to be tailored to different mission phases. "Sum alarms" can be defined that indicate a violation of any of their associated parameters. This is necessary to monitor modern spacecraft, which typically have myriads of telemetry parameters.

The database may also contain information such as flight procedures and telemetry display format definitions. Because of its central function for the manufacturer during the preparation and for the operations team during the mission execution, the database is also called mission information (data) base (MIB).

The MIB is created by the manufacturer and embedded into the spacecraft's computer software on the one hand and made available to the control center in an agreed exchange format on the other.

The operations team must validate its ground system and operations using the MIB. The operations team should ideally provide feedback to the spacecraft manufacturer on the usability of the MIB. A common problem is the naming convention of parameters and commands. Each entry, telecommand or telemetry point, is identified by a unique database identifier.

For operational safety, it is important that the naming of parameters is systematic and ergonomic and from an operator's point of view. This may differ considerably from the manufacturer perception.

For example, telemetry parameters are sometimes assigned to mnemonics that are very long and of varying length. This makes flight procedures and telemetry display systems cluttered and confusing. A good operational mnemonic has between 5 and 10 characters and follows a stringent system. An early dialog between control center and manufacturer is advisable to reach a common approach.

During a mission, the database is usually updated to include new features, correct errors, or respond to unplanned events. However, it is of utmost importance that the versions of the database on ground, on board and in the documentation are identical. Therefore, strict and formal change processes and configuration control are required.

19.3 Managing Data Handling On-board

19.3.1 General

Under normal conditions, the operation of the data handling subsystem is mostly embedded into the routine tasks of the mission. However, during mission preparation and in the case of anomalies or special campaigns, the complexity of modern spacecraft makes this a broad and challenging field. The work associated with the spacecraft database tasks described in Sect. 19.3.3 may require several people, including the data-handling expert. During mission preparation, the control center's monitoring & control system (MCS) software must be customized; a task strongly related to knowledge of the on-board data handling system. During mission execution, the functions of the MCS must be understood in order to take full advantage of the spacecraft's abilities, especially in non-nominal situations. The following sections explain the basic operational tasks associated with the OBDH.

19.3.2 Safeguard Mechanisms

Fault Detection, Isolation and Recovery (FDIR)

To protect a spacecraft from damage or mission loss, FDIR mechanisms are typically used. It is vital for safe operations to understand this mechanism and being able to interpret the telemetry correctly and quickly is essential for safe operations, as this knowledge is usually needed in critical situations.

FDIR functionality may be distributed across multiple components such as the reconfiguration module (cf. Sect. 19.2.1) and the OBC. The layout will be different for each satellite platform, using different approaches and levels of severity. Functionalities may be implemented via hardware or software, or both. Typically, errors that do not pose a threat to the health of the spacecraft will not be handled autonomously and will be left to ground (control center) interaction. Errors that can be corrected by the unit itself are normally reported but otherwise ignored by the higher-level FDIR, e.g. the error detection and correction (EDAC) of memory bitflips using available redundant information. However, this philosophy is currently changing with the availability of event action services as described in the PUS (ECSS 70-41A 2003).

Basically, FDIRs monitor system properties (telemetry values, bus voltages, hardwired signal inputs, keep-alive signal, etc.) for deviations outside the allowable limits report the incident and execute a predefined action: The surveillance defines a list of one or more health-check parameters (HCP), taken either from normal on-board telemetry buffers or from dedicated signal lines. Their values are monitored at a specific frequency (e.g., at 1 Hz) and compared with defined and possibly configurable thresholds.

Depending on the device and its criticality, a single or repeated (after n consecutive) parameter limit violation of one or more parameters (e.g., majority voting) then triggers the FDIR situation and the corresponding action, e.g., switch to a redundant device.

This event is usually recorded in some sort of error log, but in some cases must be inferred from the telemetry by the ground team: In emergencies, the memory including the log could be cleared by OBC reboots, the telemetry was not set to record, or there was no ground visibility. For this reason, geostationary satellites, for example, configure their telemetry to continuously send out the error log file. In the worst case, the error log file may be the last signal the control center sees from the spacecraft.

Usually, it is possible to configure the monitoring and actions of the safeguard mechanisms. In some cases, notifications are to be issued, but no action is to be triggered. When a redundant device is activated by an FDIR, there is usually a different and limited FDIR on the redundant device. This is usually to prevent the system from switching back to a faulty prime device in the case of a second or persisting external trigger. For certain operational tasks, is necessary to disable some FDIRs (e.g. during an orbit maneuver). This is typically covered by the appropriate flight procedures.

Recovery actions can consist of one or more discrete steps. It is common to use a "macro" functionality that triggers a sequence of discrete commands on board. This may again be implemented as software (potentially as an on-board control procedure wrapped in a PUS service, cf. Sect. 19.2.4) or hardware. The device for the latter case is called a reconfiguration module (RM). (Electro-)mechanical registers send out command signals via dedicated cables. Software macro command mechanisms usually use standard command packets (like those sent from ground).

Table 19.1 shows an example of an emergency macro command sequence for a geostationary satellite. In many emergency situations, the spacecraft is placed in a stable, safe mode by the AOCS computer. This mode is called Sun-acquisition mode (SAM) because it reorients the spacecraft to the Sun and starts a rotation about that direction. An FDIR mechanism within the OBC detects this mode and adjusts the satellite to the new situation. The steps shown are executed. The real sequence contains more than 50 discrete commands and delay instructions. If redundant commands are present, both are used to make the sequence more robust.

The recovery sequences can be reprogrammed. However, this possibility is very rarely used; usually this is the domain of the manufacturer. Changes should only be made carefully and after consultation with the spacecraft manufacturer. However, the correct content of the sequences should be checked regularly.

The operational tasks are to monitor for the occurrence of an FDIR. This daily monitoring can be done largely by automated telemetry alerts and configuration checks (concheck). Note that for LEO satellites, telemetry checks must be performed offline because large portions of the telemetry are stored on board and dumped in

Table 19.1 The main steps of a Sun-acquisition mode (SAM) macro command sequence

Step	Title	Description
1	Disable all commands triggered by ground or higher-level macro commands	Avoid contradictory actions
2	Switch off payload	Service is interrupted, avoid unnecessary power consumption, avoid uncontrolled RF output
2	Adapt battery charge control	Adapt to new electrical situation
3	Start thermal regulation	Adapt to the new thermal situation
4	Activate substitution heaters	Since the payload is off, use heaters to avoid thermal imbalance
5	Switch off payload related equipment	Reduce power consumption
6	Reconfigure the bearing and power transfer assembly (BAPTA) to the reset point	Rotate solar arrays to z-axis (= Sun-pointing) and stop rotation
7	Reconfigure telemetry	Reset to emergency configuration (display the Error log file, show basic telemetry)

compressed format. They must be unpacked and processed on ground. This is usually done in separate offline data processing systems.

Safeguard Memory (SGM)

The components and software used for spaceflight applications are very durable and well tested. The FDIR mechanisms and redundant design also contribute to this. Nevertheless, during the lifetime of a spacecraft, it can and probably will happen that the OBC has to be restarted either by ground command or after an anomaly has triggered an according FDIR.

As with a normal PC, the random access memory (RAM) is cleared on restart. The OBC software is loaded from a programmable read-only memory (PROM) or electrically erasable programmable read-only memory (EEPROM) and is configured for a default spacecraft configuration (typically the launch configuration). The configuration set includes items such as momentum wheel selection, redundant device selection, etc.

If a redundant device was activated during the mission due to a failure of the prime device, this will not be considered by the default software configuration, which can lead to re-triggering of the reboot and hence to a dangerous situation for the spacecraft. Therefore, the OBDH design includes a device called safeguard memory (SGM). This is a protected register. It is not erased during power failures and is specially protected against radiation and other environmental influences. From this memory, the OBC can take current settings that may differ from the default settings. Depending on the complexity of the spacecraft, these can be a few basic settings or extensive data ranges. Configurations that are relevant at the hardware level can be stored in electromechanical configuration relays. This may be the on/off status of the authentication mode of the TM/TC board.

The content of the SGM must be updated after each relevant reconfiguration of the spacecraft. The corresponding activity is usually part of the flight procedures. Depending on the capabilities of the OBC, it can be a tedious, time-consuming task. Nevertheless, the update should not be delayed too long. After the update, it is a good practice to dump the SGM content to ground to verify correct spelling. Also, the operations team should have enough information to interpret the contents of the memory dump. The ability to change the contents of the SGM is protected and special commands are required to allow and then disallow write access.

19.3.3 Maintenance of the On-board Software

Modifications to the Mission Information Base (MIB)
Two different aspects of the MIB must be considered: the implementation on the spacecraft and the implementation on the ground. Both must be coherent at all times. Nevertheless, it is not necessary to update the on-board version every time the ground version is changed, since the on-board part can be considered as a subset of the ground MIB, as we explain below.

The on-board side has the task to define how telecommands are unpacked from the up-link stream and to which destination device they should be forwarded. A possible change could be to reduce the list of accepted telecommands. This may be a measure to protect against unwanted commands from the ground. Changing the interpretation of telecommands is also possible, but only in conjunction with an OBC software update. For example, if the on-board software used to manage the time-tag register is updated and the new software may be able to accept new telecommands, the on-board database must be updated to accept the new command pattern.

The most common change, however, is a change in the definition of telemetry packages. This is likely to be the case for new spacecraft without sufficient operational experience, where the original definition may need to be adjusted after some time or after anomalies or failures, when the available observation capability may not be sufficient. Since telemetry packets can be very long and consume a large amount of the bandwidth, it may be better to define a small diagnostic telemetry packet for the situation at hand, which can then be sent down at high rates. In all cases, the ground database must be updated accordingly.

The on-board database is mostly implemented as an integral part of the on-board software and cannot be changed separately. Some aspects such as the definition of new TM packets can be changed via PUS service 3 (Reporting Service, cf. Sect. 12.3.4). The persistence of updates and service requests after a reboot of the OBC must be checked.

The database implementation of the ground system includes many aspects of the ground system that are adaptable and do not affect the space side:

- Inclusion of commands that are already available on board but were not originally approved for the control center

- Introducing new commands that are simply a redefinition of complex, existing commands with many specific parameters into a simple short command or that only allow a reduced parameter range
- Changing the calibration of parameters (e.g. status texts) for telemetry items or telecommands
- Adjusting telemetry ground limit entries
- Introduce or modify derived parameters or ground parameters (e.g. ground station telemetry)
- Introducing telemetry parameters for better and safer operability (e.g. defining the bits of "status words" as separate parameters that gives them text calibrations)
- Modifications of display page definitions or command sequences embedded in the MIB.

In most cases, the changes must be coordinated with and approved by the spacecraft manufacturer. Strict versioning and configuration control should be in place. Changes impact flight procedures and documentation. The effort involved can be significant. Database tools can help to keep flight procedures and the database consistent.

Whether the modification of limit entries (alerting thresholds as described in Sect. 19.2.5) in the active MCS is allowed during runtime under configuration control is controversial. Depending on the working practice established in the control center and especially for geostationary satellites, a relaxed approach is possible. Remember that the alert mechanism is primarily for situational awareness for the first-line operator. It may be useful to temporarily adjust the limits to avoid overloading the operator's attention. A compromise is to use mechanisms to temporarily overwrite the database while the MCS system is running. This preserves the original database in the configured system. This situation should be remedied during the next regular database update. Database changes are considered critical and should be tried beforehand on a test or simulation system.

OBC Software Maintenance

When the manufacturer provides a new software version for the OBC, the necessary operational tasks are uploading the software data, managing the software image on board and booting.

Typically, the software is broken down into smaller pieces and embedded into data transfer commands. All uploaded data must be checked for completeness and correctness when reassembled on board by using checksums or by dumping all data back to ground with subsequent comparison to the original. The software must then be transferred to the memory area of the OBC that will be used during the boot process. It must be ensured that the original software can still be accessed in case a problem occurs with the new software during the boot process that makes communication with the spacecraft impossible. There are several possible methods to prevent this, and the spacecraft manufacturer must provide an appropriate procedure. If the initial boot and in-orbit test of the software are successful, the new software must be configured as permanent boot software.

All of these aspects can be handled through advanced memory management services, such as those provided by the PUS, or through the use of conventional telecommands.

19.3.4 Execution Management

Time Tagging

Telecommand time tagging is used to issue preloaded commands at specific times when either the spacecraft is out of sight of a ground station, or there is a possibility of losing command link to the spacecraft, or when precise execution timing is required.

Almost all telecommands can be time-tagged and the ECSS Standard (ECSS 50-04C 2008) provides a method for this. Exceptions for this are the high-priority hardware-decoded telecommands, which are handled by the TM/TC board. When a time-tagged telecommand (TTTC) is received, it is captured by the time-tag software (or schedule handler for the PUS spacecraft) and entered into a memory table. This table contains the defined execution time and the current telecommand. If the on-board time matches the execution time, the command is released to its destination. The time-tag table (or schedule) may have limitations and special features. On some spacecraft it is possible to have more than one TTTC for the same execution time. Some spacecraft only allow time-tags to be added at the end of the list and in chronological order, forcing a complete deletion and resending of all TTTCs if a new one is to be inserted in the middle. Some designs use register numbers to store and address memory locations.

The extent and scope of use depends on the mission requirements. A geostationary satellite has almost permanent ground contact and rarely requires time-tagging:

- To put potential rescue commands a bit into the future to protect against possible uplink path dropouts (including ground failures). These TTTCs are typically cleared or updated when no failure occurs;
- To ensure nominal commanding to protect against loss of the uplink signal;
- For precise timing of commands (a rather rare thing on classical communication satellites).

Consequently, geostationary satellites need to store only a small number of time-tagged telecommands (less than 100).

Spacecraft in low Earth orbits with only few ground contacts or interplanetary missions with long signal propagation times depend heavily on this capability. Therefore, they have hundreds or thousands of slots for TTTCs. Two aspects are of particular importance here:

- To switch the ground link (telemetry) during the predicted contact times to optimize the available contact time. An example case is described in Sect. 18.5.4);
- For payload control. This can be the timing of payload operations and payload data collection.

The mission schedules of LEO satellites and interplanetary probes are quite complex. The rudimentary operations provided by previous spacecraft platforms made operations very challenging. For this reason, PUS service 11 (Scheduling, ECSS 70-41A 2003) introduced advanced methods for doing so. This standard provides for defining partial schedules than can be activated, edited, timed, and deleted separately. It is even possible to interlock schedules to make the transition to the next TTTC dependent on the success of the previous step and so on.

To keep track of schedule activity, it is recommended to use a ground tool that models the schedule and can compare this model to a schedule dump during the next visibility period. In addition, advanced ground systems allow stored and dumped command execution confirmation (or rejection) events to be compared to the command system telecommand history and indicate any discrepancies.

The time-tagged commands schedule are, of course, dependent on an available on-board time source. If the time becomes unreliable for any reason (e.g. an OBC reboot), the time-tag registers are cleared.

On-board Control Procedures

Modern spacecraft allow the storage of complete command sequences on board. This is similar to the schedule mechanism without a predefined execution time. The procedures remain in memory and can be reused and adapted. This can drastically reduce bandwidth requirements in the uplink channel if repetitive procedures are very long and only a few key parameters need to be changed. There is also a service for this in the PUS standard (PUS 18). Together with event monitoring and action services (PUS 5, 12 & 19), it gives the spacecraft a high degree of autonomy.

19.3.5 Mass Memory Management

The management of mass storage devices is especially important for missions in low-Earth orbit or for extraterrestrial missions that need to store payload data.

In the past, data were typically recorded on tape devices, such as on the Voyager mission to the outer solar system. Only in rare cases, mechanical hard disk storage drives were used (Bussinger et al. 1993). More recently, the use of solid-state storage has predominated. Various storage concepts have been developed.

Ring Buffer

A ring buffer is a robust and simple mechanism. That is ideal for continuous data streams that are expected to be erased at a predictable regular interval, and also when their handling character is first in, first out (FIFO). Here, the physically available (linear) memory is modelled in a ring-shaped manner, creating a seemingly infinite memory area (Fig. 19.2).

There are typically two or more pointers:

- *Write pointer*. This is the address of the physical storage cell that will be used for the next write operation. After writing a data set, the pointer is incremented (and,

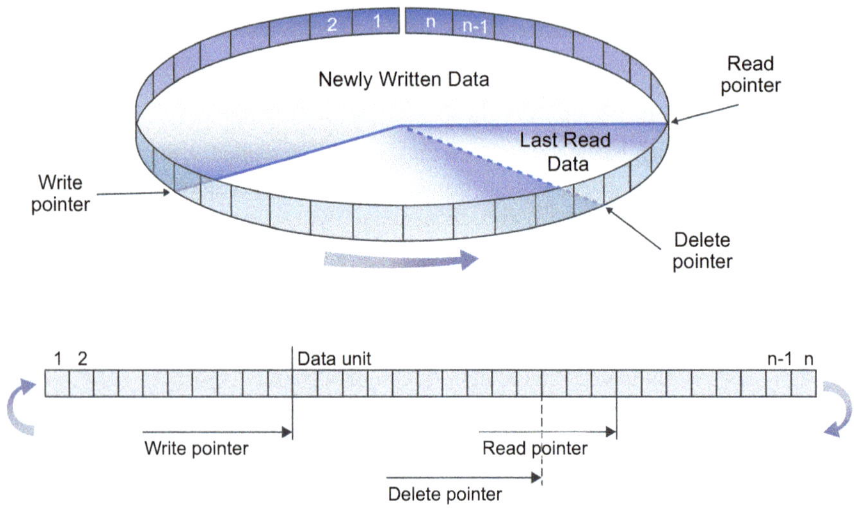

Fig. 19.2 Ring buffer model. The linear memory cells are accessed as a circular range. Each time a pointer reaches the end of physical memory it is reset to the beginning

of course, flipped to the beginning of the range at the end of the physical memory range). Whether the write operation is stopped when the delete or read pointer is reached or exceeded is a matter of mission design. Monitoring the distance between the pointers may be necessary for the ground control team. Possible measures against this situation would be to throttle the write data rate or provide additional downlink time.

- *Read pointer*. This is the address of the physical storage cell from which the next data is read. After the read operation the pointer is incremented (or flipped to the beginning of the range as described above). A possible overflow of the write pointer is not a problem, but leads to a duplication of already downloaded data, which has to be handled in the ground system. The read pointer can be pushed through the write pointer. The read operation can be triggered from ground or it can be an automatic continuous process. There can be multiple read pointers that can be used by different users.
- *Delete pointer* (optional). This can be a pointer that follows the read pointer and indicates areas that the ground control center has successfully retrieved and approved for deletion. The delete pointer may be a lock for the write operation. In other implementations, this function may be performed by the read pointer. However, if a mission cannot risk losing data, a ring buffer concept may not be the ideal solution.

The application process and the operators on the ground (or the mission planning process) need very little knowledge about the storage. The difference between read, delete and write pointer defines fill levels. Furthermore, it will usually be possible to change the read pointer, e.g. to repeat an unsuccessful dump action.

File System

A file storage system replicates data structures similar to the familiar files systems in PC computers. These files are linear data storage areas that have a certain common character. Files can be created, named, filled, closed, read, copied, modified, concatenated, and deleted on board. They can be uploaded from the ground and downloaded (dumped) to the ground. They have a unique handler or name and can be tagged with common attributes such as file creation date, last modification date, and access rights or other meta data. A very useful application for this is handling of on-board software patches or instrument measurement data (images, data-takes) that are piecewise. It is also possible to fill files with continuous data-streams such as housekeeping telemetry, but this is currently uncommon and has no particular advantage over a linear buffer or a ring-buffer. The operational tasks required can be any of the file methods mentioned above. In some cases, they can also be performed by automatisms.

Linear Storage

Linear storage is the third method for large storage devices, especially when used for housekeeping telemetry or continuous measurements. It is preferable to the ring buffer when the stored data must not be accidentally overwritten. Typically, pointers are used for reading and writing here as well, but random-access methods can also be used. If the read, write and delete management is fully automatic on board, then there is no difference to a ring-buffer. Thus, the strength of linear memory lies in random access and protection of written data, which otherwise implies manual intervention. Also, by definition, a linear memory is full at a certain point in time and a new memory must be used until the old memory has been fully read and can be flushed. Operational tasks consist of directing the data stream to a selected memory, monitoring the fill level, controlling the read operation (possibly repeating unsuccessful dumps), and setting the pointers to zero when a memory is full and has been read completely.

Advanced mass memory concepts can define multiple instances or combinations of these storage methods. These storage areas (sometimes called "stores") can be filled with data from different sources, sampled at different intervals, or targeted at different ground destinations.

The management of the data that is written to the memory usually involves the mission planning and the data handling experts. They must define what is to be stored and at what rate. This must be balanced with the available storage and the time until the next downlink opportunity. The process downloading the stored data (payload or housekeeping) is called dumping. Higher data rates can be used. To save downlink bandwidth and mass storage, it is possible to use data compression techniques (Evans and Moschini 2013).

Storage maintenance is usually not required for solid-state storage. Typically, it will be equipped with error detection and correction (EDAC) mechanisms. The reels of tape recorders had to be rewound regularly to avoid tape sticking, and the data sections had to be reformatted and rewritten (in different arrangement) regularly to avoid magnetic fading and copy effects.

19.4 Summary and Outlook

The OBDH and its operation have changed fundamentally over the past few decades as on-board storage and processing capabilities have increased dramatically. This chapter has presented the basic activities associated with the operation of the data handling subsystem.

It is obvious that some trends will intensify in the near future. There will be an even higher demand from end users for data storage and for data protection. Mission complexity and shorter development cycles will require more flexibility and autonomy on board. This will lead to more powerful computer hardware and advanced software on board and thus, as on ground, to a greater need for in-flight software maintenance and customization.

The standardization of data access and maintenance methods, as described in the Packet Utilization Standard (ECSS 70-41A 2003), serves to standardize the systems and methods. This process is not yet complete, but has already shown positive harmonization effects that ultimately benefit all parties involved.

References

Bussinger SD et al (1993) Spacecraft erasable disk mass memory (EDMM). In: 7th annual AIAA/USU conference on small satellites

ECSS 70-41A (2003) Telemetry and telecommand packet utilization standard. ECSS

ECSS 50-04C (2008) Space data links. Telecommand protocols synchronization and channel coding

ECSS-E-ST-50-12C (2003) SpaceWire. https://www.esa.int/Enabling_Support/Space_Engineering_Technology/Onboard_Computers_and_Data_Handling/SpaceWire. Cited 26 May 2021

Evans D, Moschini U (2013) Ten times more information in your real-time TM. In: Schmidhuber, Cruzen, Kehr (eds) Space operations: experience, mission systems and advanced concepts. AIAA. ISBN 978-1-62410-207-3

U.S. Department of Defense (2018) MIL-STD-1553C–interface standard: digital time division command/response multiplex data bus. https://quicksearch.dla.mil/qsDocDetails.aspx?ident_number=36973. Cited 26 May 2021

Michael Schmidhuber works in the Department for Mission Operations at the German Space Operations Center (GSOC) where he is responsible for satellite operations training. He graduated as an aerospace engineer at the Technical University of Munich in 1994. The training activities comprise not only staff training but also courses for an external audience in the annual Spacecraft Operations Course at GSOC. Also, he is involved in the organization of the bi-annual international SpaceOps Conference.

Chapter 20
Power and Thermal Operations

Kay Müller, Sebastian Löw, and Sina Scholz

Abstract This chapter provides an overview of various power sources used on satellites and the high-level physical principles they are based on. The same is done with the design elements related to maintaining proper temperature conditions in which a satellite's electrical components can function. Several approaches to operational tasks are given in the second and third section, including the handling of batteries.

20.1 PTS (Power and Thermal System) Design Aspects

20.1.1 Power System

All electrical systems require electrical power in order to function. For most stationary electrical applications on Earth this power is generated by an electric generator which basically consists of a conductor rotating in a magnetic field, using the physical effect of electromagnetic induction. The means to move the conductor are often obtained by converting heat into mechanical energy. The heat originates from chemical energy contained in combustible substances such as coal or oil or gas. Another possibility is to use mass defects of atomic nuclei.

For convenience, many appliances are physically separated from their energy supply. Energy is provided by a power plant located kilometers away. The energy is then transported to its various consumers via the electrical grid. Whenever mobility of an object is required, the means to provide power needs to become smaller and lighter in order to be able to be taken along as efficiently as possible. As a result of the mobility, the separation between the power source and consumer is no longer present.

K. Müller · S. Löw (✉)
German Space Operations Center (GSOC), German Aerospace Center (DLR), Oberpfaffenhofen, Germany
e-mail: sebastian.loew@DLR.de

S. Scholz
German Space Administration, German Aerospace Center (DLR), Bonn, Germany

F. Sellmaier et al. (eds.), *Spacecraft Operations*, Springer Aerospace Technology,
https://doi.org/10.1007/978-3-030-88593-9_20

Power plants, cars and aircraft convert chemical energy contained in fuel into thermal and then mechanical energy via combustion. Their engines are significantly scaled down versions of the large power plants. For even smaller devices such as cell phones and laptops, batteries become the power supply of choice since they can store a limited amount of energy.

For space missions, the concept of mobility is essential. Spacecraft cannot be refuelled like aeroplanes or cars nor can they be recharged by plugging them in like cell phones. Therefore, all of the required elements to provide electrical power have to be installed on board and need to be designed to be sufficient throughout the desired mission time dependent on the mission profile. All approaches to spacecraft power systems are variations of this theme with the following similarities:

- Electrical power usually needs to be generated on board (*Primary Energy*).
- It needs to be conditioned, that means to be converted and regulated, to suit the needs of the various units, specifically operating voltages, and currents. Some units might require a constant voltage for example. This usually leads to separate power busses, regulated and unregulated, which connect their respective units (*Power Management and Distribution*).
- The spacecraft might temporarily need more power than the primary power source can provide and must also be able to function when no primary power is available at all for a limited time. Therefore, the possibility to store energy must be implemented. This is done using batteries which results in charge control of some sort being necessary. Stored energy can then be fed back into the system at a later point (*Storage*).

A schematic drawing of a spacecraft's power system can be seen in Fig. 20.1.

Energy Sources and Storage

Energy supply is based on physical or chemical effects, as listed in Table 20.1.

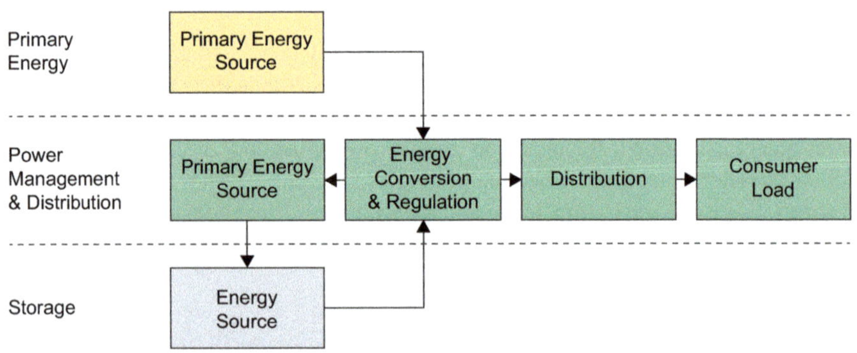

Fig. 20.1 Main functional blocks of a power supply system

Table 20.1 Physical phenomena used for the generation of electrical power in spacecraft

Physical or chemical effect	Application
Photons increase the energy level of electrons in semiconductors	Solar panel
Chemical reactions between electrodes and electrolyte move electrons from one electrode to the other	Battery, fuel cell
Temperature gradients between different electrically conductive materials create an electrical potential (see beck effect)	Radioisotope thermoelectric generator (radioactive decay as heat source)
Nuclear fission of heavy elements produces less heavy elements with lower mass per nucleon. The difference in mass is released as kinetic energy of the fission products	Nuclear fission reactor

Solar Panels

One widely used option of power generation is the photovoltaic effect. The principle of photovoltaics, literally "voltage through light", makes use of the fact that electrons in semiconductors can be excited by photons—such as solar irradiation—to a higher energy level but cannot "go back" due to a built-in junction in the semiconducting material. As such, a solar cell is a charge separator. The electrons can then move through an external circuit driven by the electrical potential created by the separation of the charges. The potential and thus the voltage are only dependent on the semiconductor materials that are used.

For initial estimates for satellites orbiting the Earth, a solar irradiation of $S_{Earth} = 1367\ \text{W/m}^2$ is usually assumed. Given a desired electrical power P_{Earth} to be generated and an efficiency μ one can approximate the required solar array area to be

$$A = \frac{P_{\text{Earth}}}{\mu S_{\text{Earth}}} \tag{20.1}$$

The solar power per square meter at other distances r from the Sun can be computed by including the square of the distance ratio:

$$A = \frac{P}{\mu S_{\text{Earth}} \left(\frac{r_{\text{Earth}}}{r}\right)^2} \tag{20.2}$$

Thus, at twice the distance from the Sun only a quarter of the power is available per square meter; therefore, the area of the solar panels would have to be increased accordingly in order to reach the required electrical power.

It was already mentioned that the chosen semiconductor material defines the voltage which is delivered by the cell. In order to generate a specific target voltage, a solar panel consisting of multiple solar cells are electrically connected in series, so that the voltages of the individual cells add up. One such series of solar cells is referred to as a "string." To reach a certain desired current output, multiple strings are connected in parallel. In order to avoid damage of the solar cells by a reversed current (which occurs when a battery is discharged), each string is usually decoupled from the bus using a diode. Modern solar cells have efficiencies of more than 30%. Since there is no atmospheric filtering of sunlight in space, they can generate more power than similar cells on ground. The panel temperature has a rather large influence on efficiency. Low temperatures yield more power output. This is the case because at lower temperatures the band gap is larger which means electrons that have been excited by incoming photons and have managed to cross the gap have received relatively more energy, which in turn results in a higher voltage. The loss of power around the power maximum is about 0.5%/K. High temperatures of up to about 130 °C cannot be avoided though, as the solar panel is naturally exposed to solar irradiation.

With increasing age, degradation will also become an issue. A loss of efficiency of 0.5% per year is assumed for crystalline cells even though the real value will usually be below that. For thin film solar cells the degradation is much faster in the first 1000 h, up to 25%. Almost no degradation occurs after that (Deng and Schiff 2002).

Electrochemical Cells—Batteries

Similarly to a solar cell, an electrochemical cell generates a voltage by separating charges. The process by which that separation occurs is, in this case, a chemical reaction. The energy needed for that charge separation is contained in chemical energy. There are electrochemical cells in which the chemical reaction can only work in one direction. The original state can thus not be restored and the cells can only be used once. These are called primary batteries. A type of battery whose chemical reaction can work in both ways is called a secondary battery. In these, the reversal of the current restores the chemical substances to their previous state (charging) so that the charge separation can start anew (discharging). The familiar term battery refers to the combination of several electrochemical cells. Like solar cells, electrochemical cells are connected in series (a string) to generate a higher voltage than each cell individually could provide. These strings are then connected in parallel in order to reach a desired current. A main figure of merit for batteries is the amount of ampere hours they can hold. Almost always, electrical energy generated by solar cells is being used to charge the batteries so that the energy that is stored in the battery can be used at a later point when the Sun is not illuminating the panels or more power is needed. A surplus of solar-electrical energy is usually designed into the power system so that the solar arrays can supply both power for the spacecraft loads as well as power for charging the battery at the same time. The battery also needs to be designed to ensure that it alone can supply enough power for the spacecraft loads when no solar irradiation is available. Peak loads that can be much higher than the usual ones need to be considered as well.

In principle, however, the charging/discharging process cannot go on indefinitely. Since the battery hardware itself, specifically the electrodes, take part in the chemical reaction, the electrode material degrades with time. Degradation can occur both chemically as well as mechanically. For instance, the electrode material dissolves in the electrolyte and the electrode materials become damaged due the constant in- and outflux of ions. Degradation occurs in all types of batteries. To some extent, measures can be taken to limit these effects. Accurate temperature control for instance can help to limit side reactions that produce no electricity and only degrade the electrodes. Performance can also be improved by making sure all the cells perform similarly (cell balancing). If some cells have a lower capacity than others, these cells are already fully charged at an earlier point in time. Since the rest of the cells are not full yet, the charging process continues while the "bad" cells are being overcharged, which in turn become warm and affect the battery as a whole.

An older type of battery that is still in use is the NiH_2 battery. It is well known for being long lived as well as rather robust over a relatively large temperature range. However, it suffers from a lower energy density (J/kg) and self-discharging while the energy efficiency is 85%. A more modern battery that does not have those downsides is the lithium-ion battery. Such a battery is comparably lightweight, has a high energy density and experiences virtually no self-discharge. However, it is sensitive to high temperatures and high depths of discharge. Its efficiency is between 80 and 90%.

Electrochemical Cells—Fuel Cells

In a conventional battery the chemical reaction takes place between the electrolyte and the electrode material. Another type of electrochemical cell is the so-called fuel cell. It can be distinguished from batteries in that the chemically reacting substances are not part of the battery structure such as the electrodes, but are added from the outside like fuel, therefore the name. Fuel cells therefore do not suffer from degradation, but they have the disadvantage that the reacting substances have to be stored in additional tanks. Almost exclusively, hydrogen and oxygen are used with hydrogen as the fuel and oxygen as the oxidizer. In this case the efficiency is up to 60%.

Radioisotope Thermoelectric Generators

While solar arrays in combination with batteries make sense when a spacecraft is relatively close to the Sun, the required area for generating the same amount of power increases with the square of the distance to the Sun, as indicated by Eq. (20.2). Therefore, most spacecraft for explorations beyond the orbit of Mars use a different power supply, the radioisotope thermoelectric generator (RTG). The device contains radioisotopes, usually Pu-238 due to its long half-life of 87.7 years. The decay of the plutonium heats up the surrounding casing, causing a temperature gradient which generates an electric voltage according to the Seebeck effect. This system has several advantages as it can provide power for decades, has no moving parts and is therefore autonomous and maintenance free. Its greatest disadvantage is the low availability of Plutonium and the associated high prices. Also, the thermal efficiency of an RTG is with at best 8% rather low. Due to large area contamination threats in case of a launch failure, RTGs are often criticized by the public. These are designed to withstand such an event, though. Famous examples of spacecraft to use RTGs are Cassini, Voyager as well as the Mars Science Laboratory Curiosity rover.

Nuclear Fission (and Fusion)

The most powerful energy source in practical use on Earth today is the process of nuclear fission. Only a few tests in the direction of nuclear-powered spacecraft were made, although a nuclear reactor as a primary power source would represent an enormous step forward for many types of space missions: The amount of power generated on board determines, among other things, what kinds of and how many payloads can be taken along. A reactor in combination with an electric propulsion system could potentially also provide an advantage when it comes to improving travel times within the solar system.

Using nuclear fission or fusion, energy is gained by making use of a physical effect that directly converts mass into energy. The mass of an atomic nucleus is slightly lower than that of the sum of its nucleons (protons and neutrons). Therefore, the average mass per nucleon is also lower. This is called the "mass defect". It depends on the number of nucleons, i.e. the element and isotope. As can be seen in Fig. 20.2, the mass per nucleon first decreases with the number of nucleons in the atom and then increases again.

It is possible to use these differences in average nucleon mass of different elements and isotopes technologically for the generation of energy. If the nuclear reaction happens in the direction from heavier to lighter nucleons, the difference in mass is

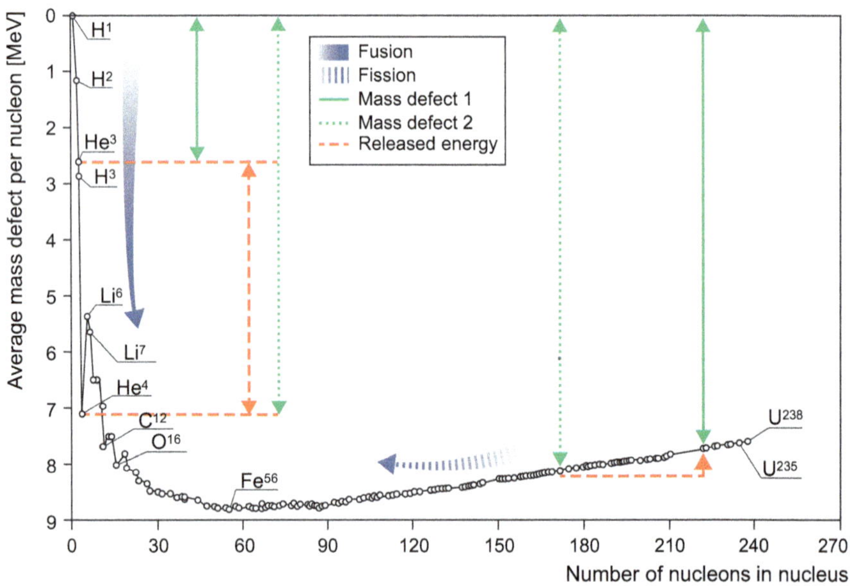

Fig. 20.2 The principle of the mass defect and the possibility to use it for energy generation. The average mass defect depending on the number of the nucleons is shown. Higher mass defect is equivalent to lower average mass per nucleon. Moving from nuclei with low mass defects to nuclei with high mass defects leads to the release of energy. Conversely, moving from nuclei with high mass defects to nuclei with low mass defects requires added energy

released as the equivalent energy $E = \Delta mc^2$. Splitting heavier elements into lighter ones is called fission and combining lighter elements to form heavier ones is termed fusion. Fusion can theoretically provide about a factor of ten more energy than fission. This is due to the fact that the mass defect increases much more rapidly for lighter elements, as seen at the very left of Fig. 20.2.

Power Regulation, Conversion and Distribution

Batteries can be charged if the power generated by the solar panels exceeds the needs of the spacecraft's systems. Once the battery is fully charged, additional power from the solar panel would actually damage it. Thus, a method to limit the charging of the battery and thus control the power generated by the solar panels must be employed. Often, a maximum power point tracker is used. This is essentially a controller with the dual purpose of, first, keeping the operational point at the maximum power point, as explained in Fig. 20.3, when the battery needs to be charged, and second, terminating the charge process of the battery by shifting the operational point towards the open circuit voltage when the battery is fully charged. This is done by increasing the resistance of the charge regulator and thus the combined regulator and bus load.

The solar array always operates where the generated power is sufficient to cover the then present bus load. When the bus load increases to values the solar array alone cannot provide, the battery is used as well and its voltage decreases. This causes the maximum power point tracker to shift the voltage of the solar array towards that of the maximum power point in order to provide as much power as possible. This goes hand in hand with the usage of the battery. As soon as the maximum power

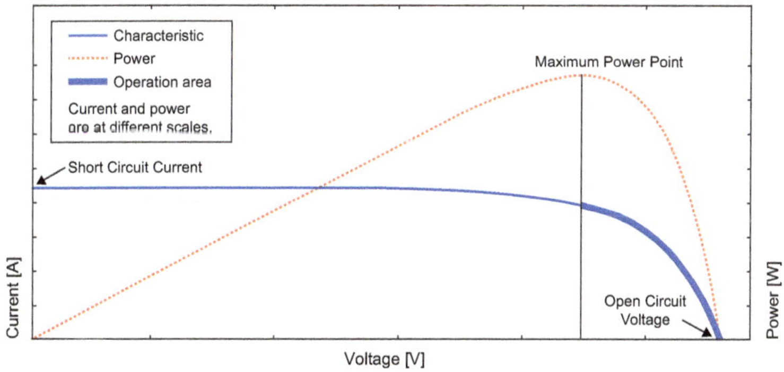

Fig. 20.3 Qualitative characteristic and power output of a solar cell. The blue graph shows, how the voltage of the solar cell depends on an external load, i.e. a resistance. The open circuit voltage on the lower right is given by the type of semiconducting material that is used and is the highest voltage that can be generated by the cell. It would appear if the resistance is infinitely high, e.g., when the circuit is open. The other extreme is a situation where the current can flow without resistance. This current is limited and called the short circuit current. When an external load is increased the voltage drops more and more quickly. From the voltage and the corresponding current, the generated power can be derived via multiplication, which results in the purple curve. The generated power reaches a maximum before dropping again and is zero at the open circuit voltage and the short circuit current. The operation area is chosen to the right of the maximum power point

point is reached, the battery provides the rest of the needed power while the solar panel remains at its maximum power point. The battery is discharged. When the bus load decreases again, power generated by the solar array is available and is used for charging the battery. Once the battery is fully charged, the charge current is decreased by gradually shifting the operational point of the solar array closer to the open circuit voltage, thus generating less power and preventing overcharging the battery.

Another cheaper but less sophisticated method is to use shunts for every string of the solar array, which can then be decoupled from the main bus individually. The unused power will then be turned into heat at shunt resistors, which are usually located in the power control and distribution unit (PCDU).

Power needs not only be provided per se, but provided at specific places in specific forms and at specific times. Spacecraft loads are usually the various on-board units such as electronics boxes, instruments or reaction wheels. Many of these units might require a specific constant voltage to operate, therefore making a voltage conversion necessary.

Consequently, most satellites have both an unregulated and a regulated voltage bus. The unregulated voltage bus usually consists of the connections between the battery and solar panels, as well as the charge controller and other associated hardware. The unregulated voltage will therefore be whatever voltage is present at the battery and can vary significantly over time. For units that require a constant voltage, converters are necessary that convert the voltage to a specific and constant one. A voltage that is often used is 28 V but lower ones are also in use such as 5 V for on board computers for instance. Usually, the functionality of controlling and distributing power described above is combined in the PCDU. The PCDU also provides means to protect against overcurrents. The input to that box is the unregulated power from the battery and solar panels. The output is both unregulated and regulated/converted power. The power is then routed to the units that require it. The PCDU is usually controlled by the on-board computer via an interface. In Fig. 20.4, schematics of the PCDU of the TerraSAR-X and TanDEM-X satellites can be seen. The spacecraft loads are comprised of the bus components such as star cameras, reaction wheels, computers, and payloads.

All the various spacecraft functions are physically located in individually placed boxes. This allows for easier design as each box can be created on its own and only the interfaces have to be determined beforehand.

20.1.2 Thermal System

Another major topic in spacecraft design are the temperatures the spacecraft and its components have to function in. For many internal units, these temperatures are desired to be at about "room temperature". Other units that are exposed directly to the space environment, such as solar panels, have to be able to withstand a wider temperature range, e.g. from -90 to $+130$ °C. Most attention needs to be drawn to the units with a very narrow operating temperature range. As can be seen in Fig. 20.5,

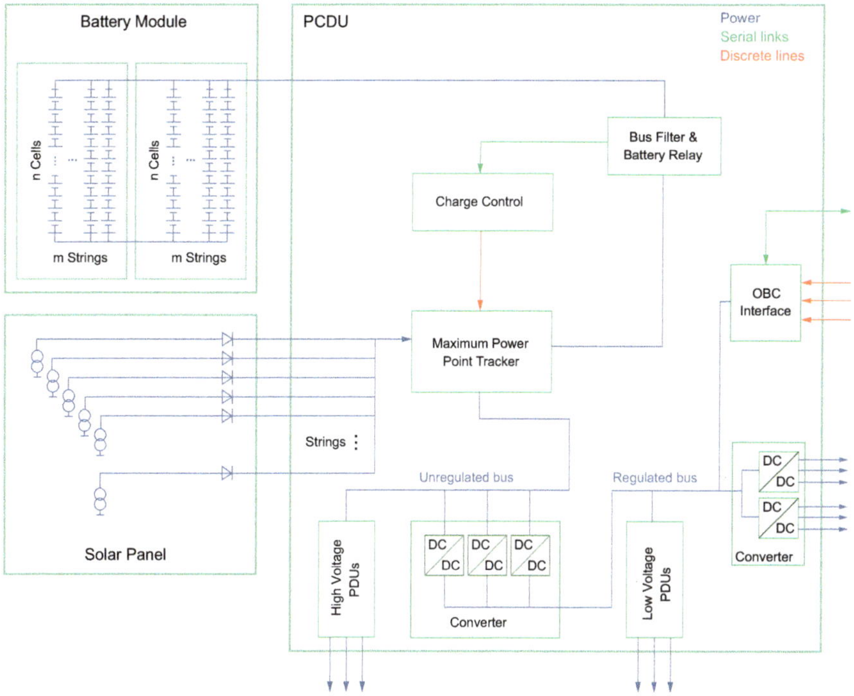

Fig. 20.4 Generic power control and distribution unit (PCDU) schematics. The PCDU is connected to the battery and the solar panel. Inside the PCDU, the connections of these two energy sources to the maximum power point tracker (MPPT) can be seen. From there the power is routed towards the unregulated bus and the regulated bus. These connections are themselves divided into various power distribution units (PDUs). The PCDU is controlled via the on-board computer interface

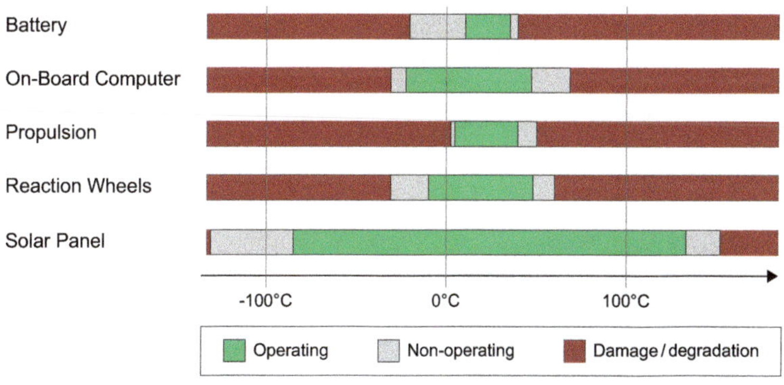

Fig. 20.5 Common operating and non-operating temperature ranges. Non-operating means that a unit is not expected to function at the low or high end of the range. However, it is expected to function when the temperature is again within the operating range

the battery and the propulsion system are often drivers for the thermal design.

The overall temperature balance of a body is determined by the heat that it produces itself and the heat it receives from the outside versus the heat that is transported away from it (see Fig. 20.6). Since internal heat sources have to be taken into account, the thermal system is closely intertwined with the electrical power system. The operation of units changes the temperature conditions and the temperature conditions determine the operability of units.

The additional heat generated on board must also be removed. It becomes therefore evident that the transfer of heat plays a fundamental role in spacecraft design.

Very generally, there are three mechanisms of heat transfer: conduction, convection and radiation. A temperature difference is always the driver with other parameters determining the effectiveness.

Conduction is the process of heat being transferred from one atom to the next within solid bodies but also within fluids and gases. This principle can be seen in Fig. 20.7a.

In physical terms, the material-dependent property of how well heat is conducted is called thermal conductivity. It is evident that the thermal conductivity is much higher in solid bodies, since there the average distance between atoms is much smaller than it is in gases, for instance. But of course, this property also varies widely between different solids.

Convection (see Fig. 20.7b) is another effective heat transfer mechanism that specifically makes use of the non-solidness of gases and fluids: They can move and

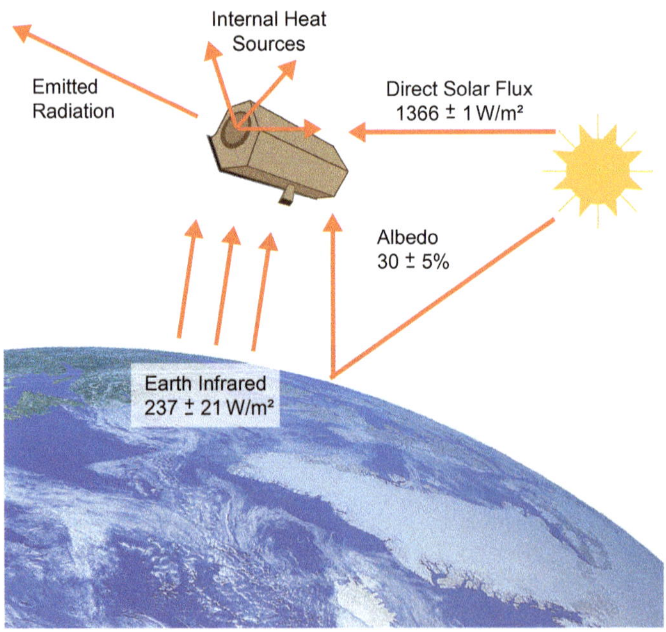

Fig. 20.6 Heat sources and radiation flux

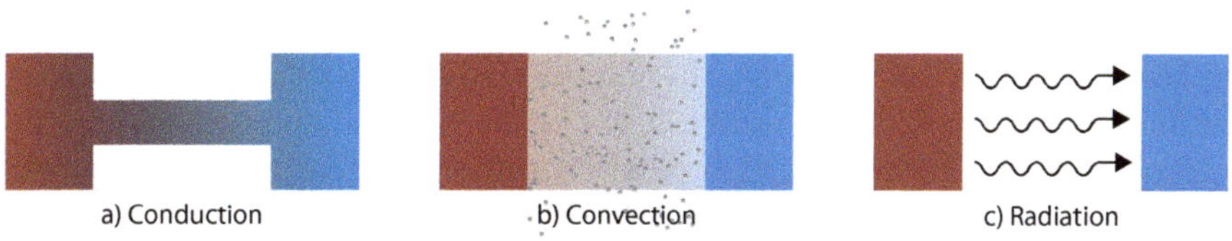

Fig. 20.7 Heat transfer mechanisms

thus carry away heat. There is free convection—where density differences caused by temperature differences result in movement in the presence of gravity and thus the heat transport—and forced convection, where the fluid moves due to an external cause, e.g. a pump or a fan.

The third mechanism is heat radiation (see Fig. 20.7c), where the heat transfer occurs by electromagnetic waves in the infrared range of the spectrum. Every physical body radiates heat. The radiated power of a body that exchanges heat with its environment exclusively via radiation is determined by the forth power of its temperature T, its emissivity ε, its radiating area A, as well as the Boltzmann Constant σ.

$$P = \varepsilon \sigma A T^4 \tag{20.3}$$

When these three principles of heat transfer are applied to spacecraft design, it becomes clear that convection cannot be used the same way it is on Earth, where it is a prominent natural and technical mechanism for temperature regulation. A technical example is a car which partially uses air from the environment to cool the engine. In free space, there is no external medium present that could allow for convection to happen. Free convection cannot occur at all due to the "lack of gravity" which could induce air movement. Within a spacecraft only forced convection is possible. In the pressurized modules of the International Space Station (and even some satellites), for instance, the air has to be constantly kept in motion artificially by a ventilation system.

The absence of convection and the absence of matter around the spacecraft leave radiation as the only means to transport heat away from the spacecraft. With only one effective cooling mechanism, it is clear that for most missions—despite the common knowledge of space being "cold"—the concern is not so much that a spacecraft becomes too cold, but rather that is becomes too hot due to the lack of effective heat rejection capability.

Of course, the prevention of heat loss can be a concern as well. But it is obviously easier to heat up a unit by switching on an electrical heater than it is to cool something by "switching on heat transfer".

Therefore, for first estimations, the assumption is made that a spacecraft needs to be able to emit all the power it generates. That means cooling needs to be "built in" at the very beginning of the design phase while heating could in theory be treated as an afterthought—provided enough electrical power is available.

Also, variations in the power consumption of a unit and the resulting changes in the heat balance are usually handled by adjusting the heating rather than the cooling function of the spacecraft. However, it is possible to build spacecraft where the thermal regulation works completely passively and autonomously. This can be done for spacecraft which do not experience large variations in their power generation and consumption. Then the radiated power can be set accurately enough for all the units to remain within their operational temperature range without the need for additional heating.

Regulating the temperature is at first dealt with by insulating the spacecraft using multi-layer insulation (MLI) foil. This serves two purposes: It prevents radiative heat loss into space and, due to its reflective golden layer, also excessive heat-up from the Sun or the Earth's infrared radiation. In order to allow for heat to escape, special radiative surfaces are employed. These should be placed strategically so that they face deep space, minimizing heat absorption from external sources which would decrease the radiator's effectiveness. As can be seen in the equation for the radiated power, see Eq. (20.3), the ability of a surface to emit heat is directly proportional to its emissivity. Consequently, radiating surfaces are usually treated with paint or coatings which have a high emissivity, for instance with white paint.

The emitted energy also is directly proportional to the forth power of the temperature. Therefore, a higher radiator temperature would significantly increase its thermal energy rejection capability and thus allow a smaller area and consequently a lower spacecraft mass. However, in a passively cooled system, the radiator needs to be cooler than the unit it is supposed to cool in order to maintain the heat flow from warm to cold. For passive heat transfer, the radiator temperature is thus rather low, often below 0 °C. A high radiator temperature is in principle possible, but this comes at the cost of having to employ active heat transfer, which can transport heat from a warm unit to an even warmer radiator. Such a system functions similarly to a common refrigerator, where heat is transported from the inside of the refrigerator to the warmer outside. Energy needs to be added to allow for the "heat pump" to work. Due to the complexity of such a system and the added mass of the active components, it is usually not employed on satellites unless absolutely needed. Examples are often scientific spacecraft that have infrared telescopes as payloads which need to be cooled to very low temperatures.

Heat transport (and prevention thereof) within the spacecraft itself is primarily realized via the usage of materials with either low or high thermal conductivity. High conductivity is used wherever heat transport is explicitly desired. This is often the case when heat shall be transported from a unit to a radiator. Low conductivity is used for parts of the spacecraft that shall be thermally insulated. Solar panels are an example, since they are by definition oriented towards the Sun and thus rather warm. Thermally insulating them shall prevent too much heat from entering the spacecraft itself via the panel structure or attachment joints.

Another common heat transfer component is the heat pipe. A heat pipe is in principle a closed pipe that contains a working fluid which evaporates at the high temperature end, the gas fills the pipe and then condenses back into liquid form at the cool end. A wicking material causes the fluid to move back to the hot end, where it can evaporate again. Heat pipes are common thermal connectors to radiators.

Thermal control predominantly makes use of electrical foil heaters. These are in principle wires encapsulated in caption foil that are being heated up by resistance heating. The other part of thermal control is the need to measure temperature. Such a temperature measurement is mostly done by thermistors, which are electrical resistors where the dependence of the resistance R on the temperature is well known, e.g., for a common type of thermistor called PT 1000 (T is here the temperature in °C without unit).

$$\frac{R}{1\,\Omega} = 1000\left(1 + 3.9083 \times 10^{-3}\left(\frac{T}{1\,°\mathrm{C}}\right) - 5.775 \times 10^{-7}\left(\frac{T}{1\,°\mathrm{C}}\right)^2\right) \quad (20.4)$$

Whether or not a heater needs to be switched on or off, is determined by the on-board computer which checks the temperature readings of the thermistors at certain intervals. These readings are then compared to on and off thresholds which can be changed from the ground. When the temperature is below the on-temperature, the heater is being switched on and when it is above the off-temperature, it is switched off again. Any temperature measurement (or any measurement in general) is originally analog and must be converted into a digital signal in order for the computer to be able to use it. This conversion usually leads to an uncertainty in the measurement. The bit range to represent a certain temperature range then determines the accuracy of the measurement. For instance, a 12-bit string can describe 4,095 values, which allows for a temperature range from -160 to $+ 160$ °C to be accurately measured to within 0.078 °C which is sufficient for temperatures. Thus, in reverse logic, the desired accuracy of the measurement determines the bit range in the design phase.

Operational Needs

The above is one example of a general principle: The design of a spacecraft must meet the operational needs. Therefore, thorough design and testing need to be employed to have accurate measurements and avoid interference among units and sensors. For instance, the magnetic field sensors on board a spacecraft, whose purpose is to measure the Earth's magnetic field for attitude determination, must not be perturbed by other magnetic fields generated on board. For this reason, magnetic fields like those generated by high currents need to be avoided by using twisted wires with opposing currents which cancel each other's magnetic fields. Unavoidable disturbances must be known beforehand and included in the on-board software; preferably in the form of commandable settings so that they can be adjusted later in the mission. Another example are batteries that might need to be operated within tight limits—especially towards the end of a mission. Here, accurate measurements of state of charge, voltage, currents and temperature are absolutely critical to ensure proper handling from the ground. It must therefore be known by the spacecraft designer what to include in the telemetry and how and where that telemetry is measured in order to allow for safe and reliable operations.

20.2 Operations

20.2.1 Preparation Phase

Satellite power and thermal system operations start several years before a spacecraft is launched. As soon as the satellite design is determined, one has to consider the consequences and necessary actions for operations. Depending on the satellite design (see previous section) and the mission requirements on the power and thermal system, an operations concept has to be developed.

A so-called power thermal model is generated for most satellite missions. This model, for example, allows estimations of how long operations, maneuvers or data takes can last without discharging (and/or heating up) the battery too much. The thermal model contains on the one hand all critical temperatures (e.g. instrument temperatures during operation) and on the other hand the thermal behavior during warm-up and cool-down. Depending on actual heat dissipation, heat capacity and operation phase, the warm-up of the units is largely linear with time whereas the cool-down in general shows an exponential temperature drop. This is caused by the fact that during heating up the conducted power (generated by a heater) is constant, whereas during cool down the dissipation of heat depends on the temperature of the thermal environment within the satellite. The battery charge control model is driven by the total power over time. It includes much information about the charge behavior of the battery, the number of allowed charge cycles and the maximum depth of discharge (DoD).

20.2.2 Launch and Early Orbit Phase (LEOP) Operations

After the launch it is necessary to bring the satellite into a nominal operating state as soon as possible in order to lower the risk of power problems. For many space missions, power is generated by the combination of solar panels and batteries (see Sect. 20.1). Several units such as the on-board computer, sensors or heaters are switched on shortly after separation. Therefore, power is drained from the battery (as long as the solar arrays do not yet generate power).

For this reason, one of the first actions after separation is the deployment of the solar panels (if they are designed to be deployable, see Fig. 20.8). Immediately prior to the deployment, the power and thermal subsystem engineer needs to evaluate the situation. Solar panels are deployed by activating pyro cutter mechanisms which require a certain operating temperature. This and other preconditions need to be checked. After deployment has been executed, the operations engineer has to confirm that the solar arrays have deployed as expected and are working properly. During Sun illumination they have to generate enough power to operate all of the satellite equipment and to recharge the battery for eclipses or times with higher power consumption.

Fig. 20.8 Deployable and undeployable solar arrays. For TerraSAR-X (left) the solar panels are mounted on the upper right of the satellite. Eu:CROPIS (right) is a compact satellite with deployable panels to increase its useful solar array surface

After this is done and the spacecraft generates enough power for operations, several units need to be activated. As soon as all activation pre-conditions are met, the nominal and redundant units are powered on by ground command to test their functionality. Also several on-board data configurations need to be checked and adapted: Since the real behavior of the vehicle in space cannot fully be predicted in the design phase, the trigger limits for autonomous on-board reactions have to possibly be adapted after launch.

Launch delays of a spacecraft by several months or even years have negative effects especially on the spacecraft battery. Even if the battery is not used before launch it degrades over the time and therefore the usable power budget decreases. The rate of degradation depends on the battery type. Li-ion batteries, for example, degrade slowly. Spacecraft batteries are already charged on ground regularly to compensate the self-discharge, which could harm or even destroy the battery cells by deep discharging. Nevertheless, a launch delay of several months or more leads to a battery with a lower capacity as desired.

If the battery has already degraded before the launch, it is desired to find possibilities to increase battery health after launch. This is called re-conditioning. It depends on the battery and satellite design whether or not it is possible to adapt the charge level in order to avoid temperatures or voltages which are outside design specifications.

The reconditioning after launch sometimes causes additional work for the power and thermal system team. Charging the battery slowly and creating a small overcharge, for example, prevents NiH_2 battery cells from degradation.

The task of the operations engineer is to derive a set of parameters which can be commanded to the satellite in order to optimize the battery charge process. Of course, it can be necessary to adapt the parameters several times over the whole mission.

20.2.3 Routine Phase

Once the satellite is in an operational state, the telemetry is continuously being monitored and the health status of the spacecraft equipment and the behavior is evaluated over large time spans. Trends are analyzed to achieve better predictions for the future subsystem behavior. For that it is indispensable to differentiate which effects are caused, e.g., by seasonal variations or operations and which are induced by ageing or degradation. An example for a suspicious behavior, which is actually explainable by normal effects, is given in Fig. 20.9.

In case telemetry readings or trends cannot be explained, it has to be clarified whether the observed behavior is critical and whether any action is required.

Battery Discharge Process

Probably the most sensitive and stressed unit within the power system is the battery. An example for its degradation is given in Fig. 20.10.

Depending on the orbit and operational concept, a battery sometimes has to go through thousands of charge cycles during its lifetime. Some experiments and payloads even need more power than the solar arrays generate. During operations, they drain power which is stored in the battery and the battery is recharged once the operations have finished. For some missions, the solar arrays generate power all the time, but payload operations are only performed over a certain amount of time (e.g.

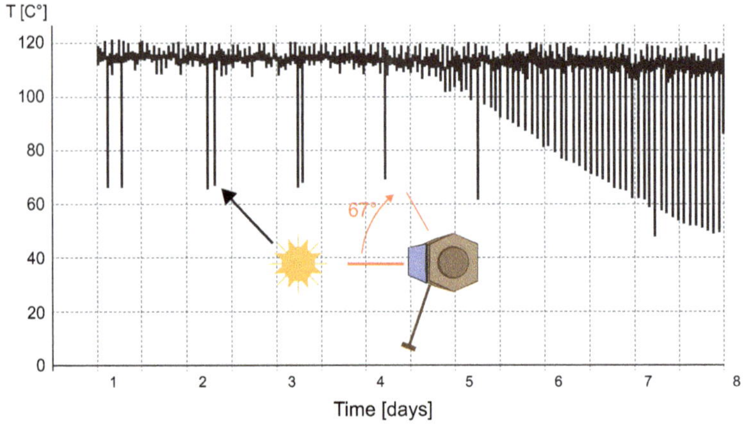

Fig. 20.9 Solar array temperature of TerraSAR-X over some days. Some irregular dips down to 65 °C can be seen as well as periodical dips intervals of 90 min with a constantly decreasing envelope. Contrary to first assumptions, this is no indication for any problem or degradation, but just shows some operational and seasonal effects on the solar panels: The irregular dips from 115 to 65 °C are caused by so called left-looking modes. The satellite is rotated by 67° around its *x*-axis. Thus, the solar arrays are no longer pointed directly towards the Sun which results in a strong temperature decrease. The periodical temperature drops are a consequence of solar eclipses. Whenever the Sun is shadowed by the Earth, the temperature of the solar arrays decreases dependent on the duration of the eclipse

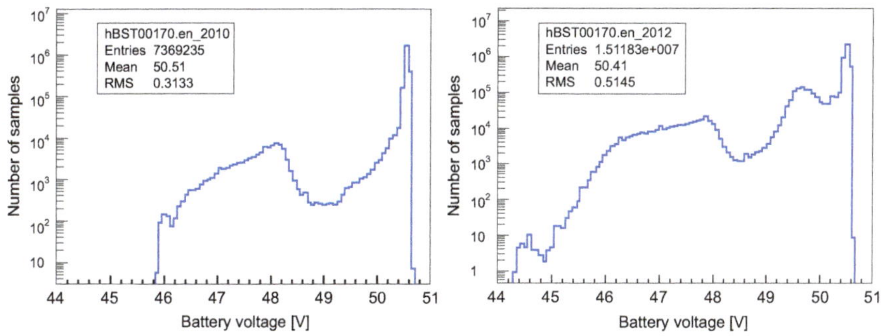

Fig. 20.10 These two histograms show the battery voltage distribution over 1 year for two subsequent years. In that time period a shift towards lower voltages can be observed; the minimum battery voltage decreased from 45.8 to 44.2 V. The cause for this behavior can be attributed to the degradation of the battery

TerraSAR-X). For this type of mission, the payload is allowed to drain more power as generated by the panels as the battery can be charged during periods without payload operations. For missions where payload operations are performed all the time (e.g. GRACE) it is necessary that the solar arrays produce more power than is used by the spacecraft.

The battery capacity is in fact the limiting resource for many missions, even more limited than propellant mass or available cycles of operations for thrusters or BUS units (see Sect. 20.1). Here, the important characteristic is the number of charge/discharge cycles at a given depth of discharge. The cycle lifetime increases as the depth of discharge decreases (Nelson 1999). If the desired lifetime for a spacecraft is five years in low Earth orbit (LEO) with 25,000 expected charge cycles, for example, it has to be ensured that the DoD is always below 15%, assuming a behavior as seen in Fig. 20.11. This usually is also part of the power thermal model. Discharge can be caused by several things such as data takes, maneuvers or eclipses. If analyses show that the DoD is already close to its limit for usual payload operations, the power and thermal engineer has to take care that no additional power-consuming operations (e.g. data takes) are done during an eclipse or a maneuver.

Battery Charge Process

Also, the charge process needs to be well-defined. Charging the battery slowly is the preferred operations concept, but need to be balanced with other requirements, like having a fully charged battery available at a certain time.

A slower charging of the battery can be achieved by rotating the satellite's solar arrays away from the Sun as the charge current depends on the effective illuminated area of the solar cells. A rotation of 30° decreases that area and therefore the charge current by approximately 14%. The smaller the charge current, the lesser heat will be produced during overcharging the battery. This makes it possible to charge the battery to a higher level of capacity without accepting an additional degradation.

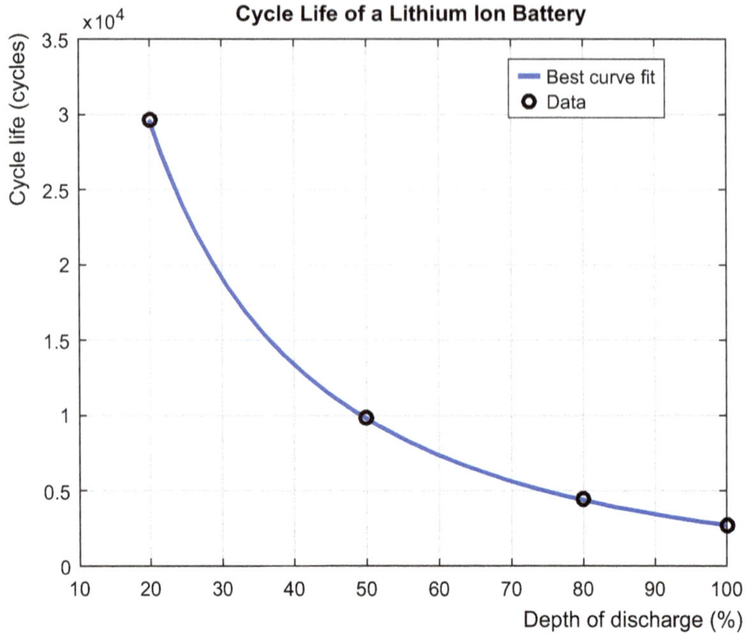

Fig. 20.11 The amount of expected charge cycles decreases with increasing depth of discharge (DoD). To achieve a certain amount of cycles, it is therefore necessary to limit the DoD (image from Mallon et al. 2017)

Another possibility to influence the charge process, is the so-called shunting of the solar array strings, which is essentially a partial deactivation of solar cells leading to a reduced current. As soon as a predefined charge level is reached, shunting would be activated by the on-board software. When the state of charge drops again below the set DoD level, shunting would be switched off and the charge process starts again.

The charge process can also be adapted to a decreased capacity by lowering the maximum charge level if it is adjustable per design: Degradation of the battery can cause the effect that it is no longer possible to charge it to its begin of life (BoL) capacity—setting the maximum charge level to 100% BoL capacity would consequently lead to an overcharge, which would be converted into heat. Therefore, the temperature within the battery would increase and possibly damage it. To avoid this, the power and thermal operations engineer has to adapt the maximum charge level to a reasonable level.

All of the scenarios mentioned above decrease the charge current which leads to the generation of less heat. Therefore, battery health can be preserved or even increased.

Battery Temperature and Performance

As already mentioned above, an important factor for battery performance is temperature. The hotter the battery, the faster chemical reactions will occur. The desired chemical reactions on which the battery depends are usually accompanied by unwanted chemical reactions which consume some of the active chemicals or impede their reactions. Even if the cell's active chemicals remain unaffected over time, cells can fail because unwanted chemical or physical changes to the seals keeping the electrolyte in place. As a rule of thumb, for every 10 °C increase in temperature the reaction rate doubles. Thus, an hour at 35 °C for lead-acid batteries is equivalent in battery life to two hours at 25 °C. Heat is the enemy of the battery and, as Arrhenius (Laidler 1987) shows, even small increases in temperature have a major influence on battery performance affecting both the desired and undesired chemical reactions.

Battery temperature has to be monitored very carefully during the whole mission. It is necessary to reach a compromise between the temperature being low enough to afford a long lifetime and being warm enough to provide good performance. It is the duty of the power and thermal engineer in cooperation with battery experts to determine when it is necessary to regulate the battery charging cycle and to determine further actions.

Thermal Subsystem

Along with the necessary operations concerning the power system there are also a set of actions that have to be performed for the thermal system. Almost every unit of a satellite has specified limits between which it has to be operated. Among the temperature sensitive units are star trackers, flow control valves for thrusters and the propulsion system. Their temperatures have to be observed over the whole mission and specific actions have to be performed if necessary. If a unit impends to cool down below its operational limit, it is necessary to switch on its heater if available. If the unit does not have heaters, sometimes it is even necessary to create new operational concepts. This can, for example, be to rotate the satellite in a way that the units are pointed towards the Earth or the Sun (if they are too cold) or to rotate them away (if they are too hot).

It is obvious that all those analyses and actions would be a huge amount of work if it all had to be done manually. To make operational life easier, automatic checks and corresponding actions can be implemented in the spacecraft. Several parameters such as battery voltage, temperatures and power statuses are monitored by the on-board software. As soon as one of these parameters is out of limit for a certain amount of time, a specific action is performed automatically (e.g. switch on heater if temperature is too low).

The degradation of units has to be analyzed periodically by the operations team and the responsible experts. The observed behavior has to be compared with the modeled one and resulting actions have to be derived from this.

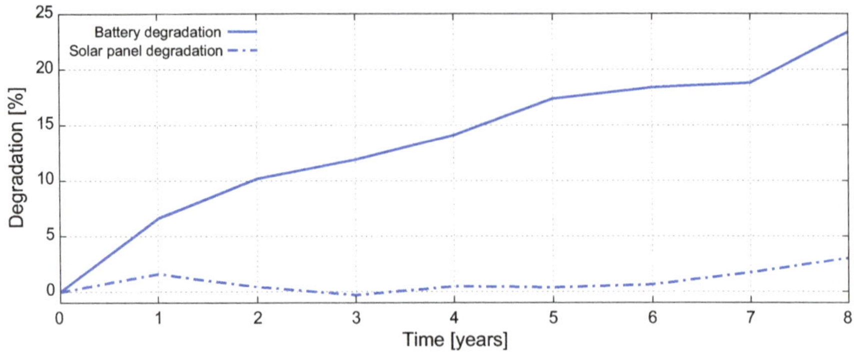

Fig. 20.12 Degradation examples of a solar panel and a lithium-ion battery. The battery is discharged at roughly 1C with an average depth of discharge of about 4%. The solar panel degradation is comparably small

Figure 20.12 shows the degradation of solar panels and a satellite battery. The causes for the reduced performance (see Sect. 20.1) can be aging, radiation or charge/discharge cycles. These effects have of course an impact on satellite operations. When solar arrays and battery degrade, this means that all in all less power is available. This has to be taken into account for all activities, monitoring and limits. Therefore, some values and parameters (e.g. heater set points, battery limits, etc.) have to be adapted periodically during the mission and actions to recover the battery cells (described in Sect. 20.2.2) can be performed. Furthermore, it has to be analyzed if payload operations have to be changed. This can result, for example, in decreased observation duration to avoid deep discharge and preserve the battery.

20.2.4 End of Life

Most of the satellite missions are extended beyond their originally planned lifetime. Once the degradation of the power equipment reaches a (mission) specific point, the satellite operations cannot be continued as before, especially during eclipse season when the satellite is mainly dependent on battery power. Solar panels also degrade, thus lengthening the battery recharge time. Ultimately, with the advancing age of batteries and solar panels daily operations become time-consuming as operations are determined by avoiding low power conditions.

The GRACE (Tapley et al. 2004) project showed very explicitly how end of life (EOL) operations used to be. During eclipse phases, which were experienced twice per year, no power could be generated by the solar panels and thus the energy needed was drained from the battery. The eclipse duration varied from some seconds at the beginning to more than half an hour in the middle of the eclipse phase. 14 years after launch the battery capacity was below 8% of its original value. As a consequence, the batteries could not provide enough power to all units during long eclipses.

There are some operational tricks to deal with already weakened batteries. If it is possible to adjust the end-of-charge voltage of the battery, this charge level can be raised at the end of an eclipse to allow an immediate maximum charging of the battery after leaving the Earth shadow. After a certain time in Sun condition, the charge level has to be reduced again to avoid an overcharge of the battery. If the battery power is no longer sufficient to maintain the nominal satellite operation during an eclipse, it can be helpful to heat up parts of the satellite such as the propellant tanks during Sun illumination. The heat dissipation can then be used afterwards to keep single units of the satellite within its operational temperature limits. Additionally, heater power can be saved by reducing the heater set points to a minimum during eclipse (for some units this cannot be done for the whole orbit duration because payload data quality is temperature dependent).

If none of these measures are successful, the only remaining option is to switch off single units or at worst the whole satellite. This implies a major impact on the mission and involves a lot of subsequent adjustments of units like the on-board computer. Potentially even software uploads are then necessary to restore the required configuration. This worst-case scenario should be avoided unless all other options have been exhausted.

Full Sun orbits with continuous power generation can also be a threat to weak battery cells because of missing discharge processes. This would then result in an irreversible cell collapse due to electrolyte bridging. Again, it is the duty of the operations team to avoid these. For the already mentioned GRACE satellites it was unfortunately neither possible just to disable some of the solar array strings as described above, nor could additional consumers be switched on. One possibility to force the battery's discharge was to generate some kind of artificial eclipse. This could be reached by rotating the satellite in a way that the solar panels were not irradiated by the Sun for some minutes. First of all, it had to be clarified how often such "eclipses" had to be generated and how long they should be. Therefore, a close coordination with the satellite support team and especially with the battery experts was required in order to find the best solution. Also the AOCS engineers needed to be involved: While the AOCS team calculates what kind of rotation should be made to save as much fuel as possible (at EOL most likely fuel budget is very low, too), it is very important to figure out if such a rotation poses a threat to the thermal system with respect to allowed unit temperatures. Usually it is required to point the solar panels towards the Sun whenever it is possible. Therefore, the spacecraft is designed in a way, so that all units which should be operated at low temperatures (e.g. star cameras) are mounted on the opposite side of the vehicle. When rotating the spacecraft to a not nominal attitude, it is possible that some units get too hot, others which are usually pointed towards the Sun could cool down. So, a compromise between thermal and power-driven attitude requirements has to be found.

These actions have to be planned, prepared and executed carefully by the operations team. Monitoring limits for temperatures or voltages during the time without payload operations have to be adapted because charge and temperature behavior change (e.g. the temperature decreases if less payloads are operated). The switching has to be timed exactly to avoid voltage or temperature variations that would violate their limits and as a consequence trigger any unwanted on-board mechanisms.

At the end of the satellite's lifetime operations become more challenging and it is necessary to know the system very well to extend the mission as long as possible. Components within the power system are the limiting factors for most missions.

20.3 Contingency Operations

The power and thermal subsystem occupies a central position for the operability of a spacecraft because all systems need power and thermal control. When a PTS contingency occurs, it is not unusual that the whole spacecraft is affected. In a multiple failure scenario, the PTS subsystem engineers have to identify and to recover electrical failures first. If there are multiple electrical failures, they have to start with the recovery of the failure that is closest to the beginning of the power generation and distribution chain and then work on failures further downstream. An example would be the failure of an on-board instrument and simultaneously voltage fluctuations of the relating power bus. Stabilizing the power bus in this case would be the precondition for a successful recovery of the failed instrument—or could even fully recover it without any further action required.

In most cases, thermal failures can be remedied after the power/electrical failure has been restored, since the thermal systems react more slowly. Some spacecraft have dedicated operation modes for a failure of a cooling circuit or for a reduced heater activity in case of a low-power situation.

In some anomaly cases, it is not directly obvious what caused the problem. If, for example, the output power of the solar panel is reduced there are a few different failures which can lead to a reduced output power displayed in the telemetry. As illustrated in Fig. 20.13 for an exemplary configuration, damage of a part of the solar panel (#1), cable break (#2), switch-off (#3a) or damage of power regulation equipment (#3b) or damage of a sensor (#4) could be the possible reasons.

The possible causes of a reduced solar panel power output are diverse, as are the effects it could have on the mission. In this example, there could be a real loss of output power of the solar panel (with possible failures #1, 2, 3a, 3b) or the loss is only displayed (failure #4). Furthermore, a truly reduced output power can be recoverable in case the power regulation equipment was switched off (failure #3a). In case of the hardware damages #1, 2 and 3b, the failures are not recoverable, therefore the power generation and probably also the mission will be impaired.

In 2004, after about 3.5 years of the GRACE mission, first a DSHL (disconnection of supplementary heater lines) and afterwards a DNEL (disconnection of non-essential loads) occurred on one of the two satellites. For this kind of spacecraft, there were four stages of DNEL. A DNEL is caused by an undervoltage situation in the battery. The deeper the voltage drops the higher the stage of DNEL. The amount of actions that are performed automatically on board increases with the DNEL stages.

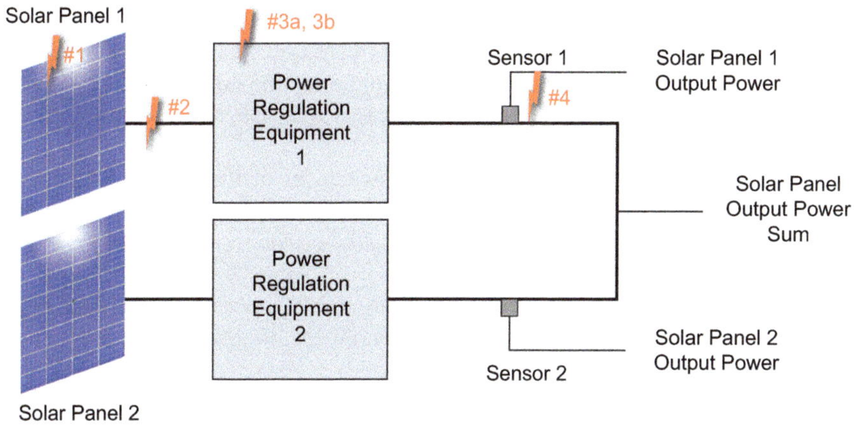

Fig. 20.13 Possible reasons causing a reduced solar panel output power

In this case, the spacecraft was found in thermal survival mode during a ground station contact. In this mode, all unnecessary heaters are switched off and the heater set points are lowered to a point, where as little power as possible is used to keep the satellite and its hardware alive. After analyzing the data, it could be seen that a battery cell was lost and the battery voltage dropped below a defined threshold for more than 30 s, which triggered the first stage of DNEL.

In addition to the change of the heater configuration, several other on-board reactions could be triggered by a DNEL depending on its configuration. The spacecraft may be sent to a coarse pointing mode which could lead to higher gas consumption, several hardware units could be switched off which are not necessary for the safety of the satellite. The transmitter could be switched to low rate to save power during a ground station contact and all commands on-board, which were saved in the timeline, could be deleted to avoid power-consuming activities.

It is obvious that a huge amount of actions may be necessary to return a satellite back to an operational state after a DNEL.

After this event in 2004, first of all, the EOC level settings of the GRACE satellite (described in Sect. 20.1) had to be changed in order to compensate for the low power situation. The EOC level was raised to get the battery stabilized. Additionally, the spacecraft was commanded back to thermal nominal mode to re-establish the desired heater configuration. As the satellite transmitter was switched to low rate by the triggered on-board failure detection, isolation and recovery (FDIR) mechanism and the ground station was still configured for high rate, a blind acquisition had to be made. In this case, the reconfiguration on board was faster than the one on the ground. This means that the subsystem engineer had to prepare a procedure to switch the transmitter back to high rate to see telemetry. As the complete timeline on-board the spacecraft was empty, they also had to ensure that the transmitter was manually switched off again at the end of the ground station contact to save power. The timeline had to be updated and reloaded. The payload equipment had to be switched on and set up again as soon as enough power was available. The telemetry had to be analyzed to find out which autonomous on-board actions were executed because an on-board action is disabled directly after triggering to avoid a continuous execution. All FDIR mechanisms had to be re-enabled after the recovery to ensure satellite safety in case of a similar problem in the future.

In the case of the GRACE mission, the switch-off of the on-board computer could be prevented by the DSHL and DNEL in 2004. Shutting down the on-board computer would have been tantamount to losing the mission, since it would not have been possible to restart.

A balanced interaction of the PTS resources and PTS operations with the other subsystems and FDIR mechanisms of a spacecraft are often of crucial importance for the success of the mission.

References

Deng X, Schiff EA (2002) Amorphous silicon-based solar cells. In: Luque A, Hegedus S (eds) Handbook of photovoltaic science and engineering. Wiley

Laidler KJ (1987) Chemical kinetics, 3rd edn. Harper & Row, p 42

Mallon KR, Assadian F, Fu B (2017) Analysis of on-board photovoltaics for a battery electric bus and their impact on battery lifespan. https://www.researchgate.net/figure/Depth-of-discharge-versus-cycle-life-of-the-lithium-ion-battery_fig4_318292540. Cited 28 May 2021

Nelson RA (1999) Spacecraft battery technology. Via Satellite

Tapley BD, Bettadpur S, Watkins M, Reigber C (2004) The gravity recovery and climate experiment: mission overview and early results. Geophys Res Lett 31(9)

Kay Müller is Project Manager for the mission operations of the TDP-1 Laser Link Mission, Flight Operations Manager for GRACE Follow-On as well as PTS Subsystem engineer for TerraSAR/TanDEM-X. He graduated in Aerospace Engineering at the University of Stuttgart in 2011 and has joined the German Space Operations Center (GSOC) in the same year.

Sebastian Löw is Project Manager for GRACE-FO and Flight Director for the mission EnMAP. He graduated in Physics from the Ludwig-Maximilian-University of Munich (Germany) in 2008. Since 2009 he is working for the German Aerospace Center at the German Space Operations Center (GSOC). He was engaged as Attitude and Orbit Control System (AOCS) as well as Power and Thermal subsystem engineer for TerraSAR-X and TanDEM-X. He was also active as AOCS subsystem engineer and Flight Director for TET-1, GRACE, and PAZ. Since 2019, he is also deputy group lead of the LEO Mission Operations Group.

Sina Scholz works for the German Space Administration and supports the National Program in the department of satellite communications. In 2007, she graduated in aerospace engineering at the Technical University of Munich and joined DLR since 2008. She worked as a sub-system engineer, flight director and project manager at the German Space Operations Center (GSOC) for LEO and GEO missions. In 2016, she changed to the German Space Administration and worked for the German Space Situational Awareness Center in Uedem before moving to the headquarter in Bonn.

Chapter 21
Propulsion Operations

Franck Chatel

Abstract This chapters gives an overview of the propulsion system operation. The principle of propulsion is briefly recalled, and the typical structure of a bi-propellant propulsion system is presented along with brief comments on other systems. Real-time and offline operation is then explained, with an emphasis on the orbit maneuvers and related activities like mass computation and lifetime estimation.

21.1 Principle of Propulsion

Satellite or rocket propulsion systems are based on the action-reaction principle: thrust arises in reaction to the expulsion of mass. The same principle is at work when a cannon is fired and moves backwards after the shot.

Some simple physics allows quantifying the force created by the expulsion of mass. Consider a satellite of mass m moving through space with a velocity vector $\vec{V}$, as shown on Fig. 21.1. The propellant mass dm to be ejected is bound to the satellite and moves with the same velocity vector. An instant dt later, the mass is expelled with an exhaust velocity $\overrightarrow{V_e}$ relatively to the satellite body, which receives a velocity increment $d\vec{V}$. Newton's second law of motion relates the change in momentum of the system to the sum of the external forces $\vec{F}$ acting on that system.

In the present case, the system consisting of the satellite is open, since its mass varies due the ejection of propellant mass. However, Newton's second law applies to closed systems, so we must consider the system consisting of the satellite and the ejected propellant mass. Before the propellant mass is ejected, the momentum of the system is $\overrightarrow{p_1} = (m + dm)\vec{V}$ whereas after ejection it becomes $\overrightarrow{p_2} = m\left(\vec{V} + d\vec{V}\right) + dm\left(\vec{V} + \overrightarrow{V_e}\right)$. Newton's second law then yields:

F. Chatel (✉)
German Space Operations Center (GSOC), German Aerospace Center (DLR), Oberpfaffenhofen, Germany
e-mail: franck.chatel@DLR.de

F. Sellmaier et al. (eds.), *Spacecraft Operations*, Springer Aerospace Technology,
https://doi.org/10.1007/978-3-030-88593-9_21

$$\vec{F} = \frac{d\vec{p}}{dt} = \frac{\overrightarrow{p_2} - \overrightarrow{p_1}}{dt} = m\frac{d\vec{V}}{dt} + \frac{dm}{dt}\overrightarrow{V_e} \tag{21.1}$$

The change of momentum is different from the usual product of mass and acceleration vector: an additional term appears because of the change of mass. This relation can be rearranged as follows:

$$m\frac{d\vec{V}}{dt} = \vec{F} + \vec{T}$$

with

$$\vec{T} = -\left(\frac{dm}{dt}\right)\overrightarrow{V_e} \tag{21.2}$$

This rearrangement shows that the situation can be considered equivalent to an ordinary closed system (term on the left), except that a new force, the thrust, has to be introduced. Of course, it must be remembered that the satellite mass m is variable. However, it can often be considered quasi-constant if the total ejected mass is small in relation to the satellite mass.

The expression obtained before shows various properties of the thrust:

- Its direction is opposite to the exhaust velocity, which shows that this force is a reaction.
- Its magnitude is proportional to the exhaust mass flow rate.
- Its magnitude is proportional to the magnitude of the exhaust velocity vector.

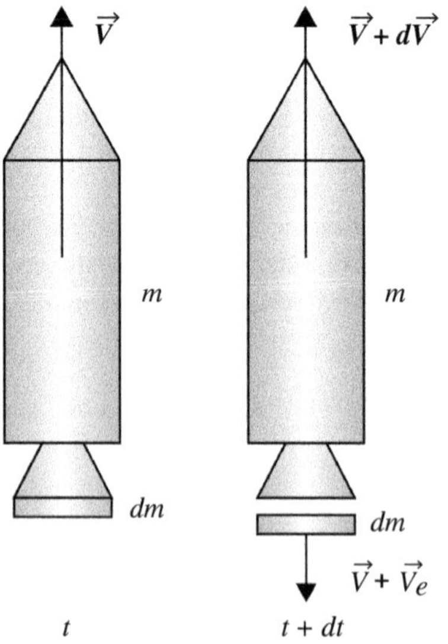

Fig. 21.1 Kinematical states of the spacecraft prior and after the expulsion of mass

The norm of the exhaust velocity vector is often written as the product of the standard gravity g_0 (whose value is 9,80665 m/s^2) and the specific impulse (noted I_{sp} and expressed in seconds). One can then write

$$\overrightarrow{V_e} = (g_0 I_{sp})\overrightarrow{u_e}$$

where $\overrightarrow{u_e}$ is a unit vector aligned with the direction of the exhaust velocity. Consequently, the expression of the thrust becomes:

$$\vec{T} = -\left(g_0 I_{sp} \frac{dm}{dt}\right)\overrightarrow{u_e} \tag{21.3}$$

It follows from this expression that the higher the specific impulse, the less the mass flow rate required to produce the same thrust level. This expression is quite general and applies to all types of engines, even if part of the propellant is not stored on board, as it is the case with aircraft.

The specific impulse is used to distinguish classes between different propulsion systems. Chemical systems use a chemical reaction to produce gases that are subsequently ejected through a nozzle. In such systems, the mass flow rate is rather high because the ejected gases are quite heavy. However, the exhaust velocity is rather low, resulting in specific impulses on the order of 250–500 s. On the other hand, an electrical system can use high power currents to generate ions that are then accelerated by a magnetic field before being ejected. In this case, the mass flow rate is rather low but the exhaust velocity is very high, resulting in specific impulses on the order of 1500–3000 s or even more.

Finally, under some assumptions, the equation of motion can be integrated into the so-called "ideal rocket equation" (named after Konstantin Tsiolkovsky, a Russian scientist who derived this relation in the late nineteenth century). Assuming that all external forces are negligible with respect to thrust and that the exhaust velocity vector is constant over the maneuver duration, we can write:

$$m\frac{d\vec{V}}{dt} = -\left(g_0 I_{sp} \frac{dm}{dt}\right)\overrightarrow{u_e} \tag{21.4}$$

Denoting the initial and final conditions with subscripts "i" and "f" respectively, we obtain:

$$\int_{\overrightarrow{V_i}}^{\overrightarrow{V_f}} d\vec{V} = -\left[g_0 I_{sp} \int_{m_i}^{m_f} \left(\frac{dm}{m}\right)\right]\overrightarrow{u_e} \tag{21.5}$$

The terms are rearranged to obtain the usual form of the ideal rocket equation:

$$\overrightarrow{V_f} - \overrightarrow{V_i} = \delta V \overrightarrow{u_e}$$

with

$$\delta V = g_0 I_{sp} ln\left(\frac{m_i}{m_f}\right) \tag{21.6}$$

21.2 Configurations of Propulsion System

Having understood the physics behind propulsion, we will use a bi-propellant propulsion system as an example to present a typical propulsion system layout and its operational configurations.

Bi-propellant systems are not the only propulsion system used for spacecraft. A mono-propellant system is very similar, except that only one propellant is used for the chemical reaction. This is the case, for example, when hydrazine is used with a catalyst to activate the decomposition into gaseous products. Propulsion systems based on cold gas or electrical systems are also in use. Their operation has some peculiarities, especially for electrical propulsion. A more detailed description can be found in Fortescue et al. (2011).

21.2.1 Layout of a Bi-propellant Propulsion System

In this kind of propulsion system, thrust is achieved by ejecting the gas produced by a chemical reaction. To achieve a high exhaust velocity, the gas is expanded through a nozzle after the combustion chamber. Propellant liquids are stored in two separate propellant tanks, and valves allow or block the flow to the combustion chamber of the various engines so that the reaction can occur on command. Generally, different engines are available depending on the intended use: apogee engines with about 400 N thrust for orbit maneuvers and reaction engines with 1–10 N thrust for attitude control or station keeping maneuvers.

Unfortunately, a system based only on the previous elements might not work because the fluids in a spacecraft are subjected to micro-gravity conditions. A system is needed to force the fluids into the outlet of the tank. A tank filled with inert gas (typically helium) at high pressure, coupled with a pressure regulator is connected to the propellant tanks. The inert gas can either be in direct contact with the propellant or separated via a bladder connected to the tank wall. The pressure regulator decreases the pressure of the inert gas to a pressure compatible with the characteristics of the propellant tanks (e.g. burst pressure, operating pressure). This allows control of the flow rate and ensures reproducible thrust conditions.

All these elements are, of course, connected by pipes so that fluids can flow between them. Apart from the valve already mentioned, which controls the flow to the combustion chambers, other valves are also needed. Some valves are used to

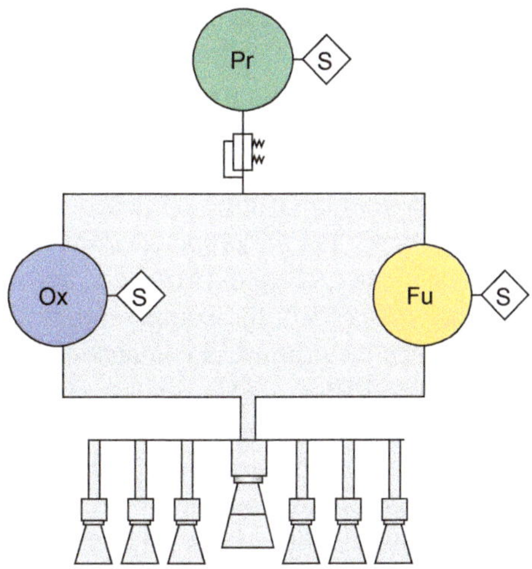

Fig. 21.2 Bi-propellant propulsion system

isolate parts of the system from the rest. Such valves can be activated only once (either to close or to open) and consist of pyrotechnic devices. Release valves, which open when the pressure exceeds a threshold, are used as a safety feature to avoid high pressures in the system. Finally, some valves are required to fill or drain the tanks and the piping at the launch pad.

The observability of the system is ensured by pressure and temperature sensors together with the related electronics. The pressure regulator and the valves are commanded by the on-board computer. The resulting system is shown on Fig. 21.2.

21.2.2 Operational Configurations

Having shown an example of propulsion system design, the first question that arises is how to operate it. Of course, operations depend on the design itself and this chapter again uses the design of a bi-propellant system as an example.

Before a spacecraft can be operated, it must be launched. The environment the spacecraft encounters during launch is drastically different (e.g. accelerations, vibrations or temperature conditions) from the one it experiences during the mission. The propulsion system is configured specifically for launch to avoid damages caused by harsh conditions: Closed valves first separate the high-pressure inert gas tanks (from 250 to 300 bar) from the propellant tanks, which are pressurized to about 18–20 bar. The propellant tanks are also isolated from the apogee engine and reaction thrusters by further closed valves to prevent accidental ignition, which would be particularly critical on top of a rocket. An inert gas at low pressure (around 3 bar) usually fills the piping between the previous valves and the engines to keep them pressurized and minimize the risk of degradation due to vibrations. The electronics controlling the propulsion system are usually shut down during ascent as a further safety measure.

After separation from the launch vehicle, the propulsion system is to be brought into an operational state (so-called "priming") before the first orbit maneuver. The logic is to put the system under increasing pressure, starting with the engines. The filler gas between the propellant tanks and the engine is vented by opening some thrusters. Then, after the thruster valves are closed again, the valve separating the propellant from the engine is opened. Finally, the valve between the high-pressure inert gas tanks and the propellant tanks is opened. The complete pressurization of the system can take up to two hours, but it can already be used under non-nominal conditions if needed, for example for attitude control purposes or in case of emergency. Once the priming is completed, the spacecraft retains its configuration until drift orbit (see Chap. 13) in the case of a geostationary satellite, or possibly until the end of the mission in the case of a low Earth orbit (LEO) satellite. These operations are either part of an automatic sequence on board or are performed manually.

The insertion into the final orbit is a delicate operation. In the case of a geostationary satellite, a box of $\pm 0.1^{\circ}$ (about 150 km wide) is allocated around the nominal longitude, and collocation of more than one spacecraft at a position is a common practice. It is therefore necessary to avoid neighboring boxes as much as possible. Reaction thrusters are better suited for such maneuvers, as well as for the further station keeping activities, because their lower thrust allows a finer orbit control. Since the apogee engine is no longer needed, it is disconnected from the propellant system by closing a pyrotechnic valve. Since the thrusters require a lower mass flow than the apogee engine, the pressure regulator is no longer needed. The inert gas tanks and the pressure regulator are isolated from the propellant tanks by closing another pyrotechnic valve. All of these activities are for the safety of the spacecraft. Once these activities are completed, the so-called blow-down phase begins, during which the pressure in the propellant tanks steadily decreases until the end of the mission. As a result, thrust performances will decrease over time. De-orbiting will be performed in this configuration. Before starting the blow-down phase, the inert gas is usually expelled as a passivation measure.

21.2.3 Electric Propulsion

The use of electric propulsion is becoming more and more widespread because of the high specific impulse of such engines. The ideal rocket equation Eq. (21.6) shows that at a higher specific impulse, a lower ratio between initial and final masses is required to achieve the same increase in velocity. This explains the observed trend, since electric propulsion allows a considerable saving in propellant mass.

Despite their high specific impulse, the mass flow rate of electric thrusters is quite low, resulting in thrusts several orders of magnitude lower than those of chemical thrusters (typically 0.01–1 N). As a result, longer run times are required to achieve the same velocity increment as chemical engines. Thus, using an electric propulsion system for a launch and early orbit phase (LEOP) would make it much longer than with a classical chemical propulsion system. An example of this is the rescue of the Artemis (Advanced Relay Technology Mission) spacecraft in 2001, which used electric propulsion to reach geostationary altitude after the rocket's upper stage failed. Despite the fact that only part of the raising was performed using the electric engine, this type of LEOP is not yet common, but still in the study phase. The duration of such a LEOP would be several months, and the concept of operations must be carefully developed to balance the cost of ground stations and operations personnel against the autonomy of the spacecraft.

The advantage of electric propulsion is often used once in geostationary orbit, for station keeping maneuvers. Because of their rather long preheating time, such engines are not used for attitude control, which relies on reaction wheels. The high-power requirements of these engines require appropriate sizing of the solar arrays so as not to interfere with the payload operation during maneuvers.

21.3 Real-Time Operations

The propulsion engineer works through the details of the propulsion system during the pre-launch preparation phase. For this purpose, he uses the manuals of the spacecraft as well as appropriate simulations. After launch comes the thrill of real-time operations.

21.3.1 Monitoring During Quiescent Periods

There is always something to do in operations, even if no activity is planned. The very first check is the limit checking. For each telemetry parameter, value ranges can be defined and automatically flagged by the telemetry processor. There are typically two types of limit ranges: a warning range, marked in yellow, indicating that the values should be monitored but no immediate action is required, and an alarm range, marked in red, indicating possible damage to a device and requiring action. The warning and alarm ranges are defined by the spacecraft manufacturer. However, operations engineers adjust these limits to suit the situation. Sometimes they are set tightly to provide early warning, sometimes they are relaxed when a situation is well understood and under control. Checking limits and on-board events is the first simple monitoring to get an overview of the system status. Under nominal conditions, no warning, alarm or non-nominal event should appear in the telemetry display system. When a warning or an alarm is triggered, it is always necessary to analyze the situation and gain a complete understanding of what is happening before acting. Many on-board automatisms, such as failure detection, isolation and recovery (FDIR), can cause warning or alarms to appear and require thorough analysis across many different subsystems.

The second monitoring concerns the pressure within the entire propulsion system. When there is no firing, the pressure should remain constant, except for the fluctuations caused by temperature changes: In fact, gas pressure responds to temperature changes according to its state equation.[1] A decrease in pressure could indicate a leakage or a thruster valve that failed to close. Although failed closure is rather obvious, leaks can be difficult to detect if they are small. In this case, only analysis over an extended period of time can reveal the problem. Temperatures should also be monitored, especially those of the thruster combustion chambers. An increase could indicate a reaction and possibly a leak.

Another value to be monitored is the time during which the thrusters are in operation (so-called "on-time"). This duration is normally generated by the on-board computer and is available in telemetry data. The more these values change, the more the thrusters are active. When an increase in on-time values is monitored, it is always important to ensure that the reason is understood. A leak or a valve that failed to close may be the root cause.

The details of monitoring will, of course, depend on the design of the spacecraft, but the previous examples provide an overview of the monitoring philosophy. Regardless of the combination of limit checking, long-term analysis, or telemetry values, the goal is to understand what is happening within the system.

21.3.2 Orbit Maneuvers

Orbit maneuvers are the most important activities performed by the propulsion system. They are carefully prepared offline (see Sect. 21.4.1 further in this chapter) and their execution is considered a critical event.

Figure 21.3 gives an overview of the pressure development during the second apogee motor firing (AMF) of a geostationary mission, while the corresponding temperature development is shown on Fig. 21.4. Both figures illustrate the main events during an orbital maneuver: The maneuver starts with the opening of the apogee engine valves, which results in a pressure drop in the propellant tanks. This pressure drop is caused by the flow of propellant toward the engine combustion chamber, which increases the space available for the inert gas and thus decreases its pressure. At the same time, the helium pressure starts to decrease as a result of the activity of the pressure regulator, which tries to compensate for the pressure drop in the propellant tanks by pumping some gas into them. After a short period of time, the pressure in the propellant tanks reaches a quasi-stationary state as an equilibrium between the two previously mentioned flows. The slight increase in propellant pressure reflects the fact that the pressure regulators are pumping helium into the propellant tanks at a higher rate than the chemical reaction is burning propellant in

[1] For example, the equation of state of an ideal gas is $PV = nRT$, where P denotes the gas pressure [Pa], V denotes the volume [m^3], n denotes the number of gas moles [mol] and T denotes the gas temperature [K]. $R = 8.314\,\mathrm{J\,mol^{-1}\,K^{-1}}$ is the ideal gas constant.

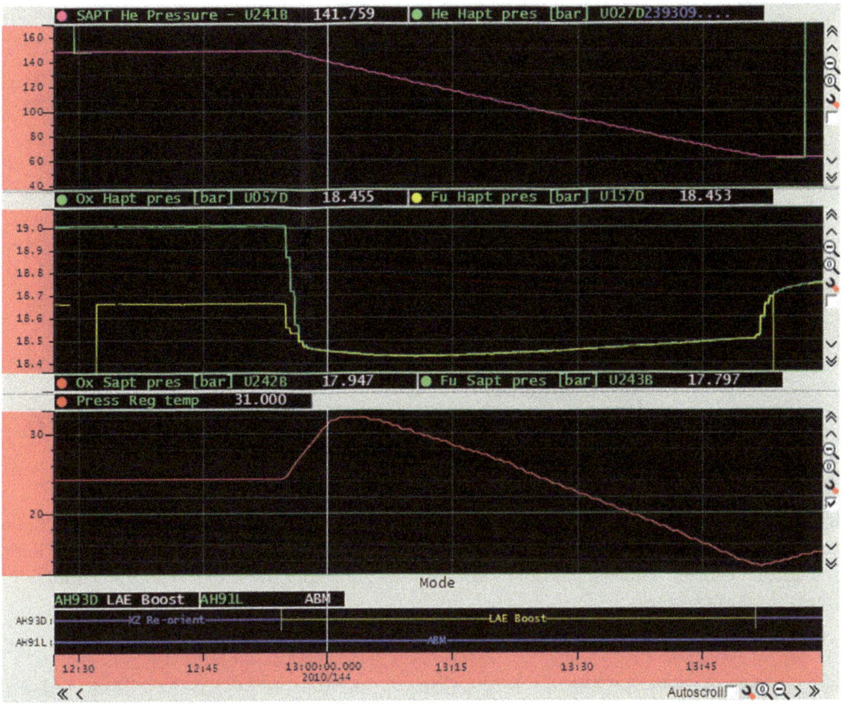

Fig. 21.3 Pressure development during AMF2. The first graph above shows the pressure in the helium tank. The second diagram shows the pressures in the two tanks for oxidizer (U057D in green) and fuel (U157D in yellow). The third graph shows the temperature of the pressure regulator and the last graph above represents the state of the attitude and orbit control system (AOCS) software and shows when the orbit maneuver takes place

the combustion chamber. When the engine valve is closed after completion of the orbit maneuver, a sharp increase in pressure is observed in the propellant tanks. This results from the propellant flow stopping while the pressure regulator is still pumping gas into the propellant tanks. Thereafter, the pressure in the tanks increases until the set point of the pressure regulator is reached.

The pressure regulator temperature increases at the beginning of the maneuver. This is due to the Joule–Thomson effect: Helium has a negative Joule–Thomson coefficient at the operating temperature, i.e. the gas heats up in the pressure regulator during expansion (from 150 to approximately 18.5 bar during this maneuver, as shown on Fig. 21.3). Later, the pressure regulator temperature decreases due to the pressure drop in the helium tank, resulting in a decrease of the helium temperature.

The temperature measured at the inlet of the combustion chamber initially drops due to the propellant flow and then reaches a steady state. As expected, the temperature increases near to the combustion chamber due to the strongly exothermic chemical reaction. More surprising is the sharp temperature rise at the end of the orbit maneuver: during the maneuver, the exhaust flow takes away some of the thermal

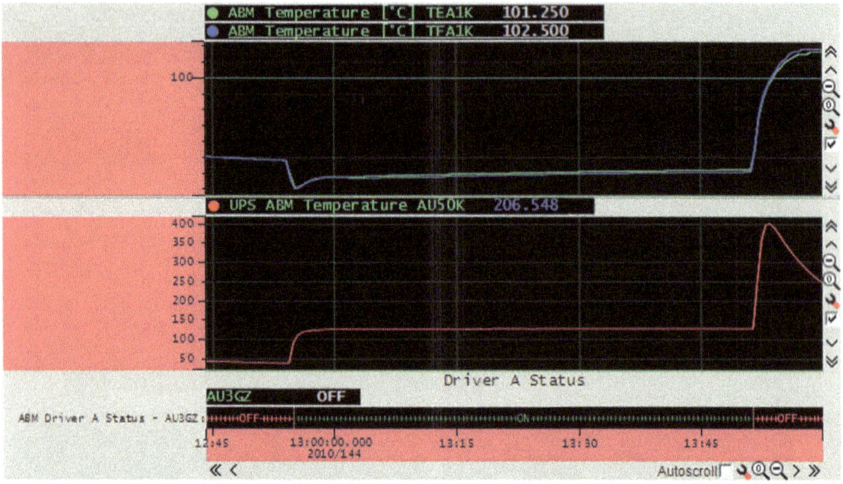

Fig. 21.4 Temperature development during AMF2. The first plot above shows the temperature at the inlet of the combustion chamber. The second plot shows the temperature measured near to the combustion chamber. The last plot gives the status of the engine valve and shows when the orbit maneuver takes place

energy generated by the chemical reaction by convection. When the engine valve is closed, the flow stops and the thermal energy can only be dissipated by radiation and conduction. Since radiation depends on temperature, the amount of heat dissipated by this mechanism initially remains relatively constant. As a result, heat begins to flow back into the relatively cold spacecraft by conduction: This is called the "heat soak-back effect" and results in a significant temperature rise (more than 350 °C) in the spacecraft around the apogee engine. Normally, a shield is installed around the engine to minimize the effects on internal equipment.

Another phenomenon that can occur is sloshing of the propellant in the tanks. The sloshing depends on the propellant mass and the space available to move freely. It is consequently reduced during AMF1 because the tanks are almost full and after AMF2 because the remaining propellant mass is low. Sloshing effects are best observed in the pointing error of the spacecraft at beginning of the maneuver.

Due to the criticality of an orbit maneuver, it is important for a propulsion engineer to quickly assess whether or not it is being performed nominally. This applies particular to the start and the end of the maneuver. The nominal start of the maneuver is first detected in the attitude system, where the rates of the spacecraft reflect the engine start. The thermal system can also confirm the correct start by monitoring the temperatures. Finally, the propulsion engineer also confirms the start by looking at the valve status. The end of the maneuver is more critical, as a failure to close the engine valve could lead to a premature end of the mission. The same systems as those previously mentioned can confirm a nominal end of maneuver, with heat soak-back and rate stabilization being the most important indicators. However, the propulsion engineer must closely monitor the pressure in the system for a period of time to ensure that no leakage has build-up.

The effects described above are quite significant for engines that produce a large amount of thrust, such as the apogee engine. For reaction thrusters, some effects might become so tiny that they cannot be observed.

Other types of engines sometimes require a preparation time, which must be considered when planning the orbit maneuvers. For example, it is necessary to preheat the catalyst of a mono-propellant system with hydrazine to avoid degradation. The same preheating time is required for the neutralizer of electric engines.

In case of electrical propulsion, the power balance and the battery status must be carefully checked before starting the maneuver. In addition, the low thrust of these engines makes a relocation dangerous. This is because several orbit maneuvers are required to achieve the necessary velocity increment. If the engine fails in the middle of the sequence, the spacecraft could be left with a drift rate still insufficient to leave the orbital slot without entering the neighboring box. This point must be considered when preparing for the relocation, possibly by contacting the operator of the neighboring satellite and exchanging maneuver plans.

21.3.3 Isolation of the Apogee Engine (Geostationary Earth Orbit—GEO)

As mentioned in Sect. 21.2.2, the apogee engine is isolated from the rest of the system after reaching the drift orbit for safety reasons. Since pyrotechnic valves can only be moved once and irreversibly, it is necessary to ensure that the apogee engine does not need to be used again. This confirmation is obtained by a precise orbit determination showing that the remaining δV required to park the spacecraft in the box can be achieved by the reaction thrusters alone. Precise orbit determination is a challenge because GPS cannot currently be used onboard geostationary satellites since their orbit is above the GPS constellation and they transmit their signals only towards the Earth. Orbit determination must then rely on ranging data (see Chap. 13, especially Figs. 13.12 and 13.13). In addition, Doppler data is of little use because the velocity in the direction of sight are small in drift orbit. In order to improve the orbit determination, ranging data is acquired alternatively from two ground stations with sufficient spacing in longitude (dual-site ranging).

The isolation closes the supply of inert gas to the propellant tanks. From this point on, the inert gas pressure in the propellant tanks decreases until the end of the mission. Sometimes the spacecraft manufacturer may decide to slightly optimize the performance of the thrusters, which depends directly on the pressure of the inert gas in the tanks. This is accomplished by turning off the tank heaters to cool the propellant and inert gas. The propellant volume changes little during this process, so the inert gas volume remains nearly constant. Consequently, a drop in temperature results in a drop in pressure, which is compensated for by the pressure regulator: This trick makes it possible to maintain a higher pressure of the inert gas in the propellant tanks after reactivation of the thermal control. Finally, it is important to carefully record pressure and temperatures in the entire propulsion system immediately after isolation for the purpose of mass calculation and lifetime estimation (see both sections under offline operations).

21.3.4 Autonomous Operations

Orbit maneuvers performed by either the apogee engine or reaction thrusters are accomplished by directly commanding the opening and closing time of the engine valves. Reaction thrusters generate a force on the spacecraft that can be used for attitude control if their direction does not pass through the current center of mass. In this case, the thruster also generates an external torque that can be used to control the attitude or angular momentum of the spacecraft.

Due to the complexity of attitude motion (see Sect. 14.5 Attitude Control), the attitude and orbit control system (AOCS) normally controls the thruster firings autonomously. When the AOCS controller aims to rotate the spacecraft, the thruster firings are combined to give a torque about the desired axis with the desired magnitude. The torque obtained with reaction thrusters can be used to control the angular momentum of the spacecraft. This is particularly interesting when a reaction wheel assembly is used to control attitude. In effect, this assembly stores angular momentum from external disturbances that cause the wheels to reach their saturation speed. Reaction thrusters aid in the desaturating of wheels in much the same way as magnetic torquers. This activity is usefully combined as part of station keeping maneuvers to take advantage of the resulting force on the spacecraft.

In such autonomous operations, monitoring is more qualitative and based on physical sense. For example, the run-up of a momentum wheel is counteracted by firing thrusters to keep the platform's orientation constant. The propulsion engineer must verify that the thruster activity indicated by telemetry data actually corresponds to a torque along the axis of the momentum wheel.

21.4 Offline Operations

The real-time operations described previously require some preparatory work and analysis after execution. Such activities are called "offline" in contrast to the actual operations.

21.4.1 Preparation and Calibration of Orbit Maneuvers

Three parties are involved in the preparation of the orbit maneuver: a flight dynamics engineer who determines the required δV and center of burn (time of apogee crossing from orbit determination), the propulsion engineer who provides the spacecraft mass, and the satellite manufacturer's propulsion expert who provides corrections based on flight experience.

The thrust direction of an apogee maneuver should be tangential to the orbit, and the spacecraft must be aligned accordingly. Flight dynamics experts provide the re-orientation commands as required by the spacecraft AOCS. On the other hand, δV and timing depend purely on the current and final orbits. These quantities are converted to start and stop times by taking the amount of thrust the apogee engine can actually deliver. Assuming the ideal rocket equation, Eq. (21.6) neglecting gravity with respect to thrust, and assuming constant specific impulse, we get:

$$M_f = M_i - \delta m = M_i e^{\frac{-\delta V}{g_0 I_{sp}}} \Rightarrow \delta m = M_i \left(1 - e^{\frac{-\delta V}{g_0 I_{sp}}}\right) \tag{21.7}$$

Knowing the initial mass provided by the propulsion engineer from his/her mass calculation protocols (see Sect. 21.4.2 below) and the specific impulse, one can calculate how much propellant mass will be consumed during the maneuver. Finally, knowledge of the expected mass flow rate $\frac{dm}{dt}$ allows calculation of how long the burn should take. Both mass flow rate and specific impulse depend on pressure and temperature within the propulsion system. An analytical approximation formula is derived from test series on the ground, e.g. as a polynomial or in form of tables. These are given in the user manual of the spacecraft. The propulsion engineer can provide the current pressure and temperature values to help calculate an estimate of the burn time. However, this estimate would not be accurate because the pressure and temperature values needed are the not current ones, but the values during the maneuver. Furthermore, Fig. 21.3 shows that the pressure in the tank drops during the maneuver and this pressure drop varies with the filling level of the tank. Finally, the pressure does not remain constant during the maneuver, but increases slightly with time, as can be seen in Fig. 21.3. At this point, the spacecraft manufacturer's propulsion expert can help by providing more accurate data based on flight experience, usually in the form of an expected average thrust level. Once the maneuver duration has been calculated, the start and stop times can be easily derived from the center of burn time.

The above calculations must be corrected because the reaction thrusters are also used for attitude control and thereby have an influence on the orbit control. The thrusters are autonomously activated by the AOCS to generate the torques required to align the apogee engine in the desired thrust direction. These torques result from forces that do not pass through the center of mass, but also these forces contribute to δV in a way that is usually predicted by simulations. Gravity, which was neglected in the calculation above, also affects the maneuver performance as it slows down the spacecraft in orbit, especially for long maneuvers. It is therefore necessary to calibrate the orbit maneuver after its performance. An accurate orbit determination gives the δV actually achieved: It is then possible to calculate a correction factor (e.g. to the mean thrust level), which is considered when preparing the next maneuver.

21.4.2 Calculation of the Propellant Mass

The previous section has clearly shown that the mass of spacecraft must be known. It can be divided into the dry mass, which includes the spacecraft equipment, and the wet mass, which includes the propellant and the various gases used in the propulsion system. The dry mass is measured accurately on ground before launch. The wet mass, on the other hand, is known at launch as the tanks are filled. The wet mass decreases as propellant is burnt and this mass cannot be easily measured directly. In fact, direct measurement of the remaining propellant mass is complicated by its distribution within the tank, which is determined by the surface tension of the propellant under micro-gravity conditions. For this reason, the remaining propellant mass is usually estimated using two complementary methods.

The first method is based on the thermodynamic properties of the inert gas used to pressurize the propellant tanks. It uses pressure and temperature measurements available in telemetry data to calculate the volume occupied by the gas from its equation of state. A simple subtraction from the tank volume yields the propellant volume and, via the propellant density, the mass from that. These quantities explain the name of PVT (pressure–volume-temperature) for this method. Although the principle is simple, several effects must be considered to obtain correct accuracy. First, the ideal gas law no longer holds accurately at pressures above 5 to 10 bars, and it is necessary to resort to more complicated equations of state. A practical way is to introduce a correction factor $Z(P, T)$ so that the equation of state is now $PV = ZnRT$. The values of the correction factors are then given for the operating range of pressure and temperature, either as a function or as table. A liquid evaporates and its vapor mixes with the gas used to pressurize the propellant, resulting in a change in tank pressure. The effect of this gas mixture on the measured tank pressure is sometimes taken into account. In addition, gases dissolve in liquids and must be considered when calculating the number of moles n. Finally, one may mention the dependency of propellant density on the temperature and the expansion of tanks under pressure. All these effects may seem so small that they could be neglected, but it will be seen that the accuracy of the thermodynamic method is closely related to end of life, justifying the consideration of these effects. Another pitfall of the PVT method is the activation of relief valves when the pressure in the system becomes too high. When this happens, it is necessary to estimate how much gas escaped from the system.

The second method is based on the knowledge of the propellant mass flow rate in all engines. As with the correction factor for real gas introduced earlier, the mass flow rate of an engine is measured on the ground and provided to the propulsion engineer as a function of pressure and temperature, whose coefficients are specified, or as a table. In either case, knowledge of the mass flow rate allows calculation of the propellant mass consumed by the engine by integration over the maneuvering time. The remaining propellant mass is then calculated simply by subtracting the mass consumed during engine activation from the mass present prior to activation. Because of this accounting of the total mass consumed during the actuations, this method is called "book-keeping".

Having two methods is a good way to cross-check calculations. However, they rarely agree and the difference can be in the range of 10 kg at end of life. The book-keeping method is more accurate in the short run than the thermodynamic method. The latter has an accuracy that is determined by the accuracies of the sensors involved and the underlying thermodynamic model. The book-keeping method, on the other hand, is applied to every maneuver and inaccuracies add up over time. Which of both methods has the better accuracy should be carefully analyzed by the satellite manufacturer, especially at end of life.

Another method based on measuring the thermal capacity of a propellant tank has emerged in the last decade. Thermal capacity depends on how much propellant remains in the tank and can be estimated by measuring the response of the tank temperature to controlled heating. However, this estimate requires modeling of the propellant tank, the location of the probes, and the location of the temperature sensors. This method has the advantage of being more accurate as the remaining propellant mass decreases.

21.4.3 Calibration of the Center of Mass

The calculation of the mass is closely related to the position of the center of mass of the spacecraft and the inertia tensor. Both are used in the attitude control algorithms and are parameters loaded into the satellite's safeguard memory. Due to the mass consumption by the propulsion system, it is often necessary to update the values of these parameters.

Some spacecrafts, such as the Spacebus 3000, overcome this problem by setting the value of a predefined index once per year. This index corresponds to a predefined position of the center of mass and is used as input by the attitude and orbit control system. The values are chosen to approximate the nominal propellant consumption over a year, but it might be necessary to choose the value differently if, for example, the spacecraft had an anomaly or had to be repositioned to a different orbital position.

The Gravity Recovery and Climate Experiment (GRACE) mission had particularly high requirements for knowledge of the center of gravity position. This mission consisted of two satellites spaced approximately 220 km apart. A microwave link between the two satellites allowed observation of the Earth's gravity field. However, it was necessary to filter out non-gravitational accelerations, which was accomplished with an accelerometer. To minimize accelerations resulting from satellite rotation, the center of mass was fixed with respect to the spacecraft structure by moving small masses around the satellite. Because of the accuracy required, special attitude maneuvers were periodically commanded about the three axes of the spacecraft to accurately calibrate the position of the center of mass. After an off-line analysis of the results, the positions of the small masses were commanded so that the center of mass remained at the assigned position.

21.4.4 Estimation of Lifetime

A mission is very often terminated not because the payload is not functioning properly, but because the propellant runs out. As debris is a growing problem, spacecraft operators are increasingly forced to leave useful orbits (LEO or GEO) and put their property either into graveyard orbits or de-orbit it into the atmosphere. The propulsion engineer plays an important role in this decision, because his/her mass estimate, considering de-orbit maneuvers, is an important input to determine the end of the mission. There is also the need to know which of the mass calculation methods to trust for this decision. However, the pressure is high to delay the end of life and "push the outside of the envelope" as far as possible.

De-orbiting maneuvers are calculated by flight dynamics experts. As propellant tanks are almost completely depleted or end-of-life is delayed longer than expected, it can be difficult to predict the amount of thrust generated. In fact, pressuring gas bubbles can affect combustion and lead to non-nominal performance.

Reference

Fortescue P, Swinerd G, Stark J (2011) Spacecraft systems engineering, 4th edn. Wiley

Franck Chatel is Flight Director for the H36w-1 and EDRS missions within the department mission operations at the German Space Operations Center (GSOC). He graduated in aerospace engineering at the Institut Supérieur de l'Aéronautique et de l'Espace (ISAE) in Toulouse (France) and joined DLR in 2001. He was involved as propulsion and AOCS sub-system engineer, later as flight director, in different LEO missions (BIRD, SAR-Lupe, GRACE) and GEO missions (EUTELSAT W24). He supported the early phase of the GALILEO project at spaceopal and DLR GfR before moving to his current position.

Chapter 22
Attitude and Orbit Control Subsystem Operations

Ralf Faller

Abstract This chapter provides a general overview of the attitude and orbit control subsystem, its components, and the operational tasks to be performed by the ground control personnel. It supplements the fundamentals of attitude and orbit control explained in Chap. 13.

After a brief introduction and an outline of its relevance for various requirements coming from other subsystems and the payload, Sect. 22.2 provides an overview of the subsystem. Its components and most common applications are described and the basic functions of the on-board control are explained. With this background, Sect. 22.3 summarizes the tasks performed by the ground control team during various mission phases. Some life experiences in various missions are reported in Sect. 22.4. Finally, the topic of the attitude and orbit control system (AOCS) is summarized in 22.5.

22.1 Introduction and Overview

The attitude and orbit control system (AOCS) is one of the most important subsystems to ensure that payload and other subsystem requirements are met.

Attitude involves two aspects: The first is the orientation of the spacecraft with respect to reference objects or directions, e.g. Earth, Sun, stars, direction of flight, etc. Attitude accuracy requirements cover the entire range, from coarse alignment of a satellite with the Sun for adequate power on the order of a few degrees, to high-precision orientation tasks on to the order of arc seconds, e.g., for pointing an astronomical telescope at specific targets. The second aspect of attitude control is its dynamic nature. Some satellites need to change their orientation from time to time,

R. Faller (✉)
German Space Operations Center (GSOC), German Aerospace Center (DLR), Oberpfaffenhofen, Germany
e-mail: ralf.faller@DLR.de

F. Sellmaier et al. (eds.), *Spacecraft Operations*, Springer Aerospace Technology,
https://doi.org/10.1007/978-3-030-88593-9_22

to rotate about specific axes (e.g. daily pitch rotation of geostationary satellites) or to stabilize their orientation after separation from the launch vehicle. Thus, rotation and rate-of-rotation control are attitude control tasks.

Orbit control encompasses all aspects of maintaining or achieving a target orbit. Geostationary (GEO) communications satellites must be positioned on the GEO and maintained in their control boxes throughout their life to adequately support the payload. Similarly, all kinds of low Earth orbit (LEO) satellite missions must acquire and maintain their target orbits, must maintain formations, or must change orbits depending on the mission purpose. Another aspect is avoiding collisions with other objects. As space becomes increasingly crowded, this issue is relevant to both manned and unmanned space missions.

Overall, maintaining attitude and orbit are tasks which are provided by the corresponding control system. The on-board control itself is realized by three main instances. Data about the current attitude and orbit are received from sensors in the form of aspect angles, rotation rates, or position measurements. These data are collected by a control unit, which derives appropriate attitude and orbit related parameter. Deviations are determined by comparison with nominal values and correction (control-) demands are generated. These control demands are sent to actuators, such as thrusters or other devices, which generate control torques and forces that correct attitude or orbit. Figure 22.1 illustrates these major components of the on-board control loop.

Attitude control is a classic on-board closed-loop control. The ground control center can adjust control settings and margins or change control strategies (see attitude modes), but the control loop requires short round-trip times of the control signals and is therefore always closed on board.

Orbit control works differently to some extent. Orbit correction maneuvers are usually performed at discrete times and not permanently. The current orbit is measured and calculated, either on board using global positioning data (e.g. GPS or Galileo) or classically on ground using station tracking information. The decision for and preparation of an orbit maneuver is made on the ground, then commanded to the satellite, and finally executed by the spacecraft at the specific maneuver time. Normally, there are no real-time requirements for the control loop. Only missions

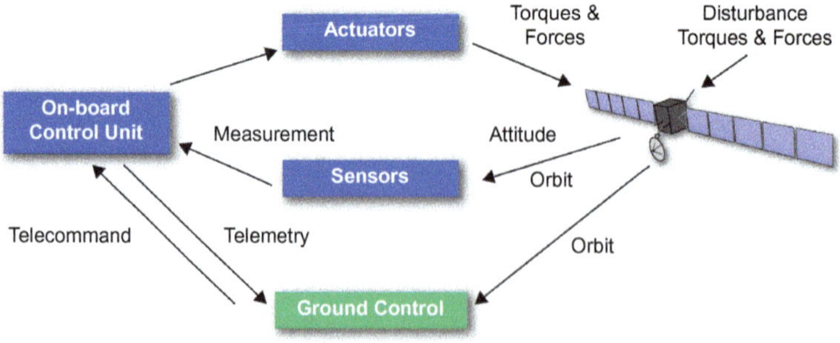

Fig. 22.1 Subsystem components and control cycles

that require permanent orbit control (e.g. rendezvous and docking situations) need to have an on-board closed loop.

22.2 Subsystem Description

With the basic understanding of the attitude and orbit control described in Sect. 22.1, this chapter gives an overview of the most common sensors and actuators and describes the typical functions of the control unit. The descriptions focus only on the most important aspects.

22.2.1 Sensors

The primary function of sensors is to measure the position, orientation, and dynamics of the spacecraft primarily with respect to celestial or other reference objects. The output of these sensors can be position and velocity vectors, attitude angles, rotation rates, or inertial orientations of the spacecraft. Whilst orbit-related data are provided with respect to Earth-related coordinate systems (e.g. the Earth-centered, Earth-fixed (ECEF) **coordinate** system), attitude measurements are provided with respect to reference axes of the sensor or spacecraft body. Depending on which type of information is used, attitude information is provided over one, two, or all three axes. The following list of typical sensors provides a brief overview.

Sun Sensor
This group of optical sensors is most commonly used in space applications. The brightest object in our solar system provides an unambiguous orientation source. Various technical solutions are available that provide directional data with an accuracy from a few degrees in case of simple safe-mode sensors to better than 0.05°. Another main feature is the large field of view of a single sensor head, which is in the order of ±60–90°. Shading of Sun light by other objects (e.g. eclipse phases by Earth or Moon) must be considered operationally. For some solar sensor models, the Earth albedo may reduce the achievable accuracy. Since the Sun is rotationally symmetric for more or less all sensor applications, only 2-axis attitude information can be obtained (Fig. 22.2).

Earth Sensor
Classic Earth sensors provide directional information by scanning the Earth's infrared disk. They are typically used for Earth related missions with moderate pointing accuracy requirements, such as geostationary communication satellites. These sensors can be interfered with by other infrared objects such as the Sun or the fully illuminated Moon, which must be considered during flight operations.

Fig. 22.2 Coarse Sun sensor (image: bradford-space.com)

Further Earth sensor applications are available with imaging systems, i.e. cameras. In addition, combinations of solar and Earth sensors have been developed. Such sensors consist of thermistor elements that provide rough directional data of the two celestial bodies.

Star Sensors

This class of optical sensors uses stars as an orientation source. Unlike Earth or Sun sensors, star sensors provide an inertial attitude reference about all three axes. The achievable accuracy is on the order of arc seconds. Interference from solar or lunar glare is possible and is usually accounted for by satellite design by implementing 2–3 sensors in different directions. See also Chap. 14 (Fig. 22.3).

Magnetometer

The Earth's magnetic field can also be used as an orientation source. Magnetometers measure the direction of local magnetic field lines and thus provide a good 2-axis attitude reference. While fairly accurate mathematical models exist, the Earth's magnetic field is affected by solar activity, so magnetometer measurements are rarely used in attitude determination calculations. Nevertheless, magnetometers are often used in combination with magnetic torque rods to provide attitude control torques.

Gyroscopes

Gyroscopes measure the rotation rates of spacecraft. Compared to the sensors described so far, gyroscopes provide relative attitude information. They are mainly used to obtain data on attitude changes and therefore allow to cover phases with reduced availability of absolute sensor data (e.g. during eclipse phases or slews

Fig. 22.3 Star sensor KM1301 (image: OHB System, formerly Kayser-Threde GmbH)

from one orientation to the next). All types of gyroscopes have a drift that causes cumulative rate integration errors over time.

Space-Based Satellite Navigation Systems

Satellite navigation systems, such as the Global Positioning System (GPS) (others: GLONASS, GALILEO, Beidou) can be used to collect position (and velocity) data from LEO missions. These data are input to on-board navigation systems that perform orbit determination tasks in space. In addition, specially equipped GPS can be used for attitude determination, preferably for larger spacecraft.

Other Sensors

In addition, these common sensor systems described above, some other sensor concepts may be in use. Radio frequency (RF) Beacon sensors may be mentioned as an example. They use RF sources on ground as an orientation reference with an accuracy of about one arcminute.

22.2.2 Actuators

The change of velocity or rotation rate of the spacecraft can be realized by different actuator systems, which generate the required forces or control torques:

Thrusters

Thrusters are the classic equipment for attitude and orbit control. Depending on the complexity of the spacecraft, they are sometimes part of a separate subsystem, as detailed in Chap. 21, but activity requests are always generated by the AOCS subsystem. Since the amount of fuel on board a spacecraft is limited, the availability of all thruster systems is correspondingly restricted.

Magnetic Torquer

These actuators use the effect of interaction with magnetic fields. Electric magnets on board a spacecraft, called magnetic torquer rods, can generate control torques by interacting with the Earth's magnetic field. Also, by mounting three torquer rods in the three body axes of the spacecraft, torques can only be generated about axes that are rectangular to the local magnetic field lines. However, this restriction to one plane can be compensated due to the change of the Earth's magnetic field vector over an orbit. Unlike thrusters, magnetic torquer are available as long as electric power is available. They are provided in two different applications, torque rods with iron core and air coils (Fig. 22.4).

Wheels

In principle, a reaction wheel consists of a rotating mass connected to an electric motor. By increasing or decreasing the rotational speed, the angular momentum of the mass increases or decreases. Due to the law of conservation of momentum, the total angular momentum of the spacecraft remains constant as long as no external torque is applied. As a result of speeding up the reaction wheel and the corresponding increase in its angular momentum, an opposite momentum is applied to the structure

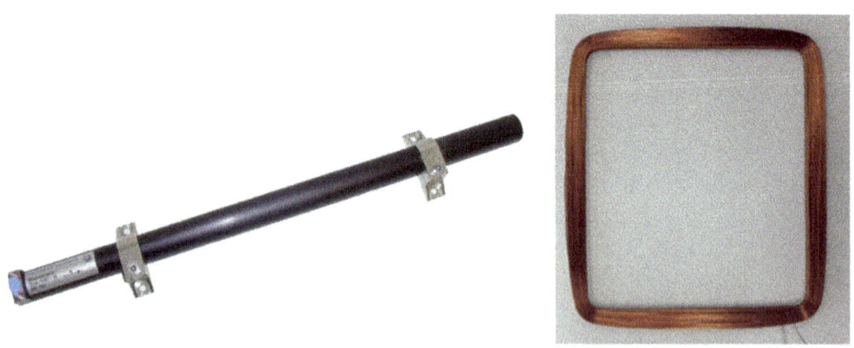

Fig. 22.4 Left: VMT-35 magnetic torquer (image: vectronic-aerospace.com); right: COMPASS-2 cubesat air coil (image: FH Aachen)

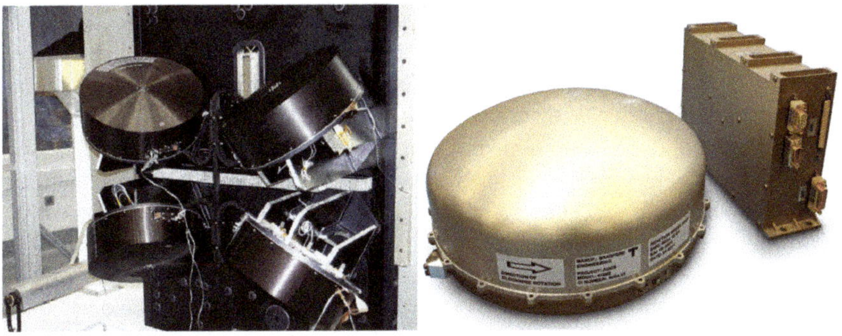

Fig. 22.5 Left: Reaction wheels on Lunar Reconnaissance Orbiter, NASA GSFC (image: NASA); right: Reaction wheel unit (image: bradford-space.com)

of the satellite. This effect can be used for attitude control. Unlike to thrusters or magnetic torquer, the angular momentum of the overall system is not changed. The angular momentum is only shifted between the spacecraft body and the wheel.

There are two different types of wheels. Reaction wheels are used for all types of satellites to generate attitude control torques. Wheels can rotate in either direction, but zero-speed transitions are usually avoided for high accuracy applications. Momentum wheels, called flywheels, run at high-speed level. They are mainly used for GEO satellites to store some of the angular momentum in the pitch axis of the spacecraft to provide some dynamic stiffness against external disturbing torques.

Due to friction effects or systematic external disturbance torques, wheels must be desaturated from time to time when their maximum speed is reached. This is called "wheel unloading" or "wheel desaturation". Therefore, purposeful torques are generated using thrusters or torquer, which can be compensated by the wheel by reducing its speed (Fig. 22.5).

22.2.3 *On-board Control Unit*

Functions

The on-board control unit is the central element of the AOCS. It can be a separate hardware or a separate process running on the on-board computer. The main tasks of the AOCS control unit are:

- *Calculation of current attitude and prediction*: From all sensor measurements, the current S/C attitude, rotation rates, and their deviations from nominal values are typically calculated using numerical filtering (e.g. Kalman filter). In addition, the evolution of the parameters for the next control grid point (see Sect. 14.3) is also predicted. Examples of such attitude parameters are the Sun aspect angle in case of Sun pointing spacecraft or errors in roll, pitch, and yaw in case of an Earth-oriented geostationary satellite.

- *Generation of control arguments for attitude correction/adjustment*: Based on the predicted attitude deviations, control requests are generated, such as accelerating or decelerating a corresponding reaction wheel, switching on/off a magnetic torquer, or firing selected thrusters in to achieve a desired attitude change.
- *Wheel unloading*: As mentioned above, wheels need to be desaturated from time to time, when they reach their maximum speed (typically a few thousand RPM). Continuous wheel unloading is also possible. This process is completely controlled by the AOCS without ground support.
- *Calculation and prediction of orbit-related data*: Spacecraft equipped with an on-board navigation system calculate the orbit using GPS measurements. Alternatively, the orbit position can be predicted using uploaded orbit reference date (e.g. two-line elements TLE).
- *Error detection and reaction*: Early detection of problems within the AOCS and autonomous response without the need for ground control intervention is a mandatory function for all types of spacecraft. This is realized by a corresponding fault detection, isolation and recovery (FDIR) mechanism.

AOCS Modes

More or less all spacecraft have different operating modes. By selecting a specific mode, different settings and pre-selections are used for the attitude control. The most relevant settings are:

- *Control strategy selection*: For example, if you switch to a Sun pointing mode, the Sun is set as the main orientation reference; the spacecraft will orient itself relative to that direction.
- *Setting the control ranges and limits*: For modes with high requirements for attitude accuracy, the control spans are kept narrow, while for modes such as a safe mode, only a coarse attitude is required and the control spans can be set to a larger interval.
- *Pre-selection of the required sensors and actuators*: The devices required for the corresponding mode are activated, others are deactivated. A star sensor, for example, would be switched on for a mode with high pointing accuracy, but could be deactivated in safe mode.
- *Selection of FDIR strategies*: This means activating or deactivating predefined reactions to possible malfunctions.

Table 22.1 shows some exemplary AOCS modes and corresponding settings.

Sub-modes

Each mode can have different states, called sub-modes, to allow for graded activities and working levels. As an example, Fig. 22.6 shows sub-modes of a simple Sun pointing mode. The lowest is a rate damping state, in which only the spacecraft rotation rates are dumped to low values to provide adequate conditions for attitude measurement. Since this is the default input state after entering the Sun pointing mode, this sub-mode is mainly defined for the first Sun orientation after separation

Table 22.1 Example AOCS mode

Mode	Orientation strategy	Control limits	Sensors/actuators
Safe mode (Sun pointing)	z-axis towards Sun slow spin around z (2-axes stabilization)	0.0° ± 0.3 in both axes 0.4°/s ± 0.01 about z	Sun sensor Gyros Thrusters
Inertial target pointing (LEO scientific satellite)	Inertial target orientation (3-axes stabilization)	± 0.002° in all 3 axes	Star camera Gyros Magnetometer Reaction wheels Magnetic torquer
Earth pointing (GEO communication satellite)	z-axis oriented to Earth, y-axis perpendicular (south) to the orbit plane, x-axis quasi in-flight direction (3-axes stabilization)	0° ± 0.01 in roll and pitch 0° ± 0.2 in yaw	Sun sensor Earth sensor Gyros Thrusters Wheels

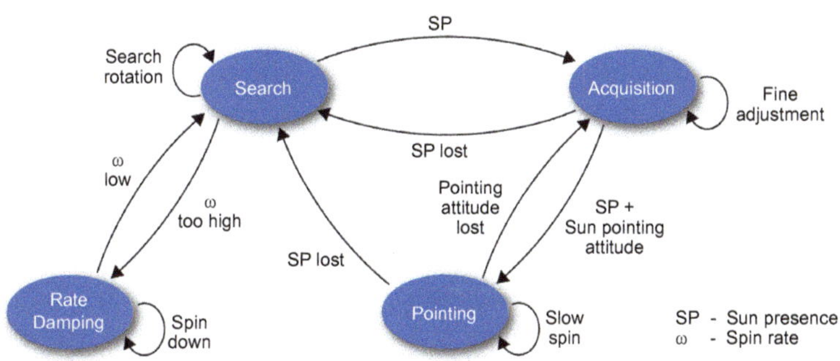

Fig. 22.6 Sub-modes and control laws for a Sun pointing mode

from the launcher, when the spacecraft has been put into high rotation. When the rates are low enough, the on-board control switches to the search sub-mode. Here, the mode logic checks if the Sun is already in the field of view of the Sun sensors (of course, this sub-mode is not useful for spacecraft with a total field of view). If the Sun sensors do not provide a Sun presence (SP) signal, the spacecraft begins a rotation about one or two body axes to bring the Sun into view. When a valid Sun presence signal is available, the acquisition sub-mode aligns the appropriate body axis of the spacecraft with the Sun. Once this is sufficiently accomplished, the highest sub-mode is invoked. Here, a slow rotation about the Sun orientation is performed to achieve passive attitude stabilization.

Fault Detection, Isolation and Recovery

FDIR routines and functions are a key element of the AOCS. While other subsystems typically have longer response times before a problem becomes severe, failures in

Table 22.2 Example of possible error levels and adequate FDIR reactions

Error level	Error/problem	FDIR reaction	Impact on payload
1	Sensor failure, indicated by equipment status bit	Switchover to redundant sensor	None Payload activities are continued Ground control has time to analyze the problem
2	Attitude error exceeds limits	Switchover to redundant set of equipment	Possible
3	Loss of attitude reference (e.g. no attitude solution available)	Switch to safe mode Switchover to redundant set of equipment	Yes Payload is shut down Ground control needs to analyze and react, until payload operations can be resumed

the AOCS control loop can immediately result in loss of proper attitude and potentially poor thermal and power conditions or loss of communication link. Therefore, autonomous on-board control response strategies are implemented for various failure conditions. A key element of these FDIR routines is the appropriateness of the reaction with respect to the problem encountered. For example, a small malfunction in a sensor system should not send the entire spacecraft into safe mode, which would result in a payload shutdown and a significant disruption to the ongoing mission. A simple switchover to the redundant sensor would be more appropriate and would allow the current activities to be maintained. Various error levels and corresponding reactions are defined for adequate responses. Table 22.2 provides some example error levels and corresponding reactions for a LEO scientific satellite mission.

22.2.4 AOCS Device Combinations and Redundancy

As described in the previous sections, there are various sensors and actuators with different technical realizations and performance levels. Table 22.3 shows sample combinations of sensors and actuators. For three different types of missions, the equipment is labeled as most suitable (++), suitable (+), or not suitable (−), but other combinations are possible and depend on mission constraints, project budget, and other S/C design factors. For example, thrusters are the main actuator for orbit control, but they could not be used for scientific missions with sensitive sensor systems because the exhaust gases from the thrusters would impinge the neighboring surfaces.

In addition to the selection of suitable sensors and actuators, the numbers of devices per category is another relevant aspect for operations. In unmanned space applications, single failure tolerance is usually achieved by using two devices, one active and in control, the other switched off and available as cold redundancy.

Table 22.3 Possible sensor/actuator combinations

Equipment	Mission		
	Commercial TV geostationary	Scientific near earth	Scientific deep space
Sun sensor	++	++	+
Earth sensor	++	±	–
Star tracker	+	++	++
Magnetometer	–	+	–
Gyros	++	++	++
Thruster	++	+±	++
Wheels	++	++	+
Magnetic torquer	–	++	–

Figure 22.7 shows an example of a functional AOCS block diagram for a geostationary communication satellite. A redundant one is available for each type of sensor. A second control unit is also provided. A redundant flywheel is available in case one of the wheels fails. The thrusters are covered by a separate subsystem in this example.

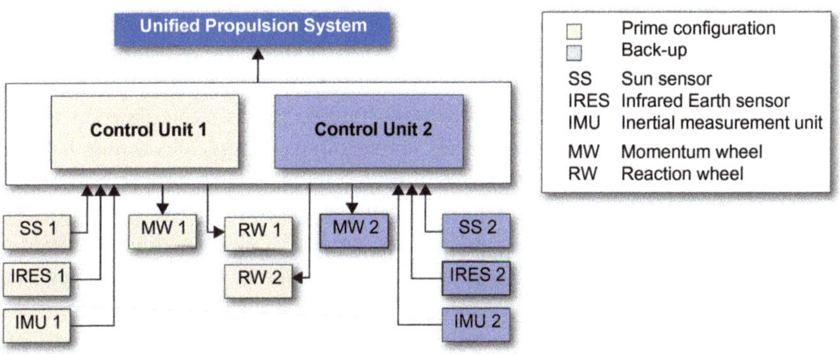

Fig. 22.7 Functional AOCS block diagram for a GEO satellite

22.3 AOCS-Related Ground Operations

22.3.1 Basic AOCS Ground Activities

This section describes general tasks and responsibilities of an AOCS engineer during all mission phases. As with all other satellite subsystems, flight procedures must be available for the AOCS to provide a reliable foundation for flight operations.

Monitoring
The most common ground activities are all types of monitoring and analysis of subsystem functions, equipment, and performance. Specifically, AOCS monitoring includes the following:

- *Current status check*: Subsystem status is provided by the AOCS mode and sub mode. Error words or flags, triggered FDIR actions, often in combination with spacecraft event messages, are indicative of problems detected by the AOCS. In case of anomalies or FDIR activities, routine activities change to contingency operations (see Sect. 22.3.3).
- *Equipment health and performance checks*: These more specific checks include on/off settings, status and health information from sensors and actuators. Direct readings of sensors and actuator duty cycles, such as wheel speeds, magnetic torquer activities, and thruster on-times provide an initial assessment of the AOCS performance.
- *Current attitude*: This expression means the attitude solution calculated by the on-board control system, considering all sensor inputs. Larger deviations between calculated and expected attitude are a direct indication of problems with the attitude control, if they have not already been triggered by FDIR mechanisms.
- *Orbit monitoring*: For spacecraft equipped with satellite navigation systems (e.g. GPS), current orbit measurements are monitored. Typical parameters to be checked are the number of satellites tracked and the availability of position and velocity solutions. If on-board routines are available to process this data, their operation will be observed. These above checks and analyses are typically performed during a ground station contact with the spacecraft by observing the incoming real-time telemetry. The main tool for real-time monitoring is the display system, which provides appropriate display pages. Figure 22.8 is an example of common sensor-specific telemetry.

Another valuable monitoring method is the analysis of recorded (offline) data. Offline checks and analysis allow the subsystem to be monitored over extended periods of time. In addition, long-time analysis of equipment telemetry can indicate

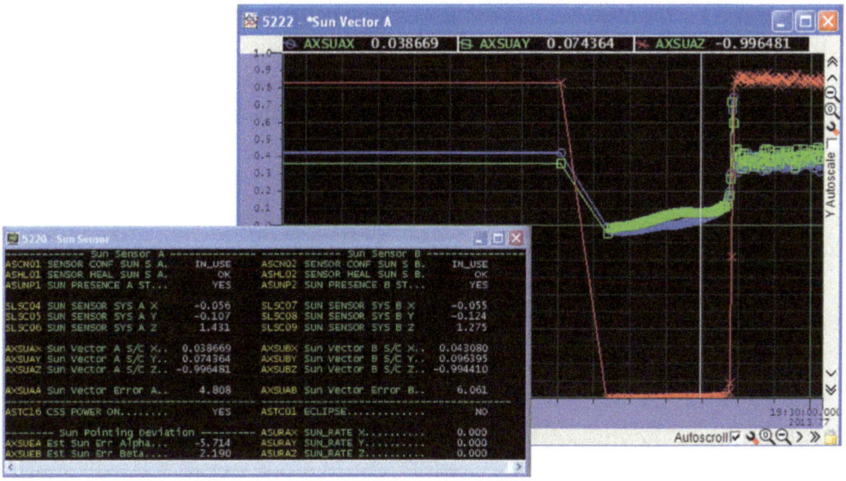

Fig. 22.8 Classic alphanumeric and graphical display pages for AOCS. While alphanumeric pages provide only a snapshot of a status, graphs allow observation of parameter behavior over time, which is especially useful for dynamic AOCS parameter, such as angles, rates, momentum, etc.

evolving problems or anomalies (e.g. increase of wheel friction due to degradation or aging effects, change in data noise of mechanical gyros, etc.).

Support of Orbit and Attitude Maneuvers

The AOCS engineer supports all types of orbit and attitude maneuvers. For maneuver preparation (e.g. station keeping maneuver for geostationary satellites or orbit correction maneuver for LEO missions), he works in close contact with the flight dynamics engineer. During routine phases, maneuvers and corresponding command lists are planned and prepared mainly by mission planning systems or flight dynamics tools.

When real-time telemetry is available, maneuver execution and corresponding AOCS performance are monitored. Otherwise, offline telemetry is evaluated retrospectively.

Various Activities

Depending on the mission and the design of the spacecraft, additional tasks are in responsibility of the AOCS engineer. Some typical tasks are mentioned here:

- *Routine updating of reference data*: On-board navigation systems for orbit determination sometimes require periodic updates of reference data (e.g. two-line elements) as fallback option in the absence of GPS measurements.
- *Sensor interferences*: For LEO missions, sensor interferences normally do not require special ground activities, but for some GEO satellites, Earth sensor interferences from the Sun or moon must be handled. The standard measure to avoid interference is to switch of the corresponding sensor head, when the Sun or moon enters its field of view. Such head switching is done either manually by time-tagged

commands (in the case of lunar events) or automatically by on-board routines (in the case of solar events). In both cases, the ground personnel monitor these events.
- *Eclipse phases*: As with sensor interferences, eclipse phases are typically only an issue for GEO missions. Some GEO satellites must be configured for each eclipse phase and are monitored by ground personnel during the event.

22.3.2 LEOP and In-Orbit Tests

The LEOP and in-orbit test (IOT) phase is the most intense phase for AOCS engineers. The engineer must perform all tasks described in the section above. In addition, he is also responsible for:

- *Initial estimate of attitude after separation from launch vehicle*: After receiving the first telemetry data from the satellite, the AOCS engineer performs a quick subsystem checkout. LEO satellites typically invoke their attitude control shortly after the board computer has started. Whether this can be monitored in real time by ground personnel depends on the visibility of the first ground station. For GEO missions, real-time telemetry is often available shortly after separation, even if they are first brought into a geostationary transfer orbit (GTO). This allows the AOCS engineer to verify attitude during separation. In this case, AOCS activation and all subsequent steps are ground-controlled.
- *Orbit and attitude maneuver support*: All orbit and attitude maneuvers performed the first time are carefully prepared, monitored in real time, and subsequently analyzed and calibrated. For GEO missions launching into the GTO, this includes performing apogee boost maneuvers.
- *In-Orbit Tests (IOT)*: Before the mission enters routine phase, the spacecraft and its payload are tested and calibrated. AOCS tests include checkouts of sensor orientation and field-of-view measurements, sensor calibration (e.g. gyroscope drift compensation), redundant equipment tests, and actuator checkouts. It depends on the mission what types of tests are performed. Customers who have ordered a satellite delivered to space often request testing of all redundant equipment (all redundant sensors and actuators) to confirm that all the components they are paying for will work. Other missions prefer to leave redundant components unused until they are really needed.
- *Payload checkouts*: Checkouts and in-orbit tests are also supported by the AOCS engineer as needed. Communication satellites often perform antenna mapping during IOT, measuring the “footprint” of the payload antennas on ground. To do this, a single ground station measures the downlink power while the entire GEO satellite is systematically rotated a few degrees about roll and pitch axes, covering the area of the ground footprint. Figure 22.9 shows an example of the footprint of a TV satellite.

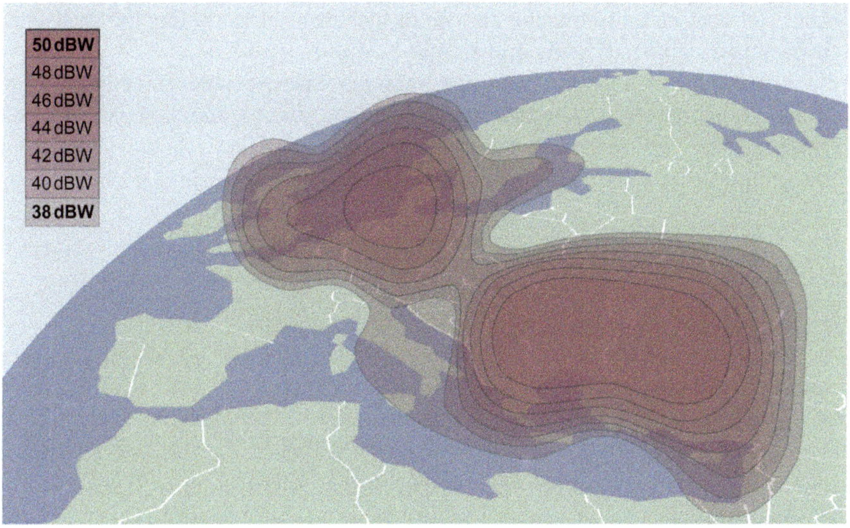

Fig. 22.9 Example of TV satellite downlink coverage

22.3.3 Contingency Operations

This type of operation generally involves handling abnormal behavior or malfunctions of the AOCS. Such anomalies are detected either by the spacecraft autonomously via FDIR tasks or by the ground personnel. The most common tasks performed by the AOCS engineer in contingency situations are:

- *Error detection and analysis*: Most contingency situations start with analysis of error bits/flags, event messages and triggered FDIRs. Since the root cause of a problem is not always clear from the beginning, an analysis of real-time and history telemetry may be required. Appropriate contingency procedures should be in place to support such analysis.
- *Reconfiguration after FDIR cases*: When a problem has been sufficiently analyzed, it may be needed to reset the device settings to nominal values (e.g., after a triggered error and automatic switchover to redundancy), if the corresponding device has been cleared OK and ready for operation. In addition, the error counters and FDIR mechanisms are reset.
- *Switching to redundant equipment*: Manual switching of sensors or actuators to the redundant equipment may be necessary if long-term analysis and specialist's assessment recommend a preventive switch before a real problem occurs.
- *AOCS recovery*: In the event of major anomalies, when a spacecraft has been placed in a safe mode, it may require extensive recovery activities until payload operations can resume. Geostationary satellites are a good example for these recovery activities. When they have been placed in a Sun pointing safe mode,

hours of work and a significant amount of fuel are required before the spacecraft is back in a stabilized 3-axis orientation.
- *Maintenance of the AOCS software*: In some cases, problem resolution may involve updating the on-board software. Loading SW upgrades and patches itself is usually not a critical task, but activating them often requires the full attention of the ground control team. Typically, some subsequent test sessions are mandatory.

22.3.4 Support Tools for AOCS

A variety of tools and software modules can be developed for AOCS flight operations to support the day-to-day operations of the engineer. The scope of these tools ranges from simple tools for quick solutions or estimates to complex and sophisticated software systems. The type and scope of tools depend on the mission, spacecraft design aspects, and operational requirements and constraints. Some typical AOCS tools are listed here:

- *Calculation of reference parameters for attitude maneuvers*: These tools are used to calculate parameters for upcoming attitude maneuvers, which serve as input for corresponding flight procedures. A typical example for GEO missions is the transition from Sun pointing to Earth pointing orientation. For this purpose, some simplified orbit parameters for the epoch of the planned mode transition are computed on ground and then sent to the satellite. For some LEO mission, lists of attitude quaternions are computed for a given period to create a dedicated attitude profile.
- *Gyro calibration*: A ground calculation tool may be needed to estimate the drift of gyroscopes and their compensation. Therefore, gyro measurements are collected over an appropriate period and a drift is estimated on ground.
- *Attitude determination*: A classic application for AOCS engineers are attitude determination tools that either provide a rough estimate of the current orientation of the spacecraft or perform highly accurate attitude determination by processing larger amounts of telemetry data.

22.4 Experience from Previous Missions

This section presents some examples of an AOCS engineer's experience.

22.4.1 AOCS Degradation

The extraordinarily successful multinational (U.S., U.K., Germany) ROSAT project is a good example of how long successful mission operations could be carried out

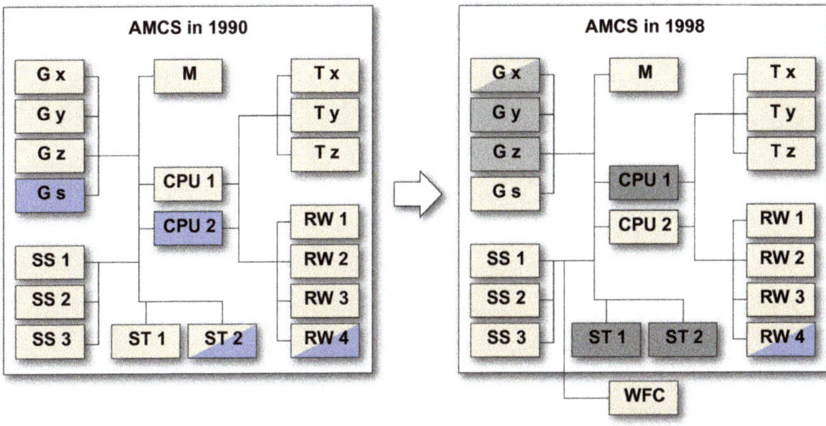

Fig. 22.10 Setup of the ROSAT AMCS in 1990 and 1998. At launch in 1990, the AMCS was equipped with two on-board computers (CPU), four gyroscopes (G), a magnetometer (M), three Sun sensor heads (SS), two-star tracker cameras (ST), three magnetic torquer (T), and four reaction wheels (RW). ROSAT did not have a propulsion system. Some equipment was lost by the end of the mission. In addition, part of the payload, a wide field camera (WFC) star tracker, was successfully used to replace the lost AMCS star trackers. Legend: prime equipment (yellow), cold redundant (blue), hot redundant (yellow/blue)

despite degradation and loss of equipment. Figure 22.10 shows the layout of the attitude measurement and control system (AMCS) at begin of the mission and at the end.

After a short LEOP and commissioning phase, the mission began in 1990 with a systematic mapping of the sky for x-ray sources, followed by a pointing phase where selected objects were observed. Over the years, some sensors degraded and eventually failed, but all these losses were overcome by updating and adjusting the AMCS software. Even after loss of the two-star tracker cameras, the mission was not over. It could be realized to bring a part of the payload instruments into the attitude control loop. For this purpose, parts of the satellite-internal data handling had to be changed as well. After the implementation of the change, some successful data takes could be performed. Of course, the performance of the system was not as good as after launch, but it could be demonstrated impressively, how flexible both the satellite configurations and the ground operation can be adapted to occurring problems.

22.4.2 Strange Wheel Behavior

This example illustrates how environmental influences can affect the attitude control. This situation occurred during the positioning of a geostationary communications satellite. On the last day of LEOP, all daily activities had been performed, including the final firing of the apogee motor, the full deployment of the solar arrays, the

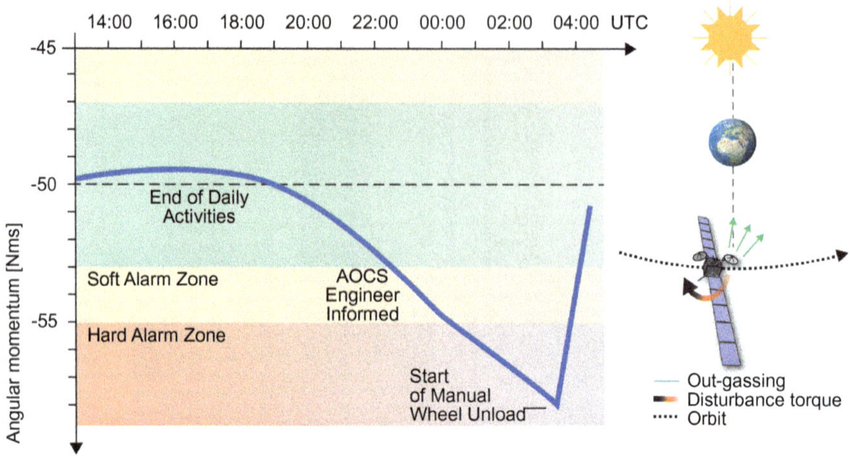

Fig. 22.11 Strange wheel behavior due to outgassing effects. The left figure shows the angular momentum telemetry of the pitch axis. The colored zones mark the soft and hard alarm zones. The right figure is a view of the orbital plane of the satellite. The Sun (yellow) shone directly into the dishes, creating a weak perturbation moment by outgassing

deployment of the two antenna dishes, and the ramp-up of the fly wheel and reaction wheels. The satellite was in a nominal configuration by the afternoon of that day. The angular momentum stored in the flywheel was nominally about 50 Nm, but developed unexpectedly during the next night, as shown in Fig. 22.11. The angular momentum and respectively the speed of the flywheel increased steadily. Analysis of the AOCS and other subsystems did not lead to a clear explanation. The effect indicated that the attitude control system was working against a weak but consistent disturbance torque about the spacecraft pitch axis (north–south axis). Fuel leakages that could cause such an effect could not be identified.

The satellite had an appropriate FDIR mechanism that would have responded by performing an autonomous wheel unload maneuver, but in order not to interfere the next day's flight operations activities, it was decided to perform a manual wheel unload procedure. During the night, the operations team performed this and the wheel speed was returned to a nominal value. This removed the accumulated momentum, but the root cause was not yet understood.

The explanation was provided by the manufacturer of the satellite the next morning. The satellite was equipped with two large payload dishes that were about the same size but made of different materials. One dish was made of metal, the other was made of composite material that tended to soak up water from the humidity at the launch site. After the dishes were deployed, the Sun shone into dishes for the first time since launch and heated them accordingly. As a result, the water evaporated from the dishes, a process known as outgassing (see Chap. 1). This created a weak but constant pressure on the dish. The disturbing torque was compensated by increasing the flywheel speed. So, there was no leakage and no AOCS failure, everything went fine. The effect was a natural one and disappeared after two days.

22.4.3 *Undocumented AOCS Function*

BIRD was a new approach in the development of a small satellite mission to detect and evaluate hot spots (forest fires, volcanic activity, burning oil wells, or coal seams) (Brieß et al. 2003). The payload and satellite were designed, built, and operated by DLR. The subsystem components for BIRD were selected with a view to a low-cost mission for a planned lifetime of one year. The attitude control subsystem (ACS) consisted of solar sensors, gyroscopes, magnetometer, star sensors, magnetic torquer, and reaction wheels. BIRD had only Sun pointing modes, which could be preset to a nadir orientation during data acquisition.

The mission started in 2001 and operated successfully for two years, collecting images of the Earth in different wavelengths. One of the first images is shown in Fig. 22.12 (*left* image). In 2003, the gyro package failed and in the following hours, three of four reaction wheels were destroyed by overload before the ground team could respond. Contact with BIRD was lost for five days. It looked like the mission was over, but telemetry could be received again on Day 6. Initial checks of the ACS showed BIRD in a stabile orientation, rotating around the Sun direction with 4°/s and an offset angle of 40°.

An initial test of a payload camera after this incident is shown in Fig. 22.12 (*right* image).

This strange mode remained stable for weeks, but it was never designed to do so. Nevertheless, it gave the ground team enough time to update the ACS software and to continue with the degraded BIRD operation.

22.5 Summary

The AOCS is one of the most complex subsystems of a spacecraft. Therefore, the ground operations team or AOCS engineers must perform various tasks ranging from simple monitoring of the subsystem to intensive maneuver preparations and executions.

It is expected that future spacecraft will have more autonomy. More and more, higher control functions, such as orbit maneuver planning tasks, can be implemented in the on-board control logic, e.g. the TanDEM-X autonomous formation flying (TAFF) control (Ardaens et al. 2008). Another issue is the availability of artificial intelligence AI methods. Precursor missions have already been flown using such algorithms for on-board navigation control (e.g. NASA's Deep Space 1). On the ground, neural networks can be used to analyze received telemetry data as part of spacecraft monitoring (O'Meara et al. 2018).

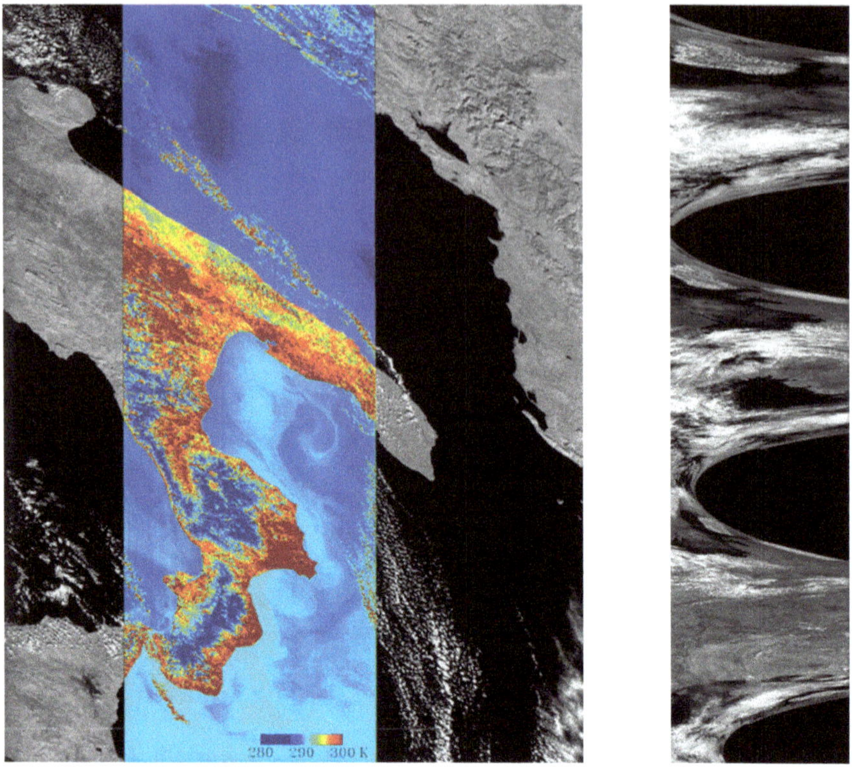

Fig. 22.12 Left image: Early BIRD data take showing a merged infrared picture of southern Italy. BIRD took the images with single-line camera detectors. By maintaining a stable attitude, the camera detectors scanned the Earth for six to eight minutes and produced such images. Right image: A test image is shown after the loss of some ACS equipment. The spacecraft had stabilized in a high spin cone-like motion, causing a rather artificial photo of the Earth

References

Ardaens J-S, D'Amico S, Ulrich D, Fischer D (2008) TanDEM-X autonomous formation flying system. In: 3rd international symposium on formation flying, missions and technology, 23–25 April 2008, ESA/ESTEC, Noordwijk

Brieß K, Bärwald W, Gill E, Halle W, Kayal H, Montenbruck O, Montenegro S, Skrbek W, Studemund H, Terzibaschian T, Venus H (2003) Technology demonstration by the BIRD mission. In: 4th IAA symposium on small satellites for earth observation, Berlin, Germany

O'Meara C, Schlag L, Wickler M (2018) Applications of deep learning neural networks to satellite telemetry monitoring. In: AIAA meeting paper, SpaceOps conference

Ralf Faller is project manager for LEO, GEO, and On-Orbit Servicing projects in the mission operations department at the German Space Operations Center (GSOC). In 1990, he received his diploma in aerospace engineering at the Technical University of Berlin, Germany. He started in 1991 at the DLR Oberpfaffenhofen in the spaceflight dynamics department, being responsible for attitude related software development and mission support and changed to the mission operations department in 1997 where he focused his work on spacecraft flight operations topics. He was involved in several missions as subsystem engineer, flight operations manager and finally project manager.

Chapter 23
Repeater Operations

Jürgen Letschnik

Abstract This chapter is dedicated to the description and operation of communications payloads of geostationary satellites. The main component of this subsystem is called a repeater, as it receives a communications signal from a ground station and simply transmits it back to a different place on Earth. Communication was the first commercial application of spaceflight.

23.1 Introduction

Sputnik 1 is known as the first satellite launched on October 4th, 1957. It was equipped with an on-board radio transmitter operating on two frequencies: 20.005 MHz and 40.002 MHz. The first satellite that could be used for communication activities was the US Army's SCORE project from 1958, which was equipped with a tape recorder to store and forward voice messages and was used to send a Christmas greeting from US President Dwight D. Eisenhower to the world. After this mission, NASA launched the Echo 1 satellite in 1960, which consisted of a balloon made of metallized polyethylene foil with a diameter of 30.5 m and a thickness of 12.7 μm. It was used to redirect transcontinental and intercontinental communications such as telephone, radio and television (Wikipedia Sputnik 1 2021).

Telstar 1 was launched into an elliptical orbit in July 1962. It was the first active, direct relay communications satellite, and the era of satellite telecommunications began. Telstar successfully transmitted television pictures, telephone calls and fax images through space and delivered the first live transatlantic television broadcast (Wikipedia Telstar 2021).

Syncom 1, launched on February 14th, 1963 on Delta B #16, was the first satellite in geosynchronous orbit. Seconds after the apogee kick motor for orbit circularization had been ignited, the spacecraft fell silent, caused by an electronics failure and

J. Letschnik (✉)
Technical University Munich, Garching, Germany
e-mail: j.letschnik@mytum.de

F. Sellmaier et al. (eds.), *Spacecraft Operations*, Springer Aerospace Technology,
https://doi.org/10.1007/978-3-030-88593-9_23

resulting in loss of the satellite. With the help of telescopic observational measurements, it could be verified that the satellite was in an orbit with a period of almost 24 h at an inclination of 33° (Wikipedia Syncom 2021).

On July 26th, 1963, the successor Syncom 2 became the first successful geosynchronous communications satellite. During Syncom 2's first year of operation, NASA conducted voice, telex and facsimile tests and 110 public demonstrations to show the satellite's capabilities and to gather feedback. In August 1963, President John F. Kennedy made a telephone call from Washington, D.C., to Nigerian Prime Minister Abubakar Balewa on board USNS Kingsport, which docked in the port of Lagos; the first live conversation between heads of state via satellite. The Kingsport served as a control and uplink station (Wikipedia Syncom 2021).

23.2 Repeater Subsystem

This chapter is dedicated to the description of the satellite communications payload (repeaters), focusing on the design principles, characteristic parameters and technologies used for the equipment.

In most commercial communications satellites, the payload consists of two separate parts with well-defined interfaces—the repeaters and the antennas (Bostian et al. 2003). The repeater usually comprises several channels (also called transponders), which are individually allocated to sub-bands within the total payload frequency band. The generic functionality and the characteristic parameters of a repeater payload are presented below.

23.2.1 Functions of a Repeater Payload

The main functions of the communication repeater payload can be divided into four main functionalities, see Fig. 23.1. A detailed description of each functionality is given in the following sections.

Receive. The repeater receives a signal through the satellite payload antenna in a specific frequency band and with a specific polarization from one or more Earth stations. The stations are located within a certain region (service zone) on Earth and are viewed from the satellite at an angle that determines the required angular width

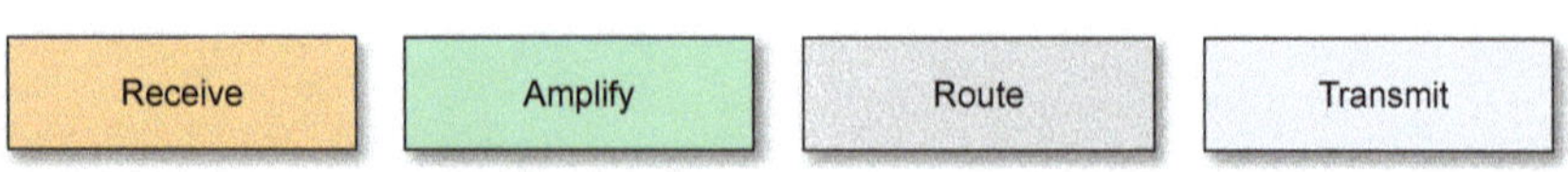

Fig. 23.1 Basic functions of the repeater payload

of the satellite antenna beam. The point of intersection of the satellite antenna beam with the Earth's surface defines the reception coverage.

Amplify. The signal strength received by the satellite's antenna system is extremely low, about -190 dBW (equivalent to 10–19 W). A direct reflection of the received signal, like it was done with ECHO 1, would result in a total loss of the signal. It is therefore imperative to amplify the signal on board the spacecraft to obtain a usable signal at the ground reception terminal. Maximizing the amplification level without creating distortion is therefore the objective.

During the amplification process, the frequency of the received signal is also converted to another frequency to avoid interference between the uplink and downlink signals. A currently typical and frequently used frequency range is the Ku band, which uses the 14 GHz band for uplink signals and the 11–12 GHz band for downlink signals (Handbook on Satellite Communications 2002; Maral and Bousquet 2002).

Route. Modern satellite systems are capable of routing signals between a variety of receiving and transmitting antenna systems on the satellite. The routing can be done at radio frequency (RF) level or at baseband level of the signal (with or without carrier frequency, as described in Sects. 23.3.2 and 23.3.3).

If the received signals are demodulated at baseband level on the satellite, the repeaters are called *regenerative repeaters*. Based on the routing configuration on board, the data streams can be routed to different spot beams. To transmit the signal, the stream is modulated back onto an RF carrier. Such systems are already available, for example on EUTELSAT 13A (HB6), EUTELSAT 7A (W3A) and on several other satellites.

If the received RF signal is forwarded without any demodulation or modulation process, the repeater is called a *transparent repeater*.

Transmit. The amplified and possibly also regenerated signal is now transmitted via a transmitter antenna system. Such antenna systems can be implemented as single-beam or multi-beam antennas, depending on the application scenarios of the satellite.

23.2.2 Overview and Layout of a Repeater/Transponder

One of the simplest transponder configurations (without routing functionality) of a dual polarization transponder (HLP—horizontal linear polarization (X), VLP—vertical linear polarization (Y)) is shown in Fig. 23.2 and refers to the functions mentioned above. The key equipment of this simple transponder is described in this section.

1. **Receive Antenna**. The typical realization of a spacecraft antenna is the parabolic antenna especially for higher frequencies (e.g. X-band, Ku-band to Ka-band). For lower frequencies like P-band (400 MHz) helix antennas are used. The sizes

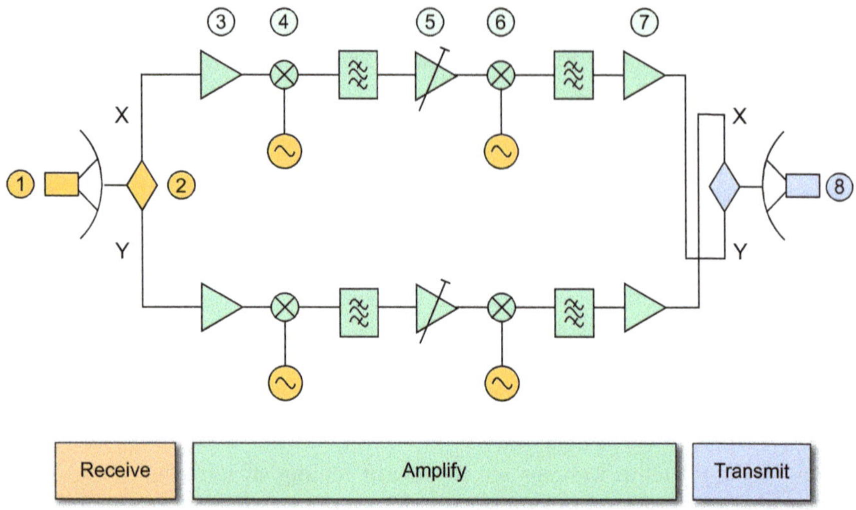

Fig. 23.2 Simplified transponder/repeater layout (Numbers refer to the text paragraph numbering)

of such antennas range from about 1–8 m (e.g. EUTELSAT 10A (W2A) uses an 8 m deployable antenna in the S-band).

The antenna system can be implemented as a fixed antenna or as a steerable antenna or a combination of both, depending on the services of the satellite provider. The use of multi-feed antennas is becoming increasingly popular, their high flux density and the possibility of frequency reuse are the driving arguments.

2. **OMT**. The OMT (orthomode transducer) is an important element of a satellite transponder system. The OMT is a waveguide element used to split the received polarizations HLP and VLP into two separate paths. The use of different polarizations on the same frequency allows an increase in the transmitted information capacity. A detailed description is given in paragraph 8.
3. **LNA**. LNA stands for low-noise amplifier. As the name suggests, this device is characterized by a very low-noise level combined with high gain. In all cases, even with high gain antennas, an LNA is necessary to obtain a good signal-to-noise ratio of the received signal.

   The LNA is a wideband RF device and is therefore able to amplify the entire reception spectrum of a dedicated frequency band (e.g. 27.5 GHz to 30.0 GHz, uplink Ka-band). In a typical communications satellite system, it is tracked in the signal path by an input multiplexer which separates the channels.
4. **Downconverter**. The frequency of the received signal is reduced to an intermediate frequency (IF) to reduce losses and simplify the components in the components following in the signal chain. On ESA's Artemis satellite, for example, the intermediate frequency is 5 GHz.

5. **Channel Amplifier**. The channel signal is then amplified again with an adjustable channel amplifier (CAMP). Depending on the detailed design of the satellite, it is possible to adjust the gain of each amplifier individually. The power of the overall connection (ground–satellite–ground) is controlled with this setting. Some satellites also have a so-called ALC (automatic level control) amplifier. This module sets the gain of the channel amplifier depending on the signal input level coming from the antenna.
6. **Upconverter**. The IF that was used for amplification and routing is then transformed to the higher frequency of the downlink. This frequency is coordinated with the ITU (International Telecommunication Union) during the design phase of the satellite. Typically, the D/L (downlink) frequency band is at a lower frequency than the U/L (uplink) frequency band. Filters located behind the converter modules eliminate frequencies generated by the mixing process itself.
7. **High Power Amplifier**. The high-power amplifier (HPA) plays an important role in a communications satellite system. In order to obtain a sufficient signal level on the ground, it is necessary to generate a high RF power level (radio frequency) at the satellite output. The generation of a signal with a high RF output level results in a high electrical power consumption which has to be provided by the satellite platform. Travelling wave tube amplifiers (TWTA) are the most commonly used amplifiers on communications satellites. Efficiencies of about 60% can be expected and the higher bandwidth compared to solid state power amplifiers (SSPA) with an efficiency of 10–20% is a positive effect. The disadvantage of a TWTA is its non-linear transmission curve, which later plays a role in the operational aspect.
8. **Transmit Antenna**. The high-power RF signal is now routed to the transmitting antenna. Here the polarizations are reversed in comparison to the receiving antenna system in order to maximize the decoupling of the uplink and downlink signal. This turns the received X-polarization into a Y-polarization in the downlink and vice versa, the Y-polarization in the uplink becomes an X-polarization in the downlink. Figure 23.3 shows an illustrated example of the principle.

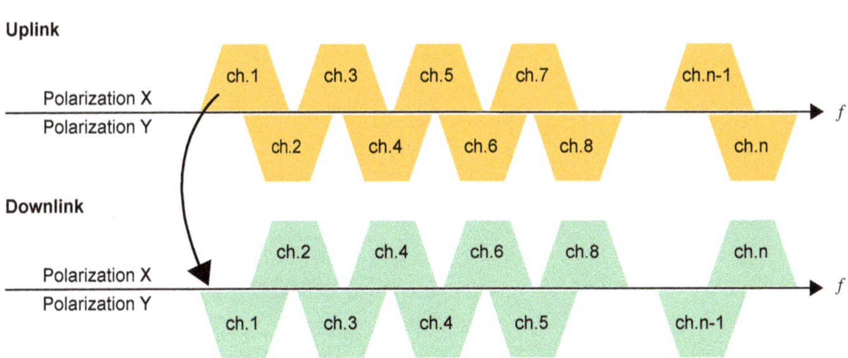

Fig. 23.3 Simplified illustration of an U/L and D/L channel allocation on a repeater subsystem for a communication satellite

23.3 Repeater Operations

The general environment for operating the repeaters of a geostationary communications satellite is very similar to other space missions. Infrastructure elements such as the ground station, the spacecraft control center and the data network between the two are necessary. The main difference is the need for a ground station with equipment to measure the in-orbit performance of the repeater payload. Such a ground station is called an IOT (in-orbit test) station. It is equipped with special RF (radio frequency) components such as power sensors, inject-pilot systems and an ASSC (automatic satellite saturation control) unit to ensure high-precision measurements of the repeater subsystem.

Repeater operations can generally be divided into communication operations and spacecraft operations as shown in Fig. 23.4, which are divided into mission phases of LEOP, IOT, and routine and emergency operations.

Communication Operations. The best-known form of satellite communication is the direct-to-home transmission of television programs via satellite. The satellite's payload (repeater) is used to broadcast television programs over a larger area. The satellite is only one element in the overall communication path. It is not the user's responsibility to supervise the spacecraft activities, since the connection time is often only rented. Typical users, e.g. playout centers of TV stations, satellite news gathering (SNG) stations are coordinated by the communication service center (CSC) of the satellite provider, e.g. SES ASTRA, EUTELSAT, INTELSAT. They are configuring the uplink channels to the satellite. The user himself has no access to the repeater configuration on board.

Spacecraft Operations. The management of the communication satellites described here is sometimes carried out by a separate team. The main task is to ensure continuous monitoring of all important parameters (housekeeping data) of the satellite, which are necessary for the proper operation of the repeater subsystem. Typical satellite parameters to be monitored are discussed in detail in the following sections.

Fig. 23.4 The structure of the repeater operating phases

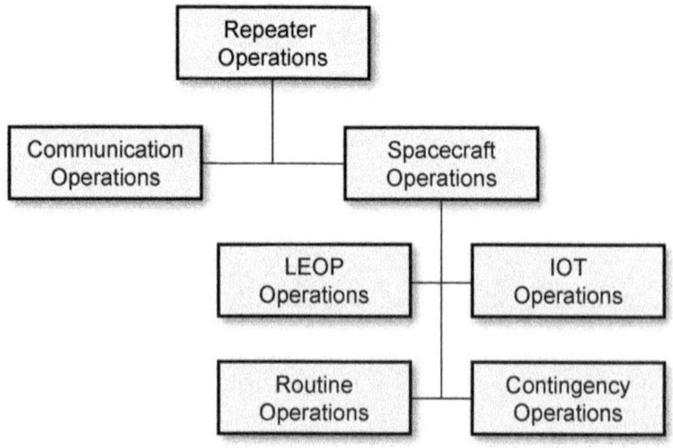

23.3.1 Launch and Early Orbit Phase (LEOP) Operations

The LEOP operations of a communications satellite focus on the stationing of the satellite and the operation of the platform subsystems; repeater operations are only a few of the payload activities that are carried out during this period. A typical task of repeater operations is the deployment of the antenna systems. In most cases, the antennas must be in a "stowed position" during launch. In many cases they are folded down because space in the launcher's envelope is limited and to protect the antenna structure from damage caused by vibrations during launch.

Before and during antenna deployment, engineers must carefully monitor the temperatures of the pyros of the hold-down mechanism, the motor currents of the antenna mechanism, the temperatures of the RF cables and other parameters defined by the satellite manufacturer.

Since the antennas are locked in their stow position during launch, they normally have to be unlocked during the LEOP phase and moved to the nominal operating position. This must be done for both axes (azimuth and elevation). Depending on the design of the antenna mechanism, the actual position of the antenna is returned either by optical encoders or potentiometers. These angular values of the antenna orientation are in the antenna's own coordinate system. In order to obtain the orientation of the antenna in relation to the coordinate system of the Earth, which is ultimately important for alignment in routine operation, these angles must be converted by means of coordinate transformations. The corresponding matrices for the conversion are supplied by the satellite manufacturer. The antenna itself is usually controlled by stepper motors. It is possible to control individual motor steps directly from the ground. Fixed limit switches in the antenna mechanism and a suitable software in the control center allow the accumulated number of commanded motor steps to be used to calculate the current orientation of the antenna; in this way optical angle encoders or potentiometers can be saved. However, if the commanded motor steps are not executed correctly or an error occurs on the motor itself, a miscalculation may occur, resulting in a misalignment of the antenna. An indicator could be the difference between the commanded motor steps and the measured antenna movement steps. Another indicator is whether the calculated change in RF signal level between the two pointing positions of the antenna is inconsistent with the expected change.

It is therefore important for the control center operators to contact the manufacturer of the mechanism before the start of the LEOP phase to ensure well-prepared routine and emergency flight procedures and the necessary software tools for troubleshooting. During the entire LEOP phase, all devices of the Repeater subsystem are switched off, especially the TWTAs. During nominal operation, the TWTAs generate waste heat, which is included as a fixed quantity in the total thermal budget of the satellite. To compensate for this lack of waste heat during launch and the LEOP phase, the satellite is equipped with so-called "substitution heaters". These heaters generate an equivalent amount of waste heat to the switched-off TWTAs and thus keep the satellite in thermal equilibrium. During the transition from the LEOP phase to the Payload IOT phase, the subsystem engineer must ensure that these substitution heaters are switched off before the TWTAs are switched on.

23.3.2 IOT Operations

In addition to the in-orbit test (IOT) of the satellite platform, the IOT phase of the payload can be considered as a single phase in itself due to its duration and complexity. The repeater subsystem has the main focus in this phase. From the customer's point of view, the communication payload is the most important satellite equipment. Money shall be earned by providing a guaranteed, defined service to the end users.

The customer and satellite manufacturer have a high interest to perform a correct and precise IOT of the repeater subsystem, to get a successful In-Orbit delivery. In order to manage a payload IOT campaign, all platform operations on the satellite must be coordinated with the RF measurement campaign, i.e., no orbital maneuvers must take place during a dedicated RF measurement. Usually two groups of engineers are involved in this process. Firstly, the IOT RF measurement engineers, who are usually located at the IOT ground station facility, and secondly the spacecraft operations engineers, who are located at the spacecraft control center. The basis for both groups is the IOT test plan for the repeater subsystem. This plan contains a very detailed schedule of all activities to be performed during the test campaign. If this plan is not meticulously prepared, this can easily lead to enormous and expensive time delays for this phase. All satellite configurations and necessary changes are part of this plan. The main task of the SCC group is to translate these configuration requirements into operational flight procedures. Depending on the number of transponders to be measured, the implementation can be complex. For this reason, the manufacturer of the repeater platform often supplies its own software, which allows to generate specific configurations and the corresponding commands. This software is typically used during the integration phase of the repeater platform; it allows to verify the correct installation. The same software is very useful in SCC for the creation of the flight procedures, which is a classic task of an engineer for the repeater subsystem, and avoids fundamental errors in the planning and development of the procedures. Special rules for repeater operations have to be considered, see Sect. 23.3.3.

In addition to the configurations on the satellite described above, this plan also specifies the orbital position at which the IOT has to be performed. The geostationary orbit is densely populated with spacecraft from a wide range of satellite operators. Their signals must be free from interference caused by RF signals from other ground stations or satellites, therefore frequency coordination is required (done by the ITU, see also Chapter 18 TM/TC telemetry/telecommand). For IOTs, this means they could be carried out at an orbital position that does not correspond to the final orbital position for operation, in order to minimize interference with other satellite operating services. For this position, the SCC must ensure command capability and reception of telemetry data, and an appropriate IOT measurement station must also be available for that orbital position. Depending on the complexity of the repeater subsystem, the IOT typically takes about three to four weeks.

The following list contains an excerpt from the large number of measurements that are performed during an IOT phase. The most important points relating to satellite operation are explained in more detail below.

- Outgassing of the TWTAs
- Antenna pointing and pattern coverage
- Gain step adjustment
- Saturation EIRP (equivalent isotropically radiated power) / IPFD (input power flux density)
- Beacon frequency, beacon EIRP and stability
- G/T (gain over noise temperature) of the Rx (receiver) antennas
- Saturation and linear gain, gain transfer (AM/AM)
- Group delay and frequency conversion.

Outgassing of TWTAs. The most common type of amplifier installed on a communications satellite is the TWTA. As the name suggests, the TWTA consists of a vacuum tube for generating an electron beam. This tube contains a certain vacuum pressure, which is different from the vacuum pressure in orbit. To allow the residual gas remaining in the tube to escape, the tube is gradually loaded with an RF signal at the input. This RF signal is generated by one or usually several ground stations and is gradually increased to the point of maximum tube power. As a result of this gradual increase, the temperature of the tube rises and the residual gas can diffuse from the tube into space. Since the tubes transmit at full RF power during this time, it is important that no other satellites or services are disturbed in the direction of the Earth. To optimize the timing of the IOT, this outgassing is often done in drift orbit or sometimes already at the end of the LEOP phase. Depending on the number of TWTAs, the outgassing may take several days. This phase offers the satellite engineer the possibility to directly observe the behavior of the helix current and the temperature. Corresponding telemetry values can be tracked in an offline table to obtain a rough estimate of the operating point of the tube load for routine operation.

Antenna Pointing and Beam Coverage. Knowledge of the actual coverage areas of the transmitting and receiving antennas is of central importance for communication operations. When the satellite is in in its IOT or final orbit position, a complete antenna diagram is generated using the attitude mode MAM (mapping antenna mode). In this mode, the satellite is rotated alternately around the roll and the pitch axis. During this entire phase, the IOT measuring station sends a signal to the satellite and receives it again. Depending on the equipment of the IOT measuring station and the satellite, the measured signal values and possibly also the values of the AGC (automatic gain control) are the generated output. This pattern can then be mapped onto the Earth's surface to obtain a corresponding beam coverage map with contour lines of equal signal strength. Such a procedure does not require a steerable antenna on the satellite and can therefore be applied generically. If a steerable antenna is available on the satellite, it is possible to limit the measurement to the 3 dB beam width,

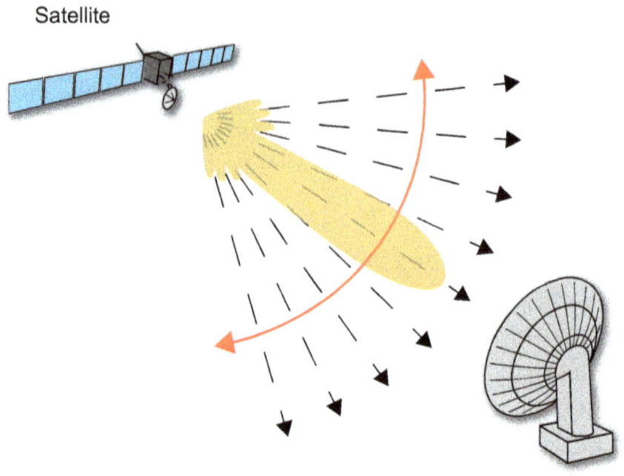

Fig. 23.5 Antenna mapping test campaign to determine the antenna pattern

which is a typical parameter for characterizing an antenna. Only the antenna, not the whole satellite, is steered step by step in azimuth and elevation. As in the previous measurement, the IOT station transmits a reference signal which the station receives again. The antenna contour can be derived from the changes in the signal level.

Depending on the number of antennas, the measurement of the antennas typically takes a considerable amount of time, up to one week, during which little other work can be done on the spacecraft (Fig. 23.5).

Gain Step Calibration. The measurement of the gain step settings is a central part of the IOT measurements. Depending on the requirements of the satellite provider, all or only part of the total adjustable gain step range is measured. This includes both the gain steps of the preamplifier (CAMP—channel amplifier) and the gain steps of the TWTAs. From a communication point of view, the actual performance of the devices at the discrete gain stages compared to their nominal values is very important as they are used for the optimal configuration of the communication link. This makes it possible to adapt to users with better or worse performing ground stations and still provide the same level of service. From the operational point of view of the SCC, it is advisable to prepare procedures in which commands can be sent to the satellite in large quantities, as some amplifiers (CAMP or TWTA) have about 40–60 gain steps that must be commanded step by step using increment-by-one instructions. Using an automated process is recommended, to avoid tedious manual work of the command operator.

23.3.3 Routine Operations

The routine operation of the repeater subsystem is based on the needs of the customers who use the repeaters for their data transmissions. The main activities in this phase are monitoring and observation. However, depending on the type of repeater (transparent or regenerative), the effort required varies.

A transparent repeater simply amplifies a signal and shifts it in frequency. The entire repeater subsystem is brought into the nominal configuration and the amplification levels of CAMP and TWTA are set. If a steerable antenna is available, it is aligned accordingly. In nominal and routine operation, the repeater itself does not require a complex analysis of the telemetry. Typical activities in routine operation on the repeater subsystem are

- the configuration of the onboard routing between different channels and different antenna spot beams,
- to change the orientation of the antenna reflectors,
- to adjust the gain stage settings of the channel amplifier,
- to change the conversion frequency of the on-board up/down converters,
- operations in strict compliance with the "repeater flight rules".

In addition to the activities mentioned above, a regenerative repeater (see Sect. 23.2.1) has considerably more configuration options (routing, modulation, etc.) and therefore requires more support both during commissioning and routine operation. The ability to route signals offers considerably more flexibility in communication operation and therefore also means considerably more operating effort in routine operation compared to the transparent transponder. Figure 23.6 shows a typical telemetry plot (tube temperature of a TWTA) of a commercial repeater payload over a period of one month, which should be monitored continuously. The plot shows over the period of one month the temperature cycle of the tube. A very clear daily temperature cycle can be seen. The little decrease of the temperature is based on the overall position an attitude of the satellite in respect to a one-year orbit duration.

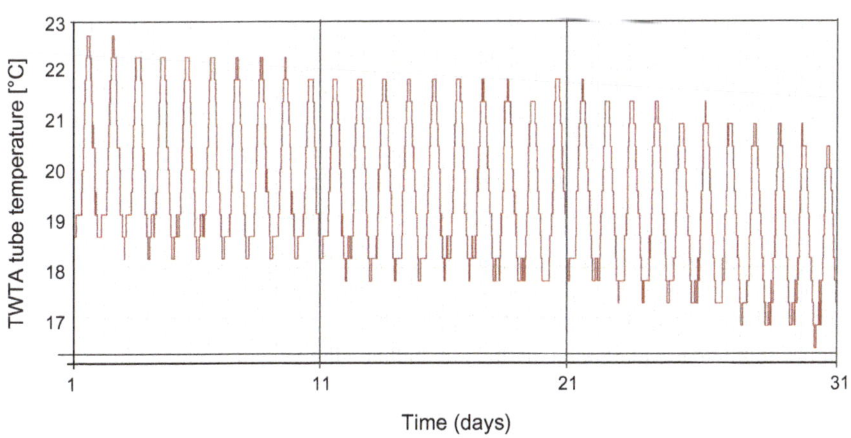

Fig. 23.6 Monitoring the tube temperature of a TWTA during one month

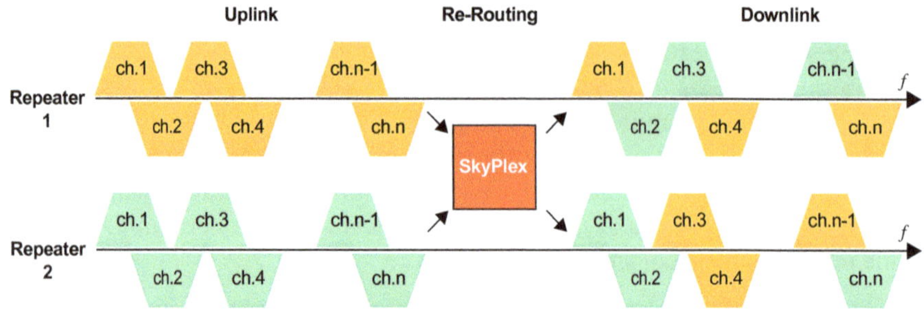

Fig. 23.7 Technical example of a routing base system (SkyPlex)

The challenge for a spacecraft engineer is to continuously observe from the large amount of telemetry parameters those from which a tendency of the degradation of an equipment can be derived. Thus, the knowledge about the value of the parameters is distributed between the manufacturer of the satellite, the manufacturer of the individual components and the space operations engineers.

Onboard Routing. The functionality to route received channels to different transmit channels is one of the most interesting activities on a telecommunication satellite. One technical example of a routing base system would be the SkyPlex technology developed by ESA and EUTELSAT (Fig. 23.7). This system allows to split and recombine DVB channels transmitted from different ground stations to one signal DVB stream onboard of the satellite.

Antenna Pointing. Some satellites are equipped with steerable antenna systems. With this antenna system it is possible to readjust the orientation of the generated spot beam to the Earth. This functionality increases flexibility in the provision of communication link services, especially point-to-point connectivity services.

The adjustment of a mechanical system in space is always associated with a higher risk than electrically based adjustments. The manufacturer of the antenna alignment mechanism and electronics typically defines a number of parameters that must be within the nominal range, such as the temperature of the motors and mechanism, to be ready for adjustments. If heaters need to be switched on to achieve this nominal range, it can take quite a long time to reach the required values. This results in a pre-setting time that must be taken into account in operations planning.

Gain Step Adjustment. Channel amplifiers, as described in the Sect. 23.2 are normally equipped with the function of gain adjustment (see also Sect. 23.3.2). This helps to optimize the communication link for one or more customers. Each channel amplifier can be adjusted separately, resulting in a high number of gain stage values that must be monitored by the SCC. In addition to the gain stages of the channel amplifiers, the high-power amplifiers can also be equipped with a gain stage setting functionality, resulting in a large number of additional parameters that need to be monitored.

Modern amplifier modules are highly resistant to cosmic rays, however, it can happen that a gain stage "reverses" its value, which has a direct and important impact on the communication link and therefore requires the immediate attention of the SCC engineers, especially to restore the correct gain stage value.

Conversion Frequency. Some satellites offer the possibility to change the conversion frequency of the up- and downconverter. When changing this frequency, it is important to be aware that all frequencies for each geostationary communications satellite are coordinated through the International Telecommunication Union (ITU). This coordination will be done in a very early design phase of the satellite mission and could be an important design driver of the communication payload. Therefore, an incorrect setting of the frequency can cause interference and disturbance on frequency bands used by other users.

Repeater Flight Rules. Flight rules are generic rules and recommendations for the operation of the satellite. This is in contrast to flight procedures, which are very specific. An overview of typical repeater flight rules is shown in Table 23.1.

The first flight in this table, "Switching coaxial and waveguide switches", is one of the most basic rules. Depending on power and frequency, RF signals can be switched with different technologies. Three main principles can be distinguished: substrate based (PIN diode switch), coaxial switch, waveguide switch. The manufacturers of the switches specifically describe the conditions under which the switches may be operated. Mechanically movable components are built into coaxial switches and waveguide switches. The actuation of mechanically movable elements is always considered very critical in space, since a mechanical defect leads to the immediate and irreversible loss of the component. In addition, RF short circuits can occur in mechanical RF switches that are switched under high RF loads. All these effects lead to the general rule that a high frequency (HF) switch should be operated with caution.

Table 23.1 Overview of typical flight rules including the recommended operational activities for a repeater subsystem

Rule	Recommendation or activity
Switching of coaxial and wave guide switches	Do not switch coaxial or waveguide switches while an RF signal is present; no "hot-switching"
Helix current of the travelling wave tube amplifier (TWTA)	Continuous monitoring of the helix current of the travelling wave tube amplifier (TWTA). A current outside the nominal range could be an indicator for a defective tube or a faulty power supply. Initiate a fault detection and backup switching sequence
Monitoring of isolator temperatures	A temperature outside the nominal range could be an indicator of abnormal RF reflection from high-power equipment. Initiate an error detection and backup switching sequence
Preheating Phases	Monitoring the preheating phases of devices such as converters or amplifiers, which can last between 2 and 20 min. No impact on spacecraft operation

A precise clarification of the conditions with the switch manufacturer is advisable at this point.

23.3.4 Contingency Operations

In general, emergency operations are carried out hand in hand with the satellite or payload manufacturer. Other very important parties involved are the end-users of the repeater service, who transmit their signals via the satellite. It is strongly recommended to involve this group in emergency operations on the repeater subsystem. Typically, all these communication business services are managed, as discussed above in "Communication Operations", by those special departments, "communication management center" (CMC) or "communication service center" (CSC).

Many contingency cases have already been investigated by the manufacturer before the launch. Flight procedures are therefore provided for rapid recovery of the system. Even though the recovery itself may only take a few hours, the entire process, starting with fault detection and analysis, can still take several days.

An example of a typical redundancy switching of a high-power amplifier module (TWTA) is described below. Starting with an observed drop in the anode voltage (see Fig. 23.8), the administrative and technical processes are triggered with all parties involved. Investigations and analyses lead to the decision to switch over to the redundant backup high-power amplifier module. Figure 23.9 shows both configurations

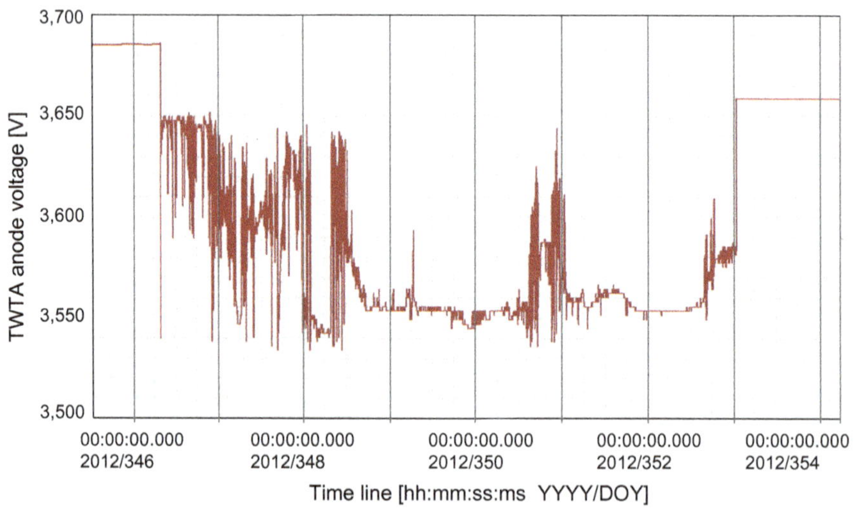

Fig. 23.8 Anode voltage drop of an travelling wave tube and self-recovery

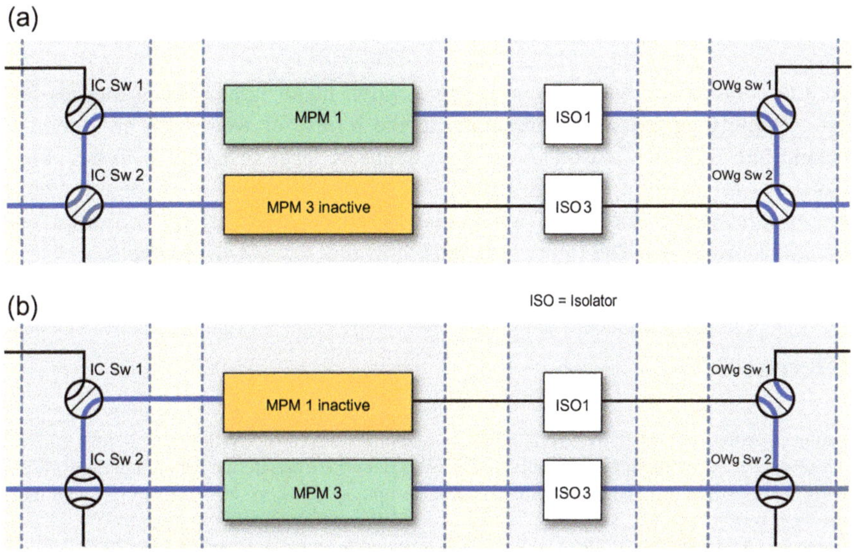

Fig. 23.9 **a** Configuration of the nominal repeater configuration, **b** Configuration of the repeater subsystem after the backup switch over

before and after the backup reconfiguration. For this reconfiguration, the following devices must be commanded:

- Switching OFF the nominal high-power amplifier module (*MPM 1*)
- Input coax switch 2 (*IC Sw 2*) from position 2 to position 3
- Output waveguide switch (*OWg SW 2*) from position 1 to position 4
- Switching ON the redundant high-power amplifier module (*MPM 3*)
- Readjustment of the amplification stages of the high-power amplifier module.

All activities shall be performed in accordance with the flight rules and based on the flight procedures supplied by the manufacturer. They must be coordinated with CSC or CMC to ensure the communication services involved. On the one hand, this switchover will lead to a total failure of all communication carriers of this faulty amplifier, but on the other hand it will restore the required quality of service.

23.4 Summary

Satellite communication has been going through major upheavals in the last few years. Rapidly advancing technologies in the terrestrial area, such as video-on-demand, are only one reason for this. In addition, large investors with new business cases for satellite communication have emerged in the space sector. Mega-constellations and "new-space" approaches are challenging established usage and operations principles. The future will show how this sector will change.

References

Bostian C, Allnutt J, Pratt T (2003) Satellite communications, 2nd ed. Wiley
International Telecommunications Union (ed) (2002) Handbook on satellite communications. Wiley
Maral G, Bousquet M (2002) Satellite communications systems: systems, techniques and technology. Wiley
Wikipedia (2021) Sputnik 1. http://en.wikipedia.org/wiki/Sputnik_1. Cited 14 June 2021
Wikipedia (2021) Syncom. https://en.wikipedia.org/wiki/Syncom. Cited 14 June 2021
Wikipedia (2021) Telstar. https://en.wikipedia.org/wiki/Telstar. Cited 14 June 2021

Jürgen Letschnik is working at Airbus Defence and Space as Head of Service Solution & Bid Support in the line of Connected Intelligence. He joined Airbus in 2018 as an Information Systems Architect. In 2008 he started his industrial career as a satellite operation engineer at LSE Space GmbH (LSE) in Weßling, Germany. He studied telecommunication engineering and did his Ph.D. in the field of tele-operations on robotic space missions at the Institute of Astronautics of Technical University Munich, Germany. Since 2016 he is lecturer at the Technical University Munich for Satellite Communications and Operations. At GSOC he was working on a geostationary telecommunication mission for the German military as a repeater payload subsystem engineer and flight control team lead. His main field of research is satellite communications engineering, in particular repeater payload communications engineering, intersatellite link communication technologies and services for specified end-user applications.

Part VII
Special Topics

Chapter 24
Human Spaceflight Operations

Jérôme Campan, Thomas Uhlig, Dennis Herrmann, and Dieter Sabath

Abstract Human spaceflight operations incorporates most of the aspects of satellite operations, but require some conceptual modifications; in addition, some new domains must be added to accommodate humans in space. This chapter briefly describes the required modifications to the classical subsystems presented earlier and introduces the additional subsystems Environmental Control and Life Support System (ECLSS), Visiting Vehicles, and extravehicular activities (EVA). Aspects specifically related to the crew, such as safety, health care, and communications are discussed, as well as the structure of the ISS flight control teams and their operational approach.

24.1 Introduction

Human Spaceflight began with Yuri Gagarin—he was the very first human in history heading towards space with his famous word "Поехали!" (Let's go!). From that historic moment with the start of the Vostok program to the current International Space Station (ISS) program and through several missions during the last decades such as Gemini, Mercury, Apollo, MIR, STS Space Shuttle (Furness et al. 2007), the concept of Human spaceflight operations has evolved into what we know today. This chapter provides an overview of the concept for human spaceflight operations currently used for the International Space Station Program (Dempsey 2017). To date, the broadest experience in human spaceflight operations could be gained in this international project.

With the first launch of the commercial U.S. crew vehicle, between three and eleven people are traveling at an average speed of 27,000 km/h at an altitude of about 400 km to conduct science in a microgravity environment. From an operational perspective, it is interesting to ask how manned versus unmanned space missions differ and what advantages an astronaut can bring to space missions.

J. Campan · T. Uhlig · D. Herrmann · D. Sabath (✉)
German Space Operations Center (GSOC), German Aerospace Center (DLR), Oberpfaffenhofen, Germany
e-mail: dieter.sabath@DLR.de

F. Sellmaier et al. (eds.), *Spacecraft Operations*, Springer Aerospace Technology,
https://doi.org/10.1007/978-3-030-88593-9_24

This chapter focuses on the deltas to what has already been presented in previous chapters regarding unmanned missions. It outlines the particularities of human spaceflight and what the major challenges are in not only operating a manned vehicle but in leading an entire human mission to success. Special processes, functions, and behaviors are required to successfully execute such a manned mission and return the astronauts safely to Earth.

24.2 Manned and Unmanned Missions

One intent in developing technology is to automate everything as much as possible. Systems should be error-free, more efficient, less time consuming, cheaper, and always produce the expected result. But with the current state of technology, some tasks that need to be performed in space cannot be automated. Consequently, there are two types of space missions: unmanned missions that go where humans could never go, and manned missions with tasks that only humans can perform. Whenever an activity is so complex that it cannot be successfully performed with an acceptable probability using current technology, the involvement of a crew member makes those tasks feasible. This is especially true for troubleshooting activities, responding to the unexpected and implementing workarounds.

24.3 From Satellite to Living Space

24.3.1 Advanced Satellite Subsystems

All the subsystems presented in the previous chapters for unmanned vehicles are also used for manned missions, but with additional functions (Messerschmitt and Bertrand 1999), which are briefly presented below (Fig. 24.1).

Thermal Control Subsystem (TCS)
This subsystem not only provides thermal cooling for the instruments and other components, but must also provide a heat sink for conditioning the cabin air. The metabolic heat produced by the people on board places an additional load on this subsystem. The allowable temperature range for crew-accessible equipment is also limited to avoid touch temperatures that are too high or too low for the astronauts.

Electrical Power Distribution Subsystem (EPDS)
In addition to generating and distributing power for the other subsystems, electrical power is also needed to provide habitable environmental conditions (e.g. light) within the space station and also power for the various electronic devices that the astronauts operate on board. The use of electrical power in an oxygen-enriched environment

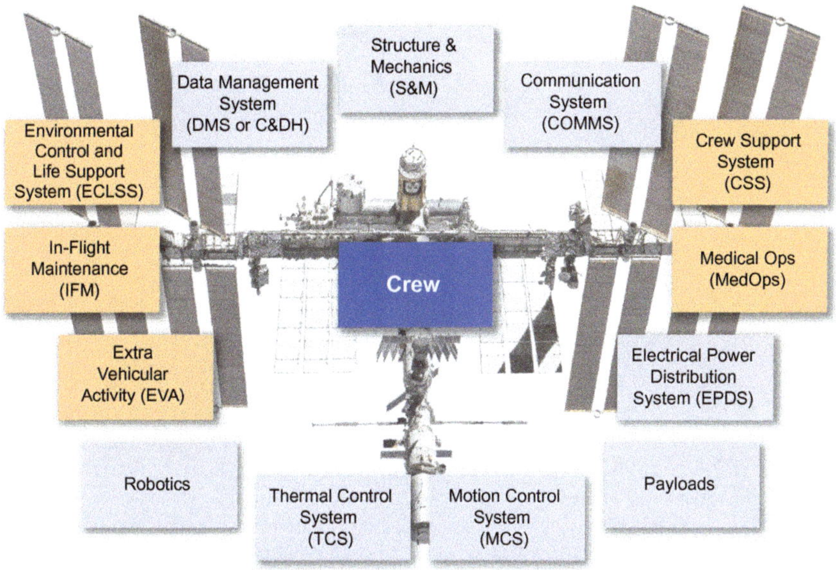

Fig. 24.1 Not only the conventional subsystems of a satellite (grey) need to be adapted, but new subsystems must also be added (orange) to enable the presence of humans on board a spacecraft

also presents a fire hazard that is not present in an unmanned vehicle. Open electrical connections pose an electrocution hazard to the crew.

Data Management Subsystem (DMS)

In terms of crew autonomy, a manned spacecraft is not always commanded only by a flight control team (FCT) on the ground. It may also be controlled by the astronauts on board, especially when critical functions are involved. Therefore, the data management system must provide a human–machine interface (HMI) that allows the crew to both send commands to the vehicle and review essential subsystem data. On the ISS, this is mainly implemented via laptops with dedicated software that provide a graphical representation of the space station and allow insight into ISS data and synoptic commanding.

Communications Subsystem (COMMS)

In human spaceflight, the communication subsystem is not limited to the transmission of "telemetry and telecommand" but also includes visual (video/image), verbal, and written communications.

Each of the ISS mission control centers, whether in Houston, Huntsville, Tsukuba, Moscow, or Munich, has the capability to interact with the crew aboard the ISS via this voice communication link, commonly called space-to-ground (S/G).

The Tracking and Data Relay Satellite System (TDRSS) is used for this link between ground and the ISS, providing near-continuous contact (Fig. 24.2).

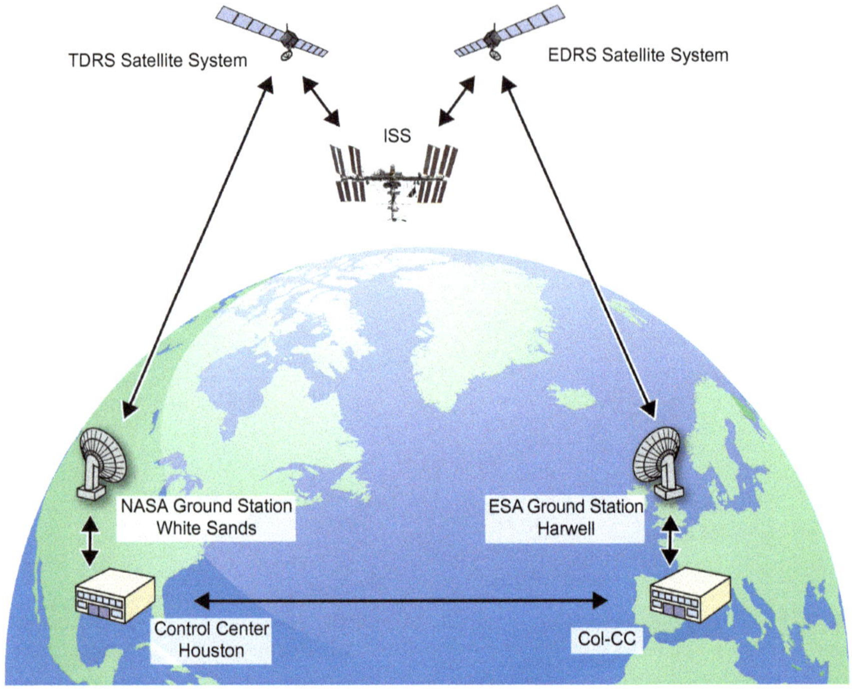

Fig. 24.2 The space station is connected to the Mission Control Center Houston (MCC-H) via the Tracking and Data Relay Satellite constellation (TDRS), which uses a ground station in White Sands, New Mexico. All other control centers are linked to Houston. Meanwhile, some control centers also use their own means of communication; for example, the Columbus Control Center (Col-CC) uses a Ka-band link via the European Data Relay System (EDRS)

Motion Control Subsystem (MCS)

Attitude and orbit control are also required for a manned spacecraft and are no different from an unmanned vehicle. For a manned station like the ISS, it is necessary that the motion control subsystem is available for years, therefore refueling is necessary. For the ISS, reboosts to counteract the continuous altitude drop due to drag forces (Sect. 1.3.5) are usually performed by docked spacecraft, as described in more detail in Sect. 24.3.3.

24.3.2 Environment Control and Life Support System (ECLSS)

Not only are modifications required to the subsystems commonly found on satellites, but at least one additional subsystem must be introduced to support human life in space: A pressurized, habitable environment, commonly referred to as shirt-sleeve environment. These functions are established by the environment control and life support system (ECLSS).

The ECLSS is a complex subsystem and is tasked with ensuring and protecting life on board. It consists of the following main functions.

Atmosphere Control and Supply (ACS)

Cabin air—a combination of oxygen and nitrogen—is provided by the ACS. It ensures that both components are supplied at the correct pressure and ratio so that the astronauts on board can breathe normally without additional equipment. The ACS provides pressure equalization between the various modules and pressure relief.

Depressurization is considered ultima ratio for extinguishing a fire or venting toxic atmosphere into space. In such a situation, the crew would be protected by closing relevant hatches to isolate the segment requiring depressurization.

Atmosphere Revitalization (AR)

Whenever astronauts breathe, they produce carbon dioxide. The AR ensures the removal of CO_2 and enables the tracing of contaminants produced by the outgassing of the various materials on board. In addition, the AR continuously analyzes the constituents of the ISS atmosphere. Because of its centralized function, the AR subsystem has a highly redundant design, and various methods of CO_2 removal are used to ensure independence in failure situations.

Temperature and Humidity Control (THC)

Temperature and humidity control (THC) aims to maintain the so-called "Crew Comfort Box" defined by a parameter range of temperature (18–27 °C), humidity (25–70%) and dew point (6–18 °C). The values in parentheses are those used for the European Columbus module. The presence of humans in the space station leads to an increase in humidity in the cabin air and to a rise in temperature due to metabolic heat. The THC must therefore remove humidity and regulate the temperature. Special air conditioning units are used for this purpose, in which the air is first cooled below the dew point, thus removing the moisture by condensation. In a second step, the cooled air is mixed with warm air to achieve the desired cabin temperature. To cool the air, the subsystem interfaces with the thermal control subsystem (TCS), which provides the cool surfaces and removes the heat via the cooling water loop.

Water Recovery and Management (WRM)

Water management is another key function of a space station, as humans consume and produce water, which consequently needs to be provided as well as collected and treated so that it can subsequently be reused. This is the task of the Water Recovery and Management subsystem. Water is collected from a variety of sources, such as condensation from cabin air, wastewater from urine, or return water after an extravehicular Activity (EVA, see Sect. 24.3.4). This water is treated to be used either as potable water or for subsystems, or it is simply vented overboard, ensuring that it does not affect any external equipment or payloads.

Fire Detection and Suppression (FDS)

The FDS function is also a part of the ECLSS system. Fire requires three basic factors to start the combustion process: fuel, oxygen and energy. Although efforts have been made not to use combustible materials aboard the ISS, energy and oxygen are present—energy in the form of electrical power and oxygen as part of the cabin air.

Therefore, a mechanism to detect a potential fire is required, as well as the ability to suppress it. The detection function is implemented using multiple smoke detectors distributed throughout the station. These detectors are connected to the on-board software, which immediately triggers automatic responses to prevent the spread of fire by shutting down active ventilation and removing any potential power source by cutting power to suspect equipment. The FDS subsystem also provides the crew with equipment for active firefighting: fire extinguishers and breathing apparatus.

24.3.3 Cargo

A manned space station will operate for years, requiring replenishment capacity for new experiments, replacement hardware, water, food, clothing, fuel, or even entire new modules. Astronauts also need to be transported up and down.

On the ISS, this is provided by an entire fleet of spacecraft (Fig. 24.3). The most famous, now retired, was the Space Shuttle—also called Space Transportation System (STS). It could carry up to 24 t of payload along with two to eight astronauts.

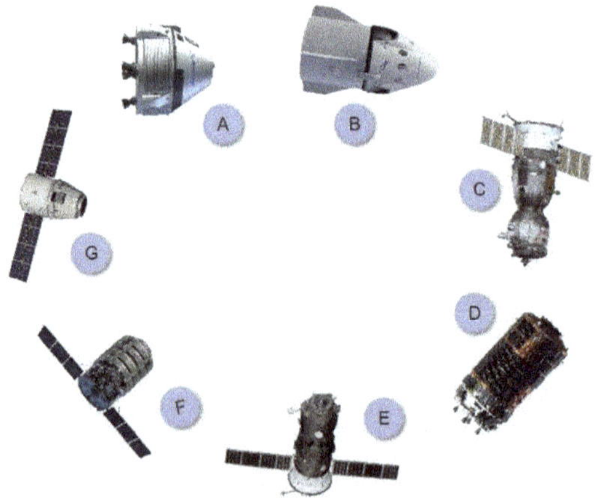

Fig. 24.3 The fleet of ISS resupply spacecraft: Three vehicles are capable to carry crew onboard: The Boeing CST-100 Starliner crew capsule (A) is scheduled to begin manned operations in 2022, the SpaceX Crew-Dragon (B) performed the first manned flight in May 2020, and the Soyuz capsule (C) has been in service for many years. The HTV (D), the Progress (E) and the Northrop–Grumman's Cygnus (F) are one-way cargo ships, while the commercial Dragon capsule (G) can also return items to Earth

The Space Shuttle was also capable of returning to the Earth's surface after its journey into space, bringing back both astronauts and cargo (about 15 t).

The main European supply vehicle was the Automated Transfer Vehicle (ATV) developed by the European Space Agency (ESA). It was designed to carry up to 8 t of cargo. Like the Progress, the ATV automatically docked with the Russian segment of the ISS. After unloading, it was used for re-boosts, filled with waste before leaving the ISS, and then destroyed during re-entry. Five ATV missions were flown between 2008 and 2015. The project was terminated after these successful missions.

The spacecraft currently used for ISS operations are listed below (the Orion spacecraft will be used in the ARTEMIS program for returning to the Moon; it is not scheduled to fly to the ISS).

- **Soyuz**: Developed by the Russian Space Agency (RSA). This is currently the vehicle with the longest legacy used for the ISS. The Soyuz was the sole means of sending astronauts to the ISS from July 2011 to May 2020. The capacity of the Soyuz is three passengers and a few dozen kilogram of cargo.
- **Progress**: Derived from RSA's Soyuz. It is designed to carry cargo of about 2.3 t and docks automatically with the station. Once docked to the Station, Progress will also be used as a re-boost vehicle to bring the ISS to a higher altitude. After completing its mission on the ISS, the Progress is filled with waste and burned in the atmosphere. A Progress is sent to the ISS two to three times a year.
- **HTV**: The H-II Transfer Vehicle was developed by the Japan Aerospace Exploration Agency (JAXA) and is also intended for cargo-only operations. The main difference from the ATV and Progress, besides the capacity (six tons for the HTV), is the way the HTV is captured by the ISS. The HTV approaches the ISS to within a few meters, then it is captured by the space station's robotic arm, which berths the HTV to the ISS. Just like the ATV and Progress, the HTV is filled with waste before being destroyed during re-entry. An improved cost reduced version known as HTV-X is planned for 2023.
- **Cygnus**: This cargo ship was built by Orbital/ATK (now Northrop–Grumman) as a commercial cargo vehicle like Dragon (below). Cygnus can carry about 3.5 t of cargo to the ISS and is berthed to ISS with the robotic arm. Cygnus is also filled with waste after unloading and burns up in the atmosphere during re-entry.
- **Cargo-Dragon**: While the above-mentioned vehicles have been operated and managed by national space agencies, Dragon is the first capsule developed by a private company, California-based SpaceX Corporation which flew to ISS. The vehicle is the only cargo spacecraft with the capability to return to Earth. In its initial version its upload and download capacity was about 6 t and 3 t, respectively. Robotic operation was required to berth the spacecraft to ISS. A second Dragon version was developed later on and can now automatically dock to the ISS. Its inaugural flight took place in December 2020 with Spx-CRS-21.
- **Crew-Dragon**: The SpaceX commercial crew capsule made its inaugural flight in March 2019 during an unmanned test flight to ISS including automatic docking and successful return to Earth. The first manned flight with two test pilots on board was already taking place during the second Crew-Dragon mission in May 2020.

Finally, the first ISS crew rotational flight of Crew Dragon launched in November 2020 with the full capacity of four crew members.
- **CST-100 Starliner**: The commercial crew capsule developed by Boeing flew for the first time in an unmanned test flight in December 2019 and successful returned to Earth, albeit without docking with the ISS. The first manned flight is planned for 2022. Like the Crew-Dragon, it can carry four crew members to the ISS.

It is planned that Soyuz, Crew Dragon and Starliner will be used for crew transport to and from ISS.

24.3.4 Extravehicular Activities (EVA)

Assembling the International Space Station in space required an enormous effort that included many flights by the Space Shuttle or unmanned Russian vehicles. A robotic arm, controlled by an astronaut from inside the station, was used to assemble the modules. But as mentioned earlier, sometimes technology just isn't good enough to do all the tasks involved robotically. This is where human help is needed, even outside the pressurized modules. Extravehicular activities (EVA) are those activities where an astronaut wearing a space suit works outside the space station. These activities are not only for construction or maintenance of the station, but are also used to install new experiments on external platforms.

The ISS program distinguishes between three types of EVAs: "scheduled" for EVAs that are planned well in advance and integrated through the nominal planning, "unscheduled" for EVAs that are not part of the nominal plan but are intended to help achieve mission objectives; and "contingency", which are also not planned in advance but are performed in a contingency situation where crew or vehicle safety is at risk.

During EVA, the astronaut can be considered as an independent spacecraft, which means that all the subsystems already discussed must be provided by the space suit: Either the extravehicular mobility unit (EMU) or the Orlan suit, with EMU developed by NASA and Orlan developed by the RSA. A space suit provides astronauts with the protection from the harsh space environment described in Chap. 1 and keeps them alive by providing oxygen and water. Although astronauts must always be tethered to the station on their well-defined translation paths along the ISS during an EVA, they also carry a rescue device just in case, which can provide the thrust and momentum needed to return to the station structure.

The longest EVA ever performed lasted 8 h and 56 min. Each "spacewalk" is much more than a walk: It is hard work for the astronauts and requires not only a lot of training and preparation on the ground, but also in orbit. For example, before the EVA, a so-called camp-out is performed to prevent the EVA crew from getting decompression sickness by having the astronauts breathe air at a lower pressure and significantly reduced amount of nitrogen to flush the nitrogen out of their blood system. This is very important because the EMUs provide only 1/3 of the pressure compared to the space station. This low pressure is needed to reduce the stiffness of the space suit in vacuum and it becomes easier for the astronaut to move around while wearing the space suit outside the ISS.

24.4 Crew—Another Subsystem to Be Operated

In the previous chapters, the spacecraft was divided into subsystems that can be treated separately and that interact with each other through well-defined interfaces. This view can be maintained if the astronauts are now also considered, treated and "operated" as one of the subsystems of the spacecraft.

24.4.1 Safety of the Crew

Whether on Earth or in space, a human life is invaluable and must therefore be protected as best as possible. This is also reflected in the basic rule of mission operations, which states that the priority order "crew—vehicle—mission" applies under all circumstance: crew safety is the highest value to be protected, the integrity of the ISS and its modules has second priority, and mission objectives rank only below (Campan et al. 2019).

Emergencies

The ISS program has defined three different emergencies (Uhlig et al. 2016): fire, rapid depression and toxic atmosphere conditions. For all emergency scenarios, there is an initial common response that the crew knows by heart and performs before executing the specific responses. These are documented in emergency procedures. The concept of operations is such that all emergency responses can be executed independently by the crew. However, the flight control teams (FCT) are trained to support the crew and act more quickly than the crew could. One reason for this is that the FCT consists of dozens of people, while the crew is only three to eleven aboard.

All emergency responses follow the same three-step approach.

1. **Warn the other crew members**. This includes triggering the ISS-wide alarm system, which creates an audible tone through the station as well as displays the emergency condition on the station computers.
2. **Gather in safe haven**. Astronauts assemble in a clearly defined safe area, near their return vehicles. There they also have access to emergency procedures, required equipment, and the station's control computers. The crew decides whether to attempt to address the emergency situation or to evacuate the ISS immediately
3. **Work emergency procedure**. Once the astronauts decide to fight the emergency, they begin working on the emergency procedures. These procedures are

written in Russian and English and are in printed form so that the crew does not have to rely on electronic aids which might fail during an emergency.

Avoidance of Hazards
Already during the design of the station and its components, a considerable effort was made to avoid possible hazards that could harm the crew during their stay in space (Musgrave et al. 2009). Special safety documentation and review processes had to be followed during the development phase to ensure that any conceivable hazards to the astronauts were avoided. These include exposure to hazardous materials, rotating equipment, extreme temperatures, powered equipment, and control of pressurized systems, oxygen enrichment, and release of shatterable material.

Any potential hazard which could be generated by a component is analyzed and demonstrated to protect the crew from it. For serious consequences of an exposure, even several independent protections measures are prescribed. If the design of the component is not capable of meeting the safety requirements, operational means of hazard control are required by the safety community. The operational products, described in more detail in Sect. 24.5.2, must include appropriate controls over the hazard (e.g. the procedure sequence first ensures the powering down of the corresponding electrical connectors before the crew could come into contact with them).

24.4.2 Crew Health

As described earlier, the spacecraft provides a habitable environment for humans; however, the human body is adapted to the conditions on the Earth's surface. Therefore, astronauts must take care of their health while in space, especially during long-duration flights (Barratt 1990).

A Flight Surgeon or a biomedical engineer (BME) is responsible for the health of the crew before, during and after the flight. They know the crew's medical history so they can provide the best possible support from ground if they have a problem on board. They are also responsible for scheduling physical exercises for the crew as well as attending private conferences such as private medical conference (PMC) and private psychological conference (PPC). They keep track of the astronaut's daily working hours so that they do not exceed the agreed limit without justification.

Astronauts on board must exercise their muscles and cardiovascular system daily to prevent degeneration due to reduced gravity. Normally, 2.5 h of training per crew member are scheduled on a daily basis on one of the various exercise devices on board, which are part of the Crew Health Care System (CHeCS). This subsystem also provides all possible means of emergency medical care, for which astronauts are trained as Crew Medical Officers.

24.4.3 Crew Communication

Communication between the crew and the ground is a key element in human spaceflight operations. There are several communication channels for the ISS project.

Space-to-Ground (S/G) Audio Communications

Voice communication with the crew are referred as Space to Ground communications (for more details, see Fortunato and Lamborelle 2012). Although radio communication to the ISS is almost continuously, thanks to the TDRS system, voice traffic is kept to a minimum for a variety of reasons, such as ensuring a crew can call the ground anytime without having to wait for a call to end. This becomes even more important in some special situations, such as an emergency. The exception, of course, is the numerous public relation activities performed by astronauts on board.

Typically, two daily planning conferences (DPC) per day are held on the ISS during the week. In the morning, for example, the crew is informed about deltas on experiments or procedures they will perform during the day. In the evening, the DPC is used to coordinate with the crew on the events of the day, the completion status of activities and the current stowage situation. On Saturdays, a so-called weekly planning conference (WPC) can also be scheduled, during which the crew receives an overview of the operations for the coming week.

In addition to these planning conferences, the audio communication link is used either to support the crew while working on a procedure or in case of off-nominal situations.

Written Communications via Operations Data File (ODF)

ODF is the term commonly used in the ISS program for procedures (for more details, see Rickerl 2009). For each activity to be performed, there is an associated ODF, whether the activity is performed by the crew or commanded from ground. Before these ODFs are released for operations, they must go through a rigorous process of review and validation involving experiment design and safety experts, as well as representatives of the astronauts and the flight control teams. The ODFs are uploaded to the ISS, and thus serve as a written means of communication.

Supporting Material

In addition to the ODFs, the crew has access to other electronic documents that provide them with additional information and that are uploaded to the ISS by the flight control teams. For example, each day the crew receives a daily summary (DS) that contains some valuable information (e.g. station configuration, potential targets for Earth observation) or some questions and answers. The astronauts also have access to so-called "crew messages" that contain, for example, a brief overview of an experiment with pictures and information that give the crew some context to better understand what they are about to do.

Timeline

The timeline is a way to communicate the daily work schedule to the astronauts on board. The daily schedule is available to the crew via the OPTIMIS Viewer. This tool provides a graphical representation of the timeline and allows the astronaut to submit a "crew note" for a specific activity in the timeline. These crew notes are used for many purposes, e.g. information on hardware used, comments on activities performed, etc.

Designing such a timeline for each day is a challenge that requires intensive international coordination. The ISS planning process is described in more details in Chap. 19.

24.5 Ground Support Operations

24.5.1 Mission Control Center (MCC)

The ISS program, spanning over four continents, is an enormous international project with five major mission control centers (MCC) working together to operate the ISS. These are the MCC-Houston (MCC-H) in Texas, the MCC-Moscow (MCC-M) in Russia, the Payload Operations and Integrations Center (POIC) in Huntsville, Alabama, the Space Station Integration and Promotion Center (SSIPC) in Tsukuba, Japan, and the Columbus Control Center (Col-CC) in Munich, Germany (for details see Kuch and Sabath 2008).

Flight Control Team (FCT)
Flight operations are conducted by dedicated flight control teams, whose core positions are usually staffed or at least available 24/7. The team works under the direction of a Flight Director, who is responsible not only for his team but also for operations in the corresponding module.

Furthermore, the flight control team includes console positions for all the different subsystems of the spacecraft mentioned above. It depends on the focus of the appropriate team and the mission phase whether the mapping between console positions and subsystems is 1:1 or whether multiple subsystems are combined into one position. Each control center also has a planning function, either as a dedicated position (e.g., Ops Plan in Houston or COMET at Col-CC) or it is merged into another console position. At each center, there is also a crew communicator position that handles all verbal communications with the astronauts and represents the crew's perspective to the FCT. In addition, some ground controllers are needed; they take care of the complicated ground infrastructure including computers and networks.

Columbus Control Center (Col-CC)
As an example of an ISS control center, the Columbus Control Center (Col-CC) located in Oberpfaffenhofen, near Munich, Germany, is briefly presented here. Col-CC is responsible for operating the Columbus module, for providing the European ground segment for human spaceflight, and for coordinating the European payloads. The actual payload operations are performed by dedicated small control centers,

called User Support and Operations Center (USOCs), distributed throughout Europe (B-USOC in Belgium, BIOTESC in Switzerland, CADMOS in France, MUSC in Germany, and E-USOC in Spain). This concept of distributed payload operations reflects the European Space Agency's (ESA) approach to actively involve all nations funding this European human spaceflight project. All of these centers are coordinated by Col-CC.

With respect to NASA, Col-CC interfaces directly with the ISS lead control center in Houston (MCC-H), as the U.S. segment provides Columbus with all the resources it needs, and with the NASA payload center in Huntsville (POIC), which operates some of its payloads within the European Columbus module.

The flight control team at Col-CC consists of two permanent positions (staffed 24/7) and three part-time positions. The two permanent positions are COL FLIGHT, who acts as the Flight Director for Columbus and is responsible for the entire Columbus module and the STRATOS (safeguarding, thermal, resources, avionics, telecommunications, operations, systems) position, introduced in 2013, which combines the DMS, COMMS, TCS, EPDS and ECLSS subsystem into one console (Sabath et al. 2014).

The planning function is performed by the EPIC (European Planning and Increment Coordination) team which is working the long-range planning in the office and the near real-time planning (less than seven days before the execution day) through their 8/5 console position called COMET.

Non-permanent positions include the EUROCOM (European spacecraft communicator), which is in charge for space-to-ground voice communications for Columbus. This position is normally filled by an astronaut who is currently on the ground or by a crew trainer. The Columbus stowage and maintenance officer (COSMO) position is responsible for on-orbit stowage, on-board logistics, waste management, up- and download coordination and all crew mechanical activities. Both positions normally work from their respective home base in Cologne (EAC) and in Turin, respectively. For special activities or in contingency cases, they come to Col-CC to collocate the core team at one location.

These five teams thus form the core of the Columbus flight control team. They are supported by the ground control team, consisting of the GSOC-GC in 24/7 operation and the Syscon in 16/5 operation. The ground control team takes care of the complex European ground segment.

Other teams also support the flight control team during office hours, such as the Engineering Support Team that interfaces with the companies that built the module and its components, and an ESA mission management team that provides programmatic guidance to the flight control team or a medical operations team.

24.5.2 Operations

Coordination

The fact that the ISS mission control centers are located on four different continents does not make coordination between them any easier. There is a need to talk to each other and to share information.

To overcome this communication problem, a voice loop system is used. Almost all positions on ground have an assigned "voice loop" which can be compared to a conference line where people can either talk and listen or just listen to what is being said. Each flight controller can then discuss with another flight controller via a headset and microphone. All relevant loops for their position are displayed on a screen and can be selected. The Space to Ground channels used for communications with the ISS are also available through this voice loop system.

A number of tools have been developed for fast, effective and well-documented information exchange and decision making. For example, the electronic flight note (EFN) system is used to exchange written information; this includes the ability to perform a review and approval process for each Flight Note. Another example is the planning product change request (PPCR) tool, which is used to coordinate and agree on changes to the schedule for coming days. Another important piece of software is the console log, which allows all decisions, events or received or delivered information to be documented in a shift diary. This is in especially helpful for the handover process between shifts.

Operations Products—OPS Products

In order to avoid conflicting policies between the different control centers and positions, it is necessary to define in some documents how decisions and interactions are made within the project. The part of this documentation that is applicable on the console is called Ops Documents and is mandatory for all control centers. The OPS products are briefly explained below.

- **Flight rule**: A flight rule documents an agreement on how to react depending on a situation. They serve to minimize the effort required for justification and coordination when off-nominal situations occur in real-time operation. A flight rule is therefore a very important document and applies as an "operational law" for all control centers and supersedes all subsequent documents in the event of a conflict.
- **Payload regulations**: They are a type of flight rule, but specific to payloads. Each MCC can have its own set of payload regulations for its respective payloads, which can be shared with other MCCs if necessary.
- **Safety documentation**: Here, some reference materials can be found for operational hazard prevention.
- **Operations data file (ODF)**: As explained in the section on crew communication, an ODF is a step-by-step procedural description of all activities on board or via ground commanding.

- **Operation interface procedure**: Depending on the level, real-time operational interfaces and standard processes are described in different documents (interactions within a control center, within Europe, or between all international partners). Depending on the task, a flight controller may need to consult three different books, if the task involves interfaces at different levels.

All ops documents require a dedicated review and approval process to change them, which can take months from initiation of the update to final release. Therefore, "real-time change processes" are defined for most products, allowing a faster turnaround on console via Electronic Flight Notes with final approval by the corresponding Flight Director.

24.6 The Future

In 2020, the ISS has been continuously manned for 20 years, reaching the second half of its total lifetime. Therefore, the focus of human spaceflight is turning to new goals. Since 2018, major spacefaring nations have been making their way to the Moon in preparation for the first manned mission to Mars. This includes, above all, the international partners (IPs) of the ISS, i.e. NASA, Roskosmos, JAXA, ESA and CSA.

The first building block is the so-called Gateway: a temporary manned space station around the Moon that will serve as a stopover for lunar landings and as a testbed for technologies for deep space missions. The core station will be built by NASA and used for the first human landing on the Moon in more than 50 years. Afterwards, the IPs will integrate their modules to add more features and capabilities to the Gateway.

Starting with this first lunar outpost, the plan is to conduct excursions on the Moon and perhaps later establish a permanent station on the lunar surface. The preferred target is the south pole on the Moon, because of the high probability of finding water ice in deep craters there, along with some permanently sunlit mountain peaks nearby.

Operations on the Moon may be accomplished in a manner similar to the ISS by direct monitoring and control from Earth. However, there are plans to test some new operational configurations that could be used for a mission to Mars. To support the astronauts in an appropriate manner, it is necessary to establish a higher level of autonomy on board—especially because of the up to 40 min signal round time for communication with Mars.

The current decade seems to be a very fascinating time for space explorers, with a high probability of leaving low Earth orbit after more than 50 years and heading for the Moon again. The next step towards Mars will be a much bigger challenge, but with a good preparation a Mars mission seems to be within reach in the next two decades—or one of the "New Space" companies finds a way to establish a Mars mission in shorter time frame—time will tell.

References

Barratt M (1990) Medical support for the International Space Station. Aviat Space Environ Med 70:155–161

Campan J, Sabath D, Söllner G (2019) Increment 56/57 ISS events put focus on safety role of the Columbus Flight Director. In: 70th international astronautical congress, IAC-19, B3.4-B6.4,7, Washington DC

Dempsey R (ed) The international space station—operating an outpost in the New Frontier. NASA-SP-2017-634. NASA

Fortunato A, Lamborelle O (2012) Space-to-ground communication: origins and development. In: Proceedings of the 63rd international astronautical congress (IAC 2012), pp 4925–4931

Furness T, Shayler D, Shayler MD (2007) Praxis manned spaceflight log 1961–2006. Springer-Verlag, Berlin Heidelberg

Kuch T, Sabath D (2008) The Columbus-CC—operating the European laboratory at ISS. Acta Astronaut 63(1–4):204–212

Messerschmitt E, Bertrand R (1999) Space station systems and utilization. Springer-Verlag, Berlin Heidelberg

Musgrave G, Larsen A, Sgobba T (2009) Safety design for space systems. Elsevier

Rickerl DA (2009) Operations data file standards. International Space Station Program SSP-50253

Sabath D, Kuch T, Söllner, G, Müller T (2014) The future of Columbus operations. In: SpaceOps 2014 conference, AIAA 2014-1618, Pasadena

Uhlig T, Roshani F, Amodio C, Rovera A, Zekusic N, Helmholz H, Fairchild M (2016) ISS emergency scenarios and a virtual training simulator for flight controllers. Acta Astronaut 128:513–520

Jérôme Campan is the head of the Columbus Flight Director Team at the Columbus Control Center in Oberpfaffenhofen. After receiving his engineering diploma from the ICAM school (Institut Catholique des Arts et Métiers) in Toulouse, he graduated from the ISAE school (Institut supérieure de l'aéronautique et de l'espace) in the "Space System engineering" field. He started his professional career as a Columbus Operations Coordinator for the first three years, then became the team lead of the European Planning and Increment Coordination team before moving to his current position.

Thomas Uhlig received his physics diploma and his Ph.D. from the University of Regensburg/Germany, where he worked on electromagnetic imaging of nanostructures. After working as scientist at the Argonne National Laboratory in Chicago he changed into the spaceflight business and to the Columbus Control Center in Oberpfaffenhofen, where he first worked as Payload expert on the Space Shuttle mission STS-122, then for several years as Columbus Flight Director. He then took over the training for the European ISS flight controllers, before he was assigned as team lead for the geostationary satellite team at GSOC. In between, he was working as expert for functional safety, first at BMW, then at Heidenhain GmbH before returning to DLR in 2022. He was awarded with the Helmholtz price 2005 of the Physikalisch-Technische Bundesanstalt. He is author of various scientific articles.

Dennis Herrmann studied aerospace engineering and received his diploma in 2007. He later joined DLR and worked as part of the Columbus team as Columbus Operations Planner (COP) and Columbus Operations Coordinator (COL OC). In 2009, he assumed the group leader role for this position before being promoted to the Columbus Flight Director group. In 2012, he acted as Lead Increment Flight Director and received a project management certification before leaving the space sector to join the automotive safety industry, where he still works for BMW today.

Dieter Sabath is group lead for Human Spaceflight at DLR's German Space Operations Center (GSOC) and working since 2004 in the project management of the Columbus Control Center. He received an aerospace diploma in 1989 and a Ph.D. from the Chair of Astronautics at TU München in 1996. After working at MAN Technologie on manned and unmanned re-entry vehicles as well as hot structures he joined DLR in 2004. As member of the project management team he is responsible for the operations preparation and for the operations of the Columbus module. He is member of the Human Space Flight Committee of the Internal Astronautical Federation and author of several scientific publications on space technology.

Chapter 25
Operations of On-Orbit Servicing Missions

Florian Sellmaier and Heike Frei

Abstract This chapter gives a comprehensive insight into the operational aspects of On-Orbit Servicing as well as rendezvous and docking missions. By means of several examples, the operational challenges are explained, and solutions are outlined. The orbit mechanics of a rendezvous mission is described in the local orbital frame. We use the Clohessy-Wiltshire equations to explain the different elements of the approach navigation. The influence of the sensor technology on the approach strategy is also discussed. In the context of robotic capture, we address the necessary changes in the communication concept, e.g., to ensure teleoperation. Finally, we describe the use of test and validation facilities for the critical maneuvers of a rendezvous and docking mission.

25.1 Introduction

25.1.1 What is On-Orbit Servicing?

On-orbit servicing (OOS) is understood here as a type of mission in which a servicing satellite (called servicer or chaser) provides service to another satellite or structure in orbit (called target or client). In principle, three OOS service classes can be distinguished: observation, motion and manipulation (see Table 25.1).

Observation of other spacecraft has so far been used mainly for military purposes. In January 2009, a first deep space inspection of the failed missiles warning satellite DSP 23 was carried out by two MITEX satellites in the geostationary Earth orbit (GEO) (see Sect. 25.2.2).

The motion of other spacecraft is used for fleet management of geo-stationary satellites, i.e. for station keeping, relocation or disposal activities after the spacecraft's fuel has been depleted. This was planned with the Orbital Life Extension Vehicle

F. Sellmaier (✉) · H. Frei
German Space Operations Center (GSOC), German Aerospace Center (DLR), Oberpfaffenhofen, Germany
e-mail: florian.sellmaier@DLR.de

F. Sellmaier et al. (eds.), *Spacecraft Operations*, Springer Aerospace Technology,
https://doi.org/10.1007/978-3-030-88593-9_25

Table 25.1 Service classes of OOS missions

Service class	Kind of service	Examples
Observation	Remote inspection	MITEX (2009)
Motion	Station keeping, relocation, disposal and de-orbiting	OLEV (study) MEV-1 (2020) DEOS (study)
Manipulation	Refueling, maintenance, repair and retrofit, docked inspection	Orbital-Express (2007) Shuttle/Hubble Space Telescope (HST) (1993)

(OLEV) and is now being demonstrated with the mission extension vehicle (MEV-1) (Sect. 25.2.3). Another application is the controlled removal of satellites from the low Earth orbit (LEO) to prevent damage on the Earth in connection with uncontrolled re-entry or in the context of space debris mitigation (SDM).

Manipulation of other spacecraft includes both maintenance and repair and is the most complex activity of OOS. In 2007, the Orbital-Express technology mission successfully demonstrated the ability to maintain a client satellite, including battery replacement and exchange of fluids, i.e. refueling (Sect. 25.2.4). The maintenance activities on the Hubble Space Telescope (HST) between 1993 and 2009 are examples of manned maintenance in orbit (Sect. 25.2.1).

25.1.2 Motivations for OOS

As Table 25.1 shows, there are several motivations for an on-orbit servicing mission. Firstly, activities such as "remote inspect", "repair" or "upgrade" should help to repair a defective spacecraft in orbit. Secondly, activities such as "station keeping" or "re-fueling" are intended to extend the lifetime of a spacecraft in orbit. All the activities mentioned so far relate to restoration and/or continuation of the service of a client spacecraft by an OOS mission. The crucial question here is whether it is worthwhile to launch a rather complex spacecraft with the aim of repairing another spacecraft. The answers will only be positive if the total value of the OOS mission is higher than the costs, e.g., if OOS becomes state-of-the-art in the future, and the servicer spacecraft would be cheaper compared to replacing an expensive geostationary communication satellite. OOS can be very effective if a service spacecraft can provide a service for more than one client satellite (fleet servicing).

The motivation for the "disposal" or "de-orbiting" of a spacecraft is either to avoid damage on Earth by uncontrolled de-orbiting or to remove a spacecraft from certain orbits to avoid further increase of space debris (e.g. discussed for ENVISAT). These activities are summarized under the term space debris mitigation (SDM), which can be seen as an important function of OOS. Especially non-operational, uncontrolled large satellites pose a high collision risk and thus an immense threat to all operational satellites in adjacent orbits.

25.1.3 Space Debris Mitigation

The amount of space debris is becoming increasingly relevant to operations, especially with the new proposed mega-constellations. Since around 100 of these satellites have been launched (as of May 2020), this will lead to a further increase in space debris. A certain percentage of operating costs is spent on collision avoidance programs. Collision avoidance maneuvers can shorten the lifetime of a mission and have an impact on the mission. Hence, the process of space debris mitigation is becoming increasingly important (cf. Chap. 1 for space debris and Chap. 13 for collision avoidance maneuvers).

Figure 25.1 shows the distribution of the catalogued resident space objects in the inclination and orbit height domain. It is remarkable that there are concentrations at orbit heights of approximately 800 km, and at inclinations at approximately 98°. The peak values exhibit orbit heights where the remaining drag is negligible resulting in stable orbits. The concentration at the inclinations results from the highly populated sun-synchronous orbits and their respective advantages. Hence, it is not surprising that the 2009 collision between Cosmos 2251 and Iridium took place at this inclination.

This situation is already quite alarming, and countermeasures must be performed immediately. As already shown in Chap. 1, mega-constellations will exacerbate the situation and lead to the well-known exponential growth, also known as Kessler syndrome (Kessler and Cour-Palais 1978). The only way to limit this increase is to

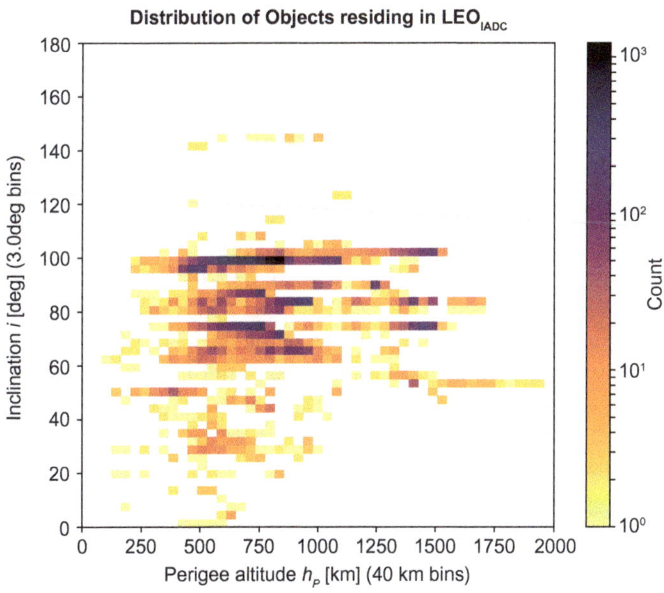

25.1 Density of particles larger 10 cm versus inclination versus mean altitude (Klinkrad 2010)

actively remove debris at the end of the mission, combined with the active removal of large resident space objects that are the source of new space debris. Current forecasts predict that in the next decades about ten large objects per year will have to be removed in order to maintain the status as is and the resource space in a sustainable way.

25.2 Examples of On-Orbit Servicing Missions

Manned OOS missions have a fairly long history and can be considered an integral part of human spaceflight. This began with the repair of the Skylab station and continued with several repair missions for the Hubble Space Telescope (HST) and maintenance missions of the International Space Station (ISS). The most complex part of the manned OOS missions is performed by people in space. Therefore, ground operations of this type of mission are quite similar to human spaceflight operations in general (see also Chap. 24). Examples of manned OOS missions are described below in Sect. 25.2.1.

On the other hand, robotic OOS is still a very new field in the space business, so there are relatively few missions so far. It is quite challenging to perform maintenance work by robots, and even more so when this activity takes place in space. Examples of robotic OOS mission are described in Sects. 25.2.2–25.2.4.

25.2.1 Manned OOS Missions

Repair of Skylab and Hubble Space Telescope

The repair of NASA's Skylab station in 1973 is the first example of a manned on-orbit servicing mission. One minute after Skylab 1 was launched, ground control received alarming telemetry signals. During launch and deployment, Skylab 1 suffered severe damage, resulting in the loss of the micrometeoroid and heat protection shield and one of the solar panels. Skylab 1 reached the planned orbit but was not usable. NASA decided to postpone the launch of Skylab 2 in order to analyze the situation. Finally, the crew of Skylab 2 and 3 successfully repaired the damage to the station.

The HST was launched in 1990. Within weeks after launch, it became clear that the images were not as sharp as expected. The reason for this problem was that the primary mirror had the wrong shape. One of the first reactions was to change the observation schedule to make fewer demanding observations such as spectroscopy. Between 1993 and 2009, the HST was serviced and maintained by five shuttle missions, starting with a corrective action for the faulty mirror (see Fig. 25.2) and including replacements of the Imaging Spectrograph, all gyroscopes and the data handling unit. As a result of this servicing mission, HST is still operational in 2021. This is a quite a respectable lifetime for a mission in LEO.

Fig. 25.2 Servicing mission SM1 (STS-61) for the Hubble Space Telescope, during which corrective optics were installed in December 1993 (*Credit* NASA)

25.2.2 Inspection Missions

MITEX

On 14 January 2009, Spaceflight Now (Covault 2009) reported:

> In a top-secret operation, the U.S. Defense Dept. is conducting the first deep space inspection of a crippled U.S. military spacecraft. To do this, it is using sensors on two covert inspection satellites that have been prowling geosynchronous orbit for nearly three years.

The failed satellite examined by the MITEX satellites was the DSP23 missile warning satellite of the Defense Support Program. Details of this inspection mission are not available, but it was reported that the radio signatures of DSP23 and one of the MITEX satellites merged, indicating a close distance between the two spacecrafts.

25.2.3 Life Extension Missions

From the Loss of TV-Sat 1 to OLEV and MEV

An early example of robotic OOS dates back to 1987, when TV-Sat 1 was launched. Unfortunately, the failure of one solar panel to deploy severely affected operations because the antenna could not be activated. After several attempts to repair the spacecraft from ground, it was finally placed in a graveyard orbit in 1989. This incident triggered the idea of a rescue satellite with TV-Sat 1 as an early candidate for OOS studies. DLR and Airbus elaborated the concept of an Experimental Servicing Satellite ESS (Settelmeyer et al. 1998), including the design of a capturing tool for the apogee thrusters of geostationary satellites (Fig. 25.3).

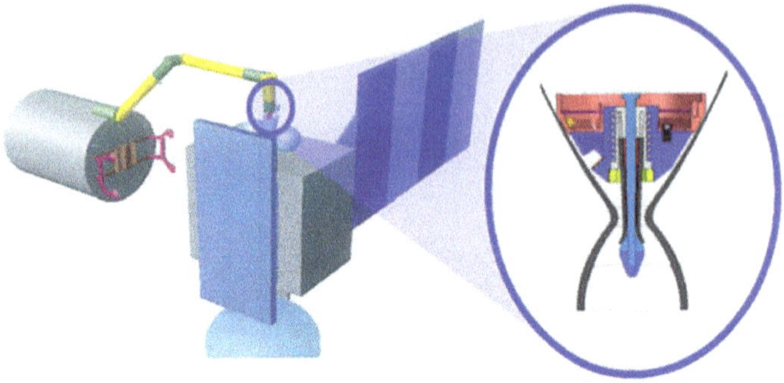

Fig. 25.3 Experimental Servicing Satellite (ESS) repairing TV-Sat 1. The study included the design of a capturing tool (*Credit* DLR)

The concept of docking to the apogees engine of communication satellites was later adopted by a commercial consortium: The Orbital Life Extension Vehicle (OLEV) aimed to extend the lifetime of communication satellites whose fuel is almost exhausted. OLEV was designed to dock on such communication satellites and take over station keeping and attitude control for about 12 years, or to undock and fly to another client and to perform re-location and disposal maneuvers (Fig. 25.4).

In 2009, the OLEV study completed phase B with a preliminary design review (PDR) and there was a customer contract for general fleet management purposes. However, in the end it was not possible to finance the non-recurring development costs

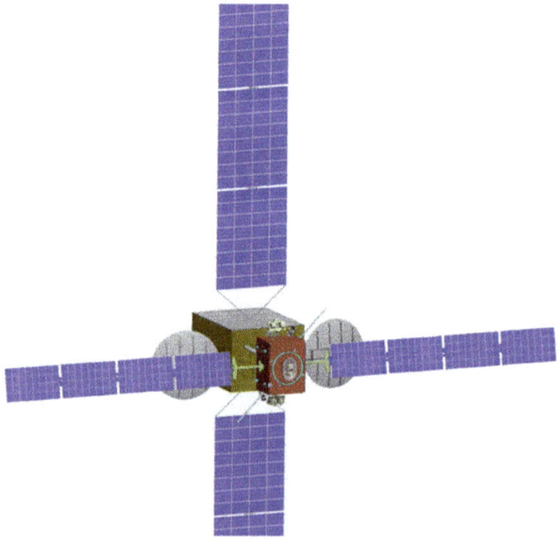

Fig. 25.4 Orbital Life Extension Vehicle (OLEV) with the horizontal solar panels docked to a communication satellite (*Credit* DLR)

for the first mission. This example also shows that commercial OOS in particular will always face the question of whether it is cheaper to repair an existing space system or to build and launch a new space system.

Robotic Servicing of Geosynchronous Satellites (RSGS)
The RSGS program of the Defense Advanced Research Projects Agency (DARPA) aims at the demonstration of robotic servicing technology in GEO or near-GEO. These technologies should enable cooperative inspection and servicing of GEO satellites. The service satellite will be based on a privately developed spacecraft designed by a commercial partner with a robotic manipulator and supporting equipment developed by DARPA. It is planned to demonstrate high resolution inspection, anomaly correction, cooperative relocation, and upgrade installation (Parrish 2019).

On-Orbit Servicing, Assembly and Manufacturing Mission 1 (OSAM-1)
The OSAM-1 mission of NASA will extend the life of a client satellite in the polar LEO orbit by a servicing spacecraft equipped with robotic arms. The client will be a LEO satellite owned by the U.S. government which will be autonomously approached and grasped followed by telerobotic refueling and relocation (Reed et al. 2016; NASA 2020). OSAM-1 was formerly known as "Restore-L", a name which emphasizes that the capabilities and functionalities of a satellite can be restored by on-orbit servicing. With the addition of another robotic payload, the mission name was changed to OSAM-1.

Mission Extension Vehicle (MEV)
Northrop Grumman (formerly Orbital ATK) developed a satellite life extension spacecraft called mission extension vehicle (MEV) with the capability to dock on a client satellite and to perform maneuvers in the coupled configuration. This will include station keeping, relocation and disposal to graveyard orbits (SpaceNews 2018). In addition to the basic MEV, Northrop Grumman is developing a Mission Robotic Vehicle (MRV) based on the MEV architecture and equipped with several modules, so-called mission extension pods (MEP), to be attached on a client satellite with a robotic arm. MEV-1 was launched on October 9, 2019 and successfully docked on a commercial customer Intelsat 901 on February 26, 2020 (Fig. 25.5).

25.2.4 OOS Technology Demonstrations

Engineering Test Satellite (ETS-VII)
The engineering test satellite No. 7 (ETS-VII) was one of the technology demonstrators related to OOS (Fig. 25.6). ETS-VII was developed by the National Space Development Agency of Japan (NASDA) and launched in 1997. It was the first satellite equipped with a robotic arm and the first unmanned spacecraft to successfully perform autonomous rendezvous and docking operations. DLR carried out experiments on ETS-VII; those are described in Sect. 25.5.2 in connection with satellite capture.

Fig. 25.5 View of target satellite in GEO from MEV-1's "far hold" position during approach from approximately 80 m with Earth in the background (*Credit* Northrop Grumman)

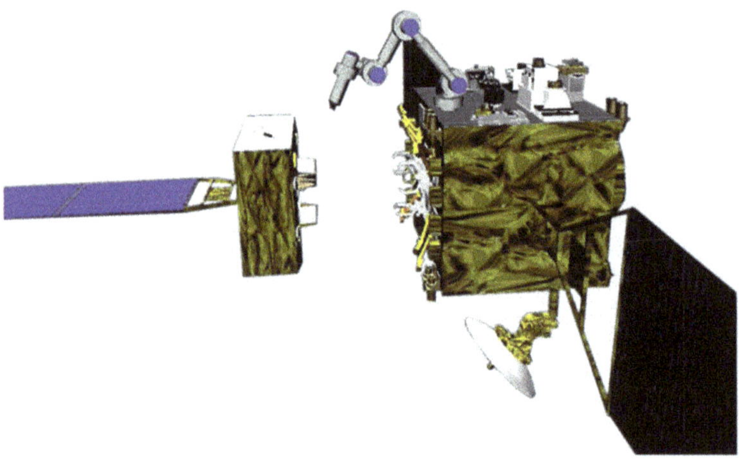

Fig. 25.6 Japanese ETS-VII servicer (right) captures target (*Source* NASDA)

Demonstration for Autonomous Rendezvous Technology (DART)

DART was a NASA mission with the goal to demonstrate a fully automatic approach to the client satellite MUBLCOM. The satellite was launched in 2005 with light detection and ranging (LIDAR) and camera-based sensors on board. DART approached its target as originally planned, when suddenly the fuel consumption became much higher than foreseen. Finally, the mission was aborted after eleven hours. The NASA investigation board declared the mission as a failure. DART had a collision shortly

Fig. 25.7 Orbital Express mission. Left image ASTRO (with two solar panels) and NextSat autonomously perform operations while unmated. Right image NextSat photographed in space by ASTRO (*Credit* Boeing Image)

before the mission abort. The target satellite MUBLCOM survived without apparent damage.

Orbital Express

A milestone for robotic OOS was set with the successful completion of DARPA's Orbital Express mission in 2007 (Mulder 2008). It comprises two satellites: ASTRO (the servicing satellite) and NextSat, a prototype of next generation serviceable satellite (client). Orbital Express successfully demonstrated the ability to autonomously perform rendezvous and docking (RvD) operations including maintenance activities. Figure 25.7 shows the two satellites ASTRO and NextSat of the Orbital Express mission.

The mission philosophy of Orbital Express was to initially crawl, walk and then run: First, both satellites performed the necessary maintenance activities while still in a mated configuration. The transfer of liquids and components like a battery and a CPU—so called orbital replacement units (ORU)—was demonstrated. After separation of the mated configuration, several rendezvous and docking maneuvers were performed including approach navigation and fly around. Both satellites were left to natural decay and subsequently re-entered the atmosphere.

25.2.5 Space Debris Mitigation

Technology Study DEOS

The technology project DEOS ("Deutsche Orbitale Servicing Mission") studied how to capture and move a target spacecraft which is no longer operational and not specially prepared for OOS.

Fig. 25.8 DEOS client (left) and servicer (*Credit* Astrium)

The primary objectives of DEOS were (1) to capture a tumbling non-supportive client satellite with a servicer spacecraft and (2) to de-orbit the coupled configuration within a pre-defined orbit corridor at end of mission (Fig. 25.8).

ESA Clean Space Initiative

ESA's clean space initiative aims to protect Earth and space by mitigating and preventing threats from space and in space, such as space debris and uncontrolled and in-operative satellites. In general, ESA intends to carry out environmentally friendly space activities (ESA 2019). Part of the initiative was the ESA program e.Deorbit with the objective to "remove a single large ESA-owned space debris from the LEO protected zone" (Biesbroek et al. 2017).

In the meantime, ESA decided to procure space debris mitigation as a service. ClearSpace-1 will be the first space mission to remove a space debris object from orbit. ClearSpace-1 is scheduled for launch in 2025.

Astroscale

Astroscale is a privately owned, global company based in Tokyo, Japan. The company develops satellite-based end-of-life and active debris removal services to contain the growing and dangerous accumulation of debris in space. Astroscale was selected as a commercial partner for the JAXA debris removal demonstration project in February 2020.

25.3 Challenges in Operating OOS Missions

The above examples of OOS missions show that there are some definite challenges when operating OOS missions. We focus here on the challenges when operating robotic OOS missions in Earth orbit.

What is the difference between a robotic OOS mission and a standard satellite mission? First of all, there are two spacecraft, one of which approaches the other. This means that approach navigation must be controlled by complex flight dynamics. When approaching the capture point, the risk of collision increases, resulting in a much shorter reaction time compared to standard operation. A collision avoidance maneuver (CAM) must be performed within seconds, whereas a standard command stack is usually time tagged hours in advance. This must be taken into accounted either by a high degree of autonomy or by an improved communication system with real-time commanding. Finally, some OOS missions require the capture of a non-cooperative target. This requires high standards regarding robotics or docking technology.

The impact of OOS on the typical ground segment system is summarized below. A more detailed discussion follows in Sects. 25.4 and 25.5 and a description of the various components of the ground segment can be found in Chaps. 2, 3 and 4.

25.3.1 Flight Operation System

The fact that there are two spacecraft, must be mapped to the flight operations team—both in terms of operational functions and responsibilities. Usually, this means that there are two sub teams, preferably integrated in one large control room (Fig. 25.9). However, sometimes it can be convenient to have two separate control rooms for client and servicer. In the worst case, operations for the two satellites are distributed to different control centers. In this case, the physical distance should be compensated by good communication connections.

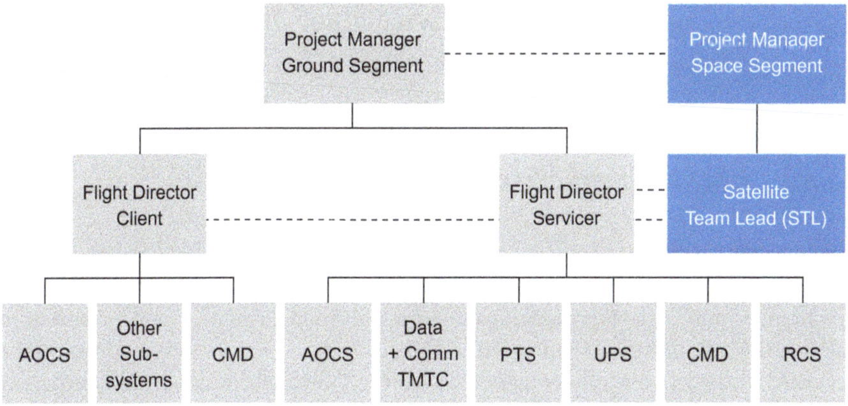

25.9 Example for an integrated flight operations team for both the servicer and the client satellite. The names of the subsystem operators in the bottom row are introduced in Chaps. 4 and 5 (see also list of abbreviations)

Also, not part of standard missions is the integration of a robotic control system (RCS) into the flight operation system. In some cases, a control device like an exoskeleton must be integrated, resulting in the reception and transmission of data in real time. The technical requirements for the so-called teleoperation are discussed in Sect. 25.5.1.

Sending commands to a spacecraft is normally the responsibility of the command operator (CMD). In most cases, this also applies to OOS missions. However, during approach and capture, the responsibility for sending attitude and orbit control system (AOCS) commands can change from CMD to RCS and back again. This situation must be carefully analyzed, also with regard to a possible collision avoidance maneuver (CAM). In some cases, it may be useful to escape forward, i.e., to try to capture instead of retreating backwards.

25.3.2 *Ground Data System*

The ground data system (GDS) connects the flight operation system with the spacecraft. For standard spacecraft operation, the GDS must ensure a stable and redundant link from ground to space and return. Since most space crafts are designed for 48 h of autonomy, the usual delay time between two and five seconds is no problem at all. The situation is different for a robotic OOS mission: Robotic activities will eventually require a telepresence communication link with real-time response (see Sect. 25.5.1).

In addition, the final approach, including a stopover on a hold point near to the client and the complete capture process should be continuously monitored and controlled with a real-time connection. In GEO this is not a problem, but in LEO the contact time is usually limited to ten minutes. For a LEO mission, the duration of the contact can be extended by a chain of ground stations (Fig. 25.23). An alternative would be to use geo-relays such as TDRS and EDRS.

Potential solutions for the above requirements are discussed in Sect. 25.5.1.

25.3.3 *Flight Dynamic System*

Any mission involving a rendezvous and docking maneuver places high demands on the flight dynamic system (FDS). One difference from standard flight dynamics for a single spacecraft is that there is a transition from absolute to relative navigation within the far range. Relative navigation will then be based on a sensor system onboard the servicer, which will eventually include systems on the client such as radio frequency transponders or reflectors for optical navigation. Relative navigation can be closed loop in space, or it can be "ground in the loop", which affects the requirements for the ground data system. After all, the maneuver strategy from far range approach until contact is very demanding for the FDS on ground.

In the next three sections, we follow the chronology of an OOS mission from the far range approach to the capture of the target satellite. In doing so, we will go into the above-mentioned challenges in detail.

25.4 Satellite Rendezvous

Approaching another spacecraft is different from launching an Earth observation satellite into a low Earth orbit. In terms of orbital mechanics, there are some similarities with the approach to a parking box in the geostationary orbit. In both cases it is advantageous to discuss the approach strategy in the local orbital frame (LOF). For the motion equations there is an analytical solution, the so called Clohessy-Wiltshire equations. Using this approximate solution, we then describe and discuss the mission phases from launch to capture. Since the sensor system plays an important role in the approach of another spacecraft, we describe the usual rendezvous sensors in the following section.

25.4.1 Orbit Mechanics in Local Orbital Frame (LOF)

In the following, we describe the movement relative to the target within LOF. The LOF is a co-moving coordinate system; the position of the target is defined as the center of the LOF (Fig. 25.10).

The relative position of the chaser with respect to the coordinate center, i.e. the target, is defined as vector s:

$$\boldsymbol{s} \equiv \boldsymbol{r} - \boldsymbol{R} \tag{25.1}$$

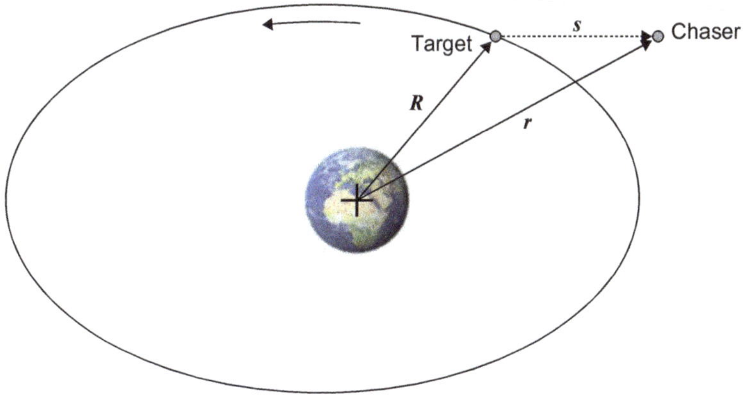

Fig. 25.10 Relative motion in LOF

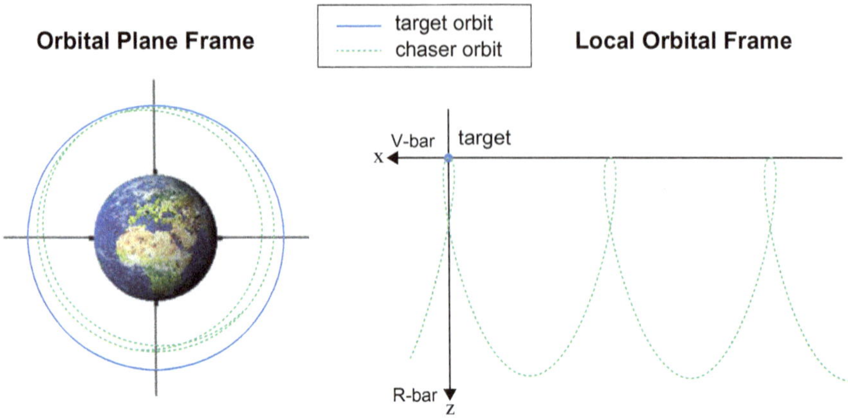

Fig. 25.11 Orbital plane frame (left) and local orbital frame (right)

The coordinates in LOF are defined as

$$s \equiv \begin{pmatrix} x \\ y \\ z \end{pmatrix} \tag{25.2}$$

This definition of (x, y, z) is often used for geostationary satellites. Alternatively, theses coordinates are also referred to as follows:

$$\begin{aligned} x &= \text{V-bar (along track)} \\ y &= \text{H-bar (south)} \\ z &= \text{R-bar (to center of Earth)} \end{aligned} \tag{25.3}$$

The advantage of using LOF is shown in Fig. 25.11. While it is not intuitive to follow the relative motion in the orbital plane frame, a clear trajectory is visible in the LOF. The LOF is a co-rotating coordinate system that is right-handed and—in case of a circular orbit—also orthogonal. However, it should be noted that it is not an inertial system, so additional pseudo forces such as the Coriolis force must be taken into account.

Using the above definition of relative coordinates Eqs. (25.1) and (25.2), an analytical set of equations can be derived from Newton's law of gravity for a central body and his 2nd law of motion (Clohessy and Wiltshire, 1960).

The following assumptions were made in deriving the Clohessy-Wiltshire equations:

- The distance of the chaser to the target is significantly smaller than its distance to the center of gravity: $|\boldsymbol{s}| \ll |\boldsymbol{r}|$
- The target is on a circular Kepler orbit with a constant angular velocity ω.

- There are no external disturbances like drag due to rest atmosphere, solar pressure or higher moments of the central body's gravitational field.

Using these assumptions, analytical equations of motion, the so-called **Clohessy-Wiltshire equations**, can be derived:

$$
\begin{aligned}
x(t) &= \left(\frac{4\dot{x}_0}{\omega} - 6z_0\right)\sin\omega t - \frac{2\dot{z}_0}{\omega}\cos\omega t + (6\omega z_0 - 3\dot{x}_0)t + \left(x_0 + \frac{2\dot{z}_0}{\omega}\right) \\
y(t) &= y_0\cos\omega t + \frac{\dot{y}_0}{\omega}\sin\omega t \\
z(t) &= \left(\frac{2\dot{x}_0}{\omega} - 3z_0\right)\cos\omega t + \frac{\dot{z}_0}{\omega}\sin\omega t + \left(4z_0 - \frac{2\dot{x}_0}{\omega}\right)
\end{aligned} \tag{25.4}
$$

where ω is the angular velocity defined as

$$\omega = \frac{2\pi}{T}$$

with T as orbital period. In addition, the Clohessy-Wiltshire equations contain the initial conditions in space and in velocity, i.e., the initial shift to the coordinate center

$$\Delta s = (x_0, y_0, z_0)$$

and the initial velocity (e.g. generated by an impulsive thrust)

$$\Delta \boldsymbol{v} = (\dot{x}_0, \dot{y}_0, \dot{z}_0)$$

In summary, the Clohessy-Wiltshire Eq. (25.4) are the equations of motion for a spacecraft in the vicinity of a target spacecraft located in the center of the coordinate system, i.e., the local orbital frame. Since additional effects due to residual atmosphere, solar pressure and higher moments of the Earth's gravitational field are not taken into account, these equations are not to be used for operational flight dynamics. However, the Clohessy-Wiltshire equations are very useful to explain and interpret the most common elements of any approach strategies to a target satellite or specific orbital position.

In the following, we analyze the orbit propagation of a body that has been displaced from the center of the local orbital frame or received an initial impulse.

(1) **Displacement: Trajectories after displacement in radial distance**

First, we study the trajectory of a body released at a radial distance Z_0 relative to the center of the coordinate system. The initial conditions for the radial displacement of this body are:

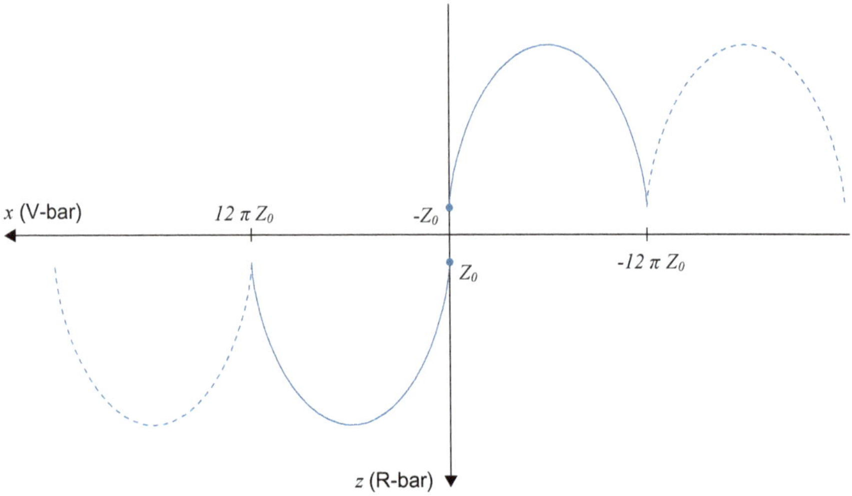

Fig. 25.12 Trajectory after displacement with a radial distance of $\pm Z_0$ (dotted line = 2nd orbit)

$$\begin{aligned} x_0 &= 0, \quad \dot{x}_0 = 0 \\ y_0 &= 0, \quad \dot{y}_0 = 0 \\ z_0 &= Z_0, \quad \dot{z}_0 = 0 \end{aligned} \tag{25.5}$$

With Eq. (25.4) the following equations of motion result:

$$\begin{aligned} x(t) &= 6Z_0[\omega t - \sin \omega t] \\ y(t) &= 0 \\ z(t) &= Z_0[4 - 3 \cos \omega t] \end{aligned} \tag{25.6}$$

Figure 25.12 shows the trajectory of a body released at a radial distance of $\pm Z_0$. This results in a cyclic motion with amplitude of $\pm 6 Z_0$. Even more relevant is that the distance in flight direction x after an orbit is $\pm 12 \pi Z_0 \approx 38 Z_0$. This propagation along or against the flight direction continues with each orbit.

The general lesson of the above consideration is that a displacement in radial direction will propagate by a factor of 38. This was the reason for Fehse (2003) to postulate a 1% rule for the accuracy of relative navigation in approach navigation:

The accuracy of a sensor system used for relative navigation should be better than 1% of the distance to the target.

This is particularly necessary for missions in LEO, which do not have permanent contact and should therefore be autonomous at least for the duration of an orbit. This rule can be relaxed if, as in GEO, a permanent connection is available, or if a closed-loop GNC system used .

(2) Displacement: Trajectories after displacement in flight direction

The situation is different for a body that is shifted in the direction of flight by X_0. The initial conditions in this case are:

$$\begin{aligned} x_0 &= X_0, \quad \dot{x}_0 = 0 \\ y_0 &= 0, \quad \dot{y}_0 = 0 \\ z_0 &= 0, \quad \dot{z}_0 = 0 \end{aligned} \tag{25.7}$$

The Clohessy-Wiltshire Eq. (25.4) result in

$$\begin{aligned} x(t) &= X_0 \\ y(t) &= 0 \\ z(t) &= 0 \end{aligned} \tag{25.8}$$

For approach navigation, this means:

Points on the target's orbit are ideal holding points with no drift relative to the target.

Now we will apply the Clohessy-Wiltshire Equations to impulsive maneuvers. We do this by setting the corresponding ΔV as the initial condition. This approach is possible if the force relations, i.e., strength and direction of the gravitational field change only insignificantly during the maneuver.

(3) Impulsive maneuver: Delta-V in flight direction (V-bar hop)

When an impulsive thrust is applied in flight direction, the initial conditions are:

$$\begin{aligned} x_0 &= 0, \quad \dot{x}_0 = \Delta V_x \\ y_0 &= 0, \quad \dot{y}_0 = 0 \\ z_0 &= 0, \quad \dot{z}_0 = 0 \end{aligned} \tag{25.9}$$

This results in the following equations of motion:

$$\begin{aligned} x(t) &= \frac{1}{\omega}\Delta V_x[4\sin(\omega t) - 3\omega t] \\ y(t) &= 0 \\ z(t) &= \frac{2}{\omega}\Delta V_x[\cos(\omega t) - 1] \end{aligned} \tag{25.10}$$

Figure 25.13 shows the corresponding trajectory. After one revolution (position 2) the impulse with ΔV has the following displacement effect Δx in the V-bar direction

$$\Delta x = -\frac{6\pi}{\omega}\Delta V_x$$

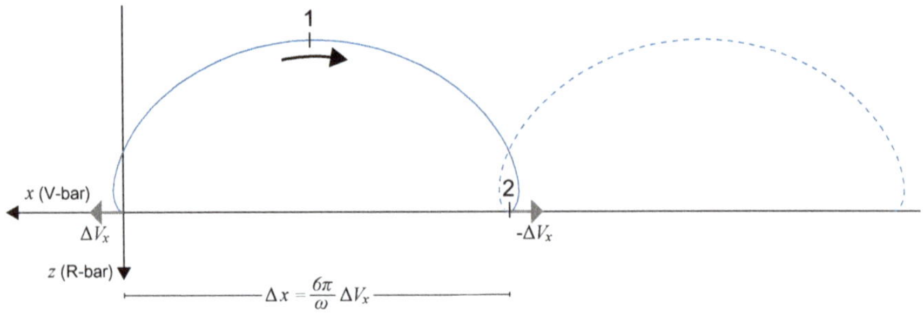

Fig. 25.13 Trajectory after ΔV in flight direction (V-bar hop)

This propagation continues on each orbit (dotted line). If the maneuver is aborted after one orbit at position 2 with an impulse of $-\Delta V$, it is called a V-bar hop. Compared to other maneuvers like the R-bar hop (see below) we find:

The V-bar hop is the most efficient maneuver to change the relative position along the orbit.

If the propagation is stopped by an impulse of $-\Delta V$ after half an orbit in position 1, we speak of a Hohmann transfer.

(4) **Impulsive maneuver: Delta-V in radial direction (R-bar hop)**

An impulsive thrust in radial direction leads to the following initial conditions:

$$\begin{aligned} x_0 &= 0, \quad \dot{x}_0 = 0 \\ y_0 &= 0, \quad \dot{y}_0 = 0 \\ z_0 &= 0, \quad \dot{z}_0 = \Delta V_z \end{aligned} \tag{25.11}$$

This results in the equations of motion for an R-bar hop:

$$\begin{aligned} x(t) &= \frac{2}{\omega}\Delta V_z[1 - \cos(\omega t)] \\ y(t) &= 0 \\ z(t) &= \frac{1}{\omega}\Delta V_z \sin(\omega t) \end{aligned} \tag{25.12}$$

Figure 25.14 shows the resulting trajectory for such an impulsive maneuver in radial direction. The efficiency of an R-bar hop after half an orbit at position 1 is

$$\Delta x = \frac{4}{\omega}\Delta V_z$$

To stop this maneuver at position 1 after half an orbit, a second impulse ΔV is required. Compared to the efficiency of a V-bar hop this is smaller by a factor of

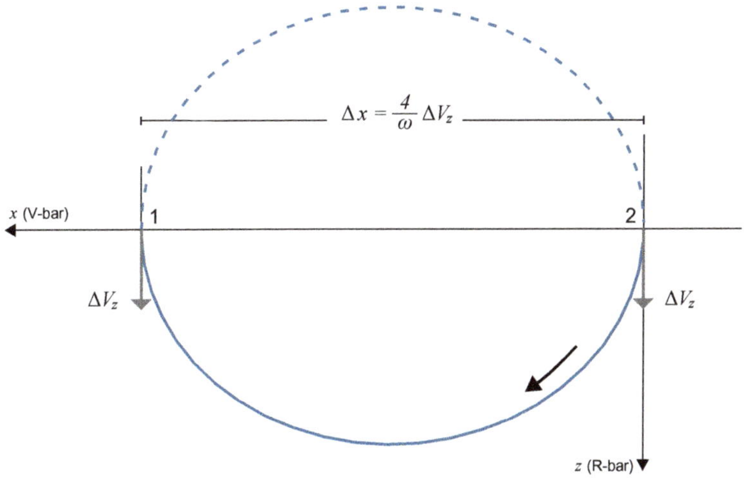

Fig. 25.14 Trajectory after Δv in radial direction (R-bar hop)

$$\frac{6\pi}{4} \approx 4.7$$

If the maneuver is not stopped at position 1, the R-bar hop continues to circle until it reaches the initial position (2) again after one orbit (dashed line), which introduces a kind of passive safety near the target:

Due to its passive safety, an R-bar hop is often used as the last two-pulse maneuver when approaching a target. This prevents a collision if the second stop maneuver fails.

With the four basic elements described above, most of the maneuvers used in a typical approach can be explained or made plausible.

25.4.2 Mission Phases from Launch to Docking

The mission phases from launch to docking are discussed in detail in Fehse (2003). Here we present one scenario each for the approach in LEO and GEO. However, the approach has to be defined and adapted for each specific mission and depends not only on the orbit but also on the target and other factors. For example, the final approach direction depends strongly on the capture strategy, in particular on the location on the target where it will be captured (location of nozzle, ring, etc.).

(1) **Typical approach in LEO**

After launch into the orbital plane, the phase angle between servicer and target is first reduced. This first phase is called "phasing" and ends at the "first aim point" S0 Fig. 25.15. This is followed by the "far range" phase from S0 to S2. During

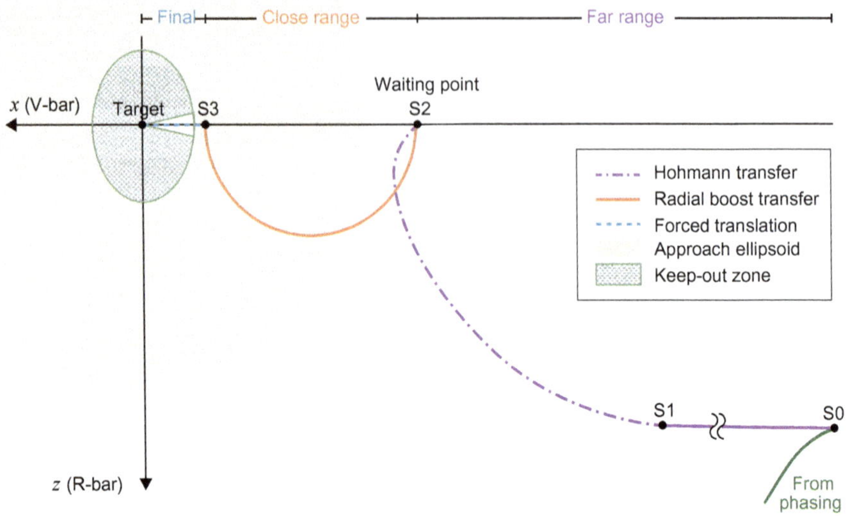

Fig. 25.15 Example of an approach strategy for a LEO mission (Fehse 2003)

this phase, absolute navigation is normally transferred to relative navigation. From S2 to S3 follows the "close range" with the last two-impulse maneuver. The "final approach" is then realized by continuous thrust maneuvers until contact and capture of the target satellite.

The approach shown in Fig. 25.15 includes most of the sub-maneuvers discussed in Sect. 25.4.1:

- After a free drift from S0 to S1, the orbit height is raised from S1 to S2 by a Hohmann transfer, which is half of a V-bar hop shown in Fig. 25.13. This is the most efficient two-pulse maneuver to reach the target orbit at S2.
- As described in Eq. (25.8), a spacecraft in the target orbit is not subject to orbital propagation relative to the target. Therefore, S2 is an ideal waiting point.
- For the close-range approach from S2 to S3, a radial boost transfer (= R-bar hop) was chosen as the last two-pulse maneuver, with the advantage that this maneuver provides passive safety in case of failure of the stop maneuver at S3 (cf. Fig. 25.14).

In summary, the advantages of this approach in LEO are (1) economical fuel consumption due to the efficient Hohmann transfer in the far range, (2) flexibility due to a waiting point in between and (3) passive safety due to the R-bar hop in the close range.

(2) **Typical approach in GEO**

The main differences between an approach in GEO and LEO are firstly the duration of the approach and secondly the required thrust Δv.

The duration of a typical two-impulse maneuver is linked to the orbital period. For the R-bar hop this means a duration of 12 h versus 45 min. This means that an approach in GEO is much slower than in LEO. In addition, the approach can be synchronized with the illumination conditions (Fig. 25.16 top).

In both V-bar and R-bar hop, the efficiency of the maneuvers is proportional to the orbital period:

$$\Delta x \sim \frac{1}{\omega}\Delta v \sim T\Delta v \tag{25.13}$$

with

$$\frac{T_{GEO}}{T_{LEO}} \approx \frac{24h}{1.5h} = 16 \tag{25.14}$$

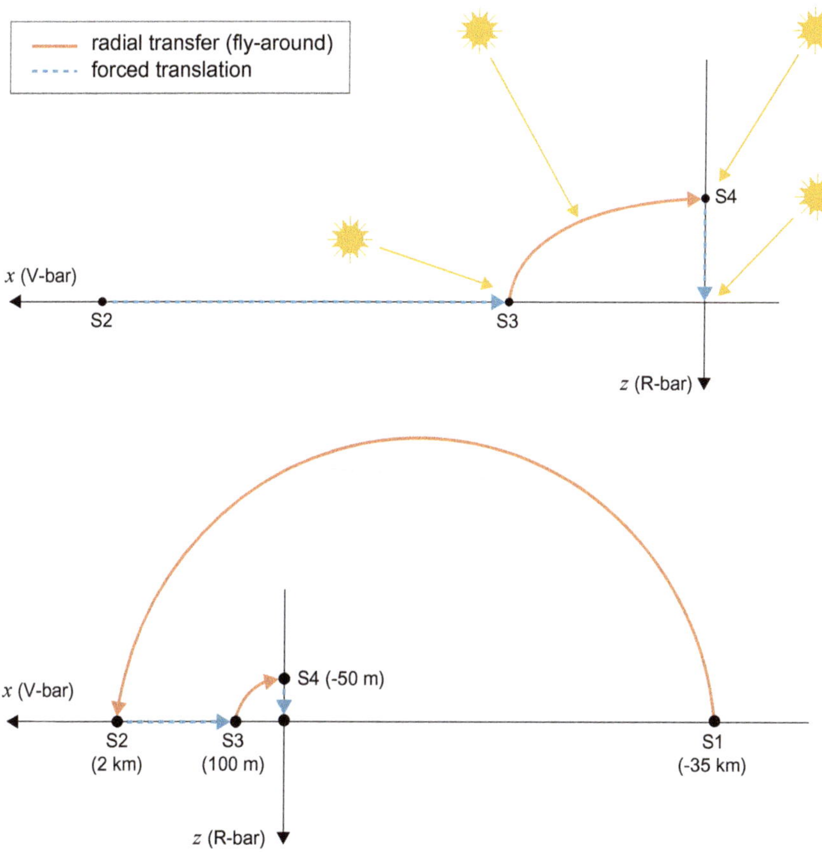

Fig. 25.16 Typical approach strategy for geostationary orbit (example OLEV)

For a given Δx, the required Δv is about 16 times smaller in GEO than in LEO.

The first part of an approach in GEO is usually the transfer from the geostationary Earth orbit to the geostationary drift orbit. In this example, the spacecraft is positioned at a waiting point S1 at a distance of 35 km from the target. The following approach strategy is shown in Fig. 25.16 (bottom):

- From S1 to S2 it includes a semi-fly around, i.e., an R-bar hop at a distance of 2 km. During this fly around, the camera-based navigation system can be calibrated to use Angles Only Navigation in the later phases (see Sect. 25.4.3).
- From S2 to S3 a forced motion maneuver with continuous thrust follows. The forced motion maneuver can be interrupted at any time as every point on the target orbit is a waiting point.
- From S3 to S4, a second fly around reduces the distance from 100 to 50 m and brings the servicer in line with the client's docking axis.
- From S4 to the client, a final forced motion maneuver in radial direction follows, as this is the docking axis for OLEV docking on the client's apogees engine.

The advantages of the approach shown in Fig. 25.16 are (1) the possibility to adjust the approach with good illumination conditions, (2) to calibrate the angles only navigation during the first fly around from S1 to S2 and (3) to use the second fly around from S3 to S4 to inspect the client.

25.4.3 Rendezvous Sensors

Basically, rendezvous sensors can be divided into (1) absolute navigation sensors, which provide the absolute position of client and servicer, and (2) relative navigation sensors, which provide the relative position and attitude of the two spacecraft.

In the following we compare the strengths and weaknesses of different sensors. For a more thorough description of rendezvous sensors, we refer to Fehse (2003).

Absolute Navigation

Absolute navigation is typically used during launch and "phasing" phase, i.e., at greater distances. In low Earth orbit (LEO), GPS can be used to determine the absolute position of the servicer and active targets. For passive targets in LEO an active radar antenna can be used (e.g., the antenna of the "Forschungsgesellschaft für angewandte Naturwissenschaften" (FGAN) in Bonn, Germany).

In Geostationary Orbits (GEO) the absolute position is usually measured using RF ranging methods (cf. Sect. 13.3.2).

Relative Navigation

The change to relative navigation starts during far range approach (cf. Fig. 25.15). Radio frequency (RF) Sensors, LIDAR and camera-type sensors are typical sensors for relative navigation, suitable for any type of orbit and target. Relative GPS (RGPS)

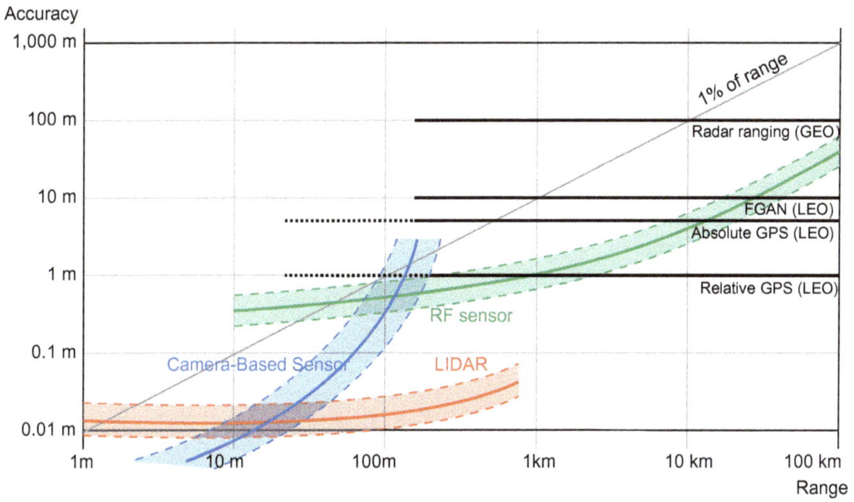

Fig. 25.17 Typical operational ranges and measurement accuracies of rendezvous sensors. The diagonal denominates an accuracy of 1% of the relative distance

is available in LEO and when the target is also prepared for RGPS. It should be noted that a camera-based sensor is a passive sensor that depends on proper illumination conditions, but on the other hand can also be used for completely passive targets.

Figure 25.17 shows a comparison of the accuracy of typical rendezvous sensors. The diagonal indicates an accuracy of 1% of the relative distance. This is the minimum accuracy normally required for approach navigation (see 1st example in Sect. 25.4.1).

Transition

When selecting rendezvous sensors, it is important to ensure a reliable transition from a distance of a few hundred kilometers to the close range. At close range, the camera-based sensors are most accurate and allow the capture or docking process to be monitored. For long distances, the absolute navigation sensors discussed above (such as GPS and RF ranging) still provide the necessary accuracy.

As shown in Fig. 25.17, there is a gap between the range where camera-based sensors are accurate enough and the range where absolute navigation can be used (in the range from 300 to 700 m). This gap can be closed by using LIDAR or RF sensors.

However, both types of sensors, LIDAR, and RF, contribute to the mass and power budget of a satellite. In the next section we describe a method that allows to extend the accuracy range of camera-based sensors to greater distances by using calibrated maneuvers.

Angles-Only Navigation

In camera-based navigation, the determination of distance is based on triangulation. The baseline for this triangulation is either the distance between two stereo cameras on the servicer, or the size of the target if object resolution is used to determine the

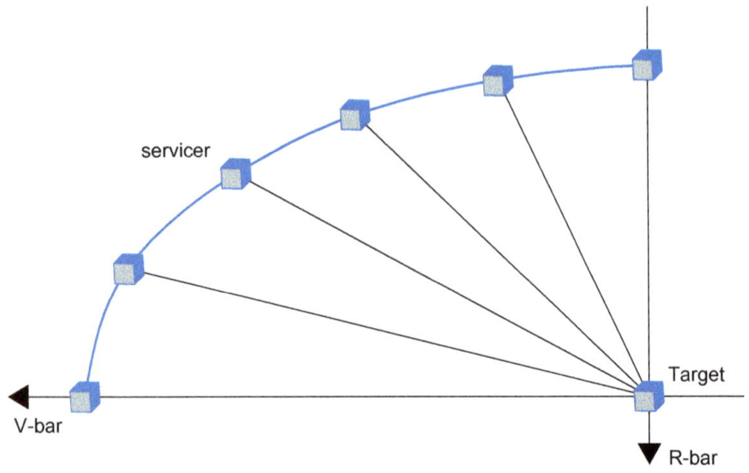

Fig. 25.18 Angles-only measurements during fly around maneuver

distance. Because the length of this baseline is limited to a few meters, the accuracy of a camera-based sensor is limited at longer distances.

Angles-only navigation is a method to extend the accuracy range of camera-based navigation by extending the baseline. It uses a calibrated maneuver to replace the baseline for triangulation with a segment of the fly around shown in Fig. 25.18. The only problem with angles-only navigation is that it requires a trajectory that is at least to some degree perpendicular to the line of sight to the target.

Figure 25.19 shows an approach strategy optimally suited for angles-only navigation as demonstrated during the ARGON experiment on PRISMA. This approach strategy combines passive safety with the possibility of using angles-only navigation. Passive safety is introduced as the servicer continues to circle around the client in case of a failed stop maneuver. Angles-only navigation is supported because the trajectory is almost perpendicular to the line of sight to the client (Sellmaier et al. 2010; Spurmann 2011).

Guidance, Navigation and Control

During the rendezvous phase, the approach of the servicer to the target object has to be controlled on the basis of rendezvous sensors (see previous Sect. 25.4.3). Further, the information about the absolute position and attitude of the servicer satellite is provided by its AOCS.

A navigation filter can be used to estimate the target's position and velocity and, in the close range, also its attitude and attitude rate based on the measurements of the sensors in combination with dynamic models for the target's orbit and attitude. It is also advantageous to use measurements from different sensors, such as camera and LIDAR, in combination (sensor fusion). Time-of-flight sensors such as LIDARs provide very accurate distance measurements, while camera systems can measure

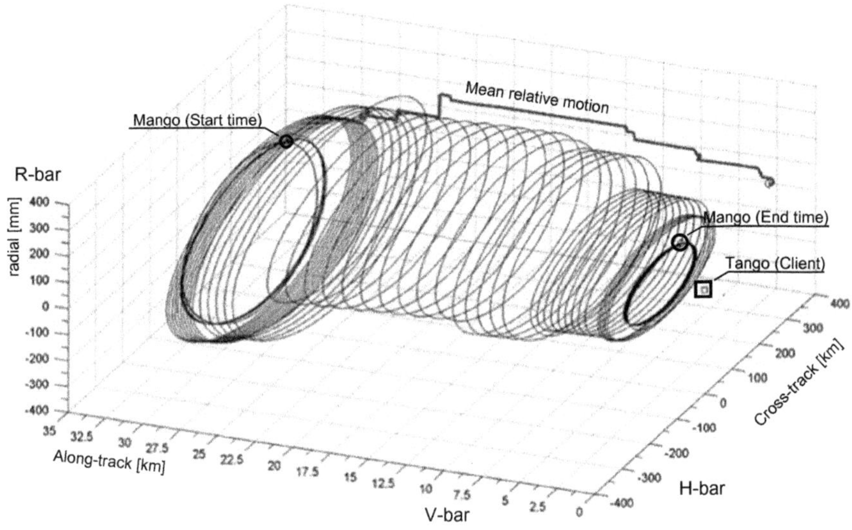

Fig. 25.19 ARGON Experiment (D'Amico et al. 2013) on PRISMA to demonstrate an approach with angles-only navigation (*Credit* DLR)

other components of the pose more accurately than time-of-flight sensors. In combination, a very accurate result of the pose estimation can be achieved. Sensor fusion at the navigation filter level can be achieved by using an extended Kalman filter or a similar filter. In the so-called filter correction step the measurements of different sensors can be included (Benninghoff et al. 2014; Rems et al. 2017).

Determining the 6D pose (i.e., 3D position and 3D orientation) from raw sensor data can be very difficult: For camera-based sensors, the illumination conditions as well as the optical properties of the target's surface materials and the target's motion, such as the rate of its tumbling, strongly influence the pose estimation. Image and data processing tools need to be developed to detect the target in the sensor data and calculate its relative position and orientation with respect to the servicer. So-called model matching algorithms are often used: Here, a 3D CAD model of the target is used and compared with the detected object edges in the images. Figure 25.20 shows an example of two images of a mono camera with the result of the image processing. With the knowledge of the optical properties of the camera, such as focal length etc., the final 6D pose can be calculated. In the case of time-of-flight sensors these directly provide a 3D point cloud with distance measurements. To calculate the full 6D pose, however, a 3D model must be used here as well.

The guidance system provides a reference trajectory for the position and attitude of the chaser. A controller compares the reference pose with the actual pose and calculates force and torque commands for the actuators. For mid and close-range rendezvous, the guidance trajectory often uses the local orbital frame (see Sect. 25.4.1). The frame and the parameterization in spherical coordinates are visualized in Fig. 25.21. Since the position and velocity of the target is calculated by the

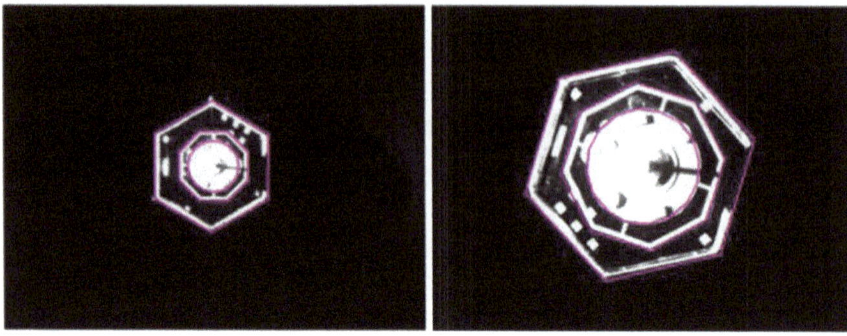

Fig. 25.20 The images from a mono camera from two different distances are overlaid with the result of the image processing (projected 3D model of the target in magenta color)

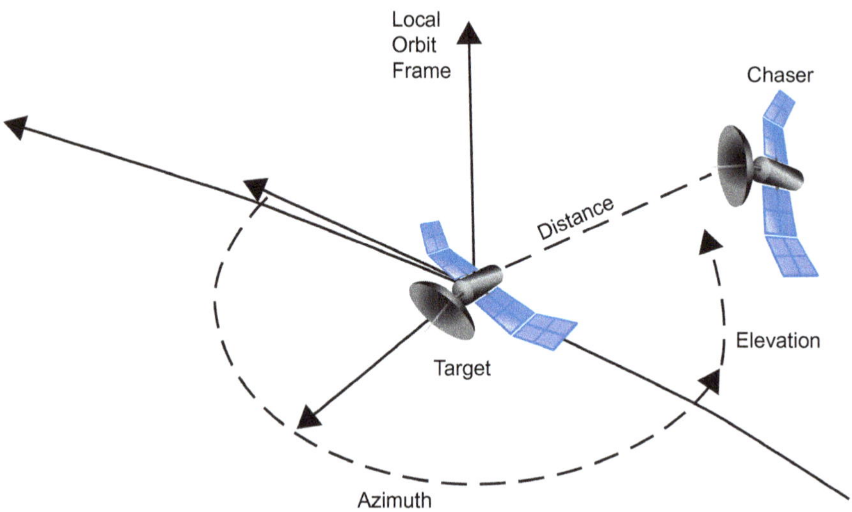

25.21 Parametrization of the guidance trajectory in the local orbital frame

navigation filter, the LOF coordinate system is known in the guidance, navigation and control (GNC) system.

From a ground console it is possible to send guidance commands to the chaser so that it performs a straight-line approach (changing the distance while keeping azimuth and elevation angles constant) or a fly-around maneuver (changing elevation or azimuth angle while keeping the distance constant). In this way, fly-around maneuvers, such as the transition from V-bar to R-bar, and straight-line approaches to reduce distance, can be realized (see Sect. 25.4.2). Not only the desired relative position but also the desired orientation, i.e., changes in the roll, pitch or yaw angle, can be commanded to the guidance system. A stand-by mode is also possible. In

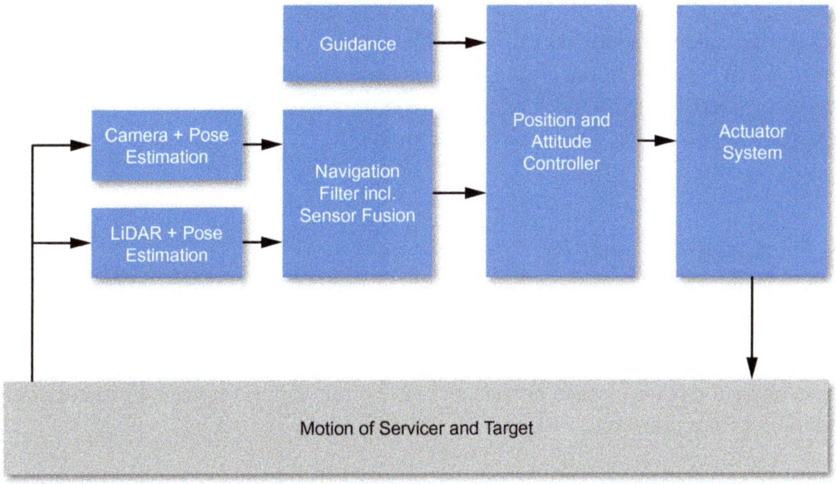

25.22 Overview of a possible guidance, navigation and control scheme for rendezvous

this case, the servicer satellite is uncontrolled, which is sometimes intended, e.g., for capturing the client (see Sect. 25.5).

Figure 25.22 shows the interaction of the different elements of the rendezvous system. The sensors acquire images of the target, followed by pose estimation. This is fed into a navigation filter. The guidance provides a reference trajectory. The controller calculates the forces and torques needed to follow the reference trajectory. Via the actuator system these forces and torques are applied to the servicer satellite, which results in a change of the relative motion between servicer and target.

When the final hold point in the immediate vicinity of the target is reached, the phase of close-range rendezvous is finished, and the satellite capture follows.

To perform the rendezvous maneuver, the operator must guide the autonomous approach by sending telecommands to its system. These commands allow the operator to change the guidance mode such as starting a straight-line approach, to select which sensors to be used by the navigation filter, etc. Constant contact is not necessary, as the main functionalities can be implemented on board and run autonomously. In addition to normal numerical telemetry, it is advantageous to provide (compressed) data of the rendezvous sensors to the rendezvous operator. The provision with such so-called "science telemetry" in combination with special rendezvous console applications such as GUIs for visualizing the approach are features developed for operating on-orbit servicing missions.

In contrast to the autonomous execution of the rendezvous on board, it is also possible to calculate some parts on the ground. However, this is only possible if there is constant contact with the spacecraft as is the case for GEO missions or with a network of ground stations (see Sect. 25.5.1). Image processing in particular could benefit from the much better computing power of standard PCs compared to on-board computers. However, the time delay in the control loop caused by sending images from cameras to the ground, processing them and then sending the result back to the satellite can lead to instabilities in the control loop. Another difficult task is the live stream of uncompressed images with sufficient image resolution and frequency. Depending on the specific mission, it must be decided which tasks are to be performed on board and which on the ground.

25.5 Satellite Capture

Capturing a target—either docking or berthing—is the most critical part of the entire OOS mission. In the following we discuss the aspects relevant to the capture process, in particular the communication concept and the interaction between the manipulator and the servicer platform.

25.5.1 Communication Concept

A stable communication link to the spacecraft is important for any spaceflight mission. For missions in the low Earth orbit, however, it is a question of mission philosophy whether the greater development effort should be put into autonomy in space or into an improved space to ground link. The two US missions DART and Orbital Express focused on autonomous rendezvous and docking. The failure and collision of DART highlighted the complexity and difficulties with autonomous RvD. However, the success of Orbital Express subsequently showed that autonomous docking can indeed be realized. Other mission concepts provide for a strategy based on an improved communication link that allows teleoperation during capture. Since the philosophy of improved autonomy only concerns the space segment, we will focus here on the requirements for improving the communication concept:

1. **Duration** of an individual contact: For the preparation and execution of a capture maneuver, sufficient time should be available to monitor and control the satellites. The time scale depends on the relative velocities during the final approach (e.g., 10–30 min in LEO and a few hours in GEO).
2. **Stability** of the connection: The link should be stable and protected against interference and shadowing from the client satellite.
3. **Teleoperation**: Whenever a manipulator is operated as "ground in the loop" there should be video streaming and/or force feedback, a short delay time (less than 100–500 ms) and low jitter.

Contact Duration

Most of the OOS missions described above are in a low Earth orbit with a maximum contact time of eight to ten minutes when passing an antenna. This time is barely sufficient to monitor and control the capture maneuver itself. However, this capture

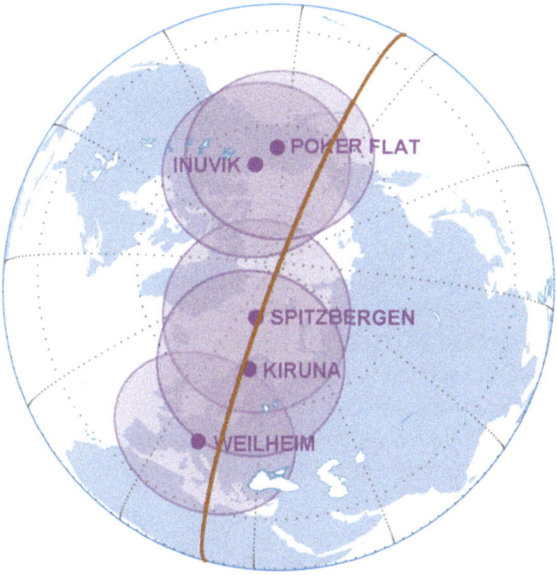

25.23 Chain of ground stations to extend contact times in LEO

maneuver should also be monitored during preparation and follow-up—especially if it involves a tumbling target satellite. Therefore, the duration of the communication link is one of the major problems for OOS mission in low Earth orbit.

One solution is to combine a chain of ground stations as shown in Fig. 25.23. This chain provides a quasi-continuous telemetry data stream in the downlink for more than 20 min. In the uplink, however, the telecommand data stream is repeatedly interrupted because a handover from one ground station to another takes a certain time. Since the Earth rotates, the chain of ground stations shown can only be used every 12 h.

At first glance, a more elegant solution for extending the possible contact time is the use of an inter-satellite link to a relay-satellite in the geostationary orbit. The additional signal delay adds up to about 500 ms roundtrip, which is tolerable. In the future, one of the planned swarms of relay satellites in low Earth orbit could also provide a continuous data link with lower delay times.

However, the use of relay services also increases the complexity of the overall system. There is an additional unit that has to cope with real-time communication and the servicer spacecraft must have additional connectivity capabilities, which often implies the use of complex antenna or laser control mechanisms. And finally, there are typically high costs for the relay services. After analyzing all aspects, it may turn out that the total cost of implementing and using relay services outweighs the potential benefits.

Interference and Shading

For a geostationary OOS mission, the situation is much better in terms of connection time. On the other hand, the communication link is often made more difficult by the fact that the approach to the target satellite is from behind, since the best docking points are located there (see Fig. 25.16 approach from S4 to the client). This means that the client satellite is in a direct line between the servicer antenna and the ground station. In this case, it is likely that the structure of the client will shadow the direct signal between the servicer and the ground station. In addition, the client's transmitter could interfere with the servicer's signals. Solutions to the shadowing problem are either to use additional antennas on the servicer or to use a network of ground station on Earth. The interference problem must be investigated by a thorough RF compatibility test and appropriate hinder analysis, which could provide potential sources of interference.

Teleoperations: Delay Time and Jitter

The requirement to minimize delay time and jitter is usually determined by the robotic operations in the vicinity of the target. To give the payload control system (PCS) on ground a realistic possibility to intervene during the capture process, the signal delay time should be less than 500 ms round trip. This delay still allows the human operator to receive the haptic feedback and react accordingly in a natural way. Similarly, a high jitter leads to a potential erroneous reaction of the human operator or to an incorrect reaction of the remote robot, making actual teleoperation impossible.

However, the communications architecture of a conventional satellite mission has a typical delay time of 2–5 s, which is mainly caused by electronic components on the ground. In addition, the automatic switching of redundant lines may cause unpredictable jitter.

One solution to these problems is to connect the PCS directly with the CORTEX of the teleoperation antenna via a dedicated, non-redundant, high rate TM/ TC connection (Fig. 25.24 dashed lines). The direct connection results in a very short delay time of 2.5 ms round trip. This solution was used for the operation of ROKVISS (Landzettel et al. 2006b), a robotic precursor experiment on the ISS. In addition, this solution used special modems that allowed the signals to be modulated directly onto the COM line, thus completely avoiding problems caused by typical protocol layers.

Of course, such a dedicated connection will not always be possible, so solutions involving UDP/IP and suitable transmission and synchronization equipment can be used to provide a jitter-free signal with an acceptable delay in the range of 10–100 ms, depending on the actual connection distance. This has the advantage that the existing IP-based telecommunications infrastructure can be used worldwide.

It should also be mentioned that with increasing distance the finite speed of light means a physical lower limit for the signal delay. So, teleoperation on the Moon or other planets will not be possible from the Earth. In these cases, robotic operations

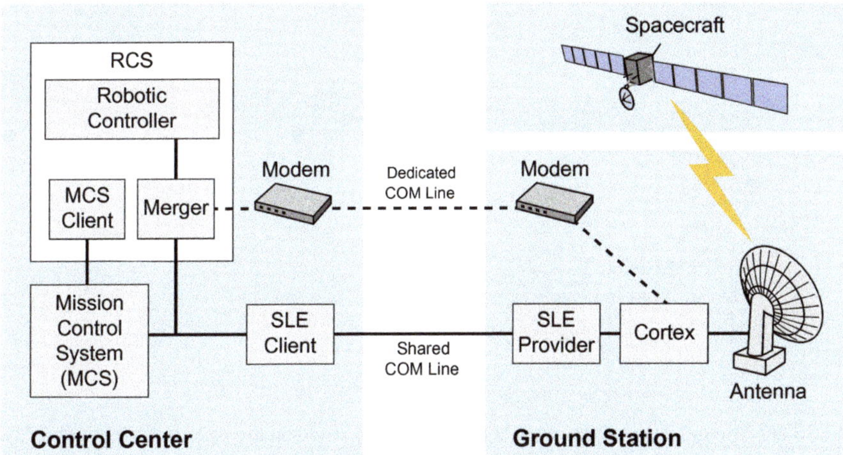

Fig. 25.24 Communication architecture to minimize delay time and jitter

must be autonomous. However, as soon as there are manned stations in their orbit, teleoperation will regain importance.

25.5.2 *Interaction of the Manipulator with the Servicer Platform*

Whenever a manipulator is used during an OOS mission, its interaction with the AOCS of the servicer platform must be taken into account. Since the total momentum and torque are maintained, the movements of the manipulator affect the dynamics of the platform. DLR has investigated this effect in the context of the ETS VII contribution GETEX (see Landzettel et al. 2006a and references within). The conclusion of this investigation was that in most cases it is better to deactivate the servicers AOCS and to pre-calculate the reaction on the free-floating platform and include it in the trajectory of the manipulator (Fig. 25.25).

25.5.3 *Other Aspects*

When capturing a satellite, there are other operationally relevant points that must be taken into account:

First, a collision avoidance strategy must be developed. The collision avoidance procedure should be independent of a standing communication link. For example, a valid collision avoidance procedure should always be loaded on board. In some

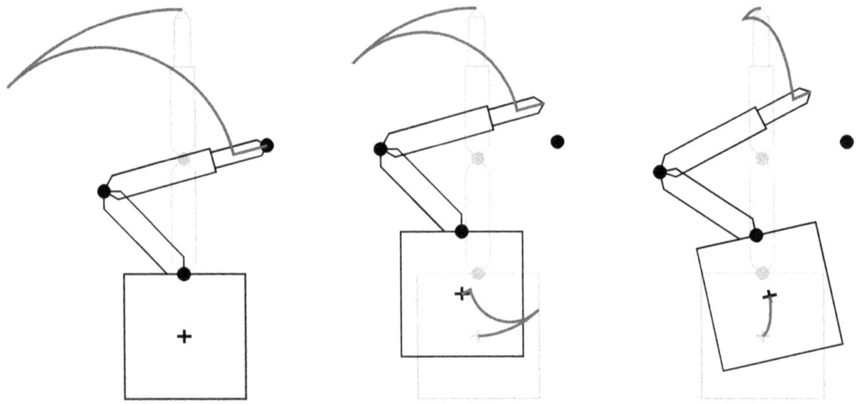

Fig. 25.25 The influence of the satellite attitude control mode on the path of the robot arm. The same joint motion is performed by a servicer with a fixed base (left); an attitude-controlled servicer (middle) and a free-floating servicer (right) (Reintsema et al. 2007)

cases, this would mean a reverse maneuver, in other cases it would lead to a forward maneuver towards docking or capture.

In addition, a strategy must be worked out for how to move or split the control authority between the robotic control system (RCS) and the servicer's AOCS in order to coordinate the overall system.

At the moment of capture, a contact voltage will occur, as it is rather unlikely that both space crafts have the same electric potential. The system will have to deal with this voltage.

After capture, a possible tumbling rate of the client has to be damped by the manipulator.

Finally, as soon as a fixed connection between the two satellites is established, the attitude control system must be modified to avoid feedback and subsequent oscillation of the two attitude control systems of client and servicer. Therefore, either the client's AOCS should be disabled or there should be a "combined control" for both spacecraft.

25.6 Verification and Test Facilities

Verification and test are of high importance in the preparation of all space missions. Especially on-orbit servicing or robotic missions which include close proximity operations between satellites require very intensive test and verification of all involved systems and sub-systems. This is due to the collision risk and hence the risk of a total loss of mission.

The requirements of OOS missions for GNC are considerably higher compared to standard spacecraft operations where for instance a communication satellite has to be positioned within a box of 70 km edge length. The critical phase of OOS missions, the RvD of two spacecraft is a very complex maneuver which requires (relative) position accuracy of a few centimeters and even millimeters close to the capturing point.

As mentioned in Sect. 25.3, new challenges arise in the context of OOS missions. All involved sub-systems and all planned maneuvers should be analyzed, simulated and verified on ground in detail before an OOS mission is launched. Numerical simulations deliver only limited results. Therefore, a test facility should be used to test the entire RvD process including the flight hardware of GNC and robotic components under realistic conditions.

We first discuss the requirements for test facilities for OOS and RvD missions. As examples we then describe two test facilities which have been developed at the DLR in Oberpfaffenhofen. Both together are capable of simulating the relevant phases of a Rendezvous and Docking mission, i.e., the approach and capture of the target.

25.6.1 Requirements for Test Facilities

Test facilities should support the verification and validation of the individual components and of the full system during all development stages. This includes test possibilities for first prototypes, for engineering models and for the final flight hardware.

The main requirement on a test facility is the ability to integrate and test on-orbit servicing hardware such as optical sensors for rendezvous (see Sect. 25.4.3) and docking/capture hardware (see Sect. 25.5). Further, also computers such as on-board computers or—in early development stages—simple standard PCs to collect sensor data need to be integrated into the tests. Therefore, mechanical, electrical and data interfaces to the hardware elements have to be defined and implemented.

In addition to the OOS payload hardware, a mission-specific target mockup is needed. This can be a model of the outer surface of a satellite or of a debris object, part of a spacecraft (like a docking port), part of a space station or just a pattern of retro-reflectors.

The environmental conditions should be as realistic as possible, depending on the items to be tested. If optical sensors are involved, realistic illumination is essential. This can be achieved by using one or more spotlights with a light spectrum similar to sunlight. If necessary, an albedo simulator could also be used. The background behind the target mockup has to be selected dependent on the scenario to test. It is often necessary that the background is black or that it shows the Earth (or part of the Earth) or the Sun.

The facility should support static and dynamic tests and the repetition of tests. Therefore, it should be possible to test various static relative positions and orientations and to test the behavior of the GNC system under dynamic conditions. It should be possible to dynamically change the relative position and orientation between the RvD sensor on the servicer and the target.

Since the accuracy requirements for the on-board navigation system of such OOS missions are in the range of centimeters or better, the requirement for the positioning accuracy of a test facility are in the millimeter range, i.e., one order of magnitude better.

Besides open-loop tests (tests with a predefined trajectory and offline post-processing of sensor data after the tests), closed-loop tests should also be supported. Closed-loop tests include sensor-in-the-loop tests, processor-in-the-loop tests or both tests. The sensor data must be processed directly and the result must be fed back into the control loop and thus back into the dynamic simulator. With such closed-loop capabilities, the entire control loop and its stability can be tested.

Both, the interaction between different systems, e.g., between GNC and robotic subsystem, and the interaction between space and ground segment is completely different in On-Orbit Servicing missions than in other mission types. Therefore, also in addition to the test of individual components, the testing of the interaction of the components is of high importance (end-to-end test capability).

Ideally, a test facility should be part of an agile development concept for the RvD and GNC system. For example, the interaction between the different components should not only take place in the later stages of development. This should be included from the beginning so that errors or missing elements are detected as early as possible.

25.6.2 European Proximity Operations Simulator

The European Proximity Operations Simulator (EPOS) is a test facility with focus on the approach phase of an RvD mission (Fig. 25.26). The EPOS facility is based on two industrial robots, each with six degrees of freedom, which simulate the 6D motion of servicer and target. One of the robots is mounted on a linear slide of 25 m length. This allows real-time simulations of the final 20–25 m of the rendezvous phase with 1:1 models, and even greater distances with scaled models (Benninghoff et al. 2017).

EPOS is designed to simulate several classes of OOS missions with both cooperative and non-cooperative and even tumbling targets. The positioning accuracy of the robots is in the sub-millimeter to millimeter range.

Since 2009, when EPOS 2.0 was built, the facility has been used for a variety of tests and test campaigns by industry, research centers and academia including test and verification of rendezvous sensors such as 2D cameras or 3D sensors like photonic mixer device (PMD) cameras or light detection and ranging sensors (LIDARs).

For realistic solar simulation (see Sect. 25.6.1), EPOS is equipped with a 12 kW HMI (hydrargyrum medium-arc iodide) lamp which can generate a luminous flux of 1.15 million lumens. The irradiance at 7 m distance to the Sun simulator is very close to the Sun's irradiance in the visual spectrum in an Earth orbit.

With its ability to move and simulate environmental conditions, rendezvous maneuvers can be simulated under different dynamic conditions (different approach

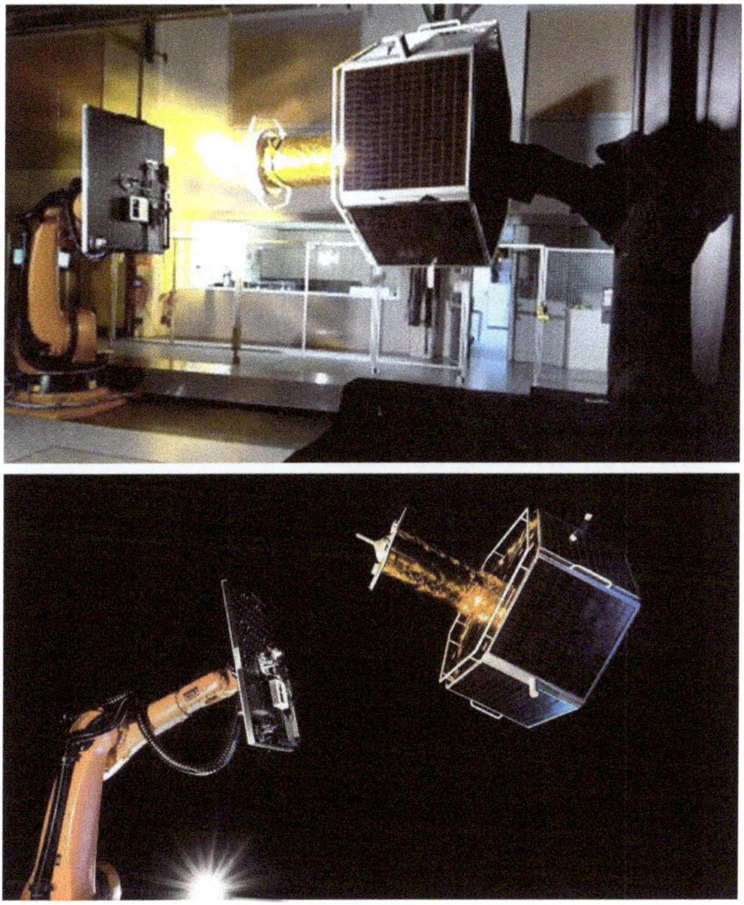

Fig. 25.26 European Proximity Operations Simulator (EPOS)—a robotic test facility for rendezvous and proximity operations located at the German Space Operation Center in Oberpfaffenhofen (*Source* DLR)

velocities and directions, different fly-arounds) and different environmental conditions (front, side, or back illumination by the solar simulator, black background or Earth in the background, etc.).

The entire guidance, navigation and control system on board can be embedded in closed hardware-in-the-loop simulations at EPOS. In addition, it is also possible to integrate not only sensors but also on-board computers and on-board software into the simulation. Real sensors and real on-board computers can be used to test whether, for example, image processing algorithms and navigation systems are robust and stable when executed on space computer hardware.

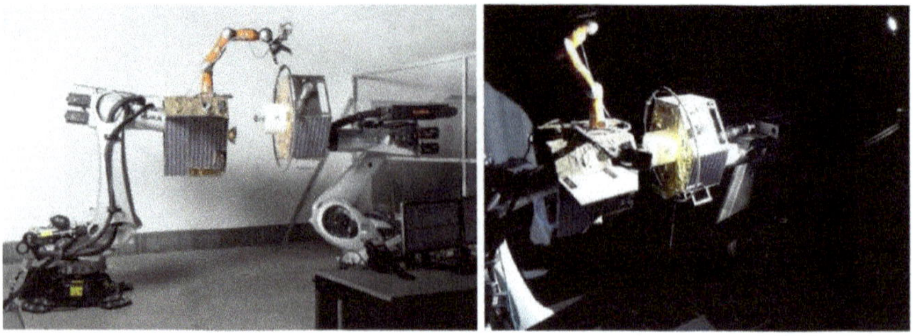

Fig. 25.27 On-orbit servicing simulator (OOS-Sim)—a test facility for capture maneuvers operations located at the Robotic and Mechatronic Center in Oberpfaffenhofen (*Source* DLR)

25.6.3 On-Orbit Servicing Simulator

The on-orbit servicing simulator (OOS-Sim) focuses on the simulation of the capture phase of an OOS mission. The facility is used for testing and validating robotic control methods. Two industrial robots with six degrees of freedom each simulate the motion of servicer and target satellite during the capture process including hardware-in-the-loop elements (Artigas et al. 2015). In the example shown in Fig. 25.27, the capture is performed by a lightweight robot as an on-board manipulator. During physical contact between the robotic on-board manipulator and the target satellite, the control forces and torques are measured by force-torque sensors and are fed back into the real-time simulation. A major challenge in the simulation of the capture process is the stability of the system when the 7D light-weight robot grasps the target. The three robots are then mechanically connected to each other, resulting in a kinematic chain of 19 degrees of freedom. To handle this situation, the lightweight robot is a torque-controlled robot and allows so-called impedance control to achieve a stable contact dynamic simulation.

For robotic capture in an autonomous way, a stereo camera is integrated to determine the relative position between the gripper and the target. Furthermore, the OOS-Sim also contains a LiDAR and an IMU (inertial measurement unit).

The illumination conditions are realized by a movable spotlight on the top of the robots. This allows the dynamic change of the illumination direction during a simulation (Fig. 25.27).

25.6.4 End-to-End Simulation

As mentioned in Sect. 25.6.1, it is not sufficient to simulate the different components and phases of an on-orbit servicing mission separately. The interaction between the on-board and the ground system also needs to be developed and tested, especially for

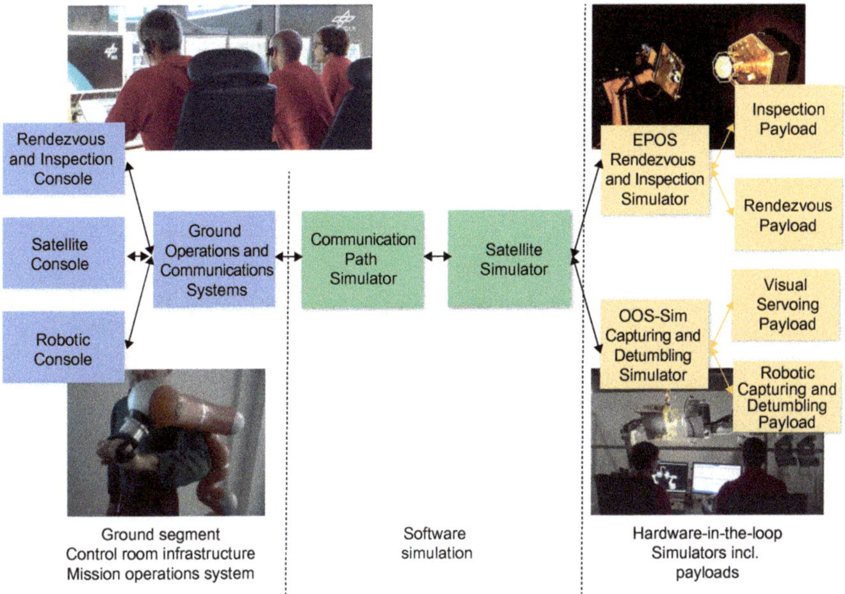

25.28 Overview on the end-to-end simulation framework for OOS

teleoperation. DLR has established such an end-to-end simulation and test environment. The phases inspection, rendezvous, capture and servicing/manipulation as well as the hand-over between the phases can be tested. It includes the on-board elements (payloads, sensors, robotic manipulator, on-board computer hardware and software) and the ground elements (control center including the mission control system and the payload control system, teleoperation elements). Those parts that cannot be represented by real systems, such as the dynamics of the satellite and the communication path, are simulated. The concept is visualized in Fig. 25.28.

25.7 Summary and Outlook

While OOS is quite common for manned space missions, it is still a new field for robotic missions. Several demonstration missions have already been carried out and the corresponding technologies developed. A commercial servicing mission has proved the concept of life extension.

For the future success of servicing missions, it will be necessary to develop interface standards for the client satellites, both for rendezvous and docking and for the exchange of so-called orbital replacement units (ORU). Current technology development projects are investigating the design of mechanical, electrical, data and heat flux interfaces for modular components.

Also, with a view to possible space debris mitigation programs, all future satellites should be equipped with a standard docking or berthing interface. In addition, it will be necessary to build a system capable of de-orbiting a series of client satellites with a single service spacecraft.

References

Artigas J, De Stefano M, Rackl W, Lampariello R, Brunner B, Bertleff W, Burger R, Porges O, Giordano A, Borst C, Albu-Schäffer A (2015) The OOS-SIM: an on-ground simulation facility for on-orbit servicing robotic operations. In: IEEE International conference on robotics and automation (ICRA), 26–30 May 2015, Seattle, Washington, USA (2015)

Benninghoff H, Rems F, Boge T (2014) Development and hardware-in-the-loop test of a guidance, navigation and control system for on-orbit servicing. Acta Astronaut 102:67–80 (Elsevier). https://doi.org/10.1016/j.actaastro.2014.05.023

Benninghoff H, Rems F, Risse EA, Mietner C (2017) European proximity operations simulator 2.0 (EPOS)—a robotic-based rendezvous and docking simulator. J Large-Scale Res Facil JLSRF. https://doi.org/10.17815/jlsrf-3-155, ISSN 2364-091X

Benninghoff H, Rems F, Risse EA, Brunner B, Stelzer M, Krenn R, Reiner M, Stangl C, Gnat M (2018) End-to-end simulation and verification of GNC and robotic systems considering both space segment and ground segment. CEAS Space J (Springer). https://doi.org/10.1007/s12567-017-0192-2, ISSN 1868-2502

Biesbroek R, Innocenti L, Wolahan A, Serrano SM (2017) e.Deorbit—ESA's active debris removal mission. In: Proceedings of the 7th European conference on space Debris, ESA Space Debris Office

Clohessy WH, Wiltshire RS (1960) Terminal guidance for satellite Rendezvous. J Aerosp Sci 27:653

Covault C (2009) Secret inspection satellites boost space intelligence ops. Spaceflight Now. https://spaceflightnow.com/news/n0901/14dsp23/, cited 1 Aug 2019

D'Amico S, Ardeans JS, Gaias G, Benninghoff H, Schlepp B, Jörgensen J (2013) Noncooperative Rendezvous using angles-only optical navigation: system design and flight results. J Guidance, Control Dyn 36(6):1576–1595 (American Institute of Aeronautics and Astronautics (AIAA)

ESA. The challenge, the clean way is the right way. https://www.esa.int/Our_Activities/Space_Safety/Clean_Space/The_Challenge, cited 1 Aug 2019

Fehse W (2003) Automated Rendezvous and docking of spacecraft. Cambridge Aerospace Series 16 (Cambridge University Press)

Kessler DJ, Cour-Palais BG (1978) Collision frequency of artificial satellites—the creation of a Debris Belt. J Geophys Res

Klinkrad H (2010) Private communication

Landzettel K, Preusche C, Albu-Schäffer A, Reintsema D, Rebele B, Hirzinger G (2006a) Robotic On-Orbit Servicing – DLR's Experience and Perspective. In: Proceedings of the 2006 IEEE/RSJ, international conference on intelligent robots and systems, 9–15 Oct. IEEE, Beijing, China. ISBN: 1-4244-0259-X

Landzettel K, Albu-Schäffer A, Brunner B, Beyer A, Gruber R, Krämer E, Preusche C, Reintsema D, Schott J, Steinmetz H-J, Hirzinger G (2006b) ROKVISS: Verification of advanced light weight robotic joints and tele-presence concepts for future space missions. ASTRA 2006—the 9th ESA workshop on advanced space technologies for robotics 7, Special Topics 413 and Automation, ESA's research and technology centre, ESTEC, Noordwijk, Netherlands, 28–30 Nov, Paper: 2.1.2.5, Proceedings ESA

Mulder TA (2008) Orbital express autonomous Rendezvous and capture flight operations; Part 1 of 2, AIAA/AAS Astrodynamics specialist conference and exhibit, 18–21 Aug 2008, Honolulu, Hawaii, AAS 08-209

NASA (2020) On-Orbit Servicing, Assembly and Manufacturing mission (OSAM-1). https://nexis.gsfc.nasa.gov/osam-1.html, cited 1 Apr 2020
Parrish J. Robotic Servicing of Geosynchronous Satellites (RSGS). https://www.darpa.mil/program/robotic-servicing-of-geosynchronous-satellites, cited 1 Aug 2019
Reed BB, Smith RC, Naasz BJ, Pellegrino JF, Bacon CE (2016) The restore-L servicing mission. AIAA Space 2016. https://doi.org/10.2514/6.2016-5478
Reintsema D, Landzettel K, Hirzinger G (2007) DLR's advanced telerobotic concepts and experiments for on-orbit servicing. Adv Telerobotics, Springer Tracts Adv Robot 31:323–345
Rems F, Risse EA, Benninghoff H (2017) Rendezvous GNC-system for autonomous orbital servicing of uncooperative targets. In: 10th international ESA Conference on guidance, navigation & control, 29 May–02 June, Salzburg, Austria
Sellmaier F, Boge T, Spurmann J, Gully S, Rupp T, Huber F (2010) On-orbit servicing missions: challenges and solutions for spacecraft operations. In: AIAA SpaceOps 2010 conference, Huntsville, Alabama, USA, p 2159
Settelmeyer E, Lehrl E, Oesterlin W, Hartmann R, Landzettel K (1998) The experimental servicing satellite—ESS. In: IARP 98-c-14, DARA, Anchorage, USA, ISBN: 0-7803-4869-9
SpaceNews (2018) Orbital ATK unveils new version of satellite servicing vehicle. https://spacenews.com/orbital-atk-unveils-new-version-of-satellite-servicing-vehicle, cited 1 Aug 2019
Spurmann J (2011) Spiraling approach for angles-only navigation within on-orbit servicing missions. Adv Aerosp Guidance Navig Control 465–472 (Springer)

Florian Sellmaier is head of business development at the German Space Operations Center (GSOC). He received his Ph.D. (Dr. rer. nat.) from the Institute of Astronomy and Astrophysics/Munich in 1996 and also spent some time at CALTECH in Pasadena/USA. After working in the field of radar and navigation and as team leader of the Siemens Competence Center for e-Business Applications he joined DLR in 2006. As project manager he led the acquisition and predevelopment of future mission like OLEV, DEOS and EDRS. He is author of several scientific publications in the field of astrophysics as well as in the space operations.

Heike Frei is head of the research team „On-Orbit Servicing and Autonomy" at the department of spaceflight technology at the German Space Operations Center (GSOC). She graduated as mathematician in 2010 and received her Ph.D. (Dr. rer. nat.) in 2015 from the University of Regensburg, Germany. Since 2010, she is a scientific employee at DLR. Her research interests comprise On-Orbit Servicing, rendezvous, optical navigation, on-board autonomy, image processing and hardware-in-the-loop simulations. She is the author of several peer-reviewed publications and conference contributions.

Chapter 26
Interplanetary Operations

Paolo Ferri

Abstract Interplanetary mission operations are characterized by extreme variability, high complexity and high vulnerability. Missions that simply fly-by their target object in the solar system, those that enter in orbit around a planet or minor body and those that attempt a landing on it have increasing levels of operations complexity. Typical challenges of interplanetary flight that impact the operations are the long signal propagation delays, the high variability of operations workload, from quiet cruise to time-critical orbit insertion phases, the scarce availability of electrical energy and communications bandwidth. Special operations like asteroids fly-by, planet orbit insertion or landing on the surface present unique operational challenges that require dedicated tools, procedures and training.

26.1 Types of Interplanetary Missions

The common definition of interplanetary mission refers to spacecraft whose trajectory leaves the Earth environment and enters a heliocentric orbit. Excluded are missions to the Sun-Earth Lagrange points L1 and L2 (see Logsdon 1997), as these remain at relatively close distances to Earth (about 1.5 million km) and do not vary significantly their distance to the Sun compared to the one of the Earth.

The heliocentric trajectory is selected such that it brings the spacecraft to the selected target, either through a pure ballistic flight or with the aid of correction maneuvers. These maneuvers are performed via the on-board propulsion system, or make use of the gravitational slingshot effect of a planet.

The objective of interplanetary missions is the scientific exploration of the Solar System. To this aim, spacecraft are flown towards the Sun, the major planets, or other minor objects like asteroids and comets. The type of trajectory and approach to the target object largely drive the complexity of the mission. We can classify interplanetary missions according to this criterion and in order of complexity as: fly-by missions, orbiting missions and landing missions.

P. Ferri (✉)
Retired, European Space Agency, ESA-ESOC, Darmstadt, Germany
e-mail: paonihon@yahoo.de

F. Sellmaier et al. (eds.), *Spacecraft Operations*, Springer Aerospace Technology,
https://doi.org/10.1007/978-3-030-88593-9_26

26.1.1 Fly-By Missions

Fly-by missions were the first historical interplanetary missions. These simply attempted to fly-by the target object, which means crossing the trajectory of the target object at the smallest possible distance, to allow observations and scientific measurements to be taken with the on-board instruments during the fly-by. After the first Lunar fly-bys (Lunik 1, USSR, and Pioneer 4, USA, 1959) various spacecraft were sent onto fly-by trajectories to the Earth's neighbour planets, Mars and Venus. Missions to the outer planets remained confined to the fly-by type until the NASA Galileo spacecraft entered in orbit around Jupiter in the 90ies, and NASA's Cassini around Saturn in 2005. Also, Mercury was only visited by a fly-by spacecraft (Mariner 10, two flybys in 1974) until Messenger entered in orbit around it in 2011. Table 26.1 shows all the solar system objects flown by by spacecraft to-date, including the date, fly-by distance and name of the spacecraft that performed the first fly-by. The fly-bys of the Jupiter and Saturn moons performed by Galileo and Cassini and of more recent small bodies have been omitted for simplicity.

Fly-by missions do not require complex maneuvers at arrival at the target and the requirements on navigation accuracy can be adapted to the available capabilities by adjusting the fly-by distance. The design of the spacecraft is therefore relatively simple and its mass does not have to include heavy propulsion systems nor the necessary fuel. Fly-by missions can further be designed so that they visit multiple targets. Perhaps the most successful typical examples of this type of mission are the two Voyager spacecraft, whose trajectory was designed such that they could visit several planets of the external solar system. Drawback of such missions is the short time spent in the proximity of the target, which is typically of the duration of a few hours, due to the large speed relative to the target object at the close encounter.

26.1.2 Orbiting Missions

Orbiting missions are those that bring the spacecraft to the target and then execute trajectory correction maneuvers to allow it to be captured by the gravity field of the target object and enter a closed orbit around it. Only a few solar system objects have been orbited to-date by spacecraft: apart from the Earth and the Moon, the planets Mars and Venus were orbited by several interplanetary probes. Mercury, Jupiter and Saturn have experienced to-date only a single orbiter each. Uranus and Neptune have not been orbited yet. Of the minor bodies, only five asteroids have been orbited (Eros, Itokawa, Vesta, Ryugu, and Bennu) one dwarf planet (Ceres) and one comet (67P/Churyumov-Gerasimenko). All the objects orbited by spacecraft to date are listed in Table 26.2.

It is obvious that orbiting missions require a higher navigation accuracy, to reach the required precision in the trajectory determination and orbit insertion maneuvers execution. The spacecraft is also normally heavier at launch, as it has to carry the

Table 26.1 Solar system objects flown by spacecraft (status August 2019—only the first fly-by is indicated)

Type	Target	Mission	Nation/agency	Date	Distance (km)
Moon	Moon	Lunik 1	USSR	02-Jan-1959	6000
Planet	Venus	Mariner 2	NASA	14-Dec-1962	34,773
Planet	Mars	Mariner 3	NASA	14-Jul-1965	9844
Planet	Jupiter	Pioneer 10	NASA	03-Dec-1973	130,000
Planet	Mercury	Mariner 10	NASA	29-Mar-1974	705
Planet	Saturn	Pioneer 11	NASA	01-Sep-1979	21,000
Planet	Uranus	Voyager 2	NASA	24-Jan-1986	81,422
Planet	Neptune	Voyager 2	NASA	25-Aug-1989	4824
Dwarf planet	Pluto	New Horizons	NASA	14-Jul-2015	12,500
Kuiper belt object	(486,958) 2014 MU_{69}	New Horizons	NASA	1-Jan-2019	3500
Asteroid	951 Gaspra	Galileo	NASA	29-Oct-1991	16,200
Asteroid	243 Isa	Galileo	NASA	28-Aug-1993	10,500
Asteroid	253 Mathilde	NEAR	NASA	27-Oct-1997	1200
Asteroid	9969 Braille	Deep space 1	NASA	29-Jul-1999	26
Asteroid	5535 Anne Frank	Stardust	NASA	02-Nov-2002	3300
Asteroid	2867 Steins	Rosetta	ESA	05-Sept-2008	800
Asteroid	21 Lutetia	Rosetta	ESA	10-Jul-2010	3160
Asteroid	4179 Toutatis	Chang'e 2	China	15-Dec-2012	3.2
Comet	21P/Giacobini-Zinner	ICE	NASA	11-Sep-1985	7800
Comet	1P/Halley	Vega 1	USSR	6-Mar-1986	8890
Comet	26P/Grigg-Skjellerup	Giotto	ESA	02-Jul-1990	200
Comet	19P/Borrelly	Deep space 1	NASA	22-Sep-2001	2200
Comet	81P/Wild2	Stardust	NASA	02-Jan-2004	240
Comet	9P/Tempel1	Deep impact	NASA	04-Jul-2005	500
Comet	103P/Hartley2	EPOXI	NASA	04-Nov-2010	700

Table 26.2 Solar system objects orbited by spacecraft (status Aug 2019)

Type	Target	Mission	Nation/agency	Year
Planet	Earth	Sputnik 1	USSR	1957
Moon	Moon	Lunik 10	USSR	1966
Planet	Mars	Mariner3	NASA	1971
Planet	Venus	Venera 9	USSR	1975
Planet	Jupiter	Galileo	NASA	1995
Asteroid	(433) Eros	Near-Shoemaker	NASA	2000
Planet	Saturn	Cassini	NASA	2004
Asteroid	(25,143) Itokawa	Hayabusa 1	JAXA	2005
Planet	Mercury	Messenger	NASA	2011
Asteroid	(4) Vesta	Dawn	NASA	2011
Comet	67P/Churyumov-Gerasimenko	Rosetta	ESA	2014
Dwarf Planet	Ceres	Dawn	NASA	2015
Asteroid	(162,173) Ryugu	Hayabusa 2	JAXA	2018
Asteroid	(101,955) Bennu	Osiris-Rex	NASA	2018

necessary propellant for the execution of the orbit insertion maneuvers, and it has to be equipped with a dedicated propulsion system. The orbit insertion operations are time critical and require intensive ground support to ensure correct and timely execution and rapid intervention in case of problems. The advantage of such missions is that the spacecraft and its scientific instruments remain in the proximity of the target object for the entire mission lifetime, which is typically of several years, allowing in-depth observations and measurements, mapping of the surface, etc.

26.1.3 Landing Missions

The next step in complexity of interplanetary missions is represented by the ones that attempt a landing on the surface of the target object. In the history of spaceflight and Solar System exploration successful landing missions have reached the Moon, Mars, Venus, Saturn's moon Titan, three asteroids and a comet. Also a special mention should be made of NASA's Near-Shoemaker spacecraft, which at the end of its successful orbiting mission around asteroid Eros, was commanded to an unplanned, graceful "landing" onto the surface of the asteroid, and managed to survive for about 16 days after the touch down. Table 26.3 lists all the solar system objects on which a successful landing has taken place, together with the first landing mission and the year of landing.

Landing missions are more complex as they normally require a separate landing module which, depending of the characteristics of the target body, in particular its gravity and atmosphere, may require complex and heavy systems like heat-shields,

Table 26.3 Solar system objects on which spacecraft successfully landed (status August 2019). For each object only the corresponding first landing mission is listed

Type	Target	Mission	Nation/agency	Year
Planet	Earth	Sputnik 5	USSR	1960
Moon	Moon	Lunik 9	USSR	1966
Planet	Venus	Venera 7	USSR	1970
Planet	Mars	Mars 2	USSR	1971
Asteroid	(433) Eros	Near-Shoemaker	NASA	2001
Moon	Titan	Cassini/Huygens	NASA—ESA	2005
Asteroid	(25,143) Itokawa	Hayabusa	JAXA	2005
Comet	67P/Churyumov-Gerasimenko	Rosetta/Philae	ESA-DLR/CNES/ASI	2014
Asteroid	(162,173) Ryugu	Hayabusa 2/Minerva/Mascot	JAXA—DLR/CNES	2018

parachutes, retro-rockets, landing gear, airbags, etc. In addition, a landing mission increases the complexity of communications to Earth, generally implying also the use of data relay orbiters around the target body. The complexity of a landing mission is therefore much larger than the one of an orbiting mission. Advantages of having a landing platform capable to carry scientific instrumentation onto the surface of the target body are, on the other hand, enormous and obvious: A lander (in case it is mobile on the surface it is called a rover) allows in situ observations and experiments that would not be possible via remote sensing from the orbit.

A special case of landing missions, which add further complexity to the mission and spacecraft design, are sample return missions. These require the capability, after having landed on the target object, to take off again, leave the object, return to Earth and deliver the samples to the surface. Apart for the manned Apollo Moon landing missions, which returned Moon samples collected by the astronauts to Earth, sample return from a solar system object was only successfully achieved by the JAXA Hayabusa 1 mission, which touched asteroid Itokawa in 2005 and collected small particles lifted by the contact, returning them to Earth in a small re-entry capsule in 2010, and more recently, in 2020, the JAXA Hayabusa 2 mission, returning samples from asteroid Ryugu, and the Chinese Chang'e-5 mission, which returned samples from the Moon. For the sake of completeness, two NASA missions, Genesis in 2004 and Stardust in 2006 managed to return some samples of solar wind particles and comet's coma dust back to Earth. These however were fly-by missions and cannot be classified as landing sample return achievements. This case and other landing missions and their complex operational implications are described in detail in Chap. 27.

26.2 The Challenges of Interplanetary Flight

The main problems of interplanetary flight are all related to energy: the enormous dynamical energy required to achieve its target orbit, the generation and delivery of electrical and thermal energy to power and keep the on-board systems warm, and the energy of the radio signals required to keep communications to Earth. All three aspects pose enormous challenges both the spacecraft design and to the operations of interplanetary probes. These are very briefly summarized in the sections below.

26.2.1 Trajectory Dynamics

Interplanetary travel means escaping from the Earth orbit into a heliocentric orbit. This requires a minimum escape velocity of 11.2 km/s that can be provided by launchers which are available today. Once the spacecraft is injected in the escape heliocentric orbit, it can start its cruise towards the target, which may last several months or years. Small trajectory corrections may be required over the cruise, or in some cases larger "deep space maneuvers", i.e., deterministic modifications of the interplanetary trajectory performed at an optimal (from the energy point of view) time to direct the spacecraft to the target. Once arrived at the target, if the objective is to enter in a closed orbit around it, the spacecraft has to perform an injection maneuver, which is typically a large burn (of the order of a few km/s) of chemical fuel over a relatively short time (of the order of one hour). Examples of direct trajectories to the target are all the historical fly-by missions (e.g., Giotto with comet Halley in 1986, or all the early missions to the planets). ESA has used this approach for its two missions to our neighbour planets, Mars Express and Venus Express.

However, the orbital energy that can be impacted by the most powerful existing launchers may be sufficient to leave the Earth's gravitational influence, but is often not sufficient to reach the mission target. This is the case, for instance, for journeys to planets beyond the main asteroids belt (Jupiter and beyond), where the orbital energy increase is enormous, or for journeys to Mercury, where in fact the spacecraft coming from the Earth has to release large amounts of orbital energy. The method adopted in this case is the use of gravity assist maneuvers. This means flying close to a planet and utilizing the gravity field of the planet to transfer some of the heliocentric orbital energy of the planet to the spacecraft. Changes to the orbital velocity of the order of a few km/s can be achieved with a single "swing-by". This is equivalent to carrying a few tons of chemical fuel, which is of course impractical for deep space probes. Table 26.4 shows the accelerations achieved by Rosetta with each of its four-planet swing-bys, compared to the total acceleration (delta-V) achievable with its own propulsion system.

With the right combination of swing-bys a spacecraft can be directed to any target in the solar system. Examples are all the missions to the external solar system, from Voyager to Galileo and Cassini. The NASA Messenger spacecraft and the ESA

Table 26.4 Rosetta achieved acceleration with planet swing-bys

	ΔV (km/s)
Earth swing-by 1	5.9
Mars swing-by	2.3
Earth swing-by 2	5.2
Earth swing-by 3	6.3
1.7 tons chemical fuel	2.2

BepiColombo mission also utilize several planet swing-bys in order to enter in orbit around Mercury, using its gravity to decelerate the spacecraft.

The operational implications of the complex trajectories and the critical activities such as planet swing-bys are explained in Sect. 26.4.

26.2.2 Energy for the On-Board Systems

All spacecraft require electrical energy to power the avionics systems and the auxiliary units like sensors and actuators and their control electronics, or to be converted into thermal energy to ensure temperature control of the various spacecraft units. Whilst for spacecraft orbiting the Earth the most common and practical source of electrical energy is a solar generator, for interplanetary spacecraft the different distance to the Sun can make this solution either difficult or impossible to utilize.

When the distance to the Sun increases, i.e. for missions to Mars and beyond, the electrical power that can be produced by a solar generator decreases with the square of the distance. Table 26.5 shows typical figures for solar electrical power at the average distances from the Sun of the planets.

Current solar generators technology is capable of providing sufficient energy to provide the required power to an interplanetary probe at distances up to the orbit of Jupiter (around 5–6 astronomical units from the Sun). Rosetta was the first solar

Table 26.5 Mean solar flux at the planets

	Min–max sun distance (AU)	Mean solar flux (W/m^2)
Mercury	0.31–0.47	9066
Venus	0.72–0.73	2601
Earth	1.0	1358
Mars	1.38–1.67	586
Jupiter	4.95–5.45	50
Saturn	9.0–10.0	15
Uranus	18.2–20.3	4
Neptune	30.0–30.3	2

powered spacecraft that reached Jupiter-like distances from the Sun (5.3 AU reached in October 2012). The maximum Sun distance record of a solar-powered spacecraft is currently hold by NASA's Juno probe, which is orbiting Jupiter since July 2016. In this decade another solar-powered spacecraft, ESA's Juice probe, will be launched to the Jupiter system, to explore the giant planet's icy moons Callisto, Europa and Ganymede.

Beyond the Jupiter orbit, the only method currently existing and utilized to power interplanetary probes is the use of radioisotope thermoelectric generators (RTGs). An RTG utilizes a radioisotope (e.g. Plutonium-238, normally utilized in the form of Plutonium Oxide, PuO_2) to produce thermal energy and then convert it into electrical energy via thermocouples.

The advantage of this type of generators is of course the independence from the Sun distance, but also from the attitude of the spacecraft with respect to the Sun. This simplifies the attitude and orbit control of the probe, and it allows higher spacecraft robustness to failures. The disadvantage is the presence of a radioactive source on-board, which has system design, integration, safety and political implications which are difficult to manage and overcome.

Contrary to what one could think, solar generators are also problematic at closer distances to the Sun. This is due to the fact that the solar cells must be—by definition—exposed to the Sun illumination, and therefore to its radiation and thermal flux. At distances of the order of the orbit of Mercury the technological challenges related to the selection of the materials and the operation of the solar array under extreme solar fluxes become a major mission driver. This was one of the main challenges in the design and operation of the ESA missions to Mercury (BepiColombo) and to the Sun (Solar Orbiter). BepiColombo, for instance, mounts very large solar arrays in order to supply electrical energy to its four solar-electric propulsion thrusters. These panels are very large since, when operating in the proximity of the Sun, they cannot be pointed efficiently such that the Sun direction is perpendicular to their surface, otherwise the Sun's radiation and heat flux would immediately destroy them. For this reason, the operational concept foresees to keep the solar panels pointed such that the sun direction is close to parallel (just a few degrees difference) to the surface of the panels. This reduces dramatically the solar flux on the panels, but also their efficiency.

The fact that just a slight mispointing, even for a few seconds, of the solar panels can cause a fatal damage of the cells by the Sun's heat flux is clearly a major design driver at system level, affecting both the spacecraft's on-board autonomy and operations concept. Hot redundancy in the on-board controllers of the solar array pointing has to be ensured, both in normal and emergency attitude control modes. On-ground, all systems and procedures required to produce the commands for the pointing of the spacecraft and the solar arrays require additional redundancy and cross-checks to eliminate any possibility of mistakes or inconsistencies.

26.2.3 Communications with Earth

Another major aspect of interplanetary flight, also related to energy, is the communications with Earth. Whilst for spacecraft orbiting the Earth even the use of a low gain, omnidirectional antenna is sufficient to maintain a bi-directional radio link with ground for telemetry and telecommanding, an interplanetary probe requires medium and high gain antennae on-board (and much larger antennae on ground) to establish the necessary communications. The problem again is the limited amount of energy available on-board for transmission of the radio signal to ground, and the enormous distances to be covered to have the signal reach the Earth stations. In order to limit the dispersion of energy the radio beam has to be concentrated to the maximum extent, and this is achieved by using highly directional antennae. This of course poses a fundamental challenge to the spacecraft design and the operations, which is the accurate and continuous pointing to Earth of the high gain antenna (also in case of contingencies). Complex mechanisms and/or pointing algorithms have to be adopted to ensure the continuity and reliability of this vital link to ground. A perfectly functioning and sun pointing interplanetary probe is totally useless if the high gain antenna is not pointed to Earth within a fraction of a degree. Figure 26.1 shows the ESA Rosetta probe and its various on-board antennae, which include low gain antennae (LGA) in S-band, medium gain antennae (MGA) in S- and X-band and a large, 2.2 m high gain antenna (HGA) supporting both S and X radio frequency bands.

Fig. 26.1 Rosetta on-board antennae

Fig. 26.2 The ESA deep space antenna (35 m diameter) in New Norcia (Western Australia)

Due to the energy limitations the achievable bitrates are also limited (typically to ranges between a few bits per second to a few hundreds of kbits per second). The achievable bitrate is also proportional to the frequency of the radio signal. For interplanetary probes the most used frequencies are in the S-band (2 GHz) and X-band (7–8 GHz). The recent use of higher frequency bands (e.g. Ka-band, 18–30 GHz) has allowed higher bitrates, also in the range of Mbits per second. However, the use of higher frequency bands carries additional operational complications, e.g., the higher sensitivity to water vapour in the Earth atmosphere and therefore to the local weather conditions.

On ground, the stations utilized for deep space communications employ large antennae, typically around 25–65 m in diameter, like for large radio-telescopes for astronomy. ESA for instance uses a network of three antennae of 35 m diameter, located in Australia, Spain and Argentina (see Fig. 26.2). NASA has also three complexes, in Australia, Spain and California. In each complex there are several 26 m and 34 m, and one 70 m antenna.

26.3 Mission Control Approach

Operations of an interplanetary mission are not only very different compared to operations of Earth-bound spacecraft, but also strongly dependent on the type of mission. On the other hand, there are common factors that characterize the operations concept and approach and are common to all types of interplanetary missions.

26.3.1 Specifics of Interplanetary Flight Operations

Long Signal Propagation Delay

Interplanetary spacecraft travel at very large distances from Earth, and the travel time of the radio signal from and to the spacecraft is not negligible like around Earth. Travelling at the speed of light the radio signal takes 8 min to cover the average distance between the Earth and the Sun (called astronomical unit, AU, equivalent to approximately 150 million km). For instance, spacecraft orbiting Mars are subject to signal propagation delays that vary between 4 and 20 min one-way (depending on the relative position of Earth and Mars in their heliocentric orbits). A spacecraft at Jupiter experiences a propagation delay of the order of 40–50 min. At Saturn, this increases to about 1.5 h. Such large delays in the radio signal propagation practically prevent any real-time interaction between the ground and the spacecraft, and in particular any activity that involves human real-time decisions based on the spacecraft response to telecommands.

Long Cruise to Target

Interplanetary mission profiles are usually characterized by long cruise periods in the interplanetary space, during which the spacecraft activity is reduced to a minimum, interrupted by short periods of intense activities, e.g. around a planet swing-by or an asteroid fly-by. An interplanetary probe is visible from a single ground station over typically 12 consecutive hours (for a spacecraft on the ecliptic this varies seasonally between typically 9 and 14 h). This is due to the fact that at large distances the motion of the spacecraft in the sky as seen from Earth is very slow (i.e., the spacecraft can be compared to a "fixed star" over the duration of one station pass[1]), thus the dominating factor constraining visibility is the rotation of the Earth. For what concerns the frequency of ground contacts, it is common practice, in order to save cost, to minimize it during the periods of quiet cruise, typically down to one contact per week. This implies of course that the spacecraft is able, under nominal circumstances but also in case of unexpected failures or external events, to survive without ground contact for a period of the order of the longest non-contact period. Also the level of activities has to be low enough to allow all communications with the spacecraft (uplink of telecommands, downlink of stored information) to be completed in a single periodic contact period.

However, the quiet cruise periods, which can last several months to years, are usually interrupted by mission critical events like planets swing-bys, each of which require intense preparation and operations execution effort over a short period of a few weeks. This variable mission profile affects the composition and profile of the mission control team, which has to be as small as possible during the quiet cruise,

---

[1] In fact, depending of the distance to Earth, the apparent motion of the spacecraft in the sky is not completely negligible, as it is typically of the order of a fraction of a degree over the duration of a ground station pass (typically 12 h). This has to be taken into account when calculating the antenna pointing angles: In particular the movement of the spacecraft over a period equivalent to twice the signal propagation delay is important, as it affects the determination of the optimal pointing.

but able to cope with sudden workload peaks during critical cruise events. The long duration of the mission, and in particular of the cruise before reaching the target and starting the actual "productive" phase of the mission, presents severe managerial challenges. First of all, enough flexibility in the size of the team has to be ensured during cruise, to cope with planned and unplanned critical events, but at the same time to minimize the cost of the operations team. This can be achieved relatively easily if resource sharing with other similar missions in different execution phases is possible within the same centre. Another challenge is keeping the motivation and the knowledge base in the team throughout the low activity cruise, to preserve it for the future science operations at the target.

Very important to tackle this problem is the organization of a plan of cross training and proficiency training activities. Furthermore, the existence of high-fidelity tools for training (e.g., software simulators or spacecraft engineering models) is essential to support this type of activities. Motivation can be stimulated by postponing some developments (e.g., ground software tools or on-board software maintenance activities) from the traditional pre-launch phase to the cruise phase. Finally, a significant problem is presented by the natural turnover in the flight control team composition. This can be mitigated by offering career opportunities within the same area (perhaps by temporarily moving people to other missions in a different role, to avoid sudden and uncontrolled losses of expertise when people decide to change job on their initiative.

Long Periods Outside Ground Contact

An interplanetary probe has to be able to autonomously sustain long periods—of the order of weeks—without ground contacts. During cruise this is often due to the decision to reduce operations cost, thereby imposing infrequent ground contacts, i.e., typically once per week to once every few months. In these periods the spacecraft has to be configured into a low activity mode and rely on its on-board autonomy to react to any anomalous situation. In any case also during normal science operations an interplanetary probe will enter periods of non-visibility, caused by the conjunctions between the Earth and the Sun in the direction of the spacecraft. As shown in Fig. 26.3, when the spacecraft, as seen from Earth, comes to an angular distance of a few degrees from the Sun, the plasma of the solar corona starts affecting the quality of the radio signal, and therefore imposes limitations to the telemetry and telecommanding activities.

There are in fact different types of conjunctions, depending on whether the spacecraft is between the Earth and the Sun (inferior conjunction) or beyond the Sun (superior conjunction). Impacts on the radio link also have to be taken into account during solar oppositions, i.e., when the Earth is between Sun and spacecraft. In general, the radio signal starts being affected below 10 degrees Sun-Earth spacecraft angle (indicated with alpha in Fig. 26.3). Below 5°, the navigation accuracy is severely affected and below 3°, the telemetry/telecommand (TM/TC) link may be totally disrupted. Interplanetary missions have therefore to be planned such that

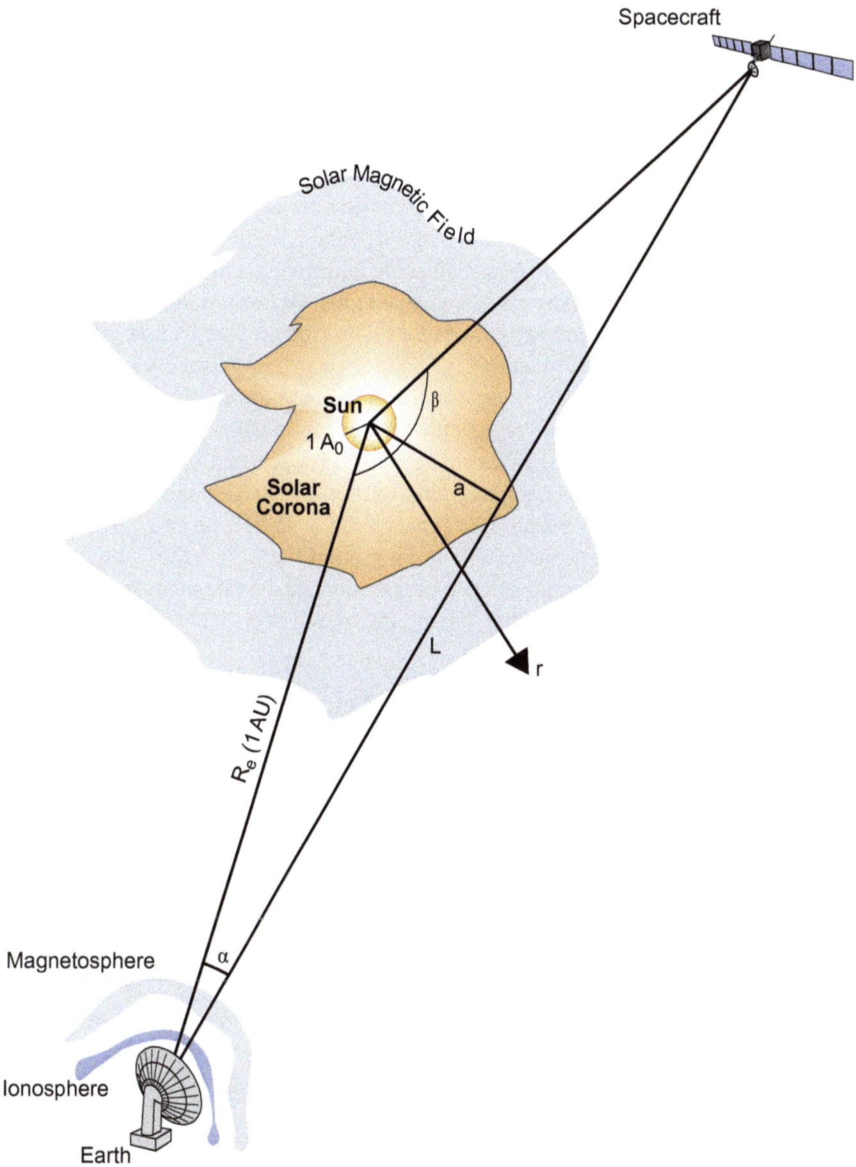

26.3 Solar conjunction (superior)

activities over these periods which, depending on the trajectory, can last several weeks, take the link interruptions and degradations into account.

High Spacecraft Vulnerability
An interplanetary spacecraft is by nature much more vulnerable to problems, whether caused by failures or operational mistakes, than Earth-bound spacecraft. All spacecraft, at least the solar-powered ones, require continuous and reliable pointing of their solar arrays to the Sun. This is normally the only vital pointing aspect the spacecraft has to ensure. Interplanetary probes require, in addition, accurate pointing of their high gain antenna to Earth. Any problem causing loss of Earth pointing will prevent any communications from and to mission control on ground, thereby possibly leading to loss of mission. Another vulnerability factor are trajectory correction maneuvers, which are often time critical and do not allow a second opportunity. Even a temporary inability of the spacecraft (e.g., due to safe mode triggering) or the ground (e.g., loss of a ground station) in a critical moment, such that a trajectory correction maneuver is missed, may directly lead to the total loss of mission. The same is true for critical mission events like planet orbit insertion or entry descent and landing. Everything on the spacecraft and on ground has to be ready and perfectly functioning at a specific time: no delay in preparation can be tolerated, and the occurrence of a single, minor problem that impacts the timing of critical activities can have catastrophic consequences.

Complex, Variable Navigation and Attitude Control Operations
Interplanetary missions require extremely complex and accurate navigation activities. Special orbit determination techniques are required to increase the accuracy, in addition to the standard radio frequency ranging and Doppler tracking: this includes delta-differential one-way ranging (Delta-DOR, see Sect. 26.3.3). For navigation in proximity of small bodies, like asteroids or comets, whose trajectory cannot be determined with sufficient accuracy via optical or radar observations from ground, relative optical navigation techniques have to be used: on-board cameras take pictures of the target body, which are then processed on ground and used in the orbit determination process.

Attitude control of an interplanetary spacecraft is also more complex than a typical Earth-bound spacecraft. This is due to the large variety of attitude modes that have to be used during the mission, to cope with geometrical constraints during cruise, to execute remote sensing observations of the target body during the science phase, and to perform special pointing to celestial objects, or automatic tracking of small objects during fly-bys. Ground control has to cope with several different attitude control modes, manage the spacecraft momentum for future complex pointing profiles (e.g., to predict and control the level of momentum loading of the reaction wheels for attitude control), design strategies over long periods to avoid unnecessary and propellant consuming frequent mode changes.

Limited Knowledge of the Target
Even after 60 years of exploration of our solar system with robotic spacecraft, the knowledge of the planets and of all the small objects that populate the space around

the Sun is still very limited. The first interplanetary spacecraft (e.g., the NASA Mariner series) launched towards the planets had to cope with the very inaccurate knowledge of the position of the planets itself (of the order of hundreds of km!). Nowadays this problem only exists with asteroids and comets, but planets maintain their challenges when it comes to landing on their surface, for which the knowledge of the atmosphere density, variations and the surface morphology is still extremely inaccurate and limited. Flying around small bodies also presents high risks due to the unknown environment in their close vicinity, which could include boulders orbiting the asteroid, or dust and gas jets around a comet.

Spacecraft design and operations approaches have to take into account this high uncertainty on the characteristics of the target. In most of the cases the operations strategy has to be re-invented or at least adapted to the actual environment during the mission, when the spacecraft reaches the target and starts with its instruments the collection of vital information about it. An extreme example is the case of the ESA mission Rosetta, which entered in orbit around a comet nucleus in August 2014. This was the first mission in the history of spaceflight that orbited and landed on a celestial object without the help of a precursor mission that observed the object at least during a fly-by. Before it became possible to enter a closed orbit around the comet nucleus, Rosetta spent several weeks at distances of 100 km first and 50 km then, collecting information about the dynamics of the nucleus, developing a model of all the forces acting on the spacecraft, and of course performing a global photographic mapping of the surface. Only after this initial, intense characterization phase it was possible to maneuver Rosetta to orbit the nucleus from closer distances, down to 10 km from the center, then to select a landing site and design the operations strategy for delivering its lander module, Philae, onto the surface. A full "engineering model" of the comet and its environment, including dust and gas density and speed, was to be built in these early phases of the comet proximity operations and maintained (another unique challenge of orbiting a comet is that the nucleus, its dynamic properties and those of its environment change continuously with the distance to the Sun) and refined throughout the two years spent by the probe around the comet.

26.3.2 Ground Contact Activities

As mentioned above, the ground station contact for an interplanetary mission is constrained by the Earth rotation rather than the spacecraft movement in the celestial sphere, which is negligible over the duration of a communication pass. This results in a theoretical visibility of 9–14 h, depending on the season (interplanetary spacecraft trajectories do typically not deviate too much from the ecliptic plane[2]).

However, if the spacecraft is in orbit around a planet, the geometrical visibility may be interrupted by occultations (i.e., when the spacecraft flies behind the planet as

---

[2] The ecliptic is the mean plane of the apparent path in the Earth's sky that the Sun follows over the course of one year.

seen from Earth) or by spacecraft constraints (e.g., transmission cannot be activated in eclipse due to power limitations, or the high gain antenna cannot be directed to Earth during scientific observations due to pointing constraints). Thus, the daily ground station contact becomes a period of short windows, of the order of tens of minutes up to hours, during which contact with the spacecraft can actually be established. The examples of Venus Express, which during its orbital lifetime was flying around Venus with a 24 h orbit period and Mars Express, flying around Mars in an orbit with about 6.5 h period, show how variable the type of contact can be. In the case of Venus Express the spacecraft oriented itself to the planet in a specific part of the orbit to take scientific measurements, and then turned the high gain antenna to Earth during the period of the orbit in which the ground station was in visibility. This resulted in a very comfortable and repeating pattern in which the full duration of the ground station geometrical visibility could be effectively used for the contact with ground without interruptions. Also Mars Express turns every orbit alternatively to the planet and the Earth for communications, but the orbit period of 6.5 h is not synchronized with the daily station visibility. Planet occultations do also occur more frequently and power constraints force a careful operation of the transmitter. This results in a very irregular pattern of short ground contacts, typically of the order of 30–120 min, interrupted by non-contact periods of the same order of magnitude. The short duration of the contacts, combined with their irregular frequency, make the planning of operations (in particular commanding and download of recorded telemetry) quite complex. Also, the rules for recovery from temporary outages or failures are not straightforward, making the replanning tasks complex and slow.

An extreme case are missions with even shorter orbit period (in the order of two hours), such as ExoMars (which is orbiting Mars since the end of its aerobraking activities in 2017 on a very low circular orbit of about 400 km altitude) or BepiColombo (which is currently cruising towards the small planet Mercury and will orbit it as of 2025 with a similar orbital period, of about two hours). For BepiColombo, for instance, the analysis shows that over the duration of a daily ground station visibility window up to ten interruptions may occur, and the duration of continuous contact between two interruptions may be as short as about 20 min.

Figure 26.4 shows an example of the BepiColombo visibility pattern during a ground station pass. The complexity of the telemetry and telecommanding activities induced by each link blockage is shown in the lower part of the figure: events on-board and on ground have to be synchronized using two-time references which differ from each other by the length of the signal propagation delay. As this delay is variable during the mission, tools are required to be able to plan and work with the two-time references and ensure synchronization. For instance, if the start of the dump of telemetry data recorded on the on-board mass memory is programmed on-board via time-tagged command, the time stamp has to take into account the signal propagation delay to reach ground at the beginning of the pass.

Other more complex examples are related, e.g., to the dump suspension when the first uplink sweep is started: shortly after the spacecraft signal is acquired at the beginning of a ground station pass, the station transmitter is activated and the radio frequency of the uplink signal is slowly changed up and down within a predefined

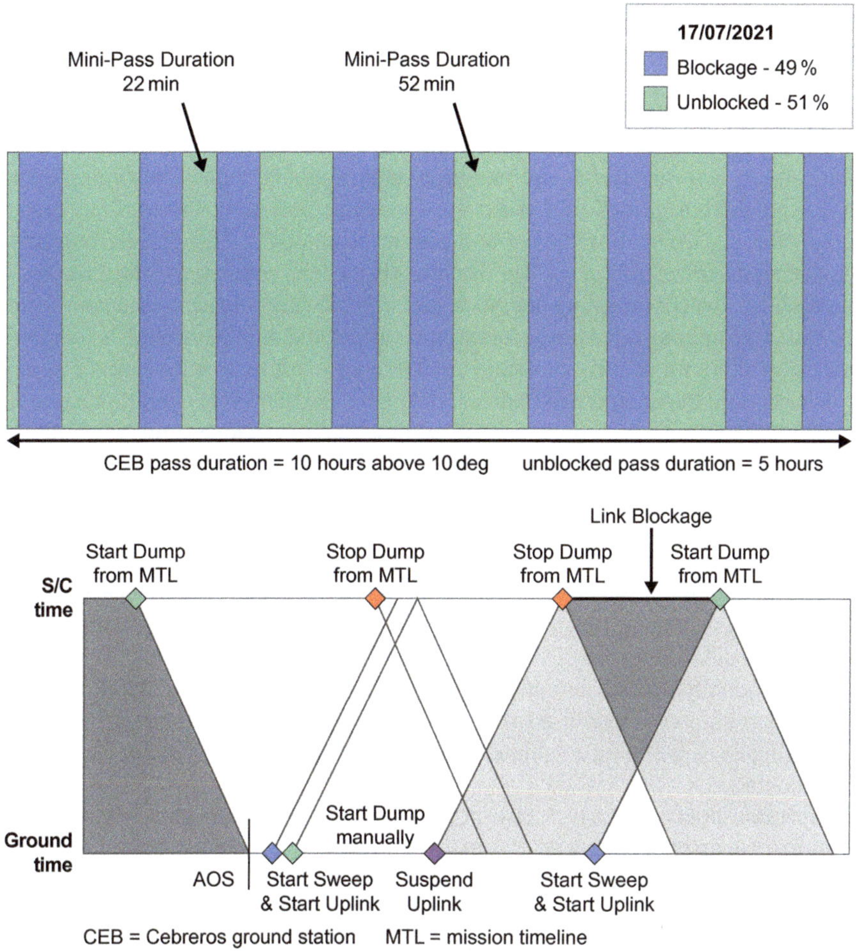

26.4 BepiColombo ground contact profile for a single pass

range (this technique is called "uplink sweep"), to facilitate the automatic lock of the on-board receiver onto the received signal. But since the transponder on board is normally configured in coherent mode (see Chap. 18), once the on-board receiver locks on the uplink radio signal, it will automatically change and move the downlink radio signal frequency. This causes a short interruption of the downlink signal reception on ground. Therefore, an interruption of the onboard memory dump has to be programmed at this point to avoid a temporary loss of data on the transmission to Earth.

The pass activities on the other hand are relatively simple, and mostly concentrated at the beginning of the contact, shortly after AOS (acquisition of signal, i.e., the moment in which the spacecraft rises over the horizon at the ground station and the first radio signal is received).

At AOS, the first engineering telemetry is received and a quick check of the spacecraft status is performed. Once the antenna has reached 10° elevation, which is the safety limit for starting the uplink signal transmission, the ground station transmitter is activated and the uplink sweep starts. It should be noted that the uplinked radio signal from ground will reach the spacecraft only after the one-way signal travel time, and the confirmation of on-board receiver lock will be received only after another signal travel time delay. This is normally too long compared to the duration of the pass. So, the commanding activities start immediately—success oriented—after the end of the uplink sweep (i.e., a few minutes after the station transmitter activation), without waiting for positive on-board receiver lock confirmation. In case the sweep is unsuccessful the ground will realize it only later and the whole operation has to be repeated. This leads to additional waste of time, but as it happens relatively rarely, the success-oriented approach is preferred over the safest but slowest one of waiting for positive uplink lock confirmation. Initial commanding is normally triggering the dump of event logs and other telemetry storage areas, which have recorded on-board the telemetry accumulated during the non-contact period. Depending on the amount of science data stored and on the length of the signal travel time, the science telemetry dump command may be stored time-tagged on-board, to increase the total dump time during the pass.

Once the initialization activities have been completed, the entire duration of the pass is normally used just to retrieve the stored telemetry. The dump is suspended automatically by on-board command for the occultations and at the end of the planned daily contact.

Operator supervision is typically only required for the initialization activities (that is around the first hours after AOS start). For the rest of the pass the dump operations do not normally require human presence in the control room. Of course, if special commanding activities are planned, this may require presence of the operator or of relevant engineering expertise. In general, though, the pass activities for an interplanetary mission are simple and relatively easy to automate. Human intervention is only required in case of anomalies, and even in this case the intervention does not have to occur within a very short time: interplanetary spacecraft are designed to survive for long periods and the rapidity of ground intervention is only driven by the need to minimize the possible interruption of the data recovery or loss of data in general, not by the safety of the spacecraft, which is ensured by the on-board autonomy.

The frequency of the ground station passes during cruise is driven by the amount of activities required on the spacecraft, but also in particular by the amount of engineering telemetry (in case of passive cruise) to be recovered. The typical useful engineering telemetry generation rate of a spacecraft in cruise should not exceed a few hundreds of bits per second, to minimize telemetry signal bandwidth occupation, in favour of more science data. However, even at this low generation rate the amount of telemetry accumulated over, say, a week out of contact is large. This fact, combined with the low downlink bitrates available to an interplanetary spacecraft traveling at large distances from Earth (normally of the order of a few tens of thousands of bits per second), means that the periodic downlink of the entire history of engineering telemetry accumulated outside the ground contact may require several hours.

For a mission like Rosetta the weekly pass frequency during the quiet cruise was mainly dictated by the wish to recover, in an average 10 h contact, the entire history of engineering telemetry. If this is not possible, either because the available downlink rate is too small, or because the frequency of passes is lower, then the spacecraft may be configured to downlink only essential telemetry, and store the rest to allow the ground to selectively downlink whatever information is deemed necessary on each occasion. This approach is more complex and requires a quite flexible storage and downlink mechanism to be implemented on the spacecraft.

Some missions may decide to downlink a simple "beacon" tone to the ground. This is a radio frequency (RF) carrier with a limited set of specific frequencies (each one is a different "tone"), which can be picked up even with relatively small antennas. The on-board autonomy decides which tone to downlink, to indicate, e.g., that an anomaly occurred or that everything is OK. The ground periodically checks the reception of the beacon and, depending on the received tone, may decide to establish a full link and acquire the complete engineering telemetry. This method has been used, e.g., by NASA's New Horizons, which was the first probe to fly-by Pluto in 2015, and another Kuiper Belt Object, 2014 MU69, in 2019. This is a complex technique and requires a full trust in the on-board autonomy, combined with a large on-board storage and a flexible store/downlink mechanism. However, for missions flying for many years at very large distances, this may be the only way to save precious ground station time and therefore mission cost.

26.3.3 Trajectory Determination

Interplanetary orbit determination is a very complex task that is often linked to critical operations and therefore directly impacting mission success. The main aspect that characterizes interplanetary orbit determination is the extremely high accuracy required, which implies high precision in all the involved measurements. For instance, an error of 1 m in the knowledge of the position of the ground station delivering the radiometric measurements directly results in an error of 750 km in the spacecraft position determination at Saturn distances. The same is true with the precision of the clocks used to measure the radio frequencies involved in the radiometric data collection. An additional complication is the accuracy of knowledge of the position of the solar system target, planet, moon, or small body like an asteroid or a comet, which becomes critical when approaching it. For instance, the position of an inner planet like Mars is known today with an accuracy of 0.5 km, whilst our knowledge of the position of the distant giant planets like Jupiter or Saturn still suffers inaccuracies of the order of 10 km. The trajectories of small bodies are known with much worse accuracy, resulting in errors of the order of 100 km.

The main methods used for the determination of the trajectory of an interplanetary probe are the same as for Earth-bound missions: ranging and Doppler measurements

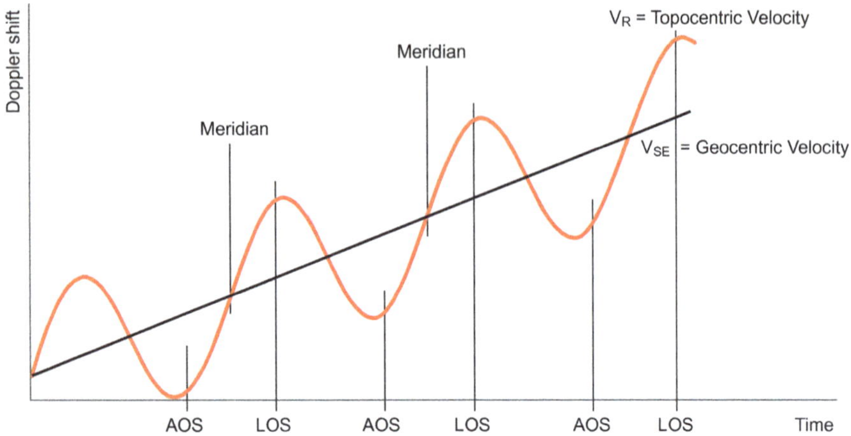

26.5 Main components of the Doppler measurement

made on the radio signal sent to and received from the spacecraft. The ranging method is based on the transmission of a sequence of tones to the spacecraft, which receives and retransmits them in its downlink signal. By measuring the time of flight of the signal the radial distance of the spacecraft from the ground station can be accurately measured. The Doppler method is based on measuring the difference between the frequency of the radio signal uplinked to the spacecraft and the frequency of the signal from the spacecraft received on ground. The latter is linked, in a coherent transponder on-board, to the uplink frequency via a fixed (known) ratio. The difference between the two frequencies is due to the Doppler effect, which allows to determine the radial velocity of the spacecraft with respect to the ground station.

The Doppler measurement can also be utilized to achieve an accurate determination of the angular position of the spacecraft in the sky (at least one component), as shown in Fig. 26.5: Given that the main component of the measured Doppler shift is due to the ground station movement during the pass due to the Earth rotation, the point in which this component is zero determines very accurately the point when the spacecraft crosses the local meridian (i.e., the local South direction, point of inversion of the radial direction of motion of the ground station).

Critical events like planet swing-bys or orbit insertions, which require extremely high spatial accuracy at a specific point in time, are nowadays supported by a new radiometric technique, called delta-differential one-way ranging (Delta-DOR). This method is based on interferometry measurements of specific spacecraft radio tones from two ground stations with simultaneous visibility of the spacecraft. The measurement is complemented by calibration of the local atmospheric delays by observing a quasar close to the direction of the spacecraft. This method, derived from radio astronomy very long base interferometry (VLBI) techniques, allows to achieve accuracy in spacecraft position determination one order of magnitude better than the one achievable with normal radiometric measurements. The principle is illustrated by Fig. 26.6.

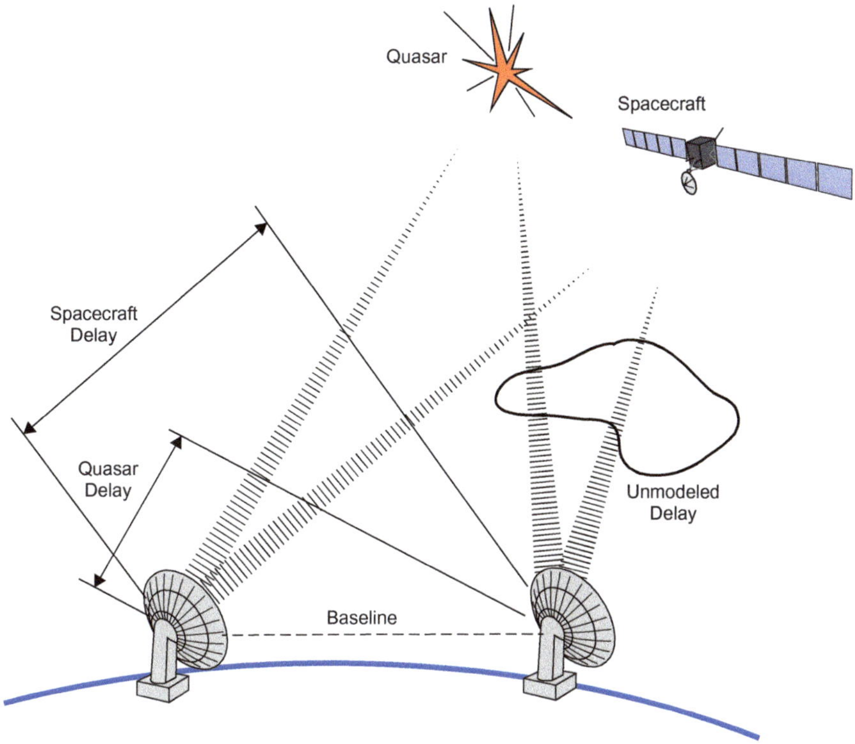

26.6 Delta-DOR measurement principle

The signal from the spacecraft is received by two ground stations. Due to the large distance between the stations on Earth, the same signal is received at a specific instant in time has a different phase at each of the two stations (which is the same as saying that the same phase is received at slightly different times at the two stations). The difference in phase gives a precise measure of the different distance between each station and the spacecraft. By knowing the exact distance between the two stations and the angle between the stations and the spacecraft a very precise measurement of the spacecraft position can be obtained. It is also very important to emphasize that this technique allows to have an independent check of the resulting trajectory, eliminating potential systematic errors in the solutions.

To resolve the inaccuracy in the known position of a small target body, optical navigation techniques are used, through the acquisition of pictures from an on-board camera, which are elaborated on ground and included in the trajectory determination process. To illustrate this technique, Fig. 26.7 shows the example of optical navigation applied to the Rosetta fly-by of asteroid Steins in September 2008.

The figure is based on the B-plane representation. The B-plane is defined as the plane perpendicular to the spacecraft trajectory (more precisely: to the asymptote to the trajectory) that contains the target object (in this case the asteroid). The figure

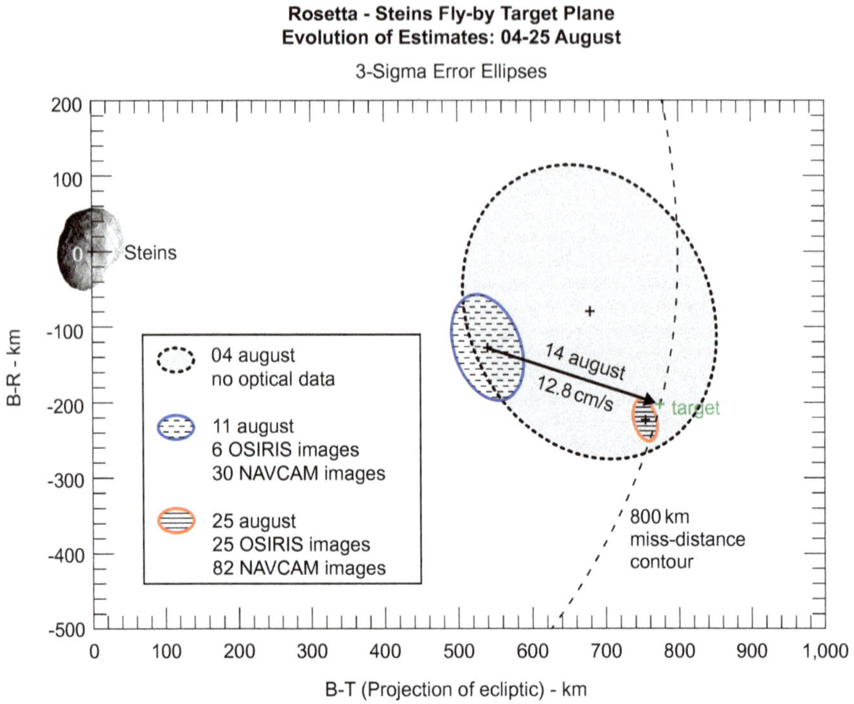

26.7 B-plane optical navigation for Rosetta asteroid fly-by

shows various ellipses with a cross in the middle of each. The cross represents the predicted point of crossing of the B-plane, i.e., when the spacecraft will be at the point of closest approach to the asteroid. The ellipse shows the bi-dimensional error (3 sigma) in the predictions. The error in the third dimension is represented by the inaccuracy in the time of closest approach. The largest error ellipse, shown in black, indicates the result of the orbit determination using only radiometric data. A sequence of pictures of the asteroid taken at 8.1–5.2 Mkm from the target shows the asteroid's apparent movement relative to the background fixed stars. By analyzing this apparent movement, the knowledge of the relative position between the spacecraft trajectory and the asteroid is enormously improved in the B-plane. Optical navigation allowed the reduction of the error ellipse (see the initial improvement of the error shown by the blue ellipse) and the planning and execution of more accurate trajectory correction maneuvers, resulting in Rosetta missing the fly-by target point by less than 0.5 km.

Optical navigation has been tried as an autonomous process on board the NASA Deep Impact impactor spacecraft. The impactor was released by the mother spacecraft to impact on a comet nucleus about 24 h before the planned impact. As the relative speed of the impactor and the comet was about 10 m/s, the error in the impact trajectory became visible only in the last two hours before impact, which made any intervention from ground practically impossible. Through its on-board camera the impactor determined its trajectory and performed three small correction maneuvers, between T-90 min and T-13 min, finally crashing on the surface of the comet as planned.

Modelling of disturbance forces on the spacecraft is a very important part of the overall orbit determination process for an interplanetary spacecraft. Over the long cruise arcs disturbance forces like solar radiation pressure can have a significant effect on the trajectory, and its accurate modelling and prediction is particularly important at the end of the arc, when the encounter with a solar system body is planned.

Even more critical and complex is the modelling of the dynamic effects of a planet's atmosphere on the descent trajectory of a landing module. Small inaccuracies in this model can lead to enormous errors in the actual landing site. For instance, error ellipses of the order of 100 km are normal for landing on Mars. The large variability and unpredictability of the density of Mars' atmosphere are one of the main problems related to designing a safe landing on the planet.

26.4 Special Operations

Interplanetary missions are characterized by a highly variable mission profile, with several complex and often critical operations which are unique of this type of missions. The sections below deal with the most important of these activities and the related operations approach.

26.4.1 Planet Swing-By

As already indicated in this chapter, planet swing-bys are the most commonly used method to acquire or release orbital energy, as they are practically almost free in terms of on-board fuel utilization. As such most of the interplanetary missions tend to make use of one or more swing-by (also called gravity assist maneuver—GAM) to achieve their target. The disadvantage of such method is first of all the complexity of the interplanetary trajectory and the duration of the cruise, which increase significantly compared to a direct flight to the target. Also, the operations around a planet swing-by are time critical and require high precision navigation and a robust fault detection, isolation and recovery (FDIR) logic on-board the spacecraft that prevents unplanned trajectory perturbations in proximity of the planet.

The activities related to a planet swing-by follow a relatively fixed pattern that begins a few months before the closest approach to the planet, with intense tracking campaigns to achieve the highest possible accuracy in orbit determination. Typically, two ground station contacts per day from different ground stations, of the duration of a few hours each, are scheduled for the acquisition of radiometric measurements. In addition, DDOR tracks are scheduled with increasing frequency (starting from once per week up to several tracks per week). Typically, 20–30 DDOR tracks are scheduled, depending on the criticality of the swing-by (e.g., the lower the swing-by altitude, the higher the criticality) in the last few months before closest approach.

Trajectory correction maneuver slots have to be planned as part of the planet swing-by phase. The first correction is normally performed at around one month before the swing-by. If the maneuver (typically of the order of tens of cm/s) is performed with high accuracy and the perturbation forces on the spacecraft are accurately modelled, there are good chances that this correction will be sufficient to perform the swing-by within the precision required by the mission. Further trajectory correction slots are anyway planned typically one week and one day before the swing-by. As a safety measure, an emergency slot at T-6 h is scheduled, to perform a pericenter raising maneuver in case of late detection of major problems with the trajectory. Such maneuver would deteriorate the performance of the fly-by, but at least save the spacecraft from the risk of collision with the planet or re-entering in its atmosphere.

As an example, the planned generic timeline of activities for each of the Solar Orbiter GAMs is shown in Table 26.6.

In order to avoid perturbations of the trajectory in the final weeks before closest approach, and in particular to minimize the chances of late disturbances to the trajectory that would force an emergency pericenter raising, the spacecraft must be designed such that its attitude control system does not create spurious forces. The "pure torques" attitude control is important when reaction wheel momentum is offloaded (required periodically to avoid saturation of the wheel momentum capacity) but in particular when the spacecraft enters into autonomous safe mode (with attitude

Table 26.6 Generic activities timeline for a solar orbiter GAM

Time (relative)	Activity
GAM—1 month	Start of phase Delta-DORs: one per week Comm passes: two to four per week
GAM—30 days	Slot for trajectory correction maneuver
GAM—16 days	Slot for trajectory correction maneuver
GAM—2 weeks	Delta-DORs: two per week comm passes: daily
GAM—1 week	Slot for trajectory correction maneuver
GAM—1 week	Delta-DORs: four per week
GAM—5 days	Comm passes: 24-support
GAM—3 days	Slot for trajectory correction maneuver
GAM—1 day	Slot for trajectory correction maneuver
GAM—3 h	Latest slot for trajectory correction maneuver
GAM	Closest approach
GAM + 1 week	Slot for trajectory correction maneuver end of phase

normally controlled with thrusters): even a small spurious acceleration of the order of 10 cm/s/day can cause, over one week (the time that may be required to bring back the spacecraft into a normal attitude control mode) major deviations (of the order of tens of km) from the nominal trajectory at the time of closest approach to the planet.

In the period around closest approach normally the ground contact is lost due to either occultation by the planet itself or, in case of Earth swing-by, due to the high velocity of the spacecraft, which may be too fast for the tracking capabilities of the ground antenna. If the spacecraft is normally keeping contact via a steerable high gain antenna, similar limitations of angular speed may require to stop the movement and configure the link over fixed, low gain antennas over the few hours around closest approach. Also, the fast dynamics of the few tens of minutes around closest approach may cause an extremely high Doppler shift in the frequency of the radio signals, which may go out of the range supported by the on-board transponder.

There are other considerations that may force special configurations of the spacecraft for the few tens of minutes around a swing-by. For instance, the gravity gradient changes may impose high torques on the reaction wheels, forcing a change to a more robust thruster-based attitude control mode. Also, the star trackers may be blinded by the planet, or other optical sensors may provide wrong inputs to the on-board software and must therefore be disabled. Finally, the spacecraft might have to fly in the shadow of the planet (eclipse), with obvious implications on its power and thermal performance and configuration.

As soon as the short period of closest approach is over, the spacecraft is reconfigured to its normal mode for interplanetary cruise. Intense tracking activities continue in order to determine the success of the maneuver. Typically, a slot for a final correction maneuver is scheduled a few days after the swing-by.

The activities described in this section are valid for a generic swing-by of a major planet, separated in time from another swing-by or critical mission activity by several months. In case of missions that operate in the Jupiter or Saturn environment, and perform frequent swing-bys of the moons (for gravity assist or scientific reasons), the entire process has to be much more time-efficient, due to the fact that subsequent fly-bys are separated by very short periods of time (of the order of weeks). New strategies are currently under analysis for the future ESA mission to the Jupiter icy moons, JUICE, which will perform about 30 swing-bys and fly-bys of Europa, Callisto and Ganymede in less than three years.

26.4.2 Asteroid Fly-By

Interplanetary probes often require to perform close fly-bys of small bodies of the solar system as part of their scientific objectives. The fly-by of a small body like an asteroid or a comet is a completely different operation from the swing-by of a major planet described above. The main difficulty in this type of operations is caused by the limited knowledge of the target position and trajectory. Asteroids ephemerides are normally known with an accuracy of the order of a few 100 km. This means that the accuracy of the trajectory determination of the spacecraft itself is not sufficient to ensure a sufficiently precise asteroid fly-by distance and time.

To improve the precision of the fly-by distance in the B-plane optical navigation is used, as described in Sect. 26.3.3 above. Several trajectory correction slots are also planned in case of an asteroid swing-by. Compared to a planetary swing-by sequence, normally the last maneuver slots are in this case always used, given the much higher requirements on the precision of the fly-by. Typically, the last maneuver is executed less than one day before the closest approach.

Having solved with optical navigation the problem of steering the trajectory to the correct fly-by distance in the B-plane is not sufficient to achieve a successful fly-by operation: in order to point the instruments to the asteroid during the short time around closest approach (the most important from a scientific point of view) the knowledge of the asteroid position along the line of flight has to be improved.

The orbit determination process on ground may estimate the time of closest approach within a few seconds. This is not good enough to steer the spacecraft instruments during closest approach, when high relative velocities (of the order of 10 km/s) combined with relatively short fly-by distances (of the order of a few 100 km) impose a very high angular rotation of the spacecraft (or its instruments) to keep the target in the field of view. Two solutions have been adopted to ensure the proper observation of the target: either the spacecraft is pre-programmed to compose a mosaic of pictures to cover an area larger than the estimated error in orbit determination (thus ensuring that at least in a few of these pictures the target object will be captured), like in the case of the asteroid fly-bys made by NASA's Galileo spacecraft; alternatively, the spacecraft uses its on-board cameras to autonomously detect the asteroid, and steers its own attitude in order to keep the asteroid in the field of view (by continuously updating its estimation of the time of closest approach). The latter technique has been used by Rosetta for its two asteroid fly-bys in 2008 and 2010, and by recent NASA spacecraft.

The operations related to asteroid fly-bys are concentrated in a few hours around closest approach. Being a one-chance activity, everything has to work perfectly and therefore a lot of tests and validation activities are performed in the months preceding the event. The combination of all the scientific requirements coming from the various remote sensing instruments involved in the fly-by into a single, robust commanding timeline is another major challenge of such an operation.

26.4.3 Planet Orbit Insertion

The operations required for a planet orbit insertion are very similar to the ones necessary for a planet swing-by. The main difference is that, at arrival at the planet, normally a large orbit injection maneuver has to be performed with the on-board propulsion system. In order to minimize gravity losses,[3] this maneuver (typically of the order of a few km/s) has to be performed as fast as possible, so the spacecraft is normally equipped with a large engine (for instance, Mars Express and Venus Express used a 400 N thruster for their insertion maneuver into a closed orbit around their target planet).

The orbit insertion maneuver normally occurs within a few tens of minutes before arrival at the planet, and lasts also several tens of minutes. Like all critical operations of interplanetary missions, it has to be executed autonomously, without any possibility for the ground to intervene in case of problems, due to the long signal propagation delays. Also, due to the pointing constraints of the spacecraft, it is normal that the high gain antenna cannot be pointed to Earth during the maneuver. If possible, a weak radio link with ground is maintained through the low gain antennae. Even if this allows only radio frequency carrier reception, this signal from the spacecraft is extremely important for the mission control team on ground to monitor the execution of the maneuver, especially in case of anomalies. The signal's Doppler shift can be used to compare it with the expected spacecraft acceleration profile, and any deviations from the expected maneuver performance can be rapidly detected, to an acceptable level of accuracy. Unfortunately, also for planet orbit insertions it is quite common that part of the maneuver is performed with the spacecraft occulted by the planet itself. This reduces the visibility and makes the moment of reappearance behind the planet extremely tense for the ground controllers: The reappearance of the spacecraft signal at the predicted time gives a first rough indication that the maneuver was performed correctly and the mission control team can start breathing again.

Given that for many planetary missions the insertion into planet orbit is a mission critical activity (i.e., if it fails, the mission is lost), all decisions at mission level give priority to the execution of the maneuver w.r.t. even the spacecraft health itself. This means that the on-board failure detection and isolation autonomous functions may be partly disabled or relaxed during the execution of the maneuver, just to minimize the chance of small problems unduly interrupting or perturbing the maneuver execution. Also all non-essential units are deactivated, also to minimize chances of failure occurrence and propagation during this mission critical period.

After orbit insertion it is normally required to perform a set of orbit correction maneuvers to move the spacecraft from the capture orbit to the final operational orbit. These maneuvers are not directly time critical, as they can be performed with the periodicity of the achieved orbit. However, since the period of the initial orbit is normally large, it is important to perform them rapidly to avoid large delays in the start of the scientific operations phase at the planet.

Planet orbit insertion maneuvers can be also less time-critical, like in the case of BepiColombo, which will inject into Mercury orbit performing a maneuver at

---

[3] Gravity losses is a term to indicate the reduced efficiency of a maneuver if the maneuver is not performed fully in the optimal point in the orbit. An instantaneous maneuver would have zero gravity losses, i.e., its fuel efficiency would be perfect. Typical chemical propulsion systems require of course a finite time to perform the maneuver: the smaller their thrust of the engine, the longer the maneuver. The longer the duration of the maneuver, the larger are the "gravity losses", i.e., the fuel efficiency of the maneuver is smaller.

the so-called "weak boundary conditions". In these cases, if the maneuver cannot be performed there is always another opportunity as the spacecraft remains in the proximity of the planet heliocentric orbit. This of course affects the considerations on priority of spacecraft health w.r.t. maneuver execution.

26.4.4 *Landing Operations*

Landing operations are described in more detail in Chap. 27.

26.5 Conclusions

Mission control for interplanetary probes involves extremely complex activities, certainly amongst the most critical and complex in the field of spaceflight. The high accuracy and criticality of navigation, the time criticality of many mission operations, the high variability of the mission profile and the fact that most operations have to be executed without the possibility for ground to intervene in real time, are the main challenges to ground operators for this type of missions.

The ground tools required for interplanetary missions are also very specialized: the large ground stations, which include extremely sensitive instrumentation and radio-astronomy techniques; the flight dynamics algorithms and models, which have to achieve extremely accurate orbit determination and support complex pointing activities under very variable conditions; the mission control approach, which is mainly based on offline interactions but has to include complex planning methods and extremely safe but flexible procedures systems.

The expertise required to perform end-to-end interplanetary mission control is consequently highly specialized and, given the relatively scarce amount of such missions, it is also rare to find and geographically concentrated in a few operation control centres in the world.

After the pioneering decades (1960–1970) and the major successes of many world space agencies in the exploration of the solar system, the next steps will become more and more complex and require more costly missions. The key to the success of future interplanetary missions is international cooperation. The final target, a deeper understanding of our solar system, is the new frontier of human exploration.

References

Clarke AC (1950) Interplanetary flight: an introduction to astronautics. Temple Press, London

Huntress W, Ya Marov M (2011) Soviet robots in the solar system: mission technologies and discoveries. Springer Praxis Books/Space Exploration

Labunsky A, Papkov OV, Sukhanov KG (1998) Multiple gravity assist interplanetary trajectories. CRC Press

Launius RD (ed) (2013) Exploring the solar system: the history and science of planetary exploration. Palgrave Studies in the History of Science and Technology (2013)

Logsdon T (1997) Orbital mechanics: theory and applications, 1st edn. Wiley

Murray C, Delmott S (2000) Solar system dynamics. Cambridge University Press

NASA Jet Propulsion Laboratory (2021) Basics of space flight. https://solarsystem.nasa.gov/basics/, cited 14 June 14 2021

Ulivi P (2007) Robotic exploration of the solar system. Springer Praxis Books/Space Exploration

Valentine L (2021) A space roadmap: mine the sky, defend the earth. Settle the Universe. http://ssi.org/reading/papers/space-studies-institute-roadmap/, cited 14 June 2021

Paolo Ferri was the long-time head of the Mission Operations Department at the European Space Operations Center (ESOC) of the European Space Agency in Darmstadt, Germany, and is now retired. After his Doctor degree in theoretical physics (University of Pavia, Italy), he started working in 1984 at ESOC as visiting scientist on the Exosat X-ray astronomy mission science operations. He moved to mission operations in 1986 and worked on different types of missions, including Eureca (microgravity, deployed and retrieved with a Space Shuttle), Cluster (space plasma physics), Rosetta (comet science), GOCE (earth gravity). In 2006, he became responsible for all ESA interplanetary mission operations, including Mars Express, Venus Express, Bepi-Colombo, ExoMars, Solar Orbiter. Since 2013, he led the Mission Operations Department, in charge of all ESA unmanned mission operations, preparation and execution.

Chapter 27
Lander Operations

Stephan Ulamec and Paolo Ferri

Abstract A special case of interplanetary missions includes devices that land on a planetary surface or descend through an atmosphere. Operations become more challenging, as a lander typically has additional restrictions in terms of visibility and mission planning. The landing and descent phase is particularly critical and needs special operational provisions, as it usually cannot be repeated or interrupted. The chapter gives insight on missions to the Moon, Mars, Venus, Titan and small bodies like asteroids or comets where landers, entry probes or rovers have been operated.

27.1 Overview

A special case of interplanetary missions is the one including landers or atmospheric entry probes. First because of the operational prerequisites to allow the landing as such, and secondly due to the particular constraints related to the operations of a vehicle positioned on the surface of a planet, moon or small body. Note that in this chapter we are not discussing human missions, where, as in the case of Apollo, landers could be commanded by astronauts directly.

There are several categories of missions, some of them to be discussed in more detail below. The design of landers differs considerably, depending on the actual target body. We distinguish between:

(a) Landing on planets or moons with a significant gravity and atmosphere (Mars, Venus, Titan); with the sub-group of entry-probes (where only the atmosphere is investigated during descent)
(b) Landing on airless bodies (Moon, also Mercury or, e.g., Ganymede)
(c) Landing on small bodies with very low gravity (asteroids, comets).

S. Ulamec (✉)
German Aerospace Center (DLR), Cologne, Germany
e-mail: stephan.ulamec@DLR.de

P. Ferri
Retired, European Space Agency, ESA-ESOC, Darmstadt, Germany

F. Sellmaier et al. (eds.), *Spacecraft Operations*, Springer Aerospace Technology,
https://doi.org/10.1007/978-3-030-88593-9_27

The descent to the surface can be performed with several strategies (Ulamec and Biele 2009), notably by:

(a) Landing the complete spacecraft, which is also used for the interplanetary cruise and remote observations (e.g., JAXA's Hayabusa, or the end of mission touch downs of NASA's Near Earth Asteroid Rendezvous (NEAR) and ESA's Rosetta orbiter spacecraft)
(b) A dedicated lander (e.g., the Viking Landers or the Rosetta Lander, Philae)
(c) Probes that can move and re-position themselves after touchdown (e.g., Luna 9, the MER Rovers or MASCOT)
(d) Penetrators.

All of these have particular requirements from an operational point of view. Some concepts are described in further detail in Sect. 27.3.

Table 27.1 gives an overview of a selection of historic landing (and some atmospheric entry) missions.

Landing phases are critical and, as for all critical activities of interplanetary missions, have to be pre-programmed and executed fully autonomously. However, the contact with mission control on Earth during these phases is essential, to follow all phases of the automatic sequence and keep the possibility to investigate failures in case of problems. This is an important lesson learned from the catastrophic failure of NASA's Mars Polar Lander, which crashed in the 90ies onto the surface of the planet, but was designed to perform the landing operations "in the blind", so that the analysis of what went wrong was severely hampered by the lack of detailed information around the time of the failure. Similar problems in failure analysis occurred after the loss of the Beagle 2 lander, which was released by the ESA Mars Express probe before entering in Mars orbit, 2003. The lander was scheduled to perform all entry, descent and landing operations without contact with the orbiting spacecraft nor with ground stations on Earth and to establish the first radio link autonomously after successful landing. This never happened and a complete explanation of what problem could have occurred could not be established. However, more than a decade later, in January 2015, high resolution pictures of the foreseen Beagle 2's landing site taken by NASA's MRO orbiter provided some information, showing an almost intact lander on the surface, in a however incompletely deployed configuration. This indicates that the atmospheric phase worked well, but a failure of the mechanisms of the lander occurred after touch-down.

More recent landing activities have therefore featured all possible "observation points", from orbit (existing orbiters), and from ground (e.g., ground stations or even radio telescopes to detect at least the radio carrier from the lander during descent in case the signal is too weak for a deep space tracking station to decode telemetry from it).

In the case of the landing of NASA's MSL Curiosity rover, three Mars orbiters [Odyssey (NASA), Mars Reconnaissance Orbiter (MRO, NASA), Mars Express (ESA)], one ground station (New Norcia, ESA) and several radio telescopes were employed to follow the Lander's UHF (ultra-high frequency) and X-band signals emitted during the entire entry, descent and landing (EDL) phase.

Table 27.1 Selected lander missions

	Country/agency	Launch date	Comment
Moon			
Luna 9	USSR	Jan 1966	First ever soft landing on an extraterrestrial body
Surveyor 1–7	USA	May 1966 to Jan. 1968	Series of five successful landings (Surveyor 1, 3, 5–7) on the Moon, in preparation of the manned Apollo missions
Luna 16	USSR	Sep 1970	First robotic sample return mission from an extraterrestrial body. Followed by the successful missions Luna 20 and 24
Luna 17	USSR	Nov 1970	Delivered Lunar rover Lunokhod 1 to the Moon. Followed by Luna 21 with Lunokhod 2. The rovers operated for several months and traversed 10.5 and 37 km on the Lunar surface
Apollo 11–17	USA	Jul 1969 to Dec 1972	Six successful manned missions to the Moon. Returned about 380 kg Lunar material to Earth and placed several scientific instruments on the Moon. Each lunar module (LM) was crewed with two astronauts. Apollo 15–17 included astronaut driven rovers
Chang'e 3 and 4	China	Dec 2013, Dec. 2018	Lunar landers with small rovers (Yutu). Chang'e 4 was the first mission ever to land on the far side of the Moon
Chang'e 5	China	Dec 2020	Robotic sample return with lunar orbiter, lander/rover/ascent vehicle on the surface and lunar orbit rendezvous between orbiter and ascent vehicle
Mars			
Mars 3	USSR	May 1971	First soft landing on Mars. However, probe ceased to transmit data only 14 s after touch-down

(continued)

Table 27.1 (continued)

	Country/agency	Launch date	Comment
Viking 1 and 2	USA	Aug/Sep 1975	Two identical, highly successful Mars stations. Operated on the Martian surface till 1980 and 1982, respectively
Mars Pathfinder	USA	Dec 1996	Fixed landing station plus small Sojourner rover. It demonstrated landing with airbag concept
Mars Exploration Rovers (MER)	USA	Jul/Jun 2003	Two rovers (Spirit and Opportunity) landed, using airbag concept. Spirit operated until 2010, Opportunity until 2018
Phoenix	USA	Aug 2007	Fixed landing platform in polar region (68°N)
Mars Science Laboratory (MSL), Curiosity	USA	Nov 2011	Successful landing of rover by "sky-crane"
Schiaparelli	ESA	March 2016	Test lander carried to Mars by the ExoMars Trace Gas Orbiter spacecraft Landing on 19 October 2016 failed in the last 50 s before planned touch down
InSight	USA	May 2018	Fixed landing platform—landed on 26 November 2018 in the near-equatorial Elysium Planitia (4.5° N, 135° E)
Mars 2020 Perseverance	USA	July 2020	Successful landing of rover by "sky-crane" From the rover deck, a helicopter (Ingenuity), which performed first powered flight on Mars, has been deployed
Tianwen 1	China	July 2020	Orbiter and lander with rover (Zhurong). Successfully landed on 14 May 2021 in Utopia Planitia
Venus			
Venera 4	USSR	Jun 1967	First successful atmospheric probe at Venus. Transmitted data down to an altitude of about 25 km

(continued)

Table 27.1 (continued)

	Country/agency	Launch date	Comment
Venera 7	USSR	Aug 1970	First successful landing on Venus. Transmitted data till 23 min after touch-down
Venera 8–14	USSR	1972–1981	Successful series of landers transmitting data from the Venusian surface
VeGa 1 & 2	USSR	Dec 1984/Jun 1985	VeGa 1 and 2 delivered successfully Landers (and balloons) to Venus and then continued their flight to comet Halley
Pioneer Venus 2	USA	Aug 1978	One large and three small entry probes to study Venusian atmosphere. (One survived impact and continued to transmit data for about 67 min.)
Jupiter			
Galileo Probe	USA	Oct 1989	Entry probe to study the Jovian atmosphere
Titan			
Huygens	ESA	Oct1997	Part of the joint NASA/ESA Cassini/ Huygens mission to the Saturnian system. Huygens entered the atmosphere of Titan in 2005 and sent data during descent; kept transmitting data after touchdown
Asteroids			
NEAR-Shoemaker	NASA	Feb 1996	Spacecraft not designed for landing. However, at the end of the orbiter mission NEAR could be safely placed at the surface of asteroid (433) Eros and continued to transmit data for about 16 days
Hayabusa (Minerva)	Japan	May 2003	Successful sample return mission from asteroid (25143) Itokawa Probe took samples by "touch & go" strategy. Attempt to deliver small Minerva lander failed

(continued)

Table 27.1 (continued)

	Country/agency	Launch date	Comment
Hayabusa 2 (MASCOT, Minerva II)	Japan	Dec 2014	Sample return mission to asteroid (162173) Ryugu. Sampling by "touch & go" in 2019. Samples successfully returned to Earth in December 2020 Successful delivery of two Minerva landers and MASCOT, a 10 kg lander provided by DLR and CNES Mission extended to rendezvous with asteroid $1998KY_{26}$ in 2031
Comet			
Rosetta Lander, Philae	DLR/CNES/ASI, ESA	Mar 2004	Mission to comet 67P/Churyumov-Gerasimenko Philae landing: 12 November 2014 End of mission Rosetta orbiter touch-down: 30 September 2016

The landing of ESA's Schiaparelli test lander was also observed in real-time from two orbiters: Mars Express, which managed to record and transmit the UHF RF carrier of the lander in real-time to ground during the descent; a ground radio telescope, the Giant Meterwave Radio Telescope (GMRT) near Pune, India also detected and recorded the UHF RF carrier in real-time; finally the ExoMars TGO orbiter itself recorded the lander radio signal including its telemetry during descent, and transmitted it later to ground, after completion of its Mars Orbit Insertion maneuver. For the post-landing phase, also other NASA Mars orbiters were scheduled to pick up the Schiaparelli signal. Unfortunately, this never materialized due to the landers crash onto the surface of the planet.

These two examples show how essential international cooperation is to enhance the monitoring capabilities of the Mars landing phase. Due to the limited communications resources, the use of all available orbiters and other ground observing facilities can make the difference between landing in the blind and full recovery of precious data over the entire entry descent and landing phase.

27.2 Lander Release

The delivery of a lander to actually descend to the surface of the target body or to enter its atmosphere can be performed by either direct insertion from the incoming hyperbolic trajectory or via a set of maneuvers, in which the spacecraft first orbits the target body prior to insertion.

The latter is energetically less favorable—particularly for planets with an atmosphere—however, it relaxes operational constraints on timing, sensitivity to possible anomalies or landing site selection/verification.

A release from orbit is the necessary choice in case the surface of the target body is not known well enough to pre-select the landing site before launch or during the cruise. For instance, the NASA Viking landers in 1976 arrived at Mars attached to the Orbiter. The assembly orbited the planet for about one month before the Landers were released and a de-orbit burn initiated the EDL (entry, descent and landing) phase. The orbit phase was used to map the planet and finally select the landing sites, which indeed resulted in a change of both originally foreseen landing sites, since the analysis of the orbiter images proved that those sites were unsafe. The Chinese Tianwen 1 mission which arrived at Mars orbit in February 2021 followed a similar approach. The lander, containing a rover, Zhurong, has been successfully delivered to the surface of Mars in May 2021.

Also, the Rosetta-Lander Philae was released from the ESA's Rosetta orbiter from an orbit around comet 67P/Churyumov-Gerasimenko after a comet nucleus mapping and characterization phase of several months, performed with the Rosetta orbiter scientific instruments (Ulamec et al. 2015).

In case of MASCOT or Minerva (the lander modules delivered by Hayabusa 2), the release took place at very low altitude (about 50 m) from a hovering mother spacecraft. This relaxed the requirements on the lander, but it dramatically increased the demand on guidance, navigation and control (GNC) of the delivering spacecraft. Hayabusa 1 and 2 have been designed to be able to operate autonomously very close to the surface, as these are sample return missions able to perform touch & go maneuvers for surface sample acquisition. After release, MASCOT fell to the surface in a planned uncontrolled manner, where it had to reorient itself upright in order to perform the scientific measurements.

The option to perform the landing via the incoming hyperbolic trajectory is more often chosen. Here, the spacecraft approaches the planet on a collision course. The lander module is released while still flying on this trajectory, and shortly after either the carrier spacecraft enters the atmosphere and is destroyed or, if the carrier spacecraft is also an orbiter, it performs the orbit insertion maneuver. In the latter case, more complex from an operational point of view, the relative timing of the lander module release and the orbit insertion of the remaining spacecraft is a crucial factor and a fundamental parameter in the overall mission and spacecraft design. In order to ensure the success of landing and orbit insertion operations also in case of contingencies, it is required to have at least a few days between the release of the lander and the start of the orbit insertion maneuver. This is due to the fact that in between, the carrier/orbiter spacecraft has to perform precise orbit determination, and also to take into account that the release operation may perturb the carrier/orbiter attitude and even cause safe mode triggering. The time for recovery from such contingency has to be accounted for planning of the orbit insertion maneuver.

27.3 Various Landing Strategies

The exact landing strategy to be applied for a surface element is strongly depending on the actual target body as well as the particular mission requirements. Entry, descent and landing (EDL), obviously, is very different in case the planet has a significant atmosphere, where, e.g., a heat shield and parachutes can be utilized, compared to airless bodies like the Moon. Small bodies like comets and asteroids are another dedicated category due to the low gravitational forces and the low velocities associated.

27.3.1 Entry, Descent and Landing Through an Atmosphere

Landing on bodies with a significant atmospheric density like Mars, Venus or Titan allows deceleration by atmospheric entry, using the drag of the atmospheric gases on the surface of the landing module. In this phase the spacecraft has to operate autonomously, since no interaction with ground is possible on the required timescales. Usually no autonomous back-up mode is designed for this phase, as recovery from major anomalies is not possible during this phase. Once released, the entry, descent and landing (EDL) module is ballistically approaching the atmosphere of the planet. The start of the EDL phase is defined as the moment in which the module senses the atmospheric drag, called Entry Interface Point. From this moment onwards the spacecraft controls autonomously its attitude (keeping the heat shield in the direction of flight) and the sequence of EDL operations, normally based on accelerometers first and radar altimeters later.

During the first phase of the descent after having entered the atmosphere the spacecraft is protected by a heat shield, which maximizes the aerodynamic drag and at the same time protects the lander from the aerodynamic pressure and the thermal load. The atmospheric drag onto the heat shield removes the majority (for instance on Mars about 90%) of the dynamical energy of the EDL module, which decelerates to a speed that allows to release the heat shield and to open a parachute (or a set of parachutes), to gradually further reduce the module's velocity. Parachutes used today for spacecraft landing on Mars can sustain supersonic speeds (up to Mach 2.2).

The final phase of the EDL scenario can use different methods, depending on the type of target body, of lander and of operations to be conducted on the surface. For landing on Mars the first successful landers used retro-rockets that slowed down the landing module during the last hundreds of meters and allowed a controlled touch down on landing legs. This was the case, e.g., of NASA's Viking landers in 1976, or the more recent Phoenix and InSight landers in 2008 and 2018, respectively. The disadvantage of this landing technique is that the lander remains on the landing spot, which may also be contaminated or at least affected by the thrust of the landing rockets.

For the Pathfinder mission, which delivered the small rover Sojourner on the surface of Mars in July 1997, NASA successfully utilized a new landing technique, based on airbags that are deployed around the lander after the parachute phase and a short burn of retro-rockets to reduce the descent speed to a few m/s. The lander module bounces on the surface until it stops; then the airbags are deflated and the protecting shell is opened. After establishment of the initial contact and a rapid systems checkout, the rover is activated and can egress the landing platform and move on the surface of the planet. This technique was further successfully used for the landing of the two Mars Exploration Rovers Spirit and Opportunity in the first decade of this millennium. The airbag technique is not practical for a fixed platform like the one of Viking or Phoenix, since the deflated airbags remain on the terrain underneath the lander shell and prevent local access to the ground for investigations. On the other hand, a landing on feet like Phoenix is not practical for rovers, since a complex egress system would have to be mounted on top of the landing gear to allow the egress of the rover after landing. Nevertheless, such approach can be adopted in case the mission includes both a fixed landing platform and a rover module. This is currently the baseline for the joint ESA-Russia ExoMars mission that is scheduled to land on Mars in June 2023 (ESA 2021c).

Finally, a new technique has been successfully employed by NASA in the landing on Mars in 2012 of the Mars Science Laboratory called Curiosity. The technique, called "sky crane" involves the use of retro-rockets to control the descent and stop the lander module at an altitude of a few tens of meters above the surface. While the platform is hovering above the surface using the retro-rockets, the rover itself is lowered with a cable, like from a crane. Once it touches the surface, the cables are cut and the sky platform flies away and crashes at a distant place on the surface. The rover Curiosity landed on its wheels and was ready for its first movements within a very short period (NASA 2014c).

This very successful strategy has been repeated in 2021, when the Perseverance rover landed at Jezero crater (NASA 2021b). An important addition to that landing concept for Perseverance was the utilization of a technique called "Terrain-relative navigation" for the very first time. Once the heat shield was ejected, exposing the rover to the Mars environment, a special camera compared the images of the terrain below to a map constructed from orbiters imaging and loaded on-board the rover. Using this technique, the rover has the capability to determine where it is heading and to steer the direction of landing to the safest place in the landing ellipse (NASA 2021a) (Fig. 27.1).

The thin atmosphere of Mars is per se a major challenge in the design of landing probes, since even small variations in its density can have dramatic effects on the efficiency of the two main tools used to dissipate orbital energy during the descent, the heat shield and the parachutes. For this reason, the altitude of the landing site and the seasonal and meteorological conditions of Mars at the time of landing play a major role in the mission and spacecraft design.

Landing on a body with a denser atmosphere is comparably simpler, as the variability of the atmospheric density plays a smaller role. For instance, the atmosphere

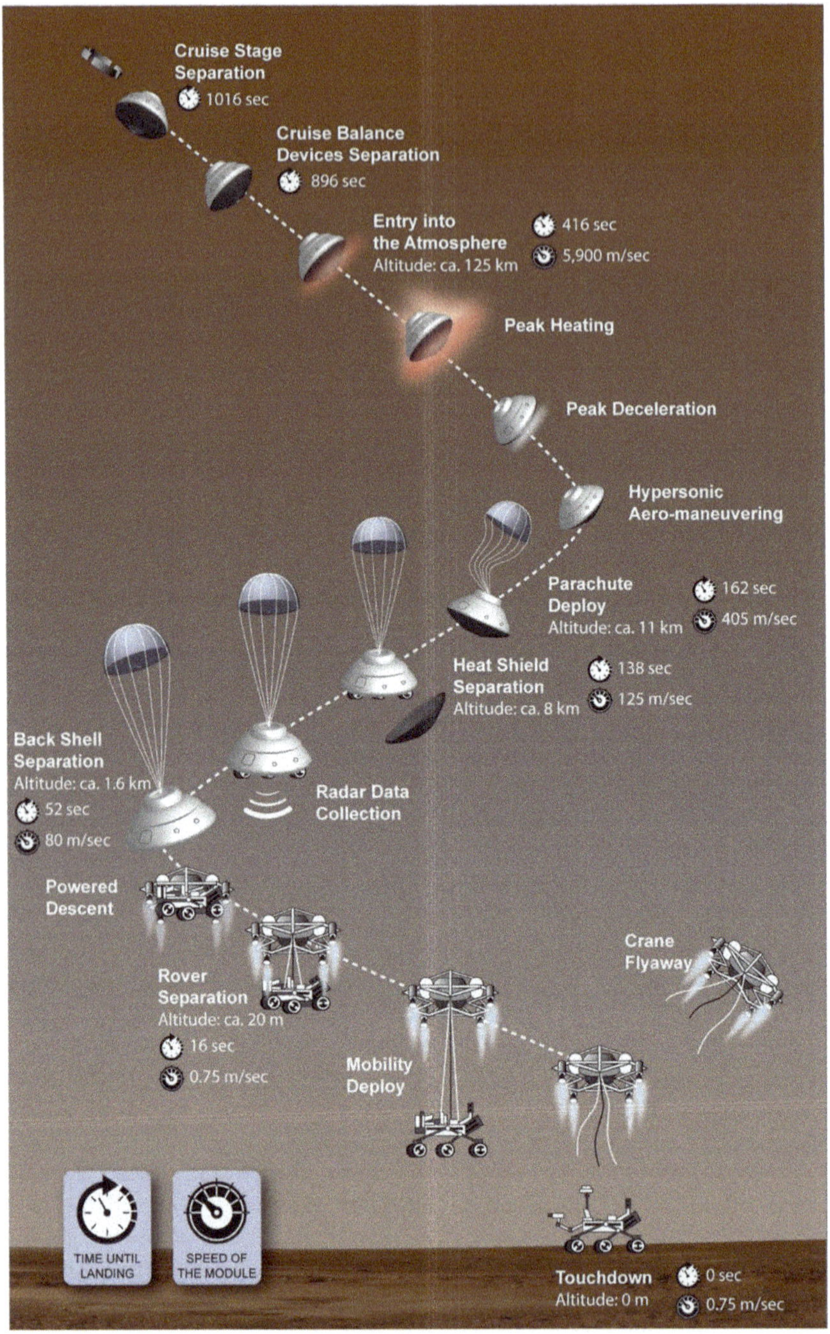

Fig. 27.1 Scheme of NASA MSL-Curiosity landing

of Venus is so dense that no retro-rockets are required for soft landing. Surface pressure at Venus is about 90 bar (mainly CO_2). Parachutes (or even a heat shield only) were sufficient to lead to surface impact velocities in the range of 7 m/s (as for Venera 9), which can be damped by the landing system.

The first successful probe to enter an extraterrestrial atmosphere was Venera 4 in 1967. After the lander was separated from its transfer vehicle it transmitted data of the Venusian atmosphere for about 96 min, until the batteries were empty while the descent module was still at an altitude of about 25 km: in this case the limited knowledge of Venus atmospheric density did have an impact on the mission: the density was estimated much lower than the one experienced by the descending spacecraft, resulting in an unexpectedly long descent and the fact that the surface could not be reached [the capsule either crashed or ran out of battery power (Wesley et al. 2011)]. All operations were performed fully autonomously.

Later, the design of the next Venera landing mission was adjusted to the correct properties of the atmosphere and Venera 7 performed a first successful landing on Venus in December 1970. After a descent of 35 min the probe touched down on the surface of Venus and continued to send data for another 23 min. The Soviet Union continued the successful series of Venera spacecraft, keeping the same basic concept but applying modifications in the design and improvements of the payload (e.g., Venera 9 sent for the first-time pictures from the surface in 1975).

The first Venera probes were transmitting data directly to Earth during descent and, in case of Venera 7 and 8, also after landing. Venera 9 to 14 included orbiters which were used to relay data from the landers. In any case ground interaction was not planned for all Venera landers after entry had started (Fig. 27.2).

Fig. 27.2 Venera 9 lander (*image* NASA/Lavochkin)

As in case of Venus, the density of the atmosphere of Saturn's largest moon, Titan, is high enough (1.6 bar at surface, mainly N_2) to allow the soft landing just with the help of a heat shield and a series of successive parachutes, as demonstrated successfully by the ESA Huygens probe in January 2005. The main purpose of the probe was to perform atmospheric investigations, and therefore it was not designed to survive a landing on the surface. Nevertheless, the probe touched down onto the surface of Titan softly enough to allow it to survive undamaged, and to continue operating as long as its on-board batteries provided power, delivering pictures and measurements from the surface. All the telemetry data were relayed via the NASA Cassini spacecraft, which had transported the Huygens probe to the Saturn system and had released it a few days earlier. The radio carrier signal of Huygens was also received directly on Earth by radio telescopes, which in real time measured its Doppler shift, allowing not only the detection of the successful landing but also complementing the atmospheric investigations done by the on-board instruments by measuring the velocity and acceleration profiles of the probe during the descent, thereby providing precious information on the winds field in Titan's atmosphere at different altitudes (ESA 2021a).

27.3.2 Landing on Large Airless Bodies: The Moon

Of course, landing on a large object with no atmosphere, like the Moon, is different, as neither a heat shield nor parachutes can be used, and the entire orbital energy has to be dissipated with retro-rockets. In comparison with Mars or Venus, objects without atmosphere have a lower gravity, so the amount of energy to be dissipated is comparatively small.

The first successful soft landing on the Moon took place on February 3rd, 1966 by USSR's Luna 9. After direct approach to the Moon the landing sequence was initiated at a distance of about 8000 km. A radar altimeter triggered the release of the cruise modules and initiated the breaking maneuver at an altitude of 75 km. Shortly before touch-down, a 5 m long boom detected contact to ground and caused the landing capsule to be ejected upward. The capsule eventually hit ground with a touchdown velocity of 15 m/s, the impact energy was damped with an airbag (Wesley et al. 2011).

After a few bounces Luna 9 came to rest, the airbag was released and four petals which covered the top half of the spacecraft were opened, thus orienting the probe on the Lunar surface. Figure 27.3 shows an image of the Luna 9 lander with open petals. The television camera was combined with a mirror system, which operated by rotating and tilting and provided panoramic images from the Lunar surface. Four panoramic views were transmitted during seven radio sessions, with a total transmission time of about 8:05 h.

The probe could be commanded from Earth and survived about three days on the surface of the Moon (allowing images with different solar aspect angle) until the batteries were depleted. It is interesting to note that the signals used the

Fig. 27.3 Luna 9 with unfolded petals and deployed antennas (*image* NASA/Lavochkin)

internationally-agreed system used by newspapers for transmitting pictures (FAX); the Jodrell Bank Observatory in England received the data and indeed, the images were first published by the Daily Express (and only one day later by Prawda). Luna 9 was followed by a very similar device with an enhanced payload, Luna 13.

Later Soviet Lunar Landers included automatic Sample Return (Luna 16, 20 and 24), even a 2.25 m depth drill-core (Luna 24) and the delivery of Rovers (Lunokhod 1 and 2, see Sect. 27.4.1). Those probes used a standard bus with retro rockets for breaking before touch-down. The landing sequence ran automatic for all these missions.

In preparation of the manned Lunar program by NASA the unmanned Surveyor landers were sent to the Moon between 1966 and 1968. Five successful landings could be performed, the probes returned a plentitude of images and scientific data back to Earth. Apollo 12 landed very close to Surveyor 3, allowing a "visit" and the return of the probe's camera back to Earth (see Fig. 27.4).

Fig. 27.4 Apollo 12 mission commander Pete Conrad at Surveyor 3; in the background the Lunar Module. (*image* NASA)

The Surveyor spacecraft approached the Moon on a direct trajectory and slowed down to about 110 m/s by a solid fuel retro rocket, which was jettisoned 11 km from the surface. The remaining descent was controlled by a Doppler radar and three vernier hydrazine engines. At an altitude of 3.4 m above ground the thrusters were switched off and the landers descended in free fall leading to an impact velocity of about 3 m/s (Ball et al. 2007).

With the successful landing of Chang'e 3 in December 2013, China became the third nation to successfully place a Lander on the surface of the Moon. The landing module carried a small (120 kg) rover, Yutu, which rolled on the lunar surface and took images. Chang'e 4, based on the design of Chang'e 3 and carrying the rover Yutu 2, was successfully landing in the South Pole-Aitken Basin in January 2019 and became the first device to land on the Lunar far side. The required relay orbiter (Queqiao) had already been launched in May 2018 and placed in a halo orbit around the Earth-Moon L2 point, to allow communications between Lander and Earth (Li et al. 2019). The next Chinese lander mission, Chang'e 5, landed in Oceanus Procellarum in December 2020. It successfully returned about 1,73 kg of lunar samples back to Earth after the ascent vehicle docked with the orbiter (Chang'e 5-T1), which then initiated the return trajectory. Further Chinese lander missions (Chang'e 6 and 7) are planned to be launched in the 2023/24 timeframe.

In this chapter we deal with the operations of robotic surface elements and, thus, do not explain the operations in the frame of the manned Apollo missions. However, the ALSEP package is worth mentioning in this context. ALSEP (Apollo Lunar Surface Experiment Packages) was installed by the Apollo astronauts near the landing sites, consisting of various individual units (including a Central Station containing a data management system, the communications unit and an antenna pointed towards the Earth by the astronauts, as well as a variety of instrument packages like, e.g., seismometers or particle detectors) all powered by a radioisotope thermoelectric generator (RTG). Those ALSEPs were operational for several years until turned off for budgetary reasons in 1977.

27.3.3 Landing on Asteroids and Comets

Landing on small objects is a very different process. Due to the extremely low gravity, the landing is just a matter of imposing a small difference in the velocity of the lander compared to the one of the asteroid or comet. The difference is of the order of cm/s, so the descent phase can last hours. The difficulty is the navigation in the proximity of the small body, and the required extreme accuracy of the involved maneuvers, as described in more detail in Chap. 26. Another difficulty is, once touched down, to avoid that the landing module bounces back into space.

So far, landing on small bodies has been attempted only a few times (Ball et al. 2007; Biele et al. 2015; Ulamec et al. 2016; Jaumann et al. 2019). In the late 1980ies the Soviet twin Phobos missions to Mars carried landers and hoppers which, unfortunately, due to early loss of the mother spacecraft, could not be delivered to the

Martian moon. Japan's Hayabusa (1) spacecraft tried to place a small lander, Minerva, onto asteroid Itokawa, but the maneuver failed and the lander missed the asteroid. However, NASA's Shoemaker-NEAR spacecraft, which orbited asteroid Eros and originally was not designed for landing, could be maneuvered to touch down on the surface of the asteroid at the end of its operations phase. This maneuver was successful and the spacecraft managed to send a weak signal from the surface to Earth for about 16 days (Dunham et al. 2002).

Rosetta and Philae's first ever landing operations on the surface of a comet's nucleus in November 2014 were even more complicated by the fact that the surrounding environment of the nucleus was much "dirtier" in terms of dust and gas compared to the one of an asteroid. This made the navigation for the landing phase particularly challenging, requiring an accurate modelling of all the forces acting on the spacecraft in this very dynamic environment (ESA 2021b). Once reached the very low gravity surface of the comet, Philae's landing system included a mechanism to shoot anchoring harpoons into the surface of the comet once its touch down is detected by sensors within the landing gear. Unfortunately, this system did not work, the harpoons did not fire and Philae bounced a few times onto the surface, traveling for two hours a distance of about 1 km before finally coming to rest. Fortunately, the damping system in the landing gear managed to eliminate part of the landing energy (landing velocity was just below 1 m/s) so that the bouncing velocity was lower than the escape velocity from the tiny gravitational field of the nucleus. Even more importantly, notwithstanding the tumbling and bouncing Philae managed to maintain a constant radio link with Rosetta, which was hovering over the surface at an altitude of a few tens of km. This allowed the continuation of the scientific operations of the lander for almost 3 days, until the energy of the batteries was fully consumed. In the end, Philae's mission was very successful, and made it the first real lander onto the surface of a small body.

After the success of Hayabusa to investigate an S-type (Barucci et al. 1987) asteroid, (25143) Itokawa, and to return samples to Earth, it was decided to send an improved version of the spacecraft to a primitive (C-type) asteroid for comparison and a better understanding of the early history of the solar system. The Hayabusa 2 spacecraft, launched in December 2014 to asteroid (162173) Ryugu, carried additional Minerva rovers as well as the "Mobile Asteroid Surface Scout", MASCOT, a small lander with the capability to upright itself with an internal torquer, after being released from about 50 m above the surface of the asteroid and an uncontrolled descent. Hayabusa 2 can hover above the surface and was also touching the surface during two maneuvers for sample acquisition. MASCOT was delivered in October 2018 and provided scientific data for about 17 h, until the batteries ran empty (as foreseen) (Jaumann et al. 2019). MASCOT is an example of a small and less complex system, as compared to, e.g., Philae, which proved to be successful in delivering images, thermal and magnetic data during two asteroid days. Hayabusa 2 successfully returned samples to Earth in December 2020, when the re-entry capsule descended at the Woomera Test Range in Australia (JAXA 2021b). The mission, however, has been extended and the spacecraft is now targeted for a flyby at asteroid $2001CC_{21}$ in 2026 and a rendezvous with $1998KY_{26}$, a fast rotator, in July 2031.

The largest difficulties described above, compared to landing on large moons or planets, are at least partially compensated, in case of small bodies, by a few advantages. First of all the landing systems do not require complex and heavy heat shields, parachutes, or retro-rockets. This is clearly a major advantage in terms of mass but also in terms of mission risk and complexity of the on-board autonomy. Another advantage is that, due to the slow relative velocity between the landing module and the target body, the landing trajectory and scenario can in general be planned such that a back-up opportunity can be guaranteed in case of last-minute contingencies that force an abort of the landing operations.

27.3.4 *Penetrators*

A special case of surface elements are penetrators (e.g., comet rendezvous asteroid flyby - CRAF), hitting ground with high velocity and being decelerated in the surface material. Penetrators have been proposed for several missions, including applications on Mars, (e.g., Mars 96 or Deep Space 2), the Moon (e.g., Lunar-A, MoonLite) or asteroids (e.g., CRAF). None of these missions has been realized successfully until now (Lorenz 2011). The advantages of such devices are the possibility to realize a relatively small and lightweight design, usually no breaking maneuver before touch-down is required and the fact that the sub-surface can be reached by definition. However, a disadvantage is the high acceleration load during impact (typically in the 10^5 m/s^2 range) that makes it challenging to implement sensitive payload elements. Another problem arises due to the uncertainty in the knowledge of the surface properties, which causes difficulties to predict the actual penetration depth. From an operational point of view, penetrators suffer from the limited possibilities to apply high quality communications systems. A summary of proposed planetary penetrator missions is given by Lorenz (2011).

27.4 Surface Operations

Operations of a landed platform are challenging since in addition to the sometimes hostile environment additional geometrical aspects have to be considered as compared to orbiters.

For instance, is the visibility not only depending on the availability of terrestrial ground stations but also on the rotational state of the target body, the particular landing site, as well as the relative position of a relay orbiter. In addition, day/night cycles have to be taken into account.

27.4.1 Operations of Stationary Surface Elements

The Viking landers were equipped to transmit information both directly to Earth with an S-band communications system, or, alternatively relayed via the orbiter with a UHF relay system. The landers also received Earth commands through the S-band system.

One of two redundant receivers used a 76 cm parabolic high-gain antenna that required pointing to the Earth. The second receiver used a fixed low-gain antenna to receive Earth commands. The UHF system was used almost exclusively during entry and the first days after landing.

The on-board computer had stored in its memory the sequence of the first 22 days after landing thus allowing autonomous operations without commanding from Earth. These sequences were updated as the communications link has been established.

Viking 1 eventually was lost in 1982 due to a faulty ground command (a software update intended to improve battery capacity erroneously overwrote the antenna positioning software) after 2245 sols (Martian days) of operations. Viking 2 ceased operations in 1980 due to a battery failure.

The NASA Phoenix lander descended to the polar region of Mars in May 2008. The landing site is at altitude of 68° North. The probe was operational during Northern summer from August to November 2008, when, eventually, solar input became insufficient. In winter, the Martian polar ice-cap was covering Phoenix. Any attempts to re-activate the lander in the following spring, in 2010, were unsuccessful.

Phoenix used UHF links, relayed through Mars orbiters during the entry, descent and landing phase and while operating on the surface, whereas during cruise an X-band system was used. The communications system on the lander used the Proximity-1 CCSDS protocol as implemented for most Mars surface missions, relaying data via orbiters and was compatible with the relay capabilities of Mars Odyssey, Mars Reconnaissance Orbiter (MRO) (both NASA) and Mars Express (ESA). This international standard is extremely useful as it allows communicating between orbiters and landers across all missions, which had implemented it, no matter which space agency they belong to.

During its 155-sol operations on Mars, the probe received regular commanding and performed a plentitude of scientific experiments, including the collection of soil samples with a grabber and analyses with a thermal evolved gas analyzer (TEGA) (NASA www.jpl.nasa.gov 2014).

InSight, a NASA mission to study the internal structure of Mars, e.g., by deploying a seismometer as well as a heat probe, landed in November 2018 in Elysium Planitia (NASA 2019). The lander system is based on Phoenix and also the communications strategy is similar.

In general, for any surface vehicle on the surface of Mars the overflights by orbiters which fly on a low circular orbit, such as NASA's MRO and Odyssey, or ESA's ExoMars TGO, are very short (of the order of minutes). During these contact periods the orbiter and lander establish the link automatically and after the lander has verified the link handshake procedure, it starts to send the data recorded by the lander during the previous non coverage period. NASA's Maven or ESA's Mars Express' elliptical orbits in principle allow longer visibility periods, but the range of distances from the planet is large, so that relay services can only be provided from the part of the orbit which is closer to the surface, due to limitations in the range of the UHF signal of the transmitters and receivers. Also, the spacecraft have to implement complex attitude profiles in order to point the UHF antenna in the optimal direction of the lander on the surface, in order to increase the efficiency of the radio signal.

The Rosetta Lander, Philae, was delivered to 67P/Churyumov-Gerasimenko applying a preloaded sequence that could be finally defined only after arrival of Rosetta at the target comet and a dedicated comet mapping and characterization phase which lasted more than two months (Ulamec et al. 2015). The instruments aboard the orbiter (most prominently the scientific camera, OSIRIS) allowed the production of a DTM (digital terrain model) of the comet nucleus, to determine the mass of the comet (including several higher harmonics of the gravitational field) and to characterize the gas and dust environment of the nucleus, together with its dynamical properties. This information was essential to allow the creation of a "comet engineering model", which was used in the navigation process of the Rosetta orbiter, to allow it to orbit around the nucleus at distances close enough to map its surface with sufficient resolution and to achieve an orbit from which the lander Philae could be released. Once initiated, the automatic sequences both on-board the Rosetta orbiter and the Philae lander had to run fully autonomously. The landing took place at a heliocentric distance of 3 AU (about 450 million km). Due to the even larger geocentric distance, the RF signal travel time to and from Earth was about 28 min one-way. During its initial bouncing on the surface until it came to rest, Philae already started a so called first science sequence (FSS), based on the use of a primary battery and lasting for about 60 h. After the end of the first contact with Rosetta there was a planned interruption of communications of about 9 h due to the nucleus rotation. This allowed the Philae operators to reassess the new situation and to prepare updates to the automatic sequence, which was uplinked to Rosetta during the period of no contact with Philae. When Rosetta appeared over the nucleus' horizon as seen by Philae, the lander initiated the two-way communications again and received and started executing the new sequence of commands. In parallel it started sending to Rosetta the scientific data recorded during the non-coverage period. This cycle of operations continued for two more days, after which the energy of the primary battery of the lander was used up and Philae switched itself off. Unfortunately, the location in which Philae had come to rest on the surface did not receive enough sunlight for its secondary battery to be re-charged by the solar cells mounted on the surface of the lander body. Also, in the dark of the comet's surface the temperature was too low to allow the charge–discharge electronic to work.

Philae managed nevertheless in the about 60 h on the surface to carry out most of its foreseen scientific measurements as part of the primary scientific sequence. In June 2015, 7 months after landing and closer to the sun, Rosetta again received signals from the Lander. Although in the following days several radio contacts could be established, due to the high activity of the comet the Rosetta probe could not be flown closer than 150 km from the nucleus, and at this distance the radio link was

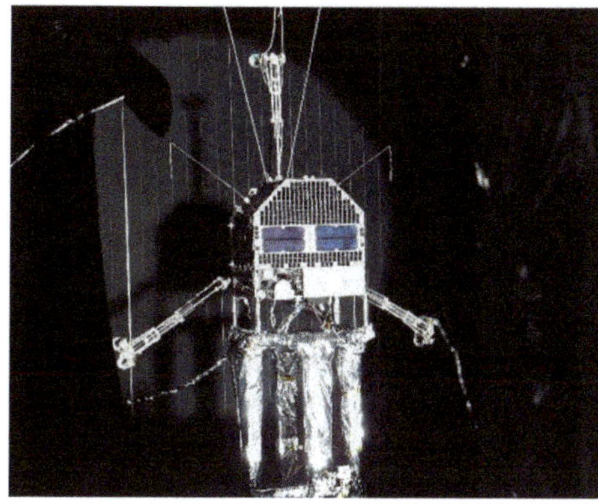

Fig. 27.5 Rosetta lander during thermal vacuum tests in chamber at IABG

too weak. Therefore, it was not possible to command another science sequence and only housekeeping data were received on-board Rosetta, before Philae eventually stopped transmitting.

Philae was operated by a Lander Control Center (LCC), located at DLR in Cologne/Germany, and a Science Operations and Navigation Center (SONC) which is located at CNES in Toulouse/France. All telemetry and telecommanding was linked via the Rosetta Mission Control Center at ESA/ESOC in Darmstadt/Germany, where also the responsibility for the overall Rosetta mission incl. the Lander delivery lied.

All Philae operations were planned and verified by using a Ground Reference Model representing the Flight Model, consisting of flight spare units and dedicated simulators. Figure 27.5 shows the Flight Model during qualification.

27.4.2 Rover Operations

One special case of operating surface elements is the one of moving devices as rovers. So far, rovers have been operated at the Moon (including those driven by astronauts in the frame of Apollo 15–17) and Mars. The small bodies hoppers Phobos PROP-F and Minerva are sometimes referred to as "rovers" as well; both could not be delivered successfully. MASCOT is a small surface package, provided by DLR and CNES to the Japanese Hayabusa 2 asteroid mission and could perform two relocation maneuvers (small "hops" of about 70 cm). In principle, also larger hops could have been performed, however, it was preferred to use the limited available time on the surface by performing science measurements rather than in performing longer hops, which in such a low gravity environment can take very long. The first successful extraterrestrial rovers were Lunokhod 1 and 2 at the Moon, launched in 1970 and 1973 in the frame of the Luna 17 and 21 missions. These devices could survive the Lunar night (thanks to radioisotopic thermal heaters), resulting in an operational lifetime on the surface of the Moon of eleven and four months, respectively.

The rovers were controlled directly from Earth by teams of five persons operating in two-hour shifts. Each crew included the commander, a driver, a navigator, a flight-engineer and an antenna-operator. There was no automated mode. The wheels of the rovers could be controlled individually, there was also an automatic fail-safe device preventing movement over extreme slopes or overheating of the motors. The control teams operated and steered the rovers by viewing through a pair of video cameras, transmitting images with a frequency of one frame per 20 s in case of Lunokhod 1 and adjustably up to one frame every 3.2 s in case of Lunokhod 2. The image rate together with the signal travel time (about 1.3 s, one-way) made operations challenging. Lunokhod 1 had only one driving speed (0.8 km/h) while Lunokhod 2 could be driven with two velocities (0.8 and 2 km/h) (Wesley et al. 2011; Aleksandrov et al. 1971). The rovers drove through rough terrain, crossed valleys and craters. Lunokhod 2 was lost when driving accidentally into a shadowed crater and scraping its lid during recovery. This eventually resulted in dust on the radiator and solar generator and consequently caused the end of the mission (Figs. 27.6 and 27.7).

On Mars, so far, six Rovers have been operated (not counting a Soviet walking robot in 1972, which never transmitted any data). The first such device, Sojourner, was part of the Mars Pathfinder mission, launched in December 1996. Sojourner had

Fig. 27.6 Lunokhod control room (*image* Aleksandrov et al., 1971)

Fig. 27.7 Lunokhod 1 (*image* NASA/Lavochkin)

Fig. 27.8 Sojourner on Mars at "Yogi rock" as imaged by central station (*image* JPL/NASA)

six wheels and a mass of 10.6 kg, it was equipped with three cameras in addition to an alpha/x-ray fluorescence (APX)-spectrometer for rock analyses. Communications with the lander were via UHF link. The rover drove off the landing platform two days after touch-down and travelled an overall distance of about 100 m but was never away more than 12 m from the station. Maximum speed was 1 cm/s (Fig. 27.8).

The next big step in terms of mobility on Mars was the Mars Exploration Rover (MER) mission including two rovers, Spirit and Opportunity. As mentioned above, they were landed with an innovative airbag system (NASA MSL Science Corner 2014). After touchdown in Meridiani Planum and inside Endurance crater, respectively (at sites where sediments and proof for early liquid water was expected and indeed found), a protective tetrahedron opened, bringing the rover in an upright position. The rovers, at the landing platform got activated and unfolded their solar generators and explored an area of many kilometers. Designed for a lifetime of about 60 days (assuming dust coverage on the solar cells) Spirit operated over six years and Opportunity for even 15 years.

After landing, Spirit did not send any science data, however, a signal could be received via Mars Global Surveyor. The problem turned out to be related to the flash memory and could be solved by reformatting it and updating the on-board software. Spirit was fully in science operations about one month after touch-down, the software update was also applied for Opportunity.

The egress phase including opening of the shell, rover stand-up, deployment of the pan-cam mast and analyzing images to select the optimum egress pass before finally driving off the platform took the first four days.

The rovers could drive with a maximum speed of 5 cm/s, however, hardly more than about 1 cm/s (on flat ground) was reached due to hazard avoidance protocols.

Both MER rovers provided a wealth of scientific results; they also demonstrated the value of moving platforms allowing detailed exploration of an area rather than a single landing spot. They operated long beyond their originally planned mission lifetime. Spirit was active until March 2010. Opportunity operated until June 2018 and holds the record for the longest distance driven by a rover vehicle on another celestial object (about 45 km).

A much larger rover operated on Mars, Curiosity, with a mass of about 900 kg, has landed successfully within Gale Crater, in August 2012 (NASA 2014b–d). After being separated from the sky-crane (see above), the system was checked and the first

Fig. 27.9 "Self Portrait" of the Curiosity Rover at the surface of Mars in Gale Crater, taken with Mars Hand Lens Imager (*image* NASA PIA16239)

images have been transmitted to Earth. Curiosity uses two communications systems: science data are primarily transferred with a UHF system (about 250 Mbit/d), while for commanding there is an X-band system available. The rover is powered by an RTG providing 125 W electric power (beginning of life). The primary mission was planned for one Martian year (i.e. 687 Earth-days), but the mission was extended indefinitely and is still operating (status May 2021) (Fig. 27.9).

In the first three months of Curiosity's operations on Mars, the operations teams were working according to "Mars time". Since a day on Mars (sol) is about 24:40 h, work shifts are not easily correlated with comfortable working hours on Earth. However, since November 2012, the teams are working in shifts linked to Earth time, with the majority of the activities from 8:00 to 20:00 local (JPL)-time. Science teams are now able to work from their institutes, while co-location was required in the first phase of the mission. For more details on science planning and operations see, e.g., NASA (2014e), MSL Science Corner.

On February 18, 2021, the Perseverance rover (with a mass of about 1026 kg even larger than Curiosity) has been landed with the same concept of a "sky crane" at Jezero Crater, a site of particular astrobiological interest as it is believed to having contained a crater lake about 3,8 billion years ago.

Entry, descent, and landing have been documented with a number of cameras, that captured e.g. the jettisoning of the heat shield, parachute deployment and sky-crane maneuvers.

The Perseverance Rover also carried a small (about 1.8 kg, rotor span about 1.2 m) helicopter, Ingenuity, which became the first device ever to actively fly in an extraterrestrial atmosphere via aerodynamic forces. It is able to fly fully autonomously. The first flight took place 19th of April 2021 (NASA 2021c).

The Curiosity and Perseverance Rovers are the most complex devices ever landed on Mars so far. They are mobile laboratories and will enhance our understanding of Mars considerably. The Perseverance Rover will sample Martian material and deposit it in cartouches, which, in a next step, are planned to be collected by a Fetching Rover (provided by ESA) and eventually returned to Earth.

In May 2021, the Chinese rover Zurong (god of fire) with a mass of about 240 kg has landed in Utopia Planitia. During descent and landing it was mounted to a platform with four landing legs and a deployable ramp to allow the rover to descend. Similar to NASA"s Perseverance, also this lander used optical imaging of the terrain during the final phase of the descent to autonomously select an optimal touch down site.

In the frame of the JAXA Mars Moon eXploration (MMX) mission it is planned to deliver a small (ca. 25 kg) rover to the surface of Phobos. The rover will analyze the properties of the surface material but also demonstrate locomotion with wheels on a low gravity body (JAXA 2021a).

27.5 Conclusions

For what concerns landing missions on solar system bodies, there is a general trend from short- to long-lived and from stationary to mobile surface elements. Accordingly, the operations get more and more demanding but there is also an increased emphasis on autonomy. In the far future, devices based on highly autonomous exploration strategies with a high degree of artificial intelligence, including intercommunications of a number of probes or rovers may be realized. Theoretical studies have proposed such concepts. However, certain decision points within the mission timeline, requiring human involvement, as well as careful planning, will stay essential elements to all future space missions.

Another future development will certainly involve combined operations between robotic and human missions on the surface of the Moon, asteroids or a planet like Mars. Combined operations may include operations of robotic surface elements by humans in orbit, and/or remote operations (from Earth or from orbit) of the robotic elements in presence and in support of humans on the surface.

References

Aleksandrov A et al (1971) Control and results of program. In: Vinogradov AP (ed) Lunokhod 1—Mobile Lunar laboratory. Nauka Publishers (original in Russian)

Ball A, Garry J, Lorenz R, Kerzhanovich V (2007) *Planetary Landers and Entry Probe;* Cambridge University Press, Cambridge

Barucci M, Capria MT, Coradini A, Fulchignoni M (1987) Classification of asteroids using G mode analysis. Icarus 72:304–324

Biele J, Ulamec S, Maibaum M et al (2015) The landing(s) of Philae and inferences about comet surface mechanical properties. Science 349(6247)

Dunham DW, Farquhar R, McAdams JV, Holdridge M, Nelson R, Whittenburg K (2002) Implementation of the First Asteroid Landing. Icarus 159:433–438

ESA (2021a) Cassini/Huygens mission. https://www.esa.int/Our_Activities/Space_Science/Cassini-Huygens, cited 15 June 2021

ESA (2021b) Rosetta Mission. https://www.esa.int/Our_Activities/Space_Science/Rosetta, cited 15 June 2021

ESA (2021c) ExoMars mission; https://www.esa.int/Our_Activities/Human_and_Robotic_Exploration/Exploration/ExoMars, cited 15 June 2021

Jaumann R, Schmitz N, Ho TM et al (2019) In-situ investigation of asteroid (162173) Ryugu by the Mobile Asteroid Surface Scout (MASCOT) Camera (MASCam). Science 365(6455):817–820

JAXA (2021a) Martian moons eXploration mission. http://mmx.jaxa.jp/en/, cited 25 Feb 2021

JAXA (2021b) Hayabusa 2 mission. https://global.jaxa.jp/projects/sas/hayabusa2/, cited 22 Dec 2021

Li C, Wang C, Wei Y, Lin Y (2019) China's present and future lunar exploration program. Science 365:238–239

Lorenz RD (2011) Planetary penetrators: their origins, history and future. Adv Space Res 48:403–431

NASA (2014a) Mars exploration rovers. http://marsrovers.jpl.nasa.gov/mission, cited 20 Jan 2014

NASA (2014b) Mars science laboratory—curiosity rover. http://marsprogram.jpl.nasa.gov/msl, cited 20 Jan 2014

NASA (2014c) Mars exploration rovers—mission timeline. http://marsrovers.jpl.nasa.gov/mission/tl_rover_egress_steps.html, cited 20 Jan 2014

NASA (2014d) Mars science laboratory. http://www.nasa.gov/mission_pages/msl/index.html, cited 20 Jan 2014

NASA (2014e) MSL science corner. http://msl-scicorner.jpl.nasa.gov/scienceoperations, cited 20 Jan 2014

NASA (2019) InSight mission overview. https://mars.nasa.gov/insight/mission/overview/, cited 10 Sept 2019

NASA (2021a) New technologies for mars exploration. https://mars.nasa.gov/mars2020/mission/technology/#Terrain-Relative-Navigation, cited 15 June 2021

NASA (2021b) Mars 2020 mission overview; https://mars.nasa.gov/mars2020/mission/overview/, cited 15 June 2021

NASA (2021c) Mars helicopter tech demo. https://mars.nasa.gov/technology/helicopter/, cited 15 June 2021

Ulamec S, Biele J (2009) Surface elements and landing strategies for small bodies missions—Philae and beyond. Adv Space Res 44:847–858

Ulamec S, Biele J, Blazquez A, Cozzoni B, Delmas C, Fantinati C, Gaudon P, Geurts K, Jurado E, Küchemann O, Lommatsch V, Maibaum M, Sierks H, Witte L (2015) Rosetta lander—philae: landing preparations. Acta Astron 107:79–86

Ulamec S, Fantinati C, Maibaum M, Geurts K, Biele J, Jansen S, Küchemann O, Cozzoni B, Finke F, Lommatsch V, Moussi-Soffys A, Delmas C, O'Rourke L (2016) Rosetta lander—landing and operations on comet 67P/Churyumov–Gerasi-menko. Acta Astron 125:80–91

Wesley T, Huntress JR, Marov MY (2011) Soviet robots in the solar system. Springer

Stephan Ulamec has more than 25 years of experience in the development and operations of space systems and instruments. After finishing his PhD at the University of Graz in 1991, he worked as a Research Fellow at ESA/ESTEC until 1993 and is since then at the German Aerospace Center, DLR in Cologne. Besides of his activities in system engineering and project management of the Rosetta Lander, Philae, he was engaged as payload manager of MASCOT for the Hayabusa 2 mission and is working as Science Lead for the MMX Rover Mission to Phobos. He was involved in numerous studies for in-situ packages for space research as well as the definition and performance of tests for various mechanisms to be operated on planetary surfaces.

Paolo Ferri was the long-time head of the Mission Operations Department at the European Space Operations Center (ESOC) of the European Space Agency in Darmstadt, Germany, and is now retired. After his Doctor degree in theoretical physics (University of Pavia, Italy), he started working in 1984 at ESOC as visiting scientist on the Exosat X-ray astronomy mission science operations. He moved to mission operations in 1986 and worked on different types of missions, including Eureca (microgravity, deployed and retrieved with a Space Shuttle), Cluster (space plasma physics), Rosetta (comet science), GOCE (earth gravity). In 2006, he became responsible for all ESA interplanetary mission operations, including Mars Express, Venus Express, BepiColombo, ExoMars, Solar Orbiter. Since 2013, he led the Mission Operations Department and is in charge of all ESA unmanned mission operations, preparation and execution.

Index

A
Acceleration profile, **5, 6**, 557, 572
Acceptance, **19, 48, 63**, 192, 221, 235
Access control, **56**, 165–167, 177
Acoustic noise, **5**
Acquisition of signal, 82, 200, 202, 260, 263, 368, 547, 548, 550
Actuators, 273, 274, 286–292, 294, 378, 380, 434, 435, 438, **440–444**, 446, 447, 515, 517, 537
Airbag technique, **569**
Air condition, 166, 169, 477
Alarm situation, 86
Albedo, 284, 402, 435, 523
Amplifier, 10, **36, 38, 40**, 198, 208, 458, 459, 463, 469
Amplitude, **34–36**, 101, 108, 200, 264, 289, 506
Angles only navigation, **512–515**
Anode voltage drop, 468
Anomalies, 9, 10, **58, 66–68**, 73, 76, 80, **82, 83, 85**, 86, 92, 95, **98**, 104, 105, **109**, 112, 114, 120, 122, 144, 146, 147, 154, 195, 215, 228, 232, 237, 239, 247, 248, 330, 337, 383, 385, 386, 414, 431, 444, 445, 447, 497, 548, 549, 557, 567, 568
Antenna, 5, 29, 36–38, 40, 46, 65, 66, 74, 76, 77, 81, 88, 91, 99, 101, 102, 104, 106, 108, 165, 178, **181–185, 191, 194, 198–200, 202, 204–209**, 224, 258, 261, 262, 266, 267, 273, 275, 313, 314, 326, 327, 358–366, 369, 371–374, 379, 446, 450, 456–459, 461, 463–466, 495, 512, 518–521, 539–541, 544, 546, 548, 549, 555, 557, 573, 574, 577, 578, 580
Antenna aspect angle, 358, 364, 369, 371, 372
Antenna gain, 36–38, 40, 184, 363
Apogee maneuver, 265, 266, 429
Apollo, 47, 83, 335, 338, 473, 535, 561, 563, 573, 574, 579
Approach, 491, 498, **499, 501–503, 505–507**, 509–518, 520, 523, 524, 531, 536, 540, 545, 548, 549, 552, 553–556, 558, 567–569, 572, 574
Argument of perigee, **247, 248**
Ascending node, **247, 248**, 259, 264
Assembly, **18**, 45, 55, 59, 64, 108, 223, 224, 342, 344, 345, 385, 428, 497, 567
Asteroid, 92, 281, 301, 371, 531–536, 541, 544, 545, 549, 551, 552, 555, 556, 561, 562, 565, 566, 568, **574–576**, 579, 583
Asynchronous, 173, 187
Atmosphere, 3, 4, **10–12**, 14, 15, 23, 27, 28, 37, 72, 78, 86, 113, 250, 251, 253, 275, 369, 432, 477, 479, 481, 499, 505, 534, 540, 545, 553, 554, 561, 565–569, 571, 572, 583
Atmosphere control and supply, **477**
Atmosphere revitalization, **477**
Atmospheric drag, **11–13, 15**, 568
Attitude and orbit control system, 23, 110, 274, 378, 425, 428, 431, **433**, 502
Attitude determination, 274, **276, 281, 283**, 290–292, 405, 436, 437, 448
Attitude dynamics, **271**

F. Sellmaier et al. (eds.), *Spacecraft Operations*, Springer Aerospace Technology,
https://doi.org/10.1007/978-3-030-88593-9

Authentication, 171, 218, 221, 222, 226, 379, 385
Automated Telemetry Health Monitoring System (ATHMoS), 154
Automated Transfer Vehicle, **479**
Automatic gain control, 100–102, 111–113, 366–369, 371, 463
Automatisms, 75, 302, 391, 423
Autonomy, 22, 80, 83, 91, 171, 185, 285, 329, 389, 392, 423, 451, 475, 487, 501, 502, 518, 538, 542, 548, 549, 576, 583

B
Bandwidth, **27–30, 32**, 171, 185, 186, 188, 191, 205, 214, 217, 218, 312, 329, 337, 339, 358, 381, 386, 389, 391, 459, 531, 548
Baseband, 28–32, 205
Battery, 74, 78, 99, 109, 111, 112, 114, 287, 292, 310, 312, 314–316, 329, 385, **393–397, 399–402, 405–415**, 427, 492, 499, 571, 572, 575, 577, 578
Battery charge, 74, 315, 385, **406, 407, 409**
Battery discharge, 111, 314, **408**
Biomedical engineer, **482**
Book-keeping method, 292, 431
B-Plane, 551–552

C
Carrier, 27–29, 32, **34–36, 38**, 40, 74, 101, 104, 178, 185, 200–202, 205, 360, 373, 457, 469, 549, 557, 562, 566, 567, 572
Cassini, 47, 397, 532, 534–536, 565, 572
Center of mass, 428, **431–432**
Central checkout system, 49, 188, **223, 224**
Channel, **29**, 40, 140, 145, 149, 150, 152, 153, 175, 186, 209, 216, 221, 227, 359, 361, 389, 456, 458, 460, 465, 466, 483, 486
Channel amplifier, 459, 464–466
Channel coding, **27–31**, 40, 186, 199
Client, 166, 169, 171, 172, 175, 177, 201–203, 226, 230, **491, 492**, 496–502, 512, 514, 515, 517, 518, 520–522, 527, 528
Clohessy-Wiltshire equations, 491, **503–505, 507**
Close range rendezvous, **517**
Cloud, 169, 178, **189–191**
Coherency, 204, **370, 371**
Collision avoidance, 77, 81, 267, 268, 493, 501, 502, 521
Columbus Control Center, 143, 152, 155, 340, **476, 484**
Columbus Stowage and Maintenance Officer (COSMO), 267, **485**
Comet, 531–536, 544, 549, 552, 553, 555, 561, 565–568, **574–576**, 578
Command error data base, 154
Command verification, 220
Commissioning, 48, 71, 72, **75**, 78, 79, 192, 208, 258, 332, 449, 465
Communication path, 181, **187, 188**, 460, 527
Communication-service-center, 460, 468, 469
Compatibility test, 59, 65, 185, **204–206**, 520
Concurrent engineering, 17, **25**
Configuration management, 22, **48**, 56, 60
Consultative Committee for Space Data Systems, 175, 185, 186, 188, 197, 199–201, 206, 209, 211, 218, 234, 379, 577
Contingency operations, 95, **414**, 444, **447**, 460, **468**
Convection, 402, 403, 426
Conversion frequency, 465, 467
Coriolis effect, **504**
Corrective action, 66, 73, 136, 494
Cosmic radiation, 103
Cosmic velocity, **249, 250**
Crew Health, **482**
Crew Health Care System, **482**
Crew Medical Officers, **482**
Crew Resource Management (CRM), **136–139**, 143, 155
Crew safety, **481**
Critical design review, **48**, 59, **62**, 192
Curiosity, 397, 562, 564, 569, 570, 581–583
Cyclic redundancy check, **31**, 216, 220

D
Daily Planning Conferences, **483**
Data bus, **380**
Data cable harness, **380**
Data flow test, **193**
Data processing, 20, 29, 165, 174, 176, 195, 326, 327, 385, 515
Data streams, 31, 32, 57, 173, 177, 178, 188, 200, 213, **216, 218**, 222, 223, 227, 389, 391, 457, 519

Data transport, 181, **216, 218**, 227
Debris, **12–15**, 267, 268, 270, 432, **492–494**, 499, 500, 523
Decision-making, 79, 80, 137, 139, **141–147**, 149, 152, 156, 157, 486
Decoding, **28, 29**, 140, 178, 198, 200, 201, 203, 379
Deep Space, 28, 46, 47, 52, 71, 86, 91, 132, 183, 184, 197, 404, 443, 451, 487, 491, 495, 533, 536, **540**, 562, 576
Degradation, 73, 75, 165, 228, 396, 397, 401, 407–412, 421, 427, 445.448, 449, 466, 544
Delay time, 502, **518–521**
Delivery, 51, 64, 76, 100, 173, 235, 236, 261, 462, 536, 566, 573, 579
Delta-Differential One-Way Ranging (Delta-DOR), 544, **550–551**
Demodulation, **39**, 178, 198, 200, 203, 457
De-orbiting, 78, 422, 432, **492**, 528
DEOS, 492, **499**, 500
Descending node, **247**
Development phase, 237, 328, 349.482
Deviation, 7, 50, 79, 97, 206, 246, 251, 273, 275, 278, 288
Docking, 6, 350, 435, 479, 480, 491, 496, 497, 499, 501, 502, 509, 512, 513, **518, 520**, 522, 523, 527, 528
Doppler, 27, 178, 184, 185, 186, **197–206**, 262, 263, 367, 368, 370, 371, 427, 544, 549, 550, 555, 557, 572, 574
Doppler shift, 27, **198**, 367, 371, 550, 555, 557, 572
Downconverter, 198, 199, **458**, 467
Dry mass, **430**

E
Earth-fixed coordinates, 90, **282**
Earth sensor, 89, 435, **436**, 441, 443, 445
Eccentricity, 15, **247, 248**, 264
Electrochemical Cell, **396, 397**
Electromagnetic Signal, **27**, 276, 358
Ellipse geometry, **246**
Emergency exit, 165, 166
Encoding, **31, 36**, 140, 176, 177, 185, 201, 203, 214, 216, 381
Encryption, 56, 177, 188, 222
End of life, 22, 412, **430–432**, 500
End-of-mission, 72, 78, 98, 303, 500, 562, 566
Engineering model, **57**, 91, 123, 156, 204, 221, 523, 542, 545, 578
Entry Interface Point, 568
Entry probe, 561, 565
Environmental Control and Life Support System, **473**
ENVISAT, 14, 47, 492
EPIC, 485
Eros, 532, 534, 535, 565, 575
EUROCOM, **152, 153**, 485
European Cooperation for Space Standardization (ECSS), 17, 47, 48, 50, 55, 60–62, 64, 65, 129, 132, 156, 185, 217, 219–221, 234, 239, 368, 379, 383, 388, 389, 392
Exhaust mass flow rate, **418**
Exhaust velocity, **417–420**
ExoMars, 546, 564, 566, 569, 577
Experimental Servicing Satellite, 108, 495, **496**
Export license, 195
Extra vehicular activity, **477**

F
Failure analysis, **102, 103, 105**, 143, 193, 562
Failure probability, 95, **98, 99**
Failure scenario, **106**, 156, 414
Far range rendezvous, **515**
Fault Detection, Isolation and Recovery, 75, 102, 292, 383, 415, 423, 440, 441, 553
File system, **391**
File transfer, 170, 173, 174, 176, **188**, 189
Final approach, 502, **509, 510**, 518
Fire detection and suppression, 478
First acquisition analysis, 257, **260**
First aim point, **509**
Fixed plan, **303**, 321
Flight control team, **46**, 71, 76, 78–82, 85, 92, 143, 144, 146, 148, 149, 154–157, 340, 349, 350, 473, 475, 481, 483–485, 542
Flight director, 53, **54**, 66, 77, 79–82, 85, 86, 142–144, 149, 150, 153, 157, 231, 351, 484, 485, 487, 501
Flight dynamics, 47, 59, 64, 66, 77, **81**, 86, 87, 97, 124, 167, 171, 202, 213, 223, **245**, 257, 258, 261, 267, 291, 292, 304, 305, 331, 370, 372, 374, 429, 432, 445, 501, 502, 505, 558
Flight dynamics system, 47, 64, 66, 83, 124, 213, **258, 261**, 305, 372
Flight model, **57**, 59, 66, 96–98, 108, 185, 204, 579

Flight operations procedures, 60, 64, 66, 82, **117–120**, 122, 126, 133, 145
Flight procedure, 56–58, 82–85, 91, 106, 113, **118–127**, 129, 131–133, 143–147, 154, 293, 294, 382, 384, 386, 387, 444, 448, 461, 462, 467–469
Flight-proven, **50**, 52
Fly-by maneuver, **554**
FORDEC, 148
Four-sides-model, 139
Frame acceptance and reporting mechanism (FARM), 221
Frame-Based Telemetry, **216, 217**
Fuel cell, 395, **397**

G
Gain, 10, 31, **36–38**, 40, 77, 91, 96, 98, 100, 101, 108, 111, 141, 157, 184, 199, 208, 291, 363, 366, 371, 423, 458, 459, 463–467, 539, 544, 546, 555, 557, 577
Gain step adjustment, 463, **466**
Galileo, 47, 86, 91, 92, 115, 224, 366, 434, 437, 532–534, 536, 556, 565
Galileo (Jupiter), 92, 366
Ganymede, 538, 555, 561
Geostationary orbit, 228, **249, 266, 268**, 380, 423, 462, 503, 511, 512, 519
Global Positioning System, 17, 20, 47, 87, 89, 199, 202, 262, 282, 283, 285, 331, 361, 362, 370, 378, 380, 427, 434, 437, 440, 444, 445, 512, 513
Graphical user interface, 223, 302, 304, 319, 327, 330, 517
Graveyard orbit, 12, 15, 72, 78, 432, 495, 497
Gravity Assist Maneuver (GAM), 553–556
Gravity Recovery and Climate Experiment (GRACE), 19, 47, 86, 87, 223, 276, 277, 291–293, 295, 302, 328, 361, 409, 412–416, 431
Ground procedures, 60, **118**
Ground Reference Model, 579
Ground segment, 17, 20, 22, 24, **45–54**, 56–65, 67–69, 97, 164, 173, 184, 223, 302, 304, 328, 329, 364, 484, 485, 501, 524, 527
Group process factors, 138, **139**
Gyroscope, 109, 110, **274**, 436, 437, 446, 448, 449, 451, 494

H
Handover, 76, 89, 102, 149, 155, 200, 204, 340, 368, 486, 519
Hayabusa, 534, 535, 562, 565, 566, 567, 575, 579, 585
Heaters, 99, 111, 114, 294, 385, 403, 405, 406, 411–415, 427, 461, 462, 466, 579
Heat Shield, 534, 568, 569, 571, 572, 576, 582
Heat Transport, 403, 404
Helix antenna, 457
Hierarchy, 79, 169, 187
High-Gain Antenna, 577
High-Power Amplifier (HPA), 198, 459, 466, 468, 469
High-priority telecommand, 378, 379, 388
High-power telecommand, 379
Hohmann Transfer, 508, 510
Horizontal linear polarization, 457
House-keeping data, 73, 220, 391, 460, 579
Hubble Space Telescope, 21, 492, 494, 495
HUDEP, 156
Human Dependability Handbook, 156
Human error, 135, 137, 155, 156, 228
Human Spaceflight, 52, 58, 71, 80, 86, 95, 117, 118, 120, 143, 148, 152, 335, 336–338, 340, 352, 473, 474, 475, 483, 484, 485, 487, 494
Huygens, 47, 535, 565, 572
Hyperbolic Trajectory, 567

I
Ideal rocket equation, **419**, 422, 429
Identifiability, 48
Inclination, 15, 86, 88, 89, 182, **247, 248**, 251, 252, 255, 256, 260, 261, 288, 290, 294, 456, 493
Incremental Scheduling, **341**
In-Flight Maintenance, **392**
Injection dispersion, **260–262**
In-orbit performance, **207**, 460
In-Orbit Test (IOT), 64, 76, 83, 96, 99, 120, 184, 198, **207, 208**, 381, 387, 446, 460, 462
Input factors, 138
Inspection, 236, 237, 333, 491, **492, 495**, 497, 527
Interference, 10, **32**, 185, 205, 264, 337, 346, 348, 358, 361, 364–366, 372, 373, 405, 436, 445, 446, 457, 462, 467, 518, 520

Intermediate frequency, **198**, 458
International Space Station, 11, 41, 126, 152, 156, 251, 287, **335**, 338, 353, 403, **473**, 480, 494
International Telecommunication Union, 195, **358**, 459, 467
Interoperability, **175**, 224, 225, 240
Iridium, 14, 267, 493
Itokawa, 532, 534, 535, 565, 575

J

Jitter, 518, **520, 521**
Joint Space Operations Center (JSpOC), **267, 268**
Joule-Thomson effect, 425
Juno, 538

K

Keplerian elements, **246, 247, 248**

L

Lander, 301, 535, 545, **561, 562, 563, 564, 565–569, 571–575, 577–579, 581, 583**, 585
Launch and Early Orbit Phase (LEOP), 22, 68, 72, **74**, 95, 120, 166, 167, 181, 198, 258, 270, 291, 298, 331, 358, 361, 406, 422, 461
Launch delay, 407
Leadership, 139, **142**
Left-Hand Circular Polarized, **198**, 364
LEOP duration, 96
LIDAR, 498, **512–514**, 517, 524, 526
Lifetime estimation, **417**, 428
Limit violation, 154, 228, 383
Linear Storage, **391**
Link-Budget-Equation, 27, **38, 39**
Local Orbital Frame, 491, **503–505**, 515, 516
Lock, 205, 262, 275, 359, 360, 365, 367, 368–370, 390, 547, 548
Long-lead items, **59**
Long-term monitoring, 86, 89
Loop Stress, **367–369**
Loss of Signal, 102, 200, 260, **368, 369**
Low noise amplifier, 38, 198, 458
Low-Earth Orbit, 69, 114, 298, 389
Luna, 562, 563, 572, 573, 579
Lunokhod, 563, 573, 579, 580

M

Magnetic torquer, 428, **438–441**, 443, 444, 449, 451
Magnetic torque rods, 274, 276, 286, 287, 290, **436**
Magnetometer, 274, 283, 284, **436**, 443, 449, 451
Magnetosphere, 8, 9, 543
Man–Machine Interface (MMI), **136**
Mars, 21, 183, 328, 397, 487, 532, 533, 534, 535, 536, 537, 541, 546, 553, 557, 559, 561–564, 566–569, 572, 574, 576, 577, 579–583, 585
Mars Express, 21, 536, 546, 557, 559, 562, 566, 577, 578, 585
Mars Polar Lander, 562
Mass Defect, 393, **398**, 399
Mass Memory Unit, **379**
Mean anomaly, **247, 248**
Mechanical Shock, **6**
Medical Ops, 80, 138, 345, 346, 485
Mercury, 335, 473, 532–534, 536, 537, 538, 546, 557, 561
Minerva, 535, 565–567, 575, 579
Mission Database, **154**
Mission director, 53, **54**, 79
Mission information base, **57**, 177, 178, 219, 386
Mission Operations Director, 205
Mission Operations Information System (MOIS), 131
Mission Planning, 64, 80, 81, 87, 124, 302, 305, 323, **326, 328, 330, 334**
Mission Requirements, **20**
Mission Statement, **17, 19**
MITEX, 491, 492, **495**
Modeling Language, **307, 313**, 316, 319, 322
Modulation, 27, 29, 30, **34–36**, 39, 40, 178, 185, 186, 198, 199, 200, 201–203, 206, 208, 216, 294, 457, 465
Monitoring, 56, 57, 64, 66, 77, 83, 84, 86, 89, 103, 119, 153, 154, 167, **173–177**, 194, 203, 204, 209, 213, 215, 220, 222, 223, 224, 226, 228, 229, 232, 239, 240, 261, 267, 284, 305, 345, 346, 357, 366, 367, 373, 383, 384, 389–391, 412, 413, 423, 424, 426, 428, 444, 451, 460, 465, 467, 487, 566
Monitoring and Control System (MCS), **215**, 216, 222
Moon, 19, 37, 46, 47, 52, 66, 91, 183, 249, 251, 267, 268, 270, 293, 296, 297,

435, 445, 479, 487, 520, 532, 533, 534, 535, 538, 549, 555, 561, 563, 568, 572–576, 579, 583
MoonLite, 576
Motion, 10, 11, 245, 251, 271, 272, 280, 281, 283, 286, 289, 403, 417, 419, 428, 452, 476, 491, 492, 503, **504–508**, 512, 515, 517, 522, 524, 541, 550
Multi-feed, **458**
Multi-mission operations, **163**
Multiplexing access point (MAP), 221

N
Network communications officer, **81**
Network Operations Project Engineer (NOPE), 53, 80, **193**
NextSat, **499**
Noise, **5**, 10, 29, 33, 36–40, 166, 184, 198, 199, 203, 208, 274, 280, 284, 366, 445, 458, 463
Nuclear fission reactor, **395**
Nyquist theorem, **29**

O
Observation, 14, 19, 21, 46, 47, 51, 52, 62, 85, 182, 259, 301, 303, 336, 361, 386, 412, 431, 445, 452, 465, 483, 491, 492, 494, 503, 532, 534, 535, 544, 546, 556, 562
Offline data, 121, 181, 188, 224, 226, **227**, 385, 444
Off-the-shelf, 127, 131, 173, 235, 321
On-board computer, 57, 185, 218, 221, 222, 274, 286, 291, 352, 360, 361, 377–380, 384–387, 389, 400, 401, 406, 413, 416, 421, 424, 439, 449, 523, 525, 527, 577
On-board control procedure, **132**, 384, 389
On-board data handling, 103, 214, 222, 360, 377, 378, 380, 381, **383**, 385, 392
On-board maintenance, 73, 386
On-orbit servicing, **491, 492, 494, 497, 517, 522–524, 526**
Operational readiness review, **48, 60**, 61, **63–65**, 67, **68**, 195
Operations data file, 145, 338, **339, 483**, 486
Operations engineering, 18, 53, 60
Operations support tools, 58
Orbital Express, 492, **499**, 518
Orbital Life Extension Vehicle (OLEV), 491, 492, **495, 496**, 511, 512, 529
Orbital period, 86, 89, **250**, 254, 260, 264, 505, 511, 546
Orbital replacement unit, **499**, 527
Orbit determination, 86, 182, 198, 202, 204, 206, **261–263**, 267, 268, 357, 362, 427, 429, 430, 437, 445, 544, 549, 552, 553, 556, 558, 567
Orbit insertion, 531, 532, 534, 544, 550, 556, 557, 566, 567
Orbit maintenance, 22, 81, 245, **263, 264**, 345
Orbit perturbations, **251**, 268
Orthomode transducer, **458**
Out of limit condition, 82, 103

P
Packet Utilization Standard, **219, 220**, 223, 234, 383, 384, 386, 388, 389, 392
Parachute, 535, **568, 569, 571, 572**, 576, 582
Passivation, 422
Pathfinder, **564**, 569, 580
Penetrator, 562, **576**
Phase lock loop, **367**, 370
Phasing, **509, 510**, 512
Philae, 535, 545, 562, 566, 567, **575**, 578, **579**
Phobos, 574, 579, 583
Phoenix, **564**, 568, 569, 577
Program for Interactive Timeline Analysis (PINTA), **319, 327**
Pioneer, 532, **533**, 565
Pitch, **272**, 273, **276–278**, 282, 283, 434, 439, 441, 446, 450, 463, 516
Pointer, **389–391**
Polarization, 66, 109, 198, 199, **358, 359**, 362–365, **371, 372**, 456–459
Power conversion, **399**
Power distribution, 74, 145, 380, 401, 474
Power management, **394**
Power system, 393, **394**, 396, 402, 408, 411, 414
Power thermal model, 406, 409
Predictive Maintenance, **231, 232**
Pre-heating, 318
Preliminary design review, **48**, 192, 496
Preventive action, 108
PRISMA, **514, 515**
Private medical conference, **482**
Private psychological conference, **482**
Procedure Language for Users in Test and Operations (PLUTO), 129–131
Processor module, 378, 379

Procurement, 18, 59, 195, 235
Project documentation, 58, 62, 192
Project manager, **53**, 54, 81, 192, 235, 501
Project office, **53, 55**
Project schedule, 51
Project structure, **53**
Propellant mass, 409, **417**, 422, 426, 429–431
Propulsion system, 21, 23, 74, 89, 109, 398, 402, 411, **417, 419–424**, 428–431, 443, 449, 531, 532, 534, 536, 557

Q
Quality assurance, **55**, 59, 60, 237
Quality of Service, 207, 220, 469
Quality standards, **55**
Quaternions, **273**, 293, 295, 448

R
Radiation, 3, **7–10**, 15, 21, 22, 103, 251, 253, 273, 274–277, 286, 358, 363, 364, 366, 371, 385, 402–404, 412, 426, 538, 553
Radiation belt, **8–10**
Radio frequency, **27, 28, 34, 36, 37**, 65, 184–186, 195, 198, 205, 357, 437, 457, 459, 460, 502, 512, 539, 544, 546, 549, 557
Radio frequency Beacon sensor, **437**
Ranging, 22, 89, 96, 169, 182, 184–186, 198–206, **263**, 285, 302, 357, 360–362, 370, 371, 377, 427, 451, 498, 512, 513, 524, 544, 549, 550
R-bar hop, **508, 509**, 511, 512
Reaction wheels, 273–275, **287**, 400, 401, 423, 428, 438–441, 443, 449–451, 544, 554, 555
Real-time, 66, 79, 88, 98, 121, 144, 145, 147, 148, 169–171, 173, 188, 201, 209, 223, 227, 293, 331, 340, 346, 350, 352, 362, 382, 423, 428, 434, 444–447, 485–487, 501, 502, 519, 524, 526, 541, 566
Receiver, 27, **29**, 32, 33, 37–40, 100, 112, 139, 140, 198–200, 202, 205, 206, 208, 223, 282, 283, 359–361, 362, 365–371, 373, 378, 380, 463, 547, 548, 577, 578
Recommendation, 68, **82, 85, 86**, 98, 109, 112, 467
Reconfiguration module, 113, **378, 384**
Recovery, 19, 58, 66, 75, 82, 102, 103, 105, 106, 107, 111, 113, 120, 276, 292, 294, 305, 378, 383, 384, 414–416, 423, 431, 440, 441, 447, 468, 477, 546, 548, 553, 566–568, 580
Redundancy, 31, 75, 102, 164–166, 168, 185, 188, 200, 216, 220, 280, 366, 378, 380, 381, 442, 447, 468, 538
Reflection, 457, 467
Regenerative repeater, **457**, 465
Regulations, 111, 137, 139, 385, 394, 399, 403, 404, 414, 415, 486
Relay satellite, 21, 181, 285, 374, 475, 476, 519
Reliability, 20, 22, 24, 91, 96, 120, 138, 142, 143, 144, 156, 157, 168, 169, 173, 188, 189, 240, 326, 539
Rendezvous, 435, 491, 497–499, 502, **503, 512–515, 517**, 523–525, 527, 562, 563, 566, 575, 576
Repeater, 84, 90, **455–470**
Repeater flight rules, 465, 467
Reporting, 68, 77, 156, 195, 220, 221, 229, 386
Requirements, 14, **17–20**, 22
Re-scheduling, 81, 352
Resonant frequency, 107, 108
Retro Rockets, 535, 568, 569, **571–574**, 576
Re-use, 21, 192, 219, 326, 328, 458
Review item discrepancy, **61, 62**
Review process, **61, 62**, 482
Right ascension of the ascending node, **247, 248**, 251
Right-hand circular polarized, 198, **359**, 364, 365, 371
Ring buffer, **389–391**
Risk, 5, 18, **50–52**, 58, 62, 68, 75, 76, 78, 81, 108, 109, 112, 123, 135–137, 147–149, 155, 157, 191, 233, 306, 373, 380, 390, 406, 421, 466, 480, 492, 501, 522, 545, 554, 576
Risk analysis, **52**
Roll, **272, 273**, 276–278, 282, 283, 290, 439, 441, 446, 463, 516
Rosetta, 21, 72, 91, 132, 533–537, 539, 545, 549, 551, 552, 556, 559, 562, 566, 567, 575, 578, **579**
Routine operations, 13, 48, 58, **76, 77**, 87, 100, 120, 121, 144, 184, 192, 193, 195, 204, 207, 239, 240, 362, 460, 461, 463–465
Routing, 168, 172, 203, **221**, 222, 457, 459, 465, 466

Rover, 302, 328, 397, 535, 561–564, 567, 569, 573–575, 579–583, 585

S
Safeguard mechanism, **383, 384**
Safeguard memory, 110, 378, **385**, 431
Sample return, 535, 563, 565–567, 573
SATCOMBw, 93
Satellite board computer, **377**
Satellite orbit, 87, 245, **246**, 262, 283
Saturation, 110, 428, 439, 460, 463, 554
Scheduling algorithm, 326
Security, 55, 56, 64, 164, 165, 169–171, 174, 179, 188, 191, 220, 231, 234, 235
Security officer, **56**
Security standard, 165
Seebeck effect, 397
Segmentation layer, 218, 222
Semi-major axis, **246–249**, 253, 255, 260, 261
Sequence of events, 58, 65, **96**, 106, 120, 301, 341, 374
Servicing, 108, 134, 452, **491**, 492, **494–497**, 499, 517, 522–524, 526, 527, 529
Shift keying, **34, 36**, 200
Shoemaker, 534, 535, 565, 575
Side band, **359, 369**
Side lobe, 34, 36, **37**, 359, 360, 369
Signal transmission, 359, 548
Simulation officer, **54**, 66–68
Single-event upset, **99**, 103
Singularity, **49**, 247, 248
Situational awareness, 139, **141**
Sky crane, 564, 569, 581, 582
Skylab, 335, 494
Software, 213, 241
Software maintenance, **387**, 392, 542
Sojourner, 564, 569, 580, 581
Solar array, 5, 7, 23, 74, 90, 99, 104–109, 111–113, 266, 378, 385, **395–397**, 399, 400, 406–410, 412, 413, 423, 449, 538, 544
Solar flux, 402, 537, 538
Solar panel, 13, 65, 76, 89, 109, 110, 112, 114, 182, 264, 274, 276, 315, 316, **395, 396**, 399–401, 404, 406–408, 412–415, 494–496, 499, 538
Solar radiation flux, **7, 8**
Solid-state memory, 379
Source coding, **27–30**
South Atlantic anomaly, **9, 10**
Space communications, **27, 28, 36**, 191, 198, 540
Spacecraft controller, 77, **80**
Spacecraft ID (SCID), 221
Spacecraft simulator, 51, **57**, 66, 100, 103
Spacecraft support team, **54**
Space debris mitigation, 492, **493**, 499, 500, 528
Space environment, **3**, 50, 74, 99, 287, 400, 480
Spirit, 564, 569, 581
Star sensor, 89, **436, 437**, 440, 451
Star tracker, **274, 280, 281, 283**, 284, 289, 293, 295–297, 411, 443, 449, 555
Storage, 22, 87, 168, 169, 174, 185, 201, 220, 226, 336, 379, **389–392**, 394, 395, 548, 549
Subsystem engineer, 20, **55**, 77, 93, 134, 241, 406, 414–416, 453, 462, 470
Suitability, **312, 313**
Sun-acquisition mode, 101, **109**, 384, 385
Sun sensor, 99, 274, **435, 436**, 441, 443, 449
Sun-synchronous orbit, 251, **252**, 264, 275, 493
Super-synchronous orbit, 89
Surveyor, 563, 573, 574, 581
Sweep, 178, **200**, 368, 369, 372, 546–548
Swing-by, 91–92, 536–537, 541, 550, **553–556**
Synthetic aperture radar, 276, 326
System contingencies, 95, 104
System engineering, **16–20**, 24, 25, 53, 54, 298, 353, 488, 585
System freeze, **68**, 69
Systems Test and Operation Language (STOL), 131

T
Tape recorder, 380, 391, 455
Team interaction, 144
Teamwork, 139, **142**
Technology readiness level, 50, 51
Telecommand, 22, 27, 28, 72, 80, 82, 83, 100, 110, 118, 119, 125, 127, 129, 145, 175–178, 181, 185, 186, 188, 198, 201, 203–206, 214–216, 218–223, 228, 302, 307, 327–329, **357**, 369, 373, 374, 377–382, 386, 388, 389, 434, 462, 475, 517, 519, 539, 541, 542, 546, 579
Telemetry and telecommand, 82, 118, 185, 186, 214, 377, 475, 539, 542, 546, 576
Teleoperation, 491, 502, 518, **520, 521**, 527

Temperature and humidity control, **477**
Terminal, 165, 169, 175, 181, 188, 189, 457
TerraSAR-X, 47, 72, 205, 223, 224, 259, 263–265, 270, 288–290, 294, 302, 305–307, 325–329, 361, 362, 378, 400, 407–409
Test data, **56**, 235
Thermal control, 22, **405**, 414, 428
Thermal control subsystem, 55, 474, 477
Time dependency, **308**
Timeline, 49, 65, 66, 80, 144, 156, 230, **301–309**, 311–313, 315–322, 325–333, 335, 336, 340, 341, 346, 349–352, 371, 415, 483, 484, 547, 554, 556, 583
Time-tagged commands, 66, 87–89, 121, 294, 331, 389, 546
Time tagging, 201–203, **388**
Titan, 534, 535, 561, 565, 568, 572
TMTC Board, **362, 363**, 378
Tolerance, **49**, 275, 294, 442
Traceability, **48**, 306, 322
Tracking speed, **183**
Training, 46, 48, 51, 54, 55, 58–60, 63–65, 67, 77, 100, 104, 117, 126, 127, 137, 143, 148, 151, **155**, 157, 195, 235, 270, 335, 345, 480, 482, 531, 541, 542
Transceiver, **361–363**, 370
Transfer orbit, 89, 97, 101, 182, 249, **255, 266**, 372, 380, 446
Transformation matrix, **271**
Transmitter, 27, 29, 31, 33, **37–40**, 83, 108, 208, 302, 321, 359–362, 366, 367, 370, 371, 373, 378, 415, 455, 457, 520, 546, 548, 578
Transponder, 104, 188, 204, 208, 370, 371, 373, **456–458**, 462, 465, 502, 547, 550, 555
Travelling Wave Tube Amplifier, **459**, 461–465, 467, 468
True anomaly, **247**
True-of-date, **282**
TV-Sat1, 104, 106–108, 495, **496**

U
Unified Propulsion System, 165, 443, 501
Upconverter, 198, 199, **459**
User Support and Operations Center, 152, 153, **485**

V
Vacuum, 3, **6**, 7, 11, 337, 463, 481, 579
Validation, 18, 46, 48, 51, 54, 56–60, **63–68**, 84, 117, 122, 123, 127, 145, 146, 195, 223, 224, 234, 237, 483, 491, 523, 556
Van Allen radiation belts, **9, 10**
V-bar Hop, **507, 508**, 510
VeGa, 533, 565
Venera, 534, 535, 564, 565, 571
Venus, 8, 21, 91, 92, 132, 532–537, 546, 557, 561, 564, 565, 568, 571, 572
Verification, **18, 19**, 45, 49, 50, 59, 60, 63, 64, 88, 194, 220, 234, 236, 237, 295, 522–524, 567
Vernal equinox, **247**, 281, 282
Vertical linear polarization, **359**, 457
Very Large Base Interferometry (VLBI), 550
Viking, 562, 564, 567–569, 577
Virtual Channel (VC), 221
Virtualization, **166, 168**, 169, 171, 179
Virtual Private Network (VPN), **167**, 169, 188
Virus scanner, **171**, 174
Visitor area, 167
Voice communication, 78, 79, **149, 150, 152**, 167, 169, 189, 475, 483, 485

W
Waiver, 50
Water recovery and management, **477**
Weekly Planning Conference, **483**
Wet mass, **430**
Wheel desaturation, **439**
Wheel unloading, **439**, 440
Work flow, 58, 85, 122, 176

Y
Yaw, **272**, 273, 277, 279, 282, 283, 288–290, 292, 294, 439, 441, 516

www.ingramcontent.com/pod-product-compliance
Ingram Content Group UK Ltd.
Pitfield, Milton Keynes, MK11 3LW, UK
UKHW022218051026
15624UKWH00001B/216

9783030885922